AF710920

NetAnalyst™ Reference Guide

Eleventh Edition

PMG
NetAnalyst
Pine Mountain Group, Inc.

PMG NetAnalyst
Pine Mountain Group, Inc.
Box 849, 20950 Ferretti Road
Groveland, CA 95321
Phone: (209) 962-6247
Fax: (209) 962-5914

Email: info@pmg.com
www.pmg.com

NetAnalyst™ Reference Guide, Eleventh Edition

Published in the United States.

ISBN: 1-58392-003-X

Published by Pine Mountain Group, Inc.
Box 849, 20950 Ferretti Rd.
Groveland, CA 95321

Table of Contents

Ethernet Packet Encapsulation

Ethernet II (See Appendix A → Ethertype)

Interfame Gap(12)	Preamble(8)	Dst(6)	Src(6)	Ethertype(2)	Layers 3-7	FCS(4)

Ethernet with 802.2 SAP (See Appendix B → SAP)

Interfame Gap(12)	Preamble(8)	Dst(6)	Src(6)	Length(2)	SAP(2) D | S	Layers 3-7	FCS(4)

Ethernet with 802.2 SNAP (See Appendix B → SAP; See Appendix A → Etype)

Interfame Gap(12)	Preamble(8)	Dst(6)	Src(6)	Length(2)	SAP(2) AA | AA	03	OUI(3)	Etype(2)	Layers 3-7	FCS(4)

Ethernet without 802.2(Novell Only)

Interfame Gap(12)	Preamble(8)	Dst(6)	Src(6)	Length(2)	FFFF(2)	Layers 3-7	FCS(4)

Interframe Gap - Minimum amount of time a station must wait after receiving the last bit of the previous frame before sending a frame. The wait period is 96 bits. For a 10BaseT LAN, this equates to 9.6 micro seconds. For a 100BaseT LAN, this equates to .96 micro seconds.

Preamble - 7 bytes of 10101010 transmitted followed by 10101011.

Length - number of bytes (not including the FCS bytes) transmitted after the length field. Maximum value of 0x5DC (1500).

Ethertype - values assigned to protocols (Appendix A) within the 802.3 (Ethernet) standard.

SAP - Service Access Points - values assigned to protocols (Appendix B) within the 802.2 (LLC) standard.

FCS(4) - Frame Check Sequence - CRC calculation on data included from the Destination field through layer 7.

Ethernet Packet Information:

- Maximum Ethernet packet size - 1518 bytes (0X5EE)
- Minimum Ethernet packet size - 64 bytes (0X040)

Ethernet LAN Errors

Collision - two or more stations transmit bits of information onto the LAN at the same time.

Local Collision - collision occurs on the same piece of wire. With 10BaseT, a station would receive bits on the RX pair at the same time it is transmitting bits on the TX pair.

Remote Collision - collision occurs between two stations on the other side of the repeater.

Valid Collision - occurs within the first 64 bytes (512 bits) of the packet. Packet is retransmitted by the Datalink layer.

Late Collision - occurs after the first 64 bytes (512 bits) of the packet. Packet is retransmitted by the Transport or Application layers.

Short (Runt) Frame - a frame which contains less than the minimum packet size of 64 bytes.

Long (Jabber) Frame - a frame which contains more than the maximum packet size of 1518 bytes.

Frame Check Sequence (FCS) - a frame which does not contain the correct FCS calculation in the FCS field.

Alignment - a frame that does not contain a number of bits which cannot be evenly divided by 8.

Token Ring Frame Encapsulation

Token Frame

SD	AC	ED

MAC Frame

SD	AC	FC	Dst(6)	Src(6)	MAC Data	ED

LLC Frame with 802.2 SAP

See Appendix B

SD	AC	FC	Dst(6)	Src(6)	SAP D | S	Layers 3-7	FCS	ED	FS

LLC Frame with 802.2 SNAP

See Appendix B See Appendix A

SD	AC	FC	Dst(6)	Src(6)	SAP AA | AA	03	OUI(3)	Etype(2)	Layers 3-7	FCS	ED	FS

MAC and LLC Frames

SD - Starting Delimiter - J K 0 J K 0 0 0

ED - Ending Delimiter - J K 1 J K 1 I E

J and K - Code Violations

I - Intermediate Frame - used when a station is sending more than one frame within the maximum token holding time of 10ms for pre-1995 IEE802.5 standards or 9.1ms in 1995 IEE802.5.

E - Error Bit - set when a workstation repeating a frame calculates a FCS which differs from the FCS field within the frame.

AC - Access Control - P P P T M R R R

PPP - Priority Bits - current level of priority used in accessing the token frame.

T - Token Bit - indicates if the frame is a token frame or LLC frame.

M - Monitor Bit - bit set by Active Monitor as the frame is repeated by the Active Monitor to ensure that the frame is not continuously sent around the ring, and by ring stations to assure the presence of Active Monitor.

RRR - Reservation Bits - set by station programmed with higher priority level to reserve use of a token by higher priority stations.

FS - Frame Status - A C R R A C R R

A - Address Recognized - workstation recognizes that the destination field of the frame contains its MAC address.

C - Frame Copied - workstation copied the contents of the frame to an upper layer of the OSI model.

R - Reserved

MAC Frames

FC - Frame Control - F F R R Z Z Z Z

FF - 00

RR - Reserved

ZZZZ - physical control field

0010 - Beacon (0X02)
0011 - Claim Token (0X03)
0100 - Ring Purge (0X04)
0101 - Active Monitor Present (0X05)
0110 - Standby Monitor Field (0X06)
0111 - Duplicate Address Test (0X07) is pre 1995 IEEE802.5 standard
0001 - Duplicate Address Test in 1995 IEEE802.5 (0X01)

MAC Data - 4th byte is the MAC vector name and indicates the type of MAC frame (often the same as the ZZZZ field in the Frame Control)

00 - Response
02 - Beacon
03 - Claim Token
04 - Ring Purge
05 - Active Monitor Present
06 - Standby Monitor Present
07 - Duplicate Address Test
08 - Lobe Test
0B - Remove Ring Station
26 - Report NAUN Change
29 - Report Error

Destination Address - MAC Frames

C00000000001 - Active Monitor
C00000000002 - Ring Parameter Server (RPS)
C00000000008 - Ring Error Monitor (REM)
C00000000010 - Configuration Report Server (CRS)
C00000000080 - NetBIOS
C00000000100 - Bridge
C000FFFFFFFF- MAC Broadcast

LLC Frames

Pre-1995 IEEE 802.5
FC - Frame Control - F F R R R Z Z Z
FF - 01
RRR - Reserved
ZZZ - LLC Priority (Usually 000)

1995 IEEE 802.5
FC - Frame Control - F F R R R Y Y Y
FF - 01
RRR - Reserved must be sent as 0
YYY - User Priority

Ethertype - values assigned to protocols (Appendix A) within the 802.3 (Ethernet) standard.

SAP - Service Access Points - values assigned to protocols (Appendix B) within the 802.2 (LLC) standard.

Token Ring Frame Information

Maximum frame size (Pre-1995 IEEE 802.5) - 4523 bytes on 4Mbps T/R, 17,851 bytes on 16Mbps T/R. Specification indicates 10ms token holding time.

Maximum frame size (1995 IEEE 802.5)- 4550 bytes on 4Mbps T/R, 18,200 bytes on 16Mbps T/R. Specification indicates 9.1ms token holding time.

Minimum frame size - 3 bytes (Token Frame).

Token Ring LAN Errors

Isolating Error Counts

Line Errors - Indicates a FCS error was detected in a frame. May also indicate a Token Ring code violation.

Internal Errors - station has detected an internal error that was recoverable.

Burst Errors - the station has not detected a signal transition for five half-bit times.

AC Errors - station has received a Standby Monitor frame prior to receiving a Active Monitor frame. Upstream neighbor unable to set AC bits.

Abort Delimiters Transmitted - station transmitted an Abort Delimiter frame prior to the completion of a Data frame.

Non-Isolating Error Counts

Lost Frame Errors - a station has not received the frame it sent before a timer expires.

Receiver Congestion - a station does not have the amount a buffer space needed to properly receive a frame.

FC Errors - a station receives a frame, but the address recognized bits have already been set.

Frequency Errors - station receives a signal outside of an acceptable range.

Token Errors - AM recognizes that a token has not passed around the ring within token holds timer value.

FDDI Frame Encapsulation

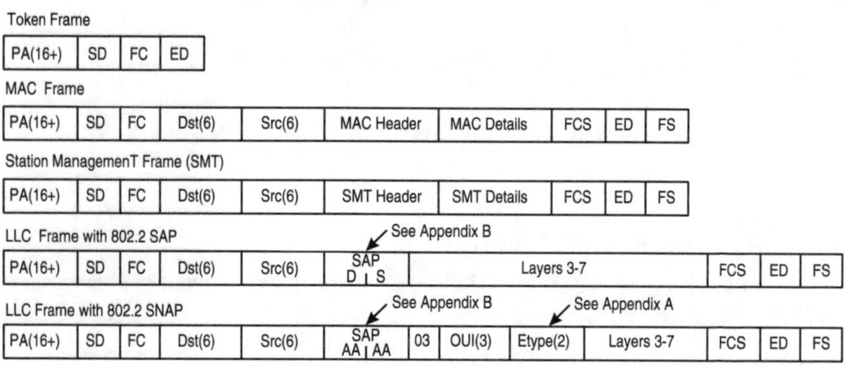

MAC, SMT and LLC Frames

PA - Idle Symbols - I(11111)
SD - Starting Delimiter - J(11000) K(10001)
ED - Ending Delimiter - T(01101)
FCS - Frequency Check Sequence - CRC calculation
FS - Frame Status - EEEEE AAAAA CCCCC

EEEEE - Error Detected - workstation has calculated a FCS value different from the value included within the frame
AAAAA - Address Recognized - workstation recognizes that the destination field of the frame contained its OUI interface address
CCCCC - Frame Copied - workstation copied the contents of the frame to an upper layer of the OSI model
00111 indicates frame status is not set
11001 indicates frame status is set

MAC Frames

FC - Frame Control - C L F F Z Z Z Z

C = 1, Class of Data Bit - 0=synchronous data, 1=asynchronous data
L = 0 or 1, Length of MAC Addresses - 0=16 bit MAC address, 1=48 bit MAC address
FF = 00
ZZZZ = 0010 - Beacon Frame
ZZZZ = 0011 - Claim Frame

SMT Frames

FC - Frame Control - C L F F Z Z Z Z

C = 0
L = 1
FF = 00
ZZZZ = 0001 - SMT frame destined to a specific address
ZZZZ = 1111 - SMT frame destined to a broadcast address

LLC Frames

FC - Frame Control - C L F F R P P P

C - Class of Data Bit - 0=synchronous data, 1=asynchronous data
L - Length of MAC Addresses - 0=16 bit MAC address, 1=48 bit MAC address
FF = 01
R - Reserved
PPP - Priority level for (001-111) the LLC frame
FDDI LAN Errors

Status Report Frame

Frame Count - Number of frames counted

Error Count - Number of frames received with CRC errors

Lost Count - Number of frames initiated, but not received correctly

FDDI Connection Rules

From	To	Validity
A	A	Legal but Undesirable
A	B	Legal
A	M	Legal if B to M is not simultaneously active
A	S	Legal but Undesirable
B	A	Legal
B	B	Illegal
B	M	Legal if A to M is not simultaneously active
B	S	Legal but Undesirable
M	A	Legal if B to M is not simultaneously active
M	B	Legal if A to M is not simultaneously active
M	M	Illegal
M	S	Legal
S	A	Legal but Undesirable
S	B	Legal but Undesirable
S	M	Legal
S	S	Legal

Media Data Encoding

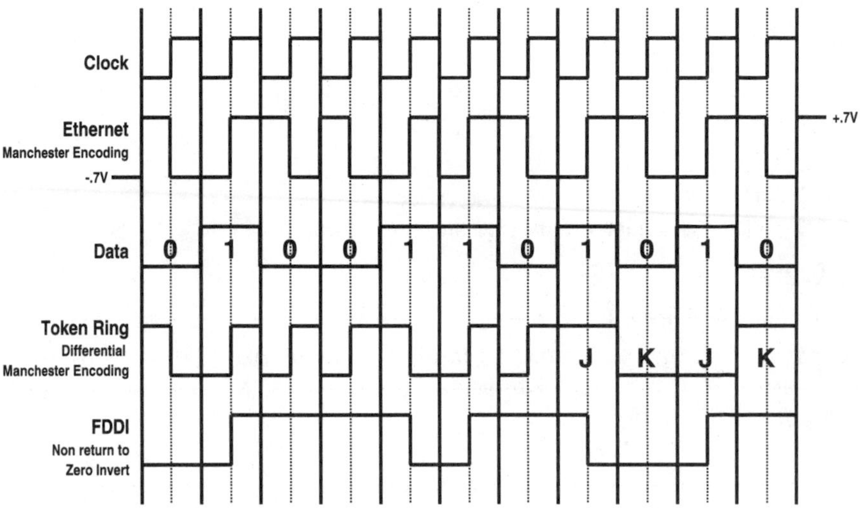

Manchester
Midpoint Low to High = 1
Midpoint High to Low = 0

NRZI
Change at Midpoint = 1
No Change at Midpoint = 0

Differential Manchester
No Midpoint Transition = J/K
Change at Start = 0
No Change at Start = 1
Violated 1 = J
Violated 0 = K

IEEE 48-Bit Address Anatomy

Canonical Format for writing addresses

08-00-5A-24-59-87

Organizationally Unique Identifier (OUI)	16,777,216 Vendor Assignments
08-00-5A	24-59-87

Ethernet Destination Address

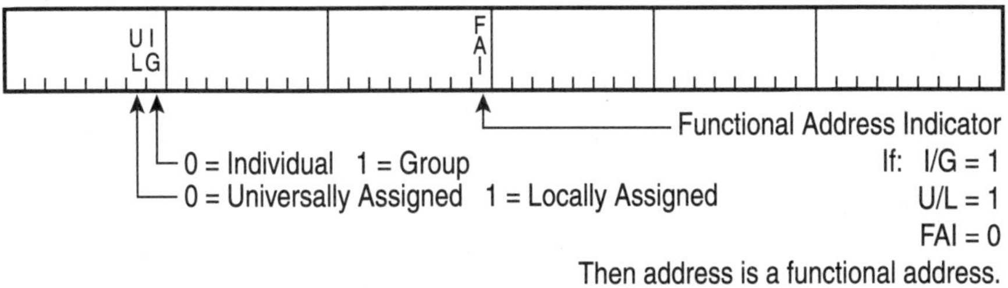

Token Ring Destination Address

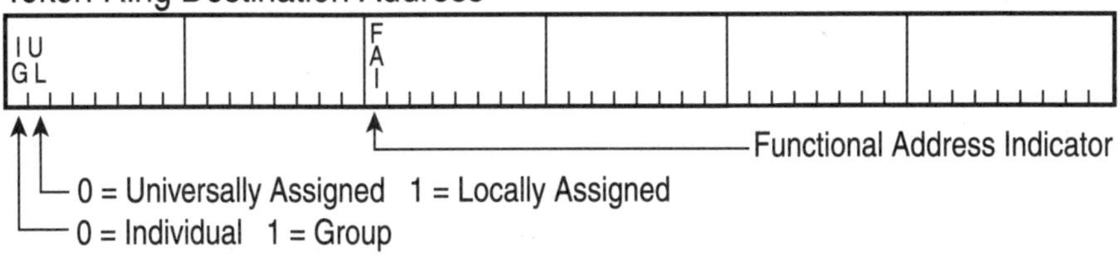

Source address I/G + UL the same as destination
if I/G = 1 then source route information is present.

IEEE mandates I/G bit transmited first on wire.
Ethernet addresses are sent LSB first.
Token Ring addresses are sent MSB first.

09	00	2B	00	00	0F
00001001	00000000	00101011	00000000	00000000	00001111

Ethernet (stored canonical)

90	00	D4	00	00	F0
10010000	00000000	11010100	00000000	00000000	11110000

Token Ring (stored non-canonical)

Bridges between Ethernet and Token Ring do bit reversal. This can cause problems if inconsistently implemented.

IEEE 48-Bit Address Anatomy Discussion

Forty-eight (48) bit LAN adapter addresses are composed of two (2) parts: 3 bytes (24 bits) are unique to a vendor—Organizationally Unique Identifier (OUI)—and are administered by the IEEE; 3 bytes (24 bits) are unique to an adapter with a particular vendors' assigned OUI, providing a vendor with the potential of 16, 777,216 unique device identifiers for each of its assigned OUIs.

ANSI/IEEE 802.1D 1993 (ISO/IEC 15802-3) defines the format for presenting these 48-bit LAN adapter address to the network: the LAN address **MUST** appear "on the wire" with the I/G (Individual/Group) bit presented first regardless of a particular technology's frame (packet) format or the storage requirements of an attaching system. For Ethernet packet encapsulation, the I/G bit is stored in the least significant bit (LSB) of the most significant byte of the address field and for Token Ring frame encapsulation, the I/G bit is stored in the most significant bit (MSB) of the most significant byte of the address field.

In memory (and sometimes mistakenly on LAN):

00	00	30	78	9A	BC
00000000	00000000	00110000	01111000	10011010	10111100

LSB (canonical)

First bit appearing on LAN (group address indicator)

↓

00000000	00000000	00001100	00011110	01011001	00111101

On LAN

In memory and on LAN:

00000000	00000000	00110000	01111000	10011010	10111100
00	00	0C	1E	59	3D

MSB Format

Unfortunately, due to the differences in frame formats for various IEEE LAN standards, the differences in storage of attaching devices as well as vendor parochialism, the LAN adapter address presented to the network may not contain the OUI actually assigned to the adapter. For example, the IEEE has assigned the OUI hexadecimal 00-00-0C to Cisco Systems, Inc. When Cisco uses this OUI with an Ethernet device, the OUI transmitted "on the wire" is the OUI assigned by IEEE; with a Token Ring device, the OUI transmitted "on the wire" is 00-00-30, which happens to be assigned by IEEE to VG Laboratory Systems Ltd.

Some vendors circumvent the issue by requesting the canonical and non-canonical forms of a given OUI (IBM with OUIs 10-00-5A and 08-00-5A and Wyle Electronics with OUIs 00—5-A8 and 00-A0-15).

See the OUIs listed in the Reference and Information Tables section—note the comments for known cases of bit-reversed LAN presentation.

IEEE 802.2 Logical Link Control (LLC)

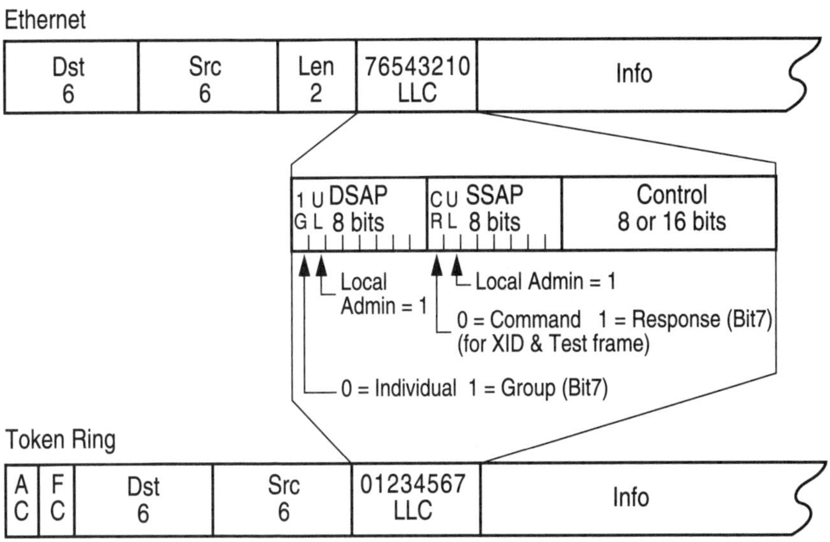

		Control Field
Type 1:	Connectionless (most NON IBM environments)	8
Type 2:	Connection Oriented Reliable Service	8/16
Type 3:	Acknowledged Connectionless Service	8

Type 1 (1 byte ctrl)	Type 2 (1 byte ctrl)	Type 2 (2 byte ctrl)
UI = Datagram	SABME = Connection Request	I = Information
XID = Exchange ID	DISC = End Connect	RR = Rcvr Ready ACK
Test = Packet Test	DM = Disc ACK	RNR = Rcvr Not Ready TACK
	FRMR = Received Invalid Pkt	
	UA = ACK DISK or SABME	
	XID = Exchange ID	
	Test = Packet Test	

LLC Frame Details

LLC Information Header

DSAP F0-Info	SSAP - F0	Bits 1-7 - N(S) Bit 8 - RR	Bits 1-7 -N(R) Bit 8 - Poll/Final
Layers 4-7			

LLC Supervisory Header

DSAP - F1	SSAP - F0	Bits 1-7 - N(S) Bit 8 - RR	Bits 1-7 -N(R) Bit 8 - Poll/Final
Layers 4-7			

LLC Unnumbered(UI) Frame Header

DSAP F0-Command F1 - Response	SSAP - F0	03	
Layers 4-7			

N(R) - Acknowledgment of the last frame number received.
N(S) - Identification number of data frame being transmitted.
Information Frame - Frame containing data beyond the LLC Layer. Session or Data Sessions Commands used with Information Frames.
Supervisory Frame - Frame acknowledging the receipt of Information Frame.
Information or Supervisory Frame Details

- **RR** - Receiver ready to accept data
- **RNR** - Receiver not ready to accept data
- **REJ** - Negative acknowledgment, go back to "N" frame
- **Poll** - Immediate frame acknowledgment requested from destination workstation
- **Final** -Acknowledgment of Poll request frame

Unnumbered Frame - N(R) and N(S) not used in the LLC UI Frame Header. Used with Management, NAME_QUERY, NAME_RECOGNIZED and Miscellaneous Commands.

- **SABME** - Connection request
- **DISC** - Disconnect connection
- **UA** - Acknowledge unnumbered command
- **DM** - Connection rejection
- **FRMR** - Reports receipt of unacceptable frame

IP

Ethernet LAN

Ethernet II*

Dst 6	Src 6	Etype 0800	Layers 3-7	FCS

802.3 w/802.2 SAP

Dst	Src	Len	SAP 06 06	Layers 3-7	FCS

802.3 w/802.2 SNAP

Dst	Src	Len	SAP AA AA	03	OUI 000000	Etype 0800	Layers 3-7	FCS

Token Ring LAN

802.5 w/802.2 SAP

AC	FC	Dst	Src	SAP 06 06	Layers 3-7	FCS

802.5 w/802.2 SNAP*

AC	FC	Dst	Src	SAP AA AA	03	OUI 000000	Etype 0800	Layers 3-7	FCS

FDDI LAN

FDDI w/802.2 SAP

FC	Dst 6	Src 6	SAP 06 06	Layers 3-7	FCS

FDDI w/802.2 SNAP*

FC	Dst 6	Src 6	SAP AA AA	03	OUI 000000	Etype 0800	Layers 3-7	FCS

* Most common use

IP Packet Structure

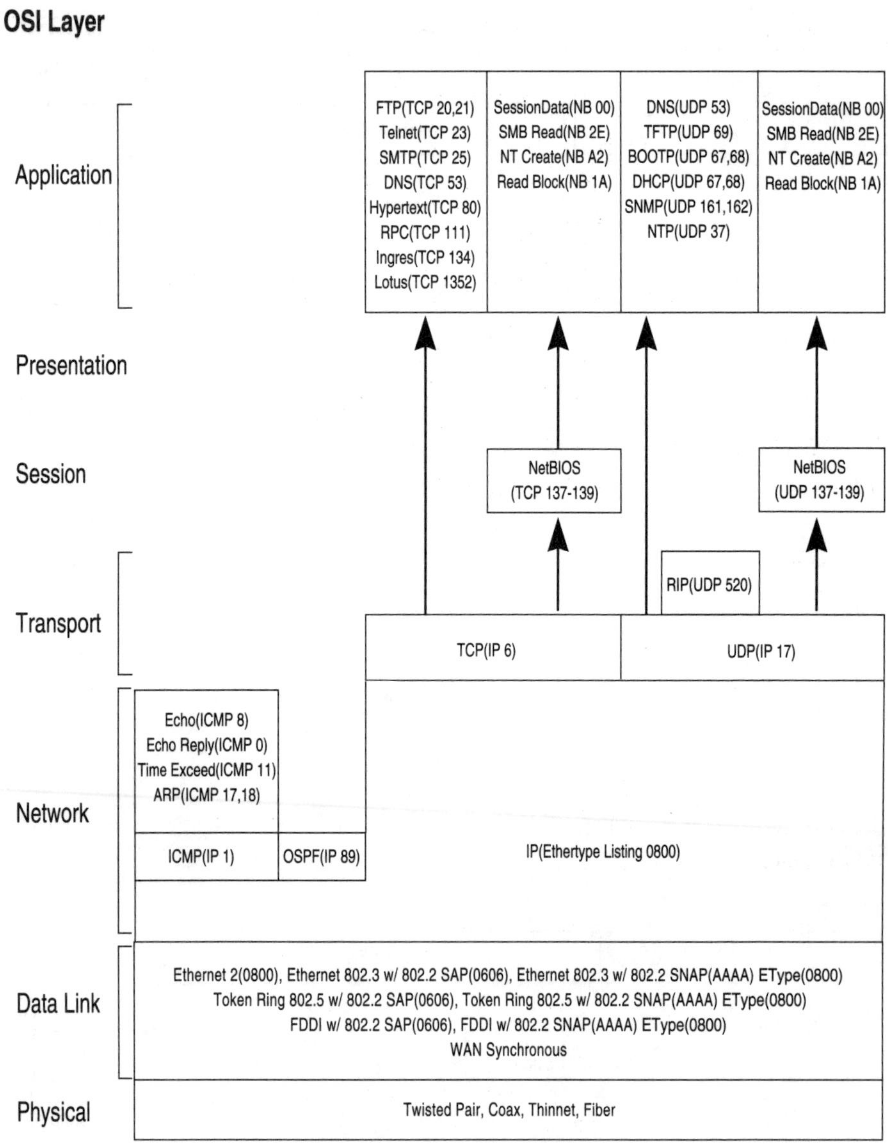

IP Packet Details

IP Header

<table>
<tr><td>Version (4)
IHL (4)</td><td>Type of Service</td><td colspan="2">Packet Length</td></tr>
<tr><td colspan="2">Packet Identifier</td><td colspan="2">Flags</td></tr>
<tr><td>Time to Live
(in Hops)</td><td>Protocol
TCP (6), UDP (17)</td><td colspan="2">IP Header Checksum</td></tr>
<tr><td colspan="4">Source IP Address</td></tr>
<tr><td colspan="4">Destination IP Address</td></tr>
<tr><td colspan="3">IP Options (Rarely Used)</td><td>Padding Options to 32 bits</td></tr>
<tr><td colspan="4">TCP, UDP, ICMP Header</td></tr>
</table>

IP Header Details

Version - Version of IP Packet used.

IHL - (IP Header Length) length of IP header in 32 bit increments.

Type of Service - xxxabcyy

a - 0 = Normal delay
1 = Minimize delay

b - 0 = Normal throughput
1 = Maximize throughput

c - 0 = Normal reliability
1 = Maximize reliability

xxxyy - Not commonly used

Packet Length - Number of bytes included in layers 3-7.

Packet Identifer - Used to identify a block of data received from an upper layer. Used when packet is fragmented.

Flags - RABY YYYY YYYY YYYY

R = Reserved

A - 0 = Can fragment
1 = Cannot fragment

B - 0 = Last fragment in a datagram
1 = Not the last fragment in a datagram (more fragments)

Y - Fragment offset in bytes from the beginning of data transfer. Number of bytes must be determined by calculating the value of Y x 8.

Time to Live - Maximum number of hops (routers) the packet can pass through (decremented)

Protocol - Identification of the Transport layer protocol

IP Header Checksum - CRC calculation for the IP header

TCP Header - Reliable IP data transport

Source Port			Destination Port (i.e. telnet = 23)	
Sequence Number (Equal to the sequence number sent in the previous packet plus the amount of data transmitted in current packet)				
Acknowledgment Number (Equal to the previous acknowledgment plus the amount of data received)				
Header Length (4)	Reserved (6)	Flags (6)	Window	
Checksum (CRC Check for TCP header)			Urgent Pointer (Points to end point in the data field considered urgent)	
Options				Padding
Application Layer or Data				

TCP Header Details

TCP Header Length - Number of bytes included in the TCP header

Flags - A B C D E F

A - 0 = No urgent data.
1 = Urgent data.

B - 0 = Acknowledgment field is not significant.
1 = Acknowledgment field is significant.

C - 0 = No push function.
1 = Push function – immediately pass the data to upper layers without waiting for additional data.

D - 0 = Packet is not requesting a reset of the session.
1 = Packet is requesting a reset of the session.

E - 0 = Not the initialization of a session.
1 = Initialization of a session.

F - 0 = The packet does not request to finish the session (synchronize).
1 = The packet does request to end the session.

Window - Number of bytes the workstation will accept for an individual connection before an acknowledgment must be sent.

Checksum - CRC check for TCP header.

Urgent Pointer - When the urgent flag bit is set, the urgent pointer field indicates where the urgent data will end within the data passed.

Options - Used to negotiate TCP options between equipment involved in a connection.
MSS (Maximum segment size) - Maximum packet size accepted by a station.

Padding - Additional bits added to make TCP layer an even increment of 32 bits.

UDP Header - Unreliable IP data transport

Source Port	Destination Port (i.e. DNS = 53)
UDP Length (Includes UDP Header and Data)	UDP Header and Data Checksum
Application Layer or Data	

ICMP Header - Internet Control Message Protocol

Type of ICMP Packet (i.e. 8=Echo Req)	Code - Detailed Packet Data	ICMP Header and Data Checksum
Optional Data		

ICMP Examples

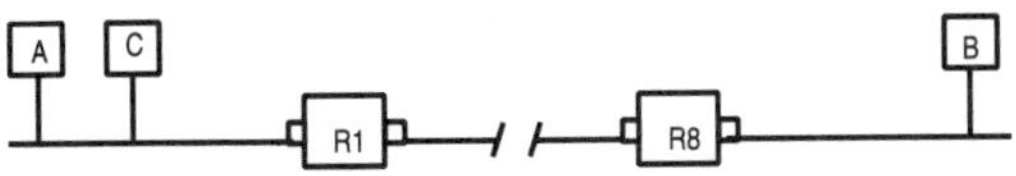

E = Packet Header Returned

				ICMP Type	ICMP Code
TTL Expired	A > B	TTL expires while B reassembles fragmented datagram	A < B	11	0 E
	A > B	TTL expired when packet arrived at R8	A < R8	11	1 E
Parameter Unintelligible	A > B	R8 can't understand IP header	A < R8	12	1/0 E
Destination Unreachable	A > B	R4's connection to R5 was lost (Network unreachable)	A < R4	3	0 E
	A > B	R8 can't find B in ARP cache and it doesn't respond	A < R8	3	1 E
	A > B	B doesn't support the TCP, UDP, ISODE higher layer protocol identified in header	A < B	3	2 E
	A > B	B doesn't support the UDP port specified	A < B	3	3 E
	A > B	R3 discards packet with no fragmentation set	A < R3	3	4 E
Source Quench	A > B	R4 has low speed link and is running out of buffers	A < R4	4	0 E
Redirect	A > C	A sends packet for C to R1 instead of directly	A < R1	5	1
	A > B	R1 sends to R2, but R2 lost connect to R3	R1< R2	5	0

IP Subnet Mask Address Table

Subnet masking is used to expand the number of networks due to the 32-bit limitation of IP's address field. When assigned an address by the NIC, the address can be further broken down to expand the single net number to many more by using host bits.

Class	Range
A	1.0.0.0 to 126.255.255.255
Loop Back	127.0.0.0 to 127.255.255.255
B	128.0.0.0 to 191.255.255.255
C	192.0.0.0 to 223.255.255.255
D	224.0.0.0 to 239.255.255.255

Subnet Bits	Bit Positions	Decimal Mask	HEX Mask	Sub/Nets Avail.	Hosts Avail.
Class A	0nnnnnnn.hhhhhhhh.hhhhhhhh.hhhhhhhh	255.0.0.0	FF-00-00-00	126	16777124
Broadcast	0nnnnnnn.11111111.11111111.11111111				
Class B	10nnnnnn.nnnnnnnn.hhhhhhhh.hhhhhhhh	255.255.0.0	FF-FF-00-00	16384	65534
2	10nnnnnn.nnnnnnnn.sshhhhhh.hhhhhhhh	255.255.192.0	FF-FF-C0-00	2	16382
3	10nnnnnn.nnnnnnnn.ssshhhhh.hhhhhhhh	255.255.224.0	FF-FF-E0-00	6	8190
4	10nnnnnn.nnnnnnnn.sssshhhh.hhhhhhhh	255.255.240.0	FF-FF-F0-00	14	4094
5	10nnnnnn.nnnnnnnn.ssssshhh.hhhhhhhh	255.255.248.0	FF-FF-F8-00	30	2046
6	10nnnnnn.nnnnnnnn.sssssshh.hhhhhhhh	255.255.252.0	FF-FF-FC-00	62	1022
7	10nnnnnn.nnnnnnnn.sssssssh.hhhhhhhh	255.255.254.0	FF-FF-FE-00	126	510
8	10nnnnnn.nnnnnnnn.ssssssss.hhhhhhhh	255.255.255.0	FF-FF-FF-00	254	254
9	10nnnnnn.nnnnnnnn.ssssssss.shhhhhhh	255.255.255.128	FF-FF-FF-80	510	126
10	10nnnnnn.nnnnnnnn.ssssssss.sshhhhhh	255.255.255.192	FF-FF-FF-C0	1022	62
11	10nnnnnn.nnnnnnnn.ssssssss.ssshhhhh	255.255.255.224	FF-FF-FF-E0	2046	30
12	10nnnnnn.nnnnnnnn.ssssssss.sssshhhh	255.255.255.240	FF-FF-FF-F0	4094	14
13	10nnnnnn.nnnnnnnn.ssssssss.ssssshhh	255.255.255.248	FF-FF-FF-F8	8190	6
13	10nnnnnn.nnnnnnnn.ssssssss.sssssshh	255.255.255.252	FF-FF-FF-FC	16382	2
Broadcast	10nnnnnn.nnnnnnnn.11111111.11111111				
Class C	110nnnnn.nnnnnnnn.nnnnnnnn.hhhhhhhh	255.255.255.0	FF-FF-FF-00	2097152	254
2	110nnnnn.nnnnnnnn.nnnnnnnn.sshhhhhh	255.255.255.192	FF-FF-FF-C0	2	62
3	110nnnnn.nnnnnnnn.nnnnnnnn.ssshhhhh	255.255.255.224	FF-FF-FF-E0	6	30
4	110nnnnn.nnnnnnnn.nnnnnnnn.sssshhhh	255.255.255.240	FF-FF-FF-F0	14	14
5	110nnnnn.nnnnnnnn.nnnnnnnn.ssssshhh	255.255.255.248	FF-FF-FF-F8	30	6
6	110nnnnn.nnnnnnnn.nnnnnnnn.sssssshh	255.255.255.252	FF-FF-FF-FC	62	2
Broadcast	110nnnnn.nnnnnnnn.nnnnnnnn.11111111				
Class D	1110xxxx.xxxxxxxx.xxxxxxxx.xxxxxxxx				
Future	11110xxx.xxxxxxxx.xxxxxxxx.xxxxxxxx				
All 0's	00000000.00000000.00000000.00000000				
All 1's	11111111.11111111.11111111.11111111				

0 = binary 0
1 = binary 1
n = network bits
h = host bits
s = subnet bits
x = other

Subnetting Class B Addresses (page 1 of 2)

Subnet/Host Bits	5/11	6/10	7/9			
Subnet Bits(1)	11111000	11111100	11111110			
Host Bits(0)	00000000	00000000	00000000			
# Subnets(x)	30	62	126			
# Hosts/Subnet	2046	1022	510			
255.255.x.y	248.0	252.0	254.0			
Subnets(x.y)	1-30	1-62	1-126	Net #	(7/9 con't)	Net #
	8.0-15.255	4.0-7.255	2.0-3.255	1	130.0-131.255	65
	16.0-23.255	8.0-11.255	4.0-5.255	2	132.0-133.255	66
	24.0-31.255	12.0-15.255	6.0-7.255	3	134.0-135.255	67
	32-0-39-255	16.0-19.255	8.0-9.255	4	136.0-137.255	68
	40.0-47.255	20.0-23.255	10.0-11.255	5	138.0-139.255	69
	48.0-55.255	24.0-27.255	12.0-13.255	6	140.0-141.255	70
	56.0-63.255	28.0-31.255	14.0-15.255	7	142.0-143.255	71
	64.0-71.255	32.0-35.255	16.0-17.255	8	144.0-145.255	72
	72.0-79.255	36.0-39.255	18.0-19.255	9	146.0-147.255	73
	80.0-87.255	40.0-43.255	20.0-21.255	10	148.0-149.255	74
	88.0-95.255	44.0-47.255	22.0-23.255	11	150.0-151.255	75
	96.0-103.255	48.0-51.255	24.0-25.255	12	152.0-153.255	76
	104.0-111.255	52.0-55.255	26.0-27.255	13	154.0-155.255	77
	112.0-119.255	56.0-59.255	28.0-29.255	14	156.0-157.255	78
	120.0-127.255	60.0-63.255	30.0-31.255	15	158.0-159.255	79
	128.0-135.255	64.0-67.255	32.0-33.255	16	160.0-161.255	80
	136.0-143.255	68.0-71.255	34.0-35.255	17	162.0-163.255	81
	144.0-151.255	72.0-75.255	36.0-37.255	18	164.0-165.255	82
	152.0-159.255	76.0-79.255	38.0-39.255	19	166.0-167.255	83
	160.0-167.255	80.0-83.255	40.0-41.255	20	168.0-169.255	84
	168.0-175.255	84.0-87.255	42.0-43.255	21	170.0-171.255	85
	176.0-183.255	88.0-91.255	44.0-45.255	22	172.0-173.255	86
	184.0-191.255	92.0-95.255	46.0-47.255	23	174.0-175.255	87
	192.0-199.255	96.0-99.255	48.0-49.255	24	176.0-177.255	88
	200.0-207.255	100.0-103.255	50.0-51.255	25	178.0-179.255	89
	208.0-215.255	104.0-107.255	52.0-53.255	26	180.0-181.255	90
	216.0-223.255	108.0-111.255	54.0-55.255	27	182.0-183.255	91
	224.0-231.255	112.0-115.255	56.0-57.255	28	184.0-185.255	92
	232.0-239.255	116.0-119.255	58.0-59.255	29	186.0-187.255	93
	240.0-247.255	120.0-123.255	60.0-61.255	30	188.0-189.255	94
		124.0-127.255	62.0-63.255	31	190.0-191.255	95
		128.0-131.255	64.0-65.255	32	192.0-193.255	96
		132.0-135.255	66.0-67.255	33	194.0-195.255	97
		136.0-139.255	68.0-69.255	34	196.0-197.255	98
		140.0-143.255	70.0-71.255	35	198.0-199.255	99
		144.0-147.255	72.0-73.255	36	200.0-201.255	100
		148.0-151.255	74.0-75.255	37	202.0-203.255	101
		152.0-155.255	76.0-77.255	38	204.0-205.255	102
		156.0-159.255	78.0-79.255	39	206.0-207.255	103
		160.0-163.255	80.0-81.255	40	208.0-209.255	104
		164.0-167.255	82.0-83.255	41	210.0-211.255	105
		168.0-171.255	84.0-85.255	42	212.0-213.255	106
		172.0-175.255	86.0-87.255	43	214.0-215.255	107
		176.0-179.255	88.0-89.255	44	216.0-217.255	108
		180.0-183.255	90.0-91.255	45	218.0-219.255	109
		184.0-187.255	92.0-93.255	46	220.0-221.255	110
		188.0-191.255	94.0-95.255	47	222.0-223.255	111
		192.0-195.255	96.0-97.255	48	224.0-225.255	112
		196.0-199.255	98.0-99.255	49	226.0-227.255	113
		200.0-203.255	100.0-101.255	50	228.0-229.255	114
		204.0-207.255	102.0-103.255	51	230.0-231.255	115
		208.0-211.255	104.0-105.255	52	232.0-233.255	116
		212.0-215.255	106.0-107.255	53	234.0-235.255	117
		216.0-219.255	108.0-109.255	54	236.0-237.255	118
		220.0-223.255	110.0-111.255	55	238.0-239.255	119
		224.0-227.255	112.0-113.255	56	240.0-241.255	120
		228.0-231.255	114.0-115.255	57	242.0-243.255	121
		232.0-235.255	116.0-117.255	58	244.0-245.255	122
		236.0-239.255	118.0-119.255	59	246.0-247.255	123
		240.0-243.255	120.0-121.255	60	248.0-249.255	124
		244.0-247.255	122.0-123.255	61	250.0-251.255	125
		248.0-251.255	124.0-125.255	62	252.0-253.255	126
			126.0-127.255	63		
			128.0-129.255	64		

Subnetting Class B Addresses (page 2 of 2)

Subnet/Host Bits	8/8	9/7	10/6	11/5	12/4	13/3	14/2	
Subnet Bits(1)	11111111	11111111	11111111	11111111	11111111	11111111	11111111	
Host Bits(0)	00000000	10000000	11000000	11100000	11110000	11111000	11111100	
# Subnets(x)	254	510	1022	2046	4094	8190	16382	
# Hosts/Subnet	254	126	62	30	14	6	2	
255.255.x.y	255.0	255.128	255.192	255.224	255.240	255.248	255.252	
Subnets(x)	1-254	1-254	1-254	1-254	1-254	1-254	1-254	Net #
Addresses(y)	.xxx.0-255	.xxx.0-127	.xxx.0-63	.xxx.0-31	.xxx.0-15	.xxx.0-7	.xxx.0-3	1
for each subnet		.xxx.128-255	.xxx.64-127	.xxx.32-63	.xxx.16-31	.xxx.8-15	.xxx.4-7	2
			.xxx.128-191	.xxx.64-95	.xxx.32-47	.xxx.16-23	.xxx.8-11	3
			.xxx.192-255	.xxx.96-127	.xxx.48-63	.xxx.24-31	.xxx.12-15	4
				.xxx.128-159	.xxx.64-79	.xxx.32-39	.xxx.16-19	5
				.xxx.160-191	.xxx.80-95	.xxx.40-47	.xxx.20-23	6
				.xxx.192-223	.xxx.96-111	.xxx.48-55	.xxx.24-27	7
				.xxx.224-255	.xxx.112-127	.xxx.56-63	.xxx.28-31	8
					.xxx.128-143	.xxx.64-71	.xxx.32-35	9
					.xxx.144-159	.xxx.72-79	.xxx.36-39	10
					.xxx.160-175	.xxx.80-87	.xxx.40-43	11
					.xxx.176-191	.xxx.88-95	.xxx.44-47	12
					.xxx.192-207	.xxx.96-103	.xxx.48-51	13
					.xxx.208-223	.xxx.104-111	.xxx.52-55	14
					.xxx.224-239	.xxx.112-119	.xxx.56-59	15
					.xxx.240-255	.xxx.120-127	.xxx.60-63	16
						.xxx.128-135	.xxx.64-67	17
						.xxx.136-143	.xxx.68-71	18
						.xxx.144-151	.xxx.72-75	19
						.xxx.152-159	.xxx.76-79	20
						.xxx.160-167	.xxx.80-83	21
						.xxx.168-175	.xxx.84-87	22
						.xxx.176-183	.xxx.88-91	23
						.xxx.184-191	.xxx.92-95	24
						.xxx.192-199	.xxx.96-99	25
						.xxx.200-207	.xxx.100-103	26
						.xxx.208-215	.xxx.104-107	27
						.xxx.216-223	.xxx.108-111	28
						.xxx.224-231	.xxx.112-115	29
						.xxx.232-239	.xxx.116-119	30
						.xxx.240-247	.xxx.120-123	31
						.xxx.248-255	.xxx.124-127	32
							.xxx.128-131	33
							.xxx.132-135	34
							.xxx.136-139	35
							.xxx.140-143	36
							.xxx.144-147	37
							.xxx.148-151	38
							.xxx.152-155	39
							.xxx.156-159	40
							.xxx.160-163	41
							.xxx.164-167	42
							.xxx.168-171	43
							.xxx.172-175	44
							.xxx.176-179	45
							.xxx.180-183	46
							.xxx.184-187	47
							.xxx.188-191	48
							.xxx.192-195	49
							.xxx.196-199	50
							.xxx.200-203	51
							.xxx.204-207	52
							.xxx.208-211	53
							.xxx.212-215	54
							.xxx.216-219	55
							.xxx.220-223	56
							.xxx.224-227	57
							.xxx.228-231	58
							.xxx.232-235	59
							.xxx.236-239	60
							.xxx.240-243	61
							.xxx.244-247	62
							.xxx.248-251	63
							.xxx.252-255	64

Subnetting a Class B Address

Subnet Mask Examples

Mask	Network	Subnet	Host	
255.255.254.0	11111111.11111111.	1111111	0.00000000	7 Bits Used for Subnet Addresses 9 Bits Used for Host Addresses
255.255.255.0	11111111.11111111.	11111111	.00000000	8 Bits Used for Subnet Addresses 8 Bits Used for Host Addresses
255.255.255.128	11111111.11111111.	11111111.1	0000000	9 Bits Used for Subnet Addresses 7 Bits Used for Host Addresses

Assigning Network Addresses Using Different Subnet Mask

Class B Address - 16 Bits		7 Bits	9 Bits	
Network Number		Subnet Num	Host Number	Subnet Mask - 255.255.254.0(126 Subnets)
8 Bits(172)	8 Bits(20)	7 Bits(93)	9 Bits(25)	
10101100	00010100	1011101	000011001	Network - 172.20, Subnet - 93, Host - 25
8 Bits(172)	8 Bits(20)	8 Bits	8 Bits	
10101100	00010100	10111010	00011001	Programmed Address(in 8 bit increments) 172.20.186.25

Class B Address - 16 Bits		8 Bits	8 Bits	
Network Number		Subnet Number	Host Number	Subnet Mask - 255.255.255.0(254 Subnets)
8 Bits(172)	8 Bits(20)	8 Bits(201)	8 Bits(43)	
10101100	00010100	11001001	00101011	Network - 172.20, Subnet - 201, Host - 43
8 Bits(172)	8 Bits(20)	8 Bits	8 Bits	
10101100	00010100	11001001	00101011	Programmed Address(in 8 bit increments) 172.20.201.43

Class B Address - 16 Bits		9 Bits	7 Bits	
Network Number		Subnet Number	Host Number	Subnet Mask - 255.255.255.128(510 Subnets)
8 Bits(172)	8 Bits(20)	9 Bits(450)	7 Bits(90)	
10101100	00010100	111000010	1011010	Network - 172.20, Subnet - 450, Host - 90
8 Bits(172)	8 Bits(20)	8 Bits	8 Bits	
10101100	00010100	11100001	01011010	Programmed Address(in 8 bit increments) 172.20.225.90

Subnet Identification and Broadcast
Address - 172.20.92.150
Subnet Mask - 255.255.255.0
Subnet ID- 172.20.92.0(lowest value within the subnet)
Subnet Broadcast - 172.20.92.255(highest value within the subnet)

Address - 172.20.92.150
Subnet Mask - 255.255.255.192
Subnet ID - 172.20.92.128
Subnet Broadcast - 172.20.92.191

Note: Cannot use the minimum(0) or maximum(255) number for subnet.

Subnetting a Class C Address

Subnet/Host Bits	0/8	1/7	2/6	3/5	4/4	5/3	6/2	
Subnet Bits(1)		1	11	111	1111	11111	111111	
Host Bits(0)	00000000	0000000	000000	00000	0000	000	00	
# Subnets(x)	1	2	4	8	16	32	64	
# Hosts/Subnet	254	126	62	30	14	6	2	
255.255.x.y	255.0	255.128	255.192	255.224	255.240	255.248	255.252	
Subnets(x)	1-254	1-255	1-255	1-255	1-255	1-255	1-255	Net #
Addresses(y)	.xxx.0-255	.xxx.0-127	.xxx.0-63	.xxx.0-31	.xxx.0-15	.xxx.0-7	.xxx.0-3	1
		.xxx.128-255	.xxx.64-127	.xxx.32-63	.xxx.16-31	.xxx.8-15	.xxx.4-7	2
			.xxx.128-191	.xxx.64-95	.xxx.32-47	.xxx.16-23	.xxx.8-11	3
			.xxx.192-255	.xxx.96-127	.xxx.48-63	.xxx.24-31	.xxx.12-15	4
				.xxx.128-159	.xxx.64-79	.xxx.32-39	.xxx.16-19	5
				.xxx.160-191	.xxx.80-95	.xxx.40-47	.xxx.20-23	6
				.xxx.192-223	.xxx.96-111	.xxx.48-55	.xxx.24-27	7
				.xxx.224-255	.xxx.112-127	.xxx.56-63	.xxx.28-31	8
					.xxx.128-143	.xxx.64-71	.xxx.32-35	9
					.xxx.144-159	.xxx.72-79	.xxx.36-39	10
					.xxx.160-175	.xxx.80-87	.xxx.40-43	11
					.xxx.176-191	.xxx.88-95	.xxx.44-47	12
					.xxx.192-207	.xxx.96-103	.xxx.48-51	13
					.xxx.208-223	.xxx.104-111	.xxx.52-55	14
					.xxx.224-239	.xxx.112-119	.xxx.56-59	15
					.xxx.240-255	.xxx.120-127	.xxx.60-63	16
						.xxx.128-135	.xxx.64-67	17
						.xxx.136-143	.xxx.68-71	18
						.xxx.144-151	.xxx.72-75	19
						.xxx.152-159	.xxx.76-79	20
						.xxx.160-167	.xxx.80-83	21
						.xxx.168-175	.xxx.84-87	22
						.xxx.176-183	.xxx.88-91	23
						.xxx.184-191	.xxx.92-95	24
						.xxx.192-199	.xxx.96-99	25
						.xxx.200-207	.xxx.100-103	26
						.xxx.208-215	.xxx.104-107	27
						.xxx.216-223	.xxx.108-111	28
						.xxx.224-231	.xxx.112-115	29
						.xxx.232-239	.xxx.116-119	30
						.xxx.240-247	.xxx.120-123	31
						.xxx.248-255	.xxx.124-127	32
							.xxx.128-131	33
							.xxx.132-135	34
							.xxx.136-139	35
							.xxx.140-143	36
							.xxx.144-147	37
							.xxx.148-151	38
							.xxx.152-155	39
							.xxx.156-159	40
							.xxx.160-163	41
							.xxx.164-167	42
							.xxx.168-171	43
							.xxx.172-175	44
							.xxx.176-179	45
							.xxx.180-183	46
							.xxx.184-187	47
							.xxx.188-191	48
							.xxx.192-195	49
							.xxx.196-199	50
							.xxx.200-203	51
							.xxx.204-207	52
							.xxx.208-211	53
							.xxx.212-215	54
							.xxx.216-219	55
							.xxx.220-223	56
							.xxx.224-227	57
							.xxx.228-231	58
							.xxx.232-235	59
							.xxx.236-239	60
							.xxx.240-243	61
							.xxx.244-247	62
							.xxx.248-251	63
							.xxx.252-255	64

Subnetting a Class C Address

Subnet Mask Examples

	Network and Subnet / Host	
255.255.255.0	11111111.11111111.11111111.00000000	0 Bits Used for Subnet Addresses 8 Bits Used for Host Addresses

	Network / Subnet/Host	
255.255.255.192	11111111.11111111.11111111.11000000	2 Bits Used for Subnet Addresses 6 Bits Used for Host Addresses

	Network / Subnet/Host	
255.255.255.224	11111111.11111111.11111111.11100000	3 Bits Used for Subnet Addresses 5 Bits Used for Host Addresses

Assigning Network Addresses Using Different Subnet Mask

Class C Address - 24 Bits			8 Bits	
Network/Subnet Number			Host Number	Subnet Mask - 255.255.255.0(1 Subnet)
8 Bits(192)	8 Bits(10)	8 Bits(15)	8 Bits(220)	
1 1 0 0 0 0 0 0	0 0 0 0 1 0 1 0	0 0 0 0 1 1 1 1	1 1 0 1 1 1 0 0	Network - 192.10.15, Host - 220
8 Bits(192)	8 Bits(10)	8 Bits(15)	8 Bits(220)	
1 1 0 0 0 0 0 0	0 0 0 0 1 0 1 0	0 0 0 0 1 1 1 1	1 1 0 1 1 1 0 0	Programmed Address(in 8 bit increments) 192.10.15.220

Class C Address - 24 Bits			8 Bits	
Network Number			Subnet and Host Number	Subnet Mask - 255.255.255.192(4 Subnets)
8 Bits(192)	8 Bits(10)	8 Bits(15)	2 Bits(2),6 Bits(52)	
1 1 0 0 0 0 0 0	0 0 0 0 1 0 1 0	0 0 0 0 1 1 1 1	1 0 1 1 0 1 0 0	Network - 192.10.15, Subnet 2, Host - 52
8 Bits(192)	8 Bits(10)	8 Bits(15)	8 Bits(116)	
1 1 0 0 0 0 0 0	0 0 0 0 1 0 1 0	0 0 0 0 1 1 1 1	1 0 1 1 0 1 0 0	Programmed Address(in 8 bit increments) 192.10.15.180

Class C Address - 24 Bits			8 Bits	
Network Number			Subnet and Host Number	Subnet Mask - 255.255.255.224(8 Subnets)
8 Bits(192)	8 Bits(10)	8 Bits(15)	3 Bits(5),5 Bits(20)	
1 1 0 0 0 0 0 0	0 0 0 0 1 0 1 0	0 0 0 0 1 1 1 1	1 0 1 1 0 1 0 0	Network - 192.10.15, Subnet 5, Host - 20
8 Bits(192)	8 Bits(10)	8 Bits(15)	8 Bits(180)	
1 1 0 0 0 0 0 0	0 0 0 0 1 0 1 0	0 0 0 0 1 1 1 1	1 0 1 1 0 1 0 0	Programmed Address(in 8 bit increments) 192.10.15.180

Subnet Identification and Broadcast

Address - 192.10.15.125
Subnet Mask - 255.255.255.192
Subnet ID- 192.10.15.64(lowest value within the subnet)
Subnet Broadcast - 192.10.15.127(highest value within the subnet)

Address - 192.10.15.125
Subnet Mask - 255.255.255.224
Subnet ID - 192.10.15.96
Subnet Broadcast - 192.10.15.127

Note: Cannot use the minimum(0) or maximum number for subnet.

Common IP Protocol Assignments

Abbreviation	IP Port Number	Description
ICMP	1	Internet Control Message
IGMP	2	Internet Group Management
TCP	6	Transmission Control Protocol
EGP	8	Exterior Gateway Protocol
IGP	9	Interior Gateway Protocol
UDP	17	User Datagram Protocol
IGRP	88	Internet Gateway Routing Protocol
OSPF	89	Open Shortest Path First

Common ICMP Port Assignments

Abbreviation	IP Port Number	Description
Echo Reply	0	Ping Reply
Destination Unreachable	3	Destination IP Address cannot be reached
Redirect	5	Router sending a packet onto the same interface it received the packet on
Echo	8	Ping
Time Exceed	11	Time to Live = 0
ARP	17,18	Address Resolution Protocols

Common TCP/UDP Port Assignments

The port numbers are divided into three ranges: the Well Known Ports, the Registered Ports, and the Dynamic and/or Private Ports. Ports are used in the TCP [RFC793] to name the ends of logical connections which carry long term conversations. For the purpose of providing services to unknown callers, a service contact port is defined. This list specifies the port used by the server process as its contact port. While the IANA can not control uses of these ports it does register or list uses of these ports as a convenience to the community. To the extent possible, these same port assignments are used with the UDP [RFC768].

WELL KNOW PORT NUMBERS

The Well Known Ports are in the range 0-1023.

The Well Known Ports are assigned by the IANA and on most systems can only be used by system (or root) processes or by programs executed by privileged users.

REGISTERED PORT NUMBERS

The Registered Ports are in the range 1024-65535.

The Registered Ports are ***not*** controlled by the IANA and on most systems can be used by ordinary user processes or programs executed by ordinary users.

DYNAMIC and/or PRIVATE PORTS

The Dynamic and/or Private Ports are in the range 49152-65535. Dynamic and/or Private Ports are a subset of Registered Ports.

Port Assignments (sorted by port number):

Abbreviation	Port-Decimal	Protocol	Description/Owner
	0	tcp/udp	Reserved—ISI
tcpmux	1	tcp/udp	TCP Port Service Multiplexer—SRI
compressnet	2	tcp/udp	Management Utility—Process
compressnet	3	tcp/udp	Compression—Process
#	4	tcp/udp	Unassigned
rje	5	tcp/udp	Remote Job Entry—ISI
#	6	tcp/udp	Unassigned
echo	7	tcp/udp	Echo—ISI
#	8	tcp/udp	Unassigned
discard	9	tcp/udp	Discard—ISI
#	10	tcp/udp	Unassigned
systat	11	tcp/udp	Active Users-ISI
#	12	tcp/udp	Unassigned
daytime	13	tcp/udp	Daytime (RFC 867)
#	14	tcp/udp	Unassigned
#	15	tcp/udp	Unassigned [was netstat]
#	16	tcp/udp	Unassigned
qotd	17	tcp/udp	Quote of the Day—ISI
msp	18	tcp/udp	Message Send Protocol
chargen	19	tcp/udp	Character Generator
ftp-data	20	tcp/udp	File Transfer [Default Data]—ISI
ftp	21	tcp/udp	File Transfer [Control]—ISI
ssh	22	tcp/udp	SSH Remote Login Protocol—Helsinki U
telnet	23	tcp/udp	Telnet—ISI
	24	tcp/udp	any private mail system—UUNET
smtp	25	tcp/udp	Simple Mail Transfer—ISI
#	26	tcp/udp	Unassigned
nsw-fe	27	tcp/udp	NSW User System FE—BBN
#	28	tcp/udp	Unassigned
msg-icp	29	tcp/udp	MSG ICP—BBN
#	30	tcp/udp	Unassigned
msg-auth	31	tcp/udp	MSG Authentication—BBN
#	32	tcp/udp	Unassigned
dsp	33	tcp/udp	Display Support Protocol—dca.mil
#	34	tcp/udp	Unassigned
	35	tcp/udp	any private printer server—ISI
#	36	tcp/udp	Unassigned
time	37	tcp/udp	Time—ISI
rap	38	tcp/udp	Route Access Protocol—The World Kiosk
rlp	39	tcp/udp	Resource Location Protocol—CMU
#	40	tcp/udp	Unassigned
graphics	41	tcp/udp	Graphics
name	42	tcp/udp	Host Name Server
nameserver	42	tcp/udp	Host Name Server
nicname	43	tcp/udp	Who Is
mpm-flags	44	tcp/udp	MPM FLAGS Protocol
mpm	45	tcp/udp	Message Processing Module [recv]
mpm-snd	46	tcp/udp	MPM [default send]—ISI
ni-ftp	47	tcp/udp	NI FTP—Isode Internet/X.400 Messaging
auditd	48	tcp/udp	Digital Audit Daemon
tacacs	49	tcp/udp	Login Host Protocol (TACACS)—BBN
re-mail-ck	50	tcp/udp	Remote Mail Checking Protocol—UIUC
la-maint	51	tcp/udp	IMP Logical Address Maintenance—Timeplex
xns-time	52	tcp/udp	XNS Time Protocol—Xerox
domain	53	tcp/udp	Domain Name Server—ISI
xns-ch	54	tcp/udp	XNS Clearinghouse—Xerox
isi-gl	55	tcp/udp	ISI Graphics Language
xns-auth	56	tcp/udp	XNS Authentication—Xerox
	57	tcp/udp	any private terminal access—ISI
xns-mail	58	tcp/udp	XNS Mail—Xerox
	59	tcp/udp	any private file service—ISI
#	60	tcp/udp	Unassigned
ni-mail	61	tcp/udp	NI MAIL—Isode Internet/X.400 Messaging
acas	62	tcp/udp	ACA Services—DEC
whois++	63	tcp/udp	whois++—Sunet
covia	64	tcp/udp	Communications Integrator (CI)—The Clout Project (Chicago)

Abbreviation	Port-Decimal	Protocol	Description/Owner
tacacs-ds	65	tcp/udp	TACACS-Database Service—BBN
sql*net	66	tcp/udp	Oracle SQL*NET
bootps	67	tcp/udp	Bootstrap Protocol Server—Stanford U
bootpc	68	tcp/udp	Bootstrap Protocol Client—Stanford U
tftp	69	tcp/udp	Trivial File Transfer—MIT
gopher	70	tcp/udp	Gopher—UMN
netrjs-1	71	tcp/udp	Remote Job Service—ISI
netrjs-2	72	tcp/udp	Remote Job Service—ISI
netrjs-3	73	tcp/udp	Remote Job Service—ISI
netrjs-4	74	tcp/udp	Remote Job Service—ISI
	75	tcp/udp	any private dial out service—ISI
deos	76	tcp/udp	Distributed External Object Store-The World Kiosk
	77	tcp/udp	any private RJE service—ISI
vettcp	78	tcp/udp	vettcp—DEC
finger	79	tcp/udp	Finger—Rutgers U
http	80	tcp/udp	World Wide Web HTTP-World Wide Web Consortium
www	80	tcp/udp	World Wide Web HTTP-World Wide Web Consortium
www-http	80	tcp/udp	World Wide Web HTTP-World Wide Web Consortium
hosts2-ns	81	tcp/udp	HOSTS2 Name Server-S1.gov
xfer	82	tcp/udp	XFER Utility—Lockheed Martin
mit-ml-dev	83	tcp/udp	MIT ML Device
ctf	84	tcp/udp	Common Trace Facility—DEC
mit-ml-dev	85	tcp/udp	MIT ML Device
mfcobol	86	tcp/udp	Micro Focus Cobol
	87	tcp/udp	any private terminal link—ISI
kerberos	88	tcp/udp	Kerberos—ISI
su-mit-tg	89	tcp/udp	SU/MIT Telnet Gateway
*	90	tcp	also being used unofficially by Pointcast
dnsix	90	tcp/udp	DNSIX Securit Attribute Token Map
mit-dov	91	tcp/udp	MIT Dover Spooler
npp	92	tcp/udp	Network Printing Protocol—UMD
dcp	93	tcp/udp	Device Control Protocol—BBN
objcall	94	tcp/udp	Tivoli Object Dispatcher
supdup	95	tcp/udp	SUPDUP—panda.com
dixie	96	tcp/udp	DIXIE Protocol Specification—Umich
swift-rvf	97	tcp/udp	Swift Remote Virtural File Protocol—U Florida
tacnews	98	tcp/udp	TAC News—ISI
metagram	99	tcp/udp	Metagram Relay
newacct	100	tcp	[unauthorized use]
hostname	101	tcp/udp	NIC Host Name Server—ISI
iso-tsap	102	tcp/udp	ISO-TSAP Class 0—Data Broadcasting Corporation
gppitnp	103	tcp/udp	Genesis Point-to-Point Trans Net
acr-nema	104	tcp/udp	ACR-NEMA Digital Imag. & Comm. 300
cso	105	tcp/udp	CCSO name server protocol—eSpecials
csnet-ns	105	tcp/udp	Mailbox Name Nameserver—UWISC
3com-tsmux	106	tcp/udp	3COM-TSMUX
rtelnet	107	tcp/udp	Remote Telnet Service—ISI
snagas	108	tcp/udp	SNA Gateway Access Server—DEC
pop2	109	tcp/udp	Post Office Protocol - Version 2—ISI
pop3	110	tcp/udp	Post Office Protocol - Version 3—Data Broadcasting Corporation
sunrpc	111	tcp/udp	SUN Remote Procedure Call
mcidas	112	tcp/udp	McIDAS Data Transmission Protocol—UCAR
ident	113	tcp	
auth	113	tcp/udp	Authentication Service—ARPA
audionews	114	tcp/udp	Audio News Multicast—Chalmers U of Technology
sftp	115	tcp/udp	Simple File Transfer Protocol—SRI
ansanotify	116	tcp/udp	ANSA REX Notify—ANSA
uucp-path	117	tcp/udp	UUCP Path Service
sqlserv	118	tcp/udp	SQL Services—DEC
nntp	119	tcp/udp	Network News Transfer Protocol-UC BERKLEY
cfdptkt	120	tcp/udp	CFDPTKT—UCOLUMBIA
erpc	121	tcp/udp	Encore Expedited Remote Pro.Call
smakynet	122	tcp/udp	SMAKYNET—Swiss Federal Institute of Technology
ntp	123	tcp/udp	Network Time Protocol—UDEL
ansatrader	124	tcp/udp	ANSA REX Trader—ANSA
locus-map	125	tcp/udp	Locus PC-Interface Net Map Ser—UCLA
nxedit	126	tcp/udp	NXEdit—Unisys
*	126	tcp	Previously assigned to application below

Abbreviation	Port-Decimal	Protocol	Description/Owner
#unitary	126	tcp/udp	Unisys Unitary Login
locus-con	127	tcp/udp	Locus PC-Interface Conn Server—UCLA
gss-xlicen	128	tcp/udp	GSS X License Verification
pwdgen	129	tcp/udp	Password Generator Protocol—ARMY.MIL
cisco-fna	130	tcp/udp	cisco FNATIVE
cisco-tna	131	tcp/udp	cisco TNATIVE
cisco-sys	132	tcp/udp	cisco SYSMAINT
statsrv	133	tcp/udp	Statistics Service—U Delaware
ingres-net	134	tcp/udp	INGRES-NET Service
epmap	135	tcp/udp	DCE endpoint resolution—HP
profile	136	tcp/udp	PROFILE Naming System—U ARIZONA
netbios-ns	137	tcp/udp	NETBIOS Name Service—ISI
netbios-dgm	138	tcp/udp	NETBIOS Datagram Service—ISI
netbios-ssn	139	tcp/udp	NETBIOS Session Service—ISI
emfis-data	140	tcp/udp	EMFIS Data Service—ISI
emfis-cntl	141	tcp/udp	EMFIS Control Service—ISI
bl-idm	142	tcp/udp	Britton-Lee IDM
imap	143	tcp/udp	Internet Message Access Protocol—U Washington
uma	144	tcp/udp	Universal Management Architecture—Power Center Software
uaac	145	tcp/udp	UAAC Protocol—Mitre
iso-tp0	146	tcp/udp	ISO-IP0—Data Broadcasting Corporation
iso-ip	147	tcp/udp	ISO-IP—Data Broadcasting Corporation
jargon	148	tcp/udp	Jargon—Bearnet
aed-512	149	tcp/udp	AED 512 Emulation Service—UPENN
sql-net	150	tcp/udp	SQL-NET
hems	151	tcp/udp	HEMS
bftp	152	tcp/udp	Background File Transfer Program—ISI
sgmp	153	tcp/udp	SGMP—NY State Ed Net (NYSER.NET)
netsc-prod	154	tcp/udp	NETSC—CSC
netsc-dev	155	tcp/udp	NETSC—CSC
sqlsrv	156	tcp/udp	SQL Service—ISI
knet-cmp	157	tcp/udp	KNET/VM Command/Message Protocol—XYLOGICS
pcmail-srv	158	tcp/udp	PCMail Server—MIT
nss-routing	159	tcp/udp	NSS-Routing—IBM
sgmp-traps	160	tcp/udp	SGMP-TRAPS—NY State Ed Net (NYSER.NET)
snmp	161	tcp/udp	SNMP—Data Broadcasting Corporation
snmptrap	162	tcp/udp	SNMPTRAP—Data Broadcasting Corporation
cmip-man	163	tcp/udp	CMIP/TCP Manager
cmip-agent	164	tcp/udp	CMIP/TCP Agent
xns-courier	165	tcp/udp	Xerox
s-net	166	tcp/udp	Sirius Systems
namp	167	tcp/udp	NAMP—NY State Ed Net (NYSER.NET)
rsvd	168	tcp/udp	RSVD—Imperial Software
send	169	tcp/udp	SEND—U Alaska
print-srv	170	tcp/udp	Network PostScript—DEC
multiplex	171	tcp/udp	Network Innovations Multiplex
cl/1	172	tcp/udp	Network Innovations CL/1
xyplex-mux	173	tcp/udp	Xyplex
mailq	174	tcp/udp	MAILQ—U Toronto
vmnet	175	tcp/udp	VMNET—Princeton U
genrad-mux	176	tcp/udp	GENRAD-MUX
xdmcp	177	tcp/udp	X Display Manager Control Protocol—MIT
nextstep	178	tcp/udp	NextStep Window Server—NEXT
bgp	179	tcp/udp	Border Gateway Protocol—Cisco
ris	180	tcp/udp	Intergraph
unify	181	tcp/udp	Unify
audit	182	tcp/udp	Unisys Audit SITP
ocbinder	183	tcp/udp	OCBinder
ocserver	184	tcp/udp	OCServer
remote-kis	185	tcp/udp	Remote-KIS
kis	186	tcp/udp	KIS Protocol
aci	187	tcp/udp	Application Communication Interface—TI
mumps	188	tcp/udp	Plus Five's MUMPS
qft	189	tcp/udp	Queued File Transport—San Diego State U
gacp	190	tcp/udp	Gateway Access Control Protocol-LANL.GOV
prospero	191	tcp/udp	Prospero Directory Service—ISI
osu-nms	192	tcp/udp	OSU Network Monitoring System—OHIO-STATE U
srmp	193	tcp/udp	Spider Remote Monitoring Protocol—SPIDER Systems UK

Abbreviation	Port-Decimal	Protocol	Description/Owner
irc	194	tcp/udp	Internet Relay Chat Protocol—Oulun Tliopisto
dn6-nlm-aud	195	tcp/udp	DNSIX Network Level Module Audit—NAVY.MIL
dn6-smm-red	196	tcp/udp	DNSIX Session Mgt Module Audit Redir—NAVY.MIL
dls	197	tcp/udp	Directory Location Service—Purdue U
dls-mon	198	tcp/udp	Directory Location Service Monitor—Purdue U
smux	199	tcp/udp	SMUX—Data Broadcasting Corporation
src	200	tcp/udp	IBM System Resource Controller
at-rtmp	201	tcp/udp	AppleTalk Routing Maintenance—CMU
at-nbp	202	tcp/udp	AppleTalk Name Binding—CMU
at-3	203	tcp/udp	AppleTalk Unused—CMU
at-echo	204	tcp/udp	AppleTalk Echo—CMU
at-5	205	tcp/udp	AppleTalk Unused—CMU
at-zis	206	tcp/udp	AppleTalk Zone Information—CMU
at-7	207	tcp/udp	AppleTalk Unused—CMU
at-8	208	tcp/udp	AppleTalk Unused—CMU
qmtp	209	tcp/udp	The Quick Mail Transfer Protocol—UC Berkeley
z39.50	210	tcp/udp	ANSI Z39.50
914c/g	211	tcp/udp	Texas Instruments 914C/G Terminal
anet	212	tcp/udp	ATEXSSTR—Kodak
ipx	213	tcp/udp	IPX—Novell
vmpwscs	214	tcp/udp	VM PWSCS
softpc	215	tcp/udp	Insignia Solutions
CAllic	216	tcp/udp	Computer Associates Int'l License Server
dbase	217	tcp/udp	dBASE Unix—Ashton-Tate
mpp	218	tcp/udp	Netix Message Posting Protocol
uarps	219	tcp/udp	Unisys ARPs
imap3	220	tcp/udp	Interactive Mail Access Protocol v3—Stanford U
fln-spx	221	tcp/udp	Berkeley rlogind with SPX auth
rsh-spx	222	tcp/udp	Berkeley rshd with SPX auth
cdc	223	tcp/udp	Certificate Distribution Center—DEC
*	222	tcp	Possible Conflict of Port 222 with "Masqdialer" below
masqdialer	224	tcp/udp	masqdialer—VillageNet
#	225-241	tcp/udp	Reserved—ISI
direct	242	tcp/udp	Direct—Peer Direct Inc.
sur-meas	243	tcp/udp	Survey Measurement—MIT
inbusiness	244	tcp/udp	inbusiness—Intel
link	245	tcp/udp	LINK
dsp3270	246	tcp/udp	Display Systems Protocol—dca.mil
subntbcst_tftp	247	tcp/udp	SUBNTBCST_TFTP—IBM
bhfhs	248	tcp/udp	bhfhs—Bell&Howell
#	249-255	tcp/udp	Reserved—ISI
rap	256	tcp/udp	RAP—IBM
set	257	tcp/udp	Secure Electronic Transaction—Pothole Industries
yak-chat	258	tcp/udp	Yak Winsock Personal Chat
esro-gen	259	tcp/udp	Efficient Short Remote Operations—Neda Communications, Inc.
openport	260	tcp/udp	Openport
nsiiops	261	tcp/udp	IIOP Name Service over TLS/SSL—Netscape
arcisdms	262	tcp/udp	Arcisdms
hdap	263	tcp/udp	HDAP—Zyxel
bgmp	264	tcp/udp	BGMP—U Michigan
x-bone-ctl	265	tcp/udp	X-Bone CTL—ISI
#	266-279	tcp/udp	Unassigned
http-mgmt	280	tcp/udp	http-mgmt—HP-UnitedKingdom
personal-link	281	tcp/udp	Personal Link—Computer Network Resources, Inc.
cableport-ax	282	tcp/udp	Cable Port A/X—Intel
rescap	283	tcp/udp	rescap—Internet Mail Consortium
corerjd	284	tcp/udp	corerjd—Core Networks
#	285-307	tcp/udp	Unassigned
novastorbakcup	308	tcp/udp	Novastor Backup
entrusttime	309	tcp/udp	EntrustTime
bhmds	310	tcp/udp	bhmds—Bell&Howell
asip-webadmin	311	tcp/udp	AppleShare IP WebAdmin
vslmp	312	tcp/udp	VSLMP
magenta-logic	313	tcp/udp	Magenta Logic-NetFusion
opalis-robot	314	tcp/udp	Opalis Robot
dpsi	315	tcp/udp	DPSI—DeskTopPaging
decauth	316	tcp/udp	decAuth
zannet	317	tcp/udp	Zannet-Access One

Abbreviation	Port-Decimal	Protocol	Description/Owner
pkix-timestamp	318	tcp/udp	PKIX TimeStamp—Entrust
ptp-event	319	tcp/udp	PTP Event—HP
ptp-general	320	tcp/udp	PTP General—HP
pip	321	tcp/udp	PIP—Activerse Inc.
rtsps	322	tcp/udp	RTSPS—Microsoft
#	323-332	tcp/udp	Unassigned
texar	333	tcp/udp	Texar Security Port
#	334-343	tcp/udp	Unassigned
pdap	344	tcp/udp	Prospero Data Access Protocol—ISI
pawserv	345	tcp/udp	Perf Analysis Workbench
zserv	346	tcp/udp	Zebra server
fatserv	347	tcp/udp	Fatmen Server
csi-sgwp	348	tcp/udp	Cabletron Management Protocol
mftp	349	tcp/udp	mftp—Microsoft
matip-type-a	350	tcp/udp	MATIP Type A—SITA Inc.
matip-type-b	351	tcp/udp	MATIP Type B—SITA Inc.
bhoetty	351	tcp/udp	bhoetty (added 5/21/97-not really assigned)—Bell&Howell
dtag-ste-sb	352	tcp/udp	DTAG (assigned long ago)—Deutsche Telekom
bhoedap4	352	tcp/udp	bhoedap4 (added 5/21/97-not really assigned)—Bell&Howell
ndsauth	353	tcp/udp	NDSAUTH—Novell
bh611	354	tcp/udp	bh611—Bell&Howell
datex-asn	355	tcp/udp	DATEX-ASN
cloanto-net-1	356	tcp/udp	Cloanto Net 1
bhevent	357	tcp/udp	bhevent—Bell&Howell
shrinkwrap	358	tcp/udp	Shrinkwrap—Green Dragon
tenebris_nts	359	tcp/udp	Tenebris Network Trace Service
scoi2odialog	360	tcp/udp	scoi2odialog
semantix	361	tcp/udp	Semantix
srssend	362	tcp/udp	SRS Send—Emergent
rsvp_tunnel	363	tcp/udp	RSVP Tunnel—UCLA
aurora-cmgr	364	tcp/udp	Aurora CMGR—Aurora Technology
dtk	365	tcp/udp	DTK—Fred Cohen & Assoc.
odmr	366	tcp/udp	ODMR—Qualcomm
mortgageware	367	tcp/udp	MortgageWare—InterLinq
qbikgdp	368	tcp/udp	QbikGDP
rpc2portmap	369	tcp/udp	rpc2portmap—FreeBSD Project
codaauth2	370	tcp/udp	codaauth2—FreeBSD Project
clearcase	371	tcp/udp	Clearcase—Atria
ulistproc	372	tcp/udp	ListProcessor—Boston U
legent-1	373	tcp/udp	Legent Corporation
legent-2	374	tcp/udp	Legent Corporation
hassle	375	tcp/udp	Hassle—U Basel
nip	376	tcp/udp	Amiga Envoy Network Inquiry Proto—I-AM Inc.
tnETOS	377	tcp/udp	NEC Corporation
dsETOS	378	tcp/udp	NEC Corporation
is99c	379	tcp/udp	TIA/EIA/IS-99 modem client—Qualcomm
is99s	380	tcp/udp	TIA/EIA/IS-99 modem server—Qualcomm
hp-collector	381	tcp/udp	hp performance data collector
hp-managed-node	382	tcp/udp	hp performance data managed node
hp-alarm-mgr	383	tcp/udp	hp performance data alarm manager
arns	384	tcp/udp	A Remote Network Server System
ibm-app	385	tcp/udp	IBM Application
asa	386	tcp/udp	ASA Message Router Object Def—Allen-Bradley
aurp	387	tcp/udp	Appletalk Update-Based Routing Pro—Novell
unidata-ldm	388	tcp/udp	Unidata LDM Version 4
ldap	389	tcp/udp	Lightweight Directory Access Protocol—U Michigan
uis	390	tcp/udp	UIS
synotics-relay	391	tcp/udp	SynOptics SNMP Relay Port
synotics-broker	392	tcp/udp	SynOptics Port Broker Port
dis	393	tcp/udp	Data Interpretation System—Metaphor
embl-ndt	394	tcp/udp	EMBL Nucleic Data Transfer—Upsala U
netcp	395	tcp/udp	NETscout Control Protocol
netware-ip	396	tcp/udp	Novell Netware over IP
mptn	397	tcp/udp	Multi Protocol Trans.Net—IBM
kryptolan	398	tcp/udp	Kryptolan—Sectra
iso-tsap-c2	399	tcp/udp	ISO Transport Class 2 Non-Control over TCP—DEC
work-sol	400	tcp/udp	Workstation Solutions
ups	401	tcp/udp	Uninterruptible Power Supply—BenaTong Softgware

Abbreviation	Port-Decimal	Protocol	Description/Owner
genie	402	tcp/udp	Genie Protocol
decap	403	tcp/udp	decap
nced	404	tcp/udp	nced
ncld	405	tcp/udp	ncld
imsp	406	tcp/udp	Interactive Mail Support Protocol—CMU
timbuktu	407	tcp/udp	Timbuktu—Netopia
prm-sm	408	tcp/udp	Prospero Resource Manager Sys. Man—ISI
prm-nm	409	tcp/udp	Prospero Resource Manager Node Man—ISI
decladebug	410	tcp/udp	DECLadebug Remote Debug Protocol
rmt	411	tcp/udp	Remote MT Protocol—Lysator
synoptics-trap	412	tcp/udp	Trap Convention Port
smsp	413	tcp/udp	SMSP
infoseek	414	tcp/udp	InfoSeek
bnet	415	tcp/udp	Bnet—Unisys
silverplatter	416	tcp/udp	Silverplatter
onmux	417	tcp/udp	Onmux-The World Kiosk
hyper-g	418	tcp/udp	Hyper-G—IICM
ariel1	419	tcp/udp	Ariel—Research Libraries Group
smpte	420	tcp/udp	SMPTE—CompuServe
ariel2	421	tcp/udp	Ariel—Research Libraries Group
ariel3	422	tcp/udp	Ariel—Research Libraries Group
opc-job-start	423	tcp/udp	IBM Operations Planning and Control Start
opc-job-track	424	tcp/udp	IBM Operations Planning and Control Track
icad-el	425	tcp/udp	ICAD
smartsdp	426	tcp/udp	smartsdp
svrloc	427	tcp/udp	Server Location—FTP Software
ocs_cmu	428	tcp/udp	OCS_CMU—US Air Force
ocs_amu	429	tcp/udp	OCS_AMU—US Air Force
utmpsd	430	tcp/udp	UTMPSD
utmpcd	431	tcp/udp	UTMPCD
iasd	432	tcp/udp	IASD—Encore
nnsp	433	tcp/udp	NNSP—UC Berkeley
mobileip-agent	434	tcp/udp	MobileIP-Agent—DEC
mobilip-mn	435	tcp/udp	MobilIP-MN—DEC
dna-cml	436	tcp/udp	DNA-CML—DEC
comscm	437	tcp/udp	comscm—DEC
dsfgw	438	tcp/udp	dsfgw
dasp	439	tcp/udp	dasp
sgcp	440	tcp/udp	sgcp—Data Broadcasting Corporation
decvms-sysmgt	441	tcp/udp	decvms-sysmgt
cvc_hostd	442	tcp/udp	cvc_hostd—Cray
https	443	tcp/udp	http protocol over TLS/SSL—Netscape
snpp	444	tcp/udp	Simple Network Paging Protocol [RFC1568]
microsoft-ds	445	tcp/udp	Microsoft-DS
ddm-rdb	446	tcp/udp	DDM-RDB—IBM
ddm-dfm	447	tcp/udp	DDM-RFM—IBM
ddm-ssl	448	tcp/udp	DDM-SSL—IBM
as-servermap	449	tcp/udp	AS Server Mapper—IBM
tserver	450	tcp/udp	Tserver—AT&T
sfs-smp-net	451	tcp/udp	Cray Network Semaphore server
sfs-config	452	tcp/udp	Cray SFS config server
creativeserver	453	tcp/udp	CreativeServer—Emotion
contentserver	454	tcp/udp	ContentServer—Emotion
creativepartnr	455	tcp/udp	CreativePartnr—Emotion
macon-tcp	456	tcp/udp	macon-tcp—Fujitsu
scohelp	457	tcp/udp	scohelp
appleqtc	458	tcp/udp	apple quick time
ampr-rcmd	459	tcp/udp	ampr-rcmd
skronk	460	tcp/udp	skronk
datasurfsrv	461	tcp/udp	DataRampSrv—MV Communications Inc.
datasurfsrvsec	462	tcp/udp	DataRampSrvSec—MV Communications Inc.
alpes	463	tcp/udp	alpes—IMAGE
kpasswd	464	tcp/udp	kpasswd—MIT
digital-vrc	466	tcp/udp	digital-vrc
mylex-mapd	467	tcp/udp	mylex-mapd
photuris	468	tcp/udp	proturis—U Michigan
rcp	469	tcp/udp	Radio Control Protocol
scx-proxy	470	tcp/udp	scx-proxy—Cray

Abbreviation	Port-Decimal	Protocol	Description/Owner
mondex	471	tcp/udp	Mondex—NatWest Offshore Ltd.
ljk-login	472	tcp/udp	ljk-login—LJK Software
hybrid-pop	473	tcp/udp	hybrid-pop
tn-tl-w1	474	tcp/udp	tn-tl-w1—ThinkNet
tcpnethaspsrv	475	tcp/udp	tcpnethaspsrv—Aladdin
tn-tl-fd1	476	tcp/udp	tn-tl-fd1—ThinkNet
ss7ns	477	tcp/udp	ss7ns—DEC
spsc	478	tcp/udp	spsc—Software Partners
iafserver	479	tcp/udp	iafserver—Solect
iafdbase	480	tcp/udp	iafdbase
ph	481	tcp/udp	Ph service—UMDAC
bgs-nsi	482	tcp/udp	bgs-nsi
ulpnet	483	tcp/udp	ulpnet—Unibol
integra-sme	484	tcp/udp	Integra Software Management Environment
powerburst	485	tcp/udp	Air Soft Power Burst
avian	486	tcp/udp	avian—Lotus
saft	487	tcp/udp	saft Simple Asynchronous File Transfer—U Stuttgart
gss-http	488	tcp/udp	gss-http—Galaxy Verticals
nest-protocol	489	tcp/udp	nest-protocol—Novell
micom-pfs	490	tcp/udp	micom-pfs
go-login	491	tcp/udp	go-login—Graphn
ticf-1	492	tcp/udp	Transport Independent Convergence for FNA-PFU Direct
ticf-2	493	tcp/udp	Transport Independent Convergence for FNA
pov-ray	494	tcp/udp	POV-Ray
intecourier	495	tcp/udp	intecourier—Intecom
pim-rp-disc	496	tcp/udp	PIM-RP-DISC—Cisco
dantz	497	tcp/udp	dantz
siam	498	tcp/udp	siam—Computer Answer Line
iso-ill	499	tcp/udp	ISO ILL Protocol—U Copenhagen
isakmp	500	tcp/udp	isakmp—NSCS-Military
stmf	501	tcp/udp	STMF—Farradyne
asa-appl-proto	502	tcp/udp	asa-appl-proto—Modicon
intrinsa	503	tcp/udp	Intrinsa
citadel	504	tcp/udp	citadel
mailbox-lm	505	tcp/udp	mailbox-lm—Stercomm
ohimsrv	506	tcp/udp	ohimsrv—Openhorizon
crs	507	tcp/udp	crs—Microsoft
xvttp	508	tcp/udp	xvttp
snare	509	tcp/udp	snare—Capres
fcp	510	tcp/udp	FirstClass Protocol—Softarc
passgo	511	tcp/udp	PassGo
exec	512	tcp	remote process execution
biff	512	udp	notify users of new mail received
login	513	tcp	remote login a la telnet;
who	513	udp	maintains data bases showing who's logged in
shell	514	tcp	"cmd like exec, but automatic authentication"
syslog	514	udp	
printer	515	tcp/udp	spooler
videotex	516	tcp/udp	videotex—Monaco Telamatique
talk	517	tcp/udp	"like tenex link, but across machines"
ntalk	518	tcp/udp	
utime	519	tcp/udp	unixtime
efs	520	tcp	extended file name server
router	520	udp	local routing process (on site)
ripng	521	tcp/udp	ripng—Ipsilon
ulp	522	tcp/udp	ULP—Microsoft
ibm-db2	523	tcp/udp	IBM-DB2
ncp	524	tcp/udp	NCP—Novell
timed	525	tcp/udp	timeserver
tempo	526	tcp/udp	newdate
stx	527	tcp/udp	Stock—IXChange
custix	528	tcp/udp	Customer—IXChange
irc-serv	529	tcp/udp	IRC-SERV—Acrux
courier	530	tcp/udp	rpc
conference	531	tcp/udp	chat
netnews	532	tcp/udp	readnews
netwall	533	tcp/udp	for emergency broadcasts

Abbreviation	Port-Decimal	Protocol	Description/Owner
mm-admin	534	tcp/udp	MegaMedia Admin—AIS-gmbh
iiop	535	tcp/udp	iiop—DEC
opalis-rdv	536	tcp/udp	opalis-rdv
nmsp	537	tcp/udp	Networked Media Streaming Protocol—Narrative
gdomap	538	tcp/udp	gdomap—Brainstorm
apertus-ldp	539	tcp/udp	Apertus Technologies Load Determination
uucp	540	tcp/udp	uucpd
uucp-rlogin	541	tcp/udp	uucp-rlogin—Wimsey
commerce	542	tcp/udp	commerce
klogin	543	tcp/udp	
kshell	544	tcp/udp	krcmd
appleqtcsrvr	545	tcp/udp	appleqtcsrvr
dhcpv6-client	546	tcp/udp	DHCPv6 Client—DEC
dhcpv6-server	547	tcp/udp	DHCPv6 Server—DEC
afpovertcp	548	tcp/udp	AFP over TCP—Apple
idfp	549	tcp/udp	IDFP—Kovi
new-rwho	550	tcp/udp	new-who
cybercash	551	tcp/udp	cybercash
deviceshare	552	tcp/udp	deviceshare—Advsyscon
pirp	553	tcp/udp	pirp—UC Berkeley
rtsp	554	tcp/udp	Real Time Stream Control Protocol—Prognet
dsf	555	tcp/udp	
remotefs	556	tcp/udp	rfs server
openvms-sysipc	557	tcp/udp	openvms-sysipc—DEC
sdnskmp	558	tcp/udp	SDNSKMP
teedtap	559	tcp/udp	TEEDTAP—GTEgsc
rmonitor	560	tcp/udp	rmonitord
monitor	561	tcp/udp	
chshell	562	tcp/udp	chcmd
nntps	563	tcp/udp	nntp protocol over TLS/SSL (was snntp)—Netscape
9pfs	564	tcp/udp	plan 9 file service
whoami	565	tcp/udp	whoami
streettalk	566	tcp/udp	streettalk
banyan-rpc	567	tcp/udp	banyan-rpc
ms-shuttle	568	tcp/udp	microsoft shuttle
ms-rome	569	tcp/udp	microsoft rome
meter	570	tcp/udp	demon
meter	571	tcp/udp	udemon
sonar	572	tcp/udp	sonar—U Tennessee Knoxville
banyan-vip	573	tcp/udp	banyan-vip
ftp-agent	574	tcp/udp	FTP Software Agent System
vemmi	575	tcp/udp	VEMMI—Monaco Telematique
ipcd	576	tcp/udp	ipcd—Ipass
vnas	577	tcp/udp	vnas—Ipass
ipdd	578	tcp/udp	ipdd—Ipass
decbsrv	579	tcp/udp	decbsrv
sntp-heartbeat	580	tcp/udp	SNTP HEARTBEAT—UUNET (owned by MCI)
bdp	581	tcp/udp	Bundle Discovery Protocol—Xylogics
scc-security	582	tcp/udp	SCC Security—Semaphorecom
philips-vc	583	tcp/udp	Philips Video-Conferencing
keyserver	584	tcp/udp	Key Server—Systemics
imap4-ssl	585	tcp/udp	IMAP4+SSL (use 993 instead)—U Washington
password-chg	586	tcp/udp	Password Change
submission	587	tcp/udp	Submission—Qualcomm
cal	588	tcp/udp	CAL—Intel
eyelink	589	tcp/udp	EyeLink—U Tornonto
tns-cml	590	tcp/udp	TNS CML—DEC
http-alt	591	tcp/udp	FileMaker, Inc.- HTTP Alternate (see Port 80)
eudora-set	592	tcp/udp	Eudora Set—Qualcomm
http-rpc-epmap	593	tcp/udp	HTTP RPC Ep Map—Microsoft
tpip	594	tcp/udp	TPIP—Platinum
cab-protocol	595	tcp/udp	CAB Protocol
smsd	596	tcp/udp	SMSD—DEC
ptcnameservice	597	tcp/udp	PTC Name Service
sco-websrvrmg3	598	tcp/udp	SCO Web Server Manager 3
acp	599	tcp/udp	Aeolon Core Protocol
ipcserver	600	tcp/udp	Sun IPC server

Abbreviation	Port-Decimal	Protocol	Description/Owner
urm	606	tcp/udp	Cray Unified Resource Manager
nqs	607	tcp/udp	nqs—Cray
sift-uft	608	tcp/udp	Sender-Initiated/Unsolicited File Transfer—Rice U
npmp-trap	609	tcp/udp	npmp-trap—CRL
npmp-local	610	tcp/udp	npmp-local—CRL
npmp-gui	611	tcp/udp	npmp-gui—CRL
hmmp-ind	612	tcp/udp	HMMP Indication—Microsoft
hmmp-op	613	tcp/udp	HMMP Operation—Microsoft
sshell	614	tcp/udp	SSLshell—Quick
sco-inetmgr	615	tcp/udp	Internet Configuration Manager
sco-sysmgr	616	tcp/udp	SCO System Administration Server
sco-dtmgr	617	tcp/udp	SCO Desktop Administration Server
dei-icda	618	tcp/udp	DEI-ICDA—Tbaytel
digital-evm	619	tcp/udp	Digital EVM
sco-websrvrmgr	620	tcp/udp	SCO WebServer Manager
escp-ip	621	tcp/udp	ESCP—Pobox
collaborator	622	tcp/udp	Collaborator—Opteamasoft
aux_bus_shunt	623	tcp/udp	Aux Bus Shunt—Intel
cryptoadmin	624	tcp/udp	Crypto Admin—Cyberus
dec_dlm	625	tcp/udp	DEC DLM
asia	626	tcp/udp	ASIA—Apple
passgo-tivoli	627	tcp/udp	PassGo Tivoli
qmqp	628	tcp/udp	QMQP
3com-amp3	629	tcp/udp	3Com AMP3
rda	630	tcp/udp	RDA—Minster
ipp	631	tcp/udp	IPP (Internet Printing Protocol)—Xerox
bmpp	632	tcp/udp	bmpp—Corvu
servstat	633	tcp/udp	Service Status update (Sterling Software)
ginad	634	tcp/udp	ginad—Cal State U
rlzdbase	635	tcp/udp	RLZ Dbase—Tyxar
ldaps	636	tcp/udp	ldap protocol over TLS/SSL (was sldap)—Xcert
lanserver	637	tcp/udp	lanserver—IBM
mcns-sec	638	tcp/udp	mcns-sec—Cablelabs
msdp	639	tcp/udp	MSDP—Cisco
entrust-sps	640	tcp/udp	entrust-sps
repcmd	641	tcp/udp	repcmd—Replicase
esro-emsdp	642	tcp/udp	ESRO-EMSDP V1.3—Neda
sanity	643	tcp/udp	SANity—OSR
dwr	644	tcp/udp	dwr—Xerox
pssc	645	tcp/udp	PSSC—DLR
ldp	646	tcp/udp	LDP—Cisco
dhcp-failover	647	tcp/udp	DHCP Failover—Process
rrp	648	tcp/udp	Registry Registrar Protocol (RRP)—Netsol
aminet	649	tcp/udp	Aminet—Adaptivemedia
obex	650	tcp/udp	OBEX—Motorola
ieee-mms	651	tcp/udp	IEEE MMS—SGI
udlr-dtcp	652	tcp/udp	UDLR_DTCP—INRIA
repscmd	653	tcp/udp	RepCmd—Tioga
aodv	654	tcp/udp	AODV—Sun
tinc	655	tcp/udp	TINC—Bigfoot
spmp	656	tcp/udp	SPMP—NoDomainName
rmc	657	tcp/udp	RMC—IBM
tenfold	658	tcp/udp	TenFold
url-rendezvous	659	tcp/udp	URL Rendezvous—Cisco
mac-srvr-admin	660	tcp/udp	MacOS Server Admin—Apple
hap	661	tcp/udp	HAP—Uroam
pftp	662	tcp/udp	PFTP
purenoise	663	tcp/udp	PureNoise—MailCity
secure-aux-bus	664	tcp/udp	Secure Aux Bus—Intel
sun-dr	665	tcp/udp	Sun DR
mdqs	666	tcp/udp	
doom	666	tcp/udp	doom Id Software
disclose	667	tcp/udp	campaign contribution disclosures - SDR Technologies
mecomm	668	tcp/udp	MeComm—Me Technology
meregister	669	tcp/udp	MeRegister—Me Technology
vacdsm-sws	670	tcp/udp	VACDSM-SWS
vacdsm-app	671	tcp/udp	VACDSM-APP
vpps-qua	672	tcp/udp	VPPS-QUA

Abbreviation	Port-Decimal	Protocol	Description/Owner
cimplex	673	tcp/udp	CIMPLEX—Cesi
acap	674	tcp/udp	ACAP—Innosoft
dctp	675	tcp/udp	DCTP—Ansa
vpps-via	676	tcp/udp	VPPS Via—Cesi
vpp	677	tcp/udp	Virtual Presence Protocol—Cobrow
ggf-ncp	678	tcp/udp	GNU Gereration Foundation NCP
mrm	679	tcp/udp	MRM—Cisco
entrust-aaas	680	tcp/udp	entrust-aaas
entrust-aams	681	tcp/udp	entrust-aams
xfr	682	tcp/udp	XFR—Ultranet
corba-iiop	683	tcp/udp	CORBA IIOP—Object Management Group
corba-iiop-ssl	684	tcp/udp	CORBA IIOP SSL—Object Management Group
mdc-portmapper	685	tcp/udp	MDC Port Mapper—Altavista
hcp-wismar	686	tcp/udp	Hardware Control Protocol Wismar—Liverpool U
asipregistry	687	tcp/udp	asipregistry—Apple
realm-rusd	688	tcp/udp	REALM-RUSD—Realminfo
nmap	689	tcp/udp	NMAP—Novonyx
vatp	690	tcp/udp	VATP—Aticasoft
msexch-routing	691	tcp/udp	MS Exchange Routing—Microsoft
hyperwave-isp	692	tcp/udp	Hyperwave-ISP
connendp	693	tcp/udp	connendp—Future-Gate
ha-cluster	694	tcp/udp	ha-cluster—Bell Labs
#	695-703	tcp	Unassigned
elcsd	704	tcp/udp	errlog copy/server daemon
agentx	705	tcp/udp	AgentX—acec
#	706	tcp/udp	Unassigned
borland-dsj	707	tcp/udp	Borland DSJ
#	708	tcp/udp	Unassigned
entrust-kmsh	709	tcp/udp	Entrust Key Management Service Handler
entrust-ash	710	tcp/udp	Entrust Administration Service Handler
cisco-tdp	711	tcp/udp	Cisco TDP
#	712-728	tcp/udp	Unassigned
netviewdm1	729	tcp/udp	IBM NetView DM/6000 Server/Client
netviewdm2	730	tcp/udp	IBM NetView DM/6000 send/tcp
netviewdm3	731	tcp/udp	IBM NetView DM/6000 receive/tcp
netgw	741	tcp/udp	netGW—netcs
netrcs	742	tcp/udp	Network based Rev. Cont. Sys
flexlm	744	tcp/udp	Flexible License Manager—Olivetti
fujitsu-dev	747	tcp/udp	Fujitsu Device Control
ris-cm	748	tcp/udp	Russell Info Sci Calendar Manager
kerberos-adm	749	tcp/udp	kerberos administration—eSpecials
rfile	750	tcp	
loadav	750	udp	
kerberos-iv	750	udp	kerberos version iv—eSpecials
pump	751	tcp/udp	
qrh	752	tcp/udp	
rrh	753	tcp/udp	
tell	754	tcp/udp	send
nlogin	758	tcp/udp	
con	759	tcp/udp	
ns	760	tcp/udp	
rxe	761	tcp/udp	
quotad	762	tcp/udp	
cycleserv	763	tcp/udp	
omserv	764	tcp/udp	
webster	765	tcp/udp	
phonebook	767	tcp/udp	phone
vid	769	tcp/udp	
cadlock	770	tcp/udp	
rtip	771	tcp/udp	
cycleserv2	772	tcp/udp	
submit	773	tcp	
rpasswd	774	tcp	
acmaint_dbd	774	udp	
entomb	775	tcp	
acmaint_transd	775	udp	
wpages	776	tcp/udp	
multiling-http	777	tcp/udp	Multiling HTTP—CTV

Abbreviation	Port-Decimal	Protocol	Description/Owner
#	778-779	tcp/udp	Unassigned
wpgs	780	tcp/udp	
concert	786	tcp/udp	Concert—IBM
qsc	787	tcp/udp	QSC—Stomped
#	788-799	tcp/udp	Unassigned
mdbs_daemon	800	tcp/udp	
device	801	tcp/udp	
#	802-809	tcp/udp	Unassigned
fcp-udp	810	tcp/udp	FCP—Softarc
#	811-827	tcp/udp	Unassigned
itm-mcell-s	828	tcp/udp	itm-mcell-s—Itmasters
pkix-3-ca-ra	829	tcp/udp	PKIX-3 CA/RA—Entrust
#	830-872	tcp/udp	Unassigned
rsync	873	tcp/udp	rsync—Australian National U
#	875-885	tcp/udp	Unassigned
iclcnet-locate	886	tcp/udp	ICL coNETion locate server
iclcnet_svinfo	887	tcp/udp	ICL coNETion server info
accessbuilder	888	tcp/udp	AccessBuilder-3Com
cddbp	888	tcp/udp	CD Database Protocol-unassigned—Moonsoft
#	889-899	tcp/udp	Unassigned
omginitialrefs	900	tcp/udp	OMG Initial Refs—Sun
#	901-910	tcp/udp	Unassigned
xact-backup	911	tcp/udp	xact-backup—Xactlabs
#	912-988	tcp/udp	Unassigned
ftps-data	989	tcp/udp	ftp protocol, data, over TLS/SSL—Consensus
ftps	990	tcp/udp	ftp protocol, control, over TLS/SSL—Consensus
nas	991	tcp/udp	Netnews Administration System—Fu-Berlin
telnets	992	tcp/udp	telnet protocol over TLS/SSL—Consensus
imaps	993	tcp/udp	imap4 protocol over TLS/SSL—Consensus
ircs	994	tcp/udp	irc protocol over TLS/SSL—Consensus
pop3s	995	tcp/udp	pop3 protocol over TLS/SSL (was spop3—Microsoft)
vsinet	996	tcp/udp	vsinet—VSI
maitrd	997	tcp/udp	
busboy	998	tcp	
puparp	998	udp	
garcon	999	tcp	
applix	999	udp	Applix ac
puprouter	999	tcp/udp	
cadlock2	1000	tcp/udp	
#	1001-1009	tcp/udp	Unassigned
#	1008	udp	Possibly used by Sun Solaris
surf	1010	tcp/udp	surf—Peapod
#	1011-1022	tcp/udp	Reserved
	1023	tcp/udp	Reserved—IANA
Reserved	1024	tcp/udp	Reserved
blackjack	1025	tcp/udp	network blackjack
iad1	1030	tcp/udp	BBN IAD
iad2	1031	tcp/udp	BBN IAD
iad3	1032	tcp/udp	BBN IAD
neod1	1047	tcp/udp	Sun's NEO Object Request Broker
neod2	1048	tcp/udp	Sun's NEO Object Request Broker
nim	1058	tcp/udp	nim—IBM
nimreg	1059	tcp/udp	nimreg—IBM
instl_boots	1067	tcp/udp	Installation Bootstrap Proto. Serv.—HP
instl_bootc	1068	tcp/udp	Installation Bootstrap Proto. Cli.—HP
socks	1080	tcp/udp	Socks—NEC
ansoft-lm-1	1083	tcp/udp	Anasoft License Manager
ansoft-lm-2	1084	tcp/udp	Anasoft License Manager
webobjects	1085	tcp/udp	Web Objects—Apple
#	1086-1096	tcp/udp	Unassigned
sunclustermgr	1097	tcp/udp	Sun Cluster Manager
rmiactivation	1098	tcp/udp	RMI Activation—Sun
rmiregistry	1099	tcp/udp	RMI Registry—Sun
nfsd-status	1110	tcp/udp	Cluster status info—HP
mini-sql	1114	tcp/udp	Mini SQL—Hughes Australia
murray	1123	tcp/udp	Murray
nfa	1155	tcp/udp	Network File Access
health-polling	1161	tcp/udp	Health Polling—Triosoftware

Abbreviation	Port-Decimal	Protocol	Description/Owner
health-trap	1162	tcp/udp	Health Trap—Triosoftware
mc-client	1180	tcp/udp	Millicent Client Proxy—DEC
hp-webadmin	1188	tcp/udp	HP Web Admin
scol	1200	tcp/udp	SCOL—Cryo-Networks
caiccipc	1202	tcp/udp	caiccipc—CAI
#	1203-1211	tcp/udp	Unassigned
lupa	1212	tcp/udp	lupa—Databus
scanstat-1	1215	tcp/udp	scanSTAT 1.0—ISI
nerv	1222	tcp/udp	SNI R&D network
#	1223-1233	tcp/udp	Unassigned
search-agent	1234	tcp/udp	Infoseek Search Agent
#	1235-1238	tcp/udp	Unassigned
nmsd	1239	tcp/udp	NMSD—PTC
#	1240-1247	tcp/udp	Unassigned
hermes	1248	tcp/udp	
#	1249-1299	tcp/udp	Unassigned
h323hostcallsc	1300	tcp/udp	H323 Host Call Secure—Intel
#	1301-1309	tcp/udp	Unassigned
husky	1310	tcp/udp	Husky—Zang
rxmon	1311	tcp/udp	RxMon—Dell
sti-envision	1312	tcp/udp	STI Envision—STIsystems
bmc_patroldb	1313	tcp/udp	BMC_PATROLDB
pdps	1314	tcp/udp	Photoscript Distributed Printing System—Compulink
#	1315-1319	tcp/udp	Unassigned
panja-axbnet	1320	tcp/udp	Panja-AXBNET
pip	1321	tcp/udp	PIP—Activerse
#	1322-1334	tcp/udp	Unassigned
digital-notary	1335	tcp/udp	Digital Notary Protocol
#	1336-1344	tcp/udp	Unassigned
vpjp	1345	tcp/udp	VPJP—America Online
alta-ana-lm	1346	tcp/udp	Alta Analytics License Manager
bbn-mmc	1347	tcp/udp	multi media conferencing
bbn-mmx	1348	tcp/udp	multi media conferencing
sbook	1349	tcp/udp	Registration Network Protocol—NEXT
editbench	1350	tcp/udp	Registration Network Protocol—NEXT
equationbuilder	1351	tcp/udp	Digital Tool Works (MIT)
lotusnote	1352	tcp/udp	Lotus Note—Iris
relief	1353	tcp/udp	Relief Consulting—PSI
rightbrain	1354	tcp/udp	RightBrain Software
intuitive edge	1355	tcp/udp	Intuitive Edge—Ohio State U
cuillamartin	1356	tcp/udp	CuillaMartin Company
pegboard	1357	tcp/udp	Electronic PegBoard
connlcli	1358	tcp/udp	CONNLCLI
ftsrv	1359	tcp/udp	FTSRV—BITNET
mimer	1360	tcp/udp	MIMER
linx	1361	tcp/udp	LinX
timeflies	1362	tcp/udp	TimeFlies—nwnexus
ndm-requester	1363	tcp/udp	Network DataMover Requester—Ricoh
ndm-server	1364	tcp/udp	Network DataMover Server——Ricoh
adapt-sna	1365	tcp/udp	Network Software Associates
netware-csp	1366	tcp/udp	Novell NetWare Comm Service Platform
dcs	1367	tcp/udp	DCS
screencast	1368	tcp/udp	ScreenCast
gv-us	1369	tcp/udp	GlobalView to Unix Shell—Fuji-Xerox
us-gv	1370	tcp/udp	Unix Shell to GlobalView—Fuji-Xerox
fc-cli	1371	tcp/udp	Fujitsu Config Protocol
fc-ser	1372	tcp/udp	Fujitsu Config Protocol
chromagrafx	1373	tcp/udp	Chromagrafx
molly	1374	tcp/udp	EPI Software Systems
bytex	1375	tcp/udp	Bytex
ibm-pps	1376	tcp/udp	IBM Person to Person Software
cichlid	1377	tcp/udp	Cichlid License Manager
elan	1378	tcp/udp	Elan License Manager
dbreporter	1379	tcp/udp	Integrity Solutions
telesis-licman	1380	tcp/udp	Telesis Network License Manager
apple-licman	1381	tcp/udp	Apple Network License Manager
udt_os	1382	tcp/udp	
gwha	1383	tcp/udp	GW Hannaway Network License Manager

Abbreviation	Port-Decimal	Protocol	Description/Owner
os-licman	1384	tcp/udp	Objective Solutions License Manager
atex_elmd	1385	tcp/udp	Atex Publishing License Manager
checksum	1386	tcp/udp	CheckSum License Manager
cadsi-lm	1387	tcp/udp	Computer Aided Design Software Inc LM
objective-dbc	1388	tcp/udp	Objective Solutions DataBase Cache
iclpv-dm	1389	tcp/udp	Document Manager—ICL
iclpv-sc	1390	tcp/udp	Storage Controller—ICL
iclpv-sas	1391	tcp/udp	Storage Access Server—ICL
iclpv-pm	1392	tcp/udp	Print Manager—ICL
iclpv-nls	1393	tcp/udp	Network Log Server—ICL
iclpv-nlc	1394	tcp/udp	Network Log Client—ICL
iclpv-wsm	1395	tcp/udp	PC Workstation Manager software—ICL
dvl-activemail	1396	tcp/udp	DVL Active Mail
audio-activmail	1397	tcp/udp	Audio Active Mail
video-activmail	1398	tcp/udp	Video Active Mail
cadkey-licman	1399	tcp/udp	Cadkey License Manager
cadkey-tablet	1400	tcp/udp	Cadkey Tablet Daemon
goldleaf-licman	1401	tcp/udp	Goldleaf License Manager
prm-sm-np	1402	tcp	Prospero Resource Manager-ISI
prm-nm-np	1403	tcp	Prospero Resource Manager—ISI
igi-lm	1404	tcp/udp	Infinite Graphics License Manager
ibm-res	1405	tcp/udp	IBM Remote Execution Starter
netlabs-lm	1406	tcp/udp	NetLabs License Manager
dbsa-lm	1407	tcp/udp	DBSA License Manager
sophia-lm	1408	tcp/udp	Sophia License Manager
here-lm	1409	tcp/udp	Here License Manager
hiq	1410	tcp/udp	HiQ License Manager
af	1411	tcp/udp	AudioFile—DEC
innosys	1412	tcp/udp	InnoSys
innosys-acl	1413	tcp/udp	Innosys-ACL
ibm-mqseries	1414	tcp/udp	IBM MQSeries
dbstar	1415	tcp/udp	DBStar
novell-lu6.2	1416	tcp/udp	Novell LU6.2
timbuktu-srv1	1417	tcp/udp	Timbuktu Service 1 Port—Netopia
timbuktu-srv2	1418	tcp/udp	Timbuktu Service 2 Port—Netopia
timbuktu-srv3	1419	tcp/udp	Timbuktu Service 3 Port—Netopia
timbuktu-srv4	1420	tcp/udp	Timbuktu Service 4 Port—Netopia
gandalf-lm	1421	tcp/udp	Gandalf License Manager
autodesk-lm	1422	tcp/udp	Autodesk License Manager
essbase	1423	tcp/udp	Essbase Arbor Software
hybrid	1424	tcp/udp	Hybrid Encryption Protocol
zion-lm	1425	tcp/udp	Zion Software License Manager
sas-1	1426	tcp/udp	Satellite-data Acquisition System 1
mloadd	1427	tcp/udp	mloadd monitoring tool
informatik-lm	1428	tcp/udp	Informatik License Manager
nms	1429	tcp/udp	Hypercom NMS
tpdu	1430	tcp/udp	Hypercom TPDU
rgtp	1431	tcp/udp	Reverse Gossip Transport
blueberry-lm	1432	tcp/udp	Blueberry Software License Manager
ms-sql-s	1433	tcp/udp	Microsoft-SQL-Server
ms-sql-m	1434	tcp/udp	Microsoft-SQL-Monitor
ibm-cics	1435	tcp/udp	IBM CISC
sas-2	1436	tcp/udp	Satellite-data Acquisition System 2
tabula	1437	tcp/udp	Tabula
eicon-server	1438	tcp/udp	Eicon Security Agent Server
eicon-x25	1439	tcp/udp	Eicon X25 SNA Gateway
eicon-slp	1440	tcp/udp	Eicon Service Location Protocol
cadis-1	1441	tcp/udp	Cadis License Management
cadis-2	1442	tcp/udp	Cadis License Management
ies-lm	1443	tcp/udp	Integrated Engineering Software
marcam-lm	1444	tcp/udp	Marcam License Management
proxima-lm	1445	tcp/udp	Proxima License Manager
ora-lm	1446	tcp/udp	Optical Research Associates License Manager
apri-lm	1447	tcp/udp	Applied Parallel Research LM
oc-lm	1448	tcp/udp	OpenConnect License Manager
peport	1449	tcp/udp	PEport
dwf	1450	tcp/udp	Tandem Distributed Workbench Facility

Abbreviation	Port-Decimal	Protocol	Description/Owner
infoman	1451	tcp/udp	IBM Information Management
gtegsc-lm	1452	tcp/udp	GTE Government Systems License Man
genie-lm	1453	tcp/udp	Genie License Manager
interhdl_elmd	1454	tcp/udp	interHDL License Manager
esl-lm	1455	tcp/udp	ESL License Manager
dca	1456	tcp/udp	DCA
valisys-lm	1457	tcp/udp	Valisys License Manager
nrcabq-lm	1458	tcp/udp	Nichols Research Corp.
proshare1	1459	tcp/udp	Proshare Notebook Application
proshare2	1460	tcp/udp	Proshare Notebook Application
ibm_wrless_lan	1461	tcp/udp	IBM Wireless LAN
world-lm	1462	tcp/udp	World License Manager
nucleus	1463	tcp/udp	Nucleus
msl_lmd	1464	tcp/udp	MSL License Manager
pipes	1465	tcp/udp	Pipes Platform
oceansoft-lm	1466	tcp/udp	Ocean Software License Manager
csdmbase	1467	tcp/udp	CSDMBASE
csdm	1468	tcp/udp	CSDM
aal-lm	1469	tcp/udp	Active Analysis Limited License Manager
uaiact	1470	tcp/udp	Universal Analytics
csdmbase	1471	tcp/udp	csdmbase
csdm	1472	tcp/udp	csdm
openmath	1473	tcp/udp	OpenMath
telefinder	1474	tcp/udp	Telefinder
taligent-lm	1475	tcp/udp	Taligent License Manager
clvm-cfg	1476	tcp/udp	clvm-cfg
ms-sna-server	1477	tcp/udp	ms-sna-server
ms-sna-base	1478	tcp/udp	ms-sna-base
dberegister	1479	tcp/udp	dberegister
pacerforum	1480	tcp/udp	PacerForum
airs	1481	tcp/udp	AIRS
miteksys-lm	1482	tcp/udp	Miteksys License Manager
afs	1483	tcp/udp	AFS License Manager
confluent	1484	tcp/udp	Confluent License Manager
lansource	1485	tcp/udp	LANSource
nms_topo_serv	1486	tcp/udp	nms_topo_serv
localinfosrvr	1487	tcp/udp	LocalInfoSrvr
docstor	1488	tcp/udp	DocStor
dmdocbroker	1489	tcp/udp	dmdocbroker
insitu-conf	1490	tcp/udp	insitu-conf
anynetgateway	1491	tcp/udp	anynetgateway
stone-design-1	1492	tcp/udp	stone-design-1
netmap_lm	1493	tcp/udp	netmap_lm
ica	1494	tcp/udp	ica
cvc	1495	tcp/udp	cvc
liberty-lm	1496	tcp/udp	liberty-lm
rfx-lm	1497	tcp/udp	rfx-lm
watcom-sql	1498	tcp/udp	Watcom-SQL
fhc	1499	tcp/udp	Federico Heinz Consultora
vlsi-lm	1500	tcp/udp	VLSI License Manager
sas-3	1501	tcp/udp	Satellite-data Acquisition System 3
shivadiscovery	1502	tcp/udp	Shiva
imtc-mcs	1503	tcp/udp	Databeam
evb-elm	1504	tcp/udp	EVB Software Engineering License Manager
funkproxy	1505	tcp/udp	Funk Software,Inc.
utcd	1506	tcp/udp	Universal Time daemon (utcd)
symplex	1507	tcp/udp	symplex
diagmond	1508	tcp/udp	diagmond
robcad-lm	1509	tcp/udp	Robcad, Ltd. License Manager
mvx-lm	1510	tcp/udp	Midland Valley Exploration Ltd. Lic. Man.
3L-L1	1511	tcp/udp	3L-L1
wins	1512	tcp/udp	Microsoft's Windows Internet Name Service
fujitsu-dtc	1513	tcp/udp	Fujitsu DTC
fujitsu-dtcns	1514	tcp/udp	Fujitsu DTCNS
ifor-protocol	1515	tcp/udp	ifor-protocol
vpad	1516	tcp/udp	Virtual Places Audio data
vpac	1517	tcp/udp	Virtual Places Audio control
vpvd	1518	tcp/udp	Virtual Places Video data

Abbreviation	Port-Decimal	Protocol	Description/Owner
vpvc	1519	tcp/udp	Virtual Places Video control
atm-zip-office	1520	tcp/udp	atm zip office
ncube-lm	1521	tcp/udp	nCube License Manager
rna-lm	1522	tcp/udp	Ricardo North America License Manager
cichild-lm	1523	tcp/udp	cichild
ingreslock	1524	tcp/udp	ingres
orasrv	1525	tcp/udp	oracle
prospero-np	1525	tcp/udp	Prospero Directory Service non-priv
pdap-np	1526	tcp/udp	Prospero Data Access Prot non-priv
tlisrv	1527	tcp/udp	oracle
mciautoreg	1528	tcp/udp	micautoreg
coauthor	1529	tcp/udp	oracle
rap-service	1530	tcp/udp	rap-service
rap-listen	1531	tcp/udp	rap-listen
miroconnect	1532	tcp/udp	miroconnect
virtual-places	1533	tcp/udp	Virtual Places Software
micromuse-lm	1534	tcp/udp	micromuse-lm
ampr-info	1535	tcp/udp	ampr-info
ampr-inter	1536	tcp/udp	ampr-inter
sdsc-lm	1537	tcp/udp	isi-lm
3ds-lm	1538	tcp/udp	3ds-lm
intellistor-lm	1539	tcp/udp	Intellistor License Manager
rds	1540	tcp/udp	rds
rds2	1541	tcp/udp	rds2
gridgen-elmd	1542	tcp/udp	gridgen-elmd
simba-cs	1543	tcp/udp	simba-cs
aspeclmd	1544	tcp/udp	aspeclmd
vistium-share	1545	tcp/udp	vistium-share
abbaccuray	1546	tcp/udp	abbaccuray
laplink	1547	tcp/udp	laplink—Traveling Software
axon-lm	1548	tcp/udp	Axon License Manager
shivahose	1549	tcp	Shiva Hose
shivasound	1549	udp	Shiva Sound
3m-image-lm	1550	tcp/udp	Image Storage license manager 3M Company
hecmtl-db	1551	tcp/udp	HECMTL-DB
pciarray	1552	tcp/udp	pciarray—AVL
sna-cs	1553	tcp/udp	sna-cs—Data Connection Ltd.
caci-lm	1554	tcp/udp	CACI Products Company License Manager
livelan	1555	tcp/udp	livelan—Pictel
ashwin	1556	tcp/udp	AshWin CI Tecnologies
arbortext-lm	1557	tcp/udp	ArborText License Manager
xingmpeg	1558	tcp/udp	xingmpeg—xingtech.com
web2host	1559	tcp/udp	web2host—Mindspring
asci-val	1560	tcp/udp	asci-val—Advanced Systems
facilityview	1561	tcp/udp	facilityview—Particle Measuring Systems Inc.
pconnectmgr	1562	tcp/udp	pconnectmgr—Veritas
cadabra-lm	1563	tcp/udp	Cadabra License Manager
pay-per-view	1564	tcp/udp	Pay-Per-View—ISI
winddlb	1565	tcp/udp	WinDD—Tektronix
corelvideo	1566	tcp/udp	CORELVIDEO
jlicelmd	1567	tcp/udp	jlicelmd
tsspmap	1568	tcp/udp	tsspmap
ets	1569	tcp/udp	ets
orbixd	1570	tcp/udp	orbixd—Iona Technologies
rdb-dbs-disp	1571	tcp/udp	Oracle Remote Data Base
chip-lm	1572	tcp/udp	Chipcom License Manager
itscomm-ns	1573	tcp/udp	itscomm-ns—IBM
mvel-lm	1574	tcp/udp	mvel-lm—Midland Valley
oraclenames	1575	tcp/udp	oraclenames
moldflow-lm	1576	tcp/udp	moldflow-lm
hypercube-lm	1577	tcp/udp	hypercube-lm
jacobus-lm	1578	tcp/udp	Jacobus License Manager
ioc-sea-lm	1579	tcp/udp	ioc-sea-lm
tn-tl-r2	1580	tcp/udp	tn-tl-r2—Thinknet
mil-2045-47001	1581	tcp/udp	MIL-2045-47001—ITT Industries
msims	1582	tcp/udp	MSIMS
simbaexpress	1583	tcp/udp	simbaexpress

Abbreviation	Port-Decimal	Protocol	Description/Owner
tn-tl-fd2	1584	tcp/udp	tn-tl-fd2—ThinkNet Inc.
intv	1585	tcp/udp	intv—Claddagh Films
ibm-abtact	1586	tcp/udp	ibm-abtact—Stanford U
pra_elmd	1587	tcp/udp	pra_elmd—Platte River Associates, Inc.
triquest-lm	1588	tcp/udp	triquest-lm
vqp	1589	tcp/udp	VQP—Cisco
gemini-lm	1590	tcp/udp	gemini-lm—OpenGVS
ncpm-pm	1591	tcp/udp	ncpm-pm—HP
commonspace	1592	tcp/udp	commonspace—Within Technology
mainsoft-lm	1593	tcp/udp	mainsoft-lm
sixtrak	1594	tcp/udp	sixtrak—Wizvax Communications
radio	1595	tcp/udp	radio—ISI
radio-sm	1596	tcp	radio-sm—ISI
radio-bc	1596	udp	radio-bc—ISI
orbplus-iiop	1597	tcp/udp	orbplus-iiop—HP (Apollo)
picknfs	1598	tcp/udp	picknfs—Pick Systems
simbaservices	1599	tcp/udp	simbaservices
issd	1600	tcp/udp	
aas	1601	tcp/udp	aas—Lachman
inspect	1602	tcp/udp	inspect—DEC
picodbc	1603	tcp/udp	pickodbc—picknfs
icabrowser	1604	tcp/udp	icabrowser—Citrix
slp	1605	tcp/udp	Salutation Manager (Salutation Protocol)—IBM
slm-api	1606	tcp/udp	Salutation Manager (SLM-API)—IBM
stt	1607	tcp/udp	stt—Microsoft
smart-lm	1608	tcp/udp	Smart Corp. License Manager
isysg-lm	1609	tcp/udp	isysg-lm
taurus-wh	1610	tcp/udp	taurus-wh
ill	1611	tcp/udp	Inter Library Loan—University College Dublin
netbill-trans	1612	tcp/udp	NetBill Transaction Server—CMU
netbill-keyrep	1613	tcp/udp	NetBill Key Repository—CMU
netbill-cred	1614	tcp/udp	NetBill Credential Server—CMU
netbill-auth	1615	tcp/udp	NetBill Authorization Server—CMU
netbill-prod	1616	tcp/udp	NetBill Product Server—CMU
nimrod-agent	1617	tcp/udp	nimrod Inter-Agent Communication—BBN
skytelnet	1618	tcp/udp	skytelnet—Bell South
xs-openstorage	1619	tcp/udp	xs-openstorage—XuiSis Software Ltd.
faxportwinport	1620	tcp/udp	faxportwinport—LanSource
softdataphone	1621	tcp/udp	softdataphone—IBM
ontime	1622	tcp/udp	ontime
jaleosnd	1623	tcp/udp	jaleosnd
udp-sr-port	1624	tcp/udp	udp-sr-port—ITT Industries
svs-omagent	1625	tcp/udp	sys-omagent—IBM
shockwave	1626	tcp/udp	Shockwave—Macromedia
t128-gateway	1627	tcp/udp	T.128 Gateway—Data Connection Ltd.
lontalk-norm	1628	tcp/udp	Lontalk normal—Cisco
lontalk-urgnt	1629	tcp/udp	Lontalk urgent—Cisco
oraclenet8cman	1630	tcp/udp	Oracle Net8 Cman
visitview	1631	tcp/udp	Visit view—U Wisconsin
pammratc	1632	tcp/udp	PAMMRATC—Peak Audio
pammrpc	1633	tcp/udp	PAMMRPC—Peak Audio
loaprobe	1634	tcp/udp	Log On America Probe
edb-server1	1635	tcp/udp	EDB Server 1—Simple Software Solutions
cncp	1636	tcp/udp	CableNet Control Protocol
cnap	1637	tcp/udp	CableNet Admin Protocol
cnip	1638	tcp/udp	CableNet Info Protocol
cert-initiator	1639	tcp/udp	cert-initiator—Sun
cert-responder	1640	tcp/udp	cert-responder—Sun
invision	1641	tcp/udp	InVision—CSK Software
isis-am	1642	tcp/udp	isis-am
isis-ambc	1643	tcp/udp	isis-ambc
isis-ambc	1643	tcp/udp	isis-ambc
saiseh	1644	tcp	Satellite-data Acquisition System 4—U Wisconsin
datametrics	1645	tcp/udp	datametrics
sa-msg-port	1646	tcp/udp	sa-msg-port—ITT Industries
rsap	1647	tcp/udp	rsap—TU Illmenau (Tech University)
concurrent-lm	1648	tcp/udp	concurrent-lm

Abbreviation	Port-Decimal	Protocol	Description/Owner
kermit	1649	tcp/udp	kermit—Columbia U
nkd	1650	tcp/udp	
shiva_confsrvr	1651	tcp/udp	shiva_confsrvr
xnmp	1652	tcp/udp	xnmp—At&T Cerfnet
alphatech-lm	1653	tcp/udp	alphatech-lm
stargatealerts	1654	tcp/udp	stargatealerts—Intel
dec-mbadmin	1655	tcp/udp	dec-mbadmin
dec-mbadmin-h	1656	tcp/udp	dec-mbadmin-h
fujitsu-mmpdc	1657	tcp/udp	fujitsu-mmpdc
sixnetudr	1658	tcp/udp	sixnetudr—Wizvax Communications
sg-lm	1659	tcp/udp	Silicon Grail License Manager
skip-mc-gikreq	1660	tcp/udp	skip-mc-gikreq—Sun
netview-aix-1	1661	tcp/udp	netview-aix-1—IBM
netview-aix-2	1662	tcp/udp	netview-aix-2—IBM
netview-aix-3	1663	tcp/udp	netview-aix-3—IBM
netview-aix-4	1664	tcp/udp	netview-aix-4—IBM
netview-aix-5	1665	tcp/udp	netview-aix-5—IBM
netview-aix-6	1666	tcp/udp	netview-aix-6—IBM
netview-aix-7	1667	tcp/udp	netview-aix-7—IBM
netview-aix-8	1668	tcp/udp	netview-aix-8—IBM
netview-aix-9	1669	tcp/udp	netview-aix-9—IBM
netview-aix-10	1670	tcp/udp	netview-aix-10—IBM
netview-aix-11	1671	tcp/udp	netview-aix-11—IBM
netview-aix-12	1672	tcp/udp	netview-aix-12—IBM
proshare-mc-1	1673	tcp/udp	Intel Proshare Multicast
proshare-mc-2	1674	tcp/udp	Intel Proshare Multicast
pdp	1675	tcp/udp	Pacific Data Products
netcomm1	1676	tcp	netcomm1—Symantec
netcomm2	1676	udp	netcomm2—Symantec
groupwise	1677	tcp/udp	groupwise—Novell
prolink	1678	tcp/udp	prolink—Tektronix
darcorp-lm	1679	tcp/udp	darcorp-lm
microcom-sbp	1680	tcp/udp	microcom-sbp
sd-elmd	1681	tcp/udp	sd-elmd—SoftDesk
lanyon-lantern	1682	tcp/udp	lanyon-lantern
ncpm-hip	1683	tcp/udp	ncpm-hip—HP
snaresecure	1684	tcp/udp	SnareSecure—Copenhagen Applied Research
n2nremote	1685	tcp/udp	n2nremote—Net2Net
cvmon	1686	tcp/udp	cvmon—HP
nsjtp-ctrl	1687	tcp/udp	nsjtp-ctrl—HP
nsjtp-data	1688	tcp/udp	nsjtp-data—HP
firefox	1689	tcp/udp	firefox
ng-umds	1690	tcp/udp	ng-umds
empire-empuma	1691	tcp/udp	empire-empuma—Empire Technology
sstsys-lm	1692	tcp/udp	sstsys-lm—Netcom
rrirtr	1693	tcp/udp	rrirtr—RRI Inc.
rrimwm	1694	tcp/udp	rrimwm—RRI Inc.
rrilwm	1695	tcp/udp	rrilwm—RRI Inc.
rrifmm	1696	tcp/udp	rrifmm—RRI Inc.
rrisat	1697	tcp/udp	rrisat—RRI Inc.
rsvp-encap-1	1698	tcp/udp	RSVP-ENCAPSULATION-1—ISI
rsvp-encap-2	1699	tcp/udp	RSVP-ENCAPSULATION-2—ISI
mps-raft	1700	tcp/udp	mps-raft—AOL
l2f	1701	tcp/udp	l2f—Cisco
l2tp	1701	tcp/udp	l2tp—Cisco
deskshare	1702	tcp/udp	deskshare
deskshare	1702	tcp/udp	deskshare
hb-engine	1703	tcp/udp	hb-engine—Zoom Inc.
bcs-broker	1704	tcp/udp	bcs-broker—Knoware BV
slingshot	1705	tcp/udp	slingshot—CSK Software
jetform	1706	tcp/udp	jetform
vdmplay	1707	tcp/udp	vdmplay
gat-lmd	1708	tcp/udp	gat-lmd—global-tech.com
centra	1709	tcp/udp	centra—Centra Inc.
impera	1710	tcp/udp	impera
pptconference	1711	tcp/udp	pptconference—Microsoft
registrar	1712	tcp/udp	resource monitoring service—HP

Abbreviation	Port-Decimal	Protocol	Description/Owner
conferencetalk	1713	tcp/udp	ConferenceTalk—Videoserver
sesi-lm	1714	tcp/udp	sesi-lm—Side Effects SW
houdini-lm	1715	tcp/udp	houdini-lm—Side Effects SW
xmsg	1716	tcp/udp	xmsg—Xantel Connex
fj-hdnet	1717	tcp/udp	fj-hdnet—Fujitsu
h323gatedisc	1718	tcp/udp	h323gatedisc—Intel
h323gatestat	1719	tcp/udp	h323gatestat—Intel
h323hostcall	1720	tcp/udp	h323hostcall—Intel
caicci	1721	tcp/udp	caicci—Computer Associates
hks-lm	1722	tcp/udp	HKS License Manager
pptp	1723	tcp/udp	pptp—Microsoft
csbphonemaster	1724	tcp/udp	csbphonemaster
iden-ralp	1725	tcp/udp	iden-ralp—Motorola Inc.
iberiagames	1726	tcp/udp	IBERIAGAMES
winddx	1727	tcp/udp	winddx—Tektronix
telindus	1728	tcp/udp	TELINDUS
citynl	1729	tcp/udp	CityNL License Management
roketz	1730	tcp/udp	roketz—Bluemoon Interactive
msiccp	1731	tcp/udp	MSICCP—Microsoft
proxim	1732	tcp/udp	proxim
siipat	1733	tcp/udp	SIMS - SIIPAT Protocol for Alarm Transmission—SimWare
cambertx-lm	1734	tcp/udp	Camber Corporation License Management
privatechat	1735	tcp/udp	PrivateChat
street-stream	1736	tcp/udp	street-stream—Netcom
ultimad	1737	tcp/udp	ultimad—Ultima Interconnect Technology
gamegen1	1738	tcp/udp	GameGen1—MultiGen-Paradigm Inc.
webaccess	1739	tcp/udp	webaccess—Asymetrix Learning Systems
encore	1740	tcp/udp	encore—PRI Automation
cisco-net-mgmt	1741	tcp/udp	cisco-net-mgmt
3Com-nsd	1742	tcp/udp	3Com-nsd
cinegrfx-lm	1743	tcp/udp	Cinema Graphics License Manager—RFX Systems Integration
ncpm-ft	1744	tcp/udp	ncpm-ft—HP
remote-winsock	1745	tcp/udp	remote-winsock—Microsoft
ftrapid-1	1746	tcp/udp	ftrapid-1—Unisys
ftrapid-2	1747	tcp/udp	ftrapid-2—Unisys
oracle-em1	1748	tcp/udp	oracle-em1
aspen-services	1749	tcp/udp	aspen-services—Aspen Graphics Technical Analysis
sslp	1750	tcp/udp	Simple Socket Library's PortMaster—NASA
swiftnet	1751	tcp/udp	"SwiftNet-Pentek, Inc."
lofr-lm	1752	tcp/udp	Leap of Faith Research License Manager
translogic-lm	1753	tcp/udp	Translogic License Manager
oracle-em2	1754	tcp/udp	oracle-em2
ms-streaming	1755	tcp/udp	ms-streaming
capfast-lmd	1756	tcp/udp	capfast-lmd—Phase Three Logic Inc.
cnhrp	1757	tcp/udp	cnhrp—Virata Semiconductor
tftp-mcast	1758	tcp/udp	tftp-mcast—LanWorks
spss-lm	1759	tcp/udp	SPSS License Manager
www-ldap-gw	1760	tcp/udp	www-ldap-gw—DEC
cft-0	1761	tcp/udp	cft-0
cft-1	1762	tcp/udp	cft-1
cft-2	1763	tcp/udp	cft-2
cft-3	1764	tcp/udp	cft-3
cft-4	1765	tcp/udp	cft-4
cft-5	1766	tcp/udp	cft-5
cft-6	1767	tcp/udp	cft-6
cft-7	1768	tcp/udp	cft-7
bmc-net-adm	1769	tcp/udp	bmc-net-adm—BMC Software, Inc.
bmc-net-svc	1770	tcp/udp	bmc-net-svc—BMC Software, Inc.
vaultbase	1771	tcp/udp	vaultbase
essweb-gw	1772	tcp/udp	EssWeb Gateway—Hyperion Inc.
kmscontrol	1773	tcp/udp	KMSControl—KMSystems Inc.
global-dtserv	1774	tcp/udp	global-dtserv—Global Communications
femis	1776	tcp/udp	Federal Emergency Management Information System—pl.gov
powerguardian	1777	tcp/udp	powerguardian—BenaTong Software
prodigy-intrnet	1778	tcp/udp	prodigy-internet
pharmasoft	1779	tcp/udp	pharmasoft
dpkeyserv	1780	tcp/udp	dpkeyserv—Omron Software

Abbreviation	Port-Decimal	Protocol	Description/Owner
answersoft-lm	1781	tcp/udp	answersoft-lm
hp-hcip	1782	tcp/udp	hp-hcip
fjris	1783	tcp/udp	Fujitsu Remote Install Service
finle-lm	1784	tcp/udp	Finle License Manager
windlm	1785	tcp/udp	Wind River Systems License Manager
funk-logger	1786	tcp/udp	funk-logger
funk-license	1787	tcp/udp	funk-license
psmond	1788	tcp/udp	psmond—HP
hello	1789	tcp/udp	hello—U Illinois Chicago
nmsp	1790	tcp/udp	Narrative Media Streaming Protocol
ea1	1791	tcp/udp	EA1
ibm-dt-2	1792	tcp/udp	ibm-dt-2
rsc-robot	1793	tcp/udp	rsc-robot
cera-bcm	1794	tcp/udp	cera-bcm—IBM
dpi-proxy	1795	tcp/udp	dpi-proxy—NETsilicon
vocaltec-admin	1796	tcp/udp	Vocaltec Server Administration
uma	1797	tcp/udp	UMA—The Open Group
etp	1798	tcp/udp	Event Transfer Protocol—HP
netrisk	1799	tcp/udp	NETRISK—Trident Data Systems
ansys-lm	1800	tcp/udp	ANSYS-License manager
msmq	1801	tcp/udp	Microsoft Message Que
concomp1	1802	tcp/udp	ConComp1
hp-hcip-gwy	1803	tcp/udp	HP-HCIP-GWY
enl	1804	tcp/udp	ENL—Seagate Software
enl-name	1805	tcp/udp	ENL-Name—Seagate Software
musiconline	1806	tcp/udp	Musiconline—SysSpace
fhsp	1807	tcp/udp	Fujitsu Hot Standby Protocol
oracle-vp2	1808	tcp/udp	Oracle-VP2
oracle-vp1	1809	tcp/udp	Oracle-VP1
jerand-lm	1810	tcp/udp	Jerand License Manager
scientia-sdb	1811	tcp/udp	Scientia-SDB
radius	1812	tcp/udp	RADIUS—Livingston
radius-acct	1813	tcp/udp	RADIUS Accounting—Livingston
tdp-suite	1814	tcp/udp	TDP Suite—Motorola Inc.
mmpft	1815	tcp/udp	MMPFT
harp	1816	tcp/udp	HARP—Portland State U
rkb-oscs	1817	tcp/udp	RKB-OSCS
etftp	1818	tcp/udp	Enhanced Trivial File Transfer Protocol—Mitre
plato-lm	1819	tcp/udp	Plato License Manager—Ermuk
mcagent	1820	tcp/udp	mcagent—IBM
donnyworld	1821	tcp/udp	donnyworld
es-elmd	1822	tcp/udp	es-elmd
unisys-lm	1823	tcp/udp	Unisys Natural Language License Manager
metrics-pas	1824	tcp/udp	metrics-pas
direcpc-video	1825	tcp/udp	DirecPC Video—Hughes Network Systems
ardt	1826	tcp/udp	ARDT—Ardent
asi	1827	tcp/udp	ASI—The ABB Group
itm-mcell-u	1828	tcp/udp	itm-mcell-u—ITmasters
optika-emedia	1829	tcp/udp	Optika eMedia
net8-cman	1830	tcp/udp	Oracle Net8 CMan Admin
myrtle	1831	tcp/udp	Myrtle—Genscan
tht-treasure	1832	tcp/udp	ThoughtTreasure—Signiform
udpradio	1833	tcp/udp	udpradio—Warande
ardusuni	1834	tcp/udp	ARDUS Unicast
ardusmul	1835	tcp/udp	ARDUS Multicast
ste-smsc	1836	tcp/udp	ste-smsc—ST-Electronics
csoft1	1837	tcp/udp	csoft1
talnet	1838	tcp/udp	TALNET-Taltrade
netopia-vo1	1839	tcp/udp	netopia-vo1
netopia-vo2	1840	tcp/udp	netopia-vo2
netopia-vo3	1841	tcp/udp	netopia-vo3
netopia-vo4	1842	tcp/udp	netopia-vo4
netopia-vo5	1843	tcp/udp	netopia-vo5
direcpc-dll	1844	tcp/udp	DirecPC-DLL—Hughes Network Systems
#	1845-1849	tcp/udp	Unassigned
gsi	1850	tcp/udp	GSI
ctcd	1851	tcp/udp	ctcd—Cybertrace
#	1852-1859	tcp/udp	Unassigned

Abbreviation	Port-Decimal	Protocol	Description/Owner
sunscalar-svc	1860	tcp/udp	SunSCALAR Services
lecroy-vicp	1861	tcp/udp	LeCroy VICP
techra-server	1862	tcp/udp	techra-server—Maxware
msnp	1863	tcp/udp	MSNP—Microsoft
paradym-31port	1864	tcp/udp	Paradym 31 Port—Wizdom
entp	1865	tcp/udp	ENTP—Epson
#	1866-1869	tcp/udp	Unassigned
sunscalar-dns	1870	tcp/udp	SunSCALAR DNS Service
canocentral0	1871	tcp/udp	Cano Central 0—Canon
canocentral1	1872	tcp/udp	Cano Central 1—Canon
fjmpjps	1873	tcp/udp	jps—Fujitsu
fjswapsnp	1874	tcp/udp	swapsnp—Fujitsu
#	1875-1880	tcp/udp	Unassigned
ibm-mqseries2	1881	tcp/udp	IBM MQSeries
#	1882-1894	tcp/udp	Unassigned
vista-4gl	1895	tcp/udp	vista 4GL—Vistacomp
#	1896-1898	tcp/udp	Unassigned
mc2studios	1899	tcp/udp	MC2Studios—TheCube
ssdp	1900	tcp/udp	SSDP—Microsoft
fjicl-tep-a	1901	tcp/udp	Fujitsu ICL Terminal Emulator Program A
fjicl-tep-b	1902	tcp/udp	Fujitsu ICL Terminal Emulator Program B
linkname	1903	tcp/udp	Local Link Name Resolution—Lucent
fjicl-tep-c	1904	tcp/udp	Fujitsu ICL Terminal Emulator Program C
sugp	1905	tcp/udp	Secure UP.Link Gateway Protocol—UpLANet
tpmd	1906	tcp/udp	TPortMapperReq—IBM
intrastar	1907	tcp/udp	IntraSTAR—IntraSTAR
dawn	1908	tcp/udp	Dawn—Travsoft
global-wlink	1909	tcp/udp	Global World Link—Global Communications
ultrabac	1910	tcp/udp	ultrabac—CNW
mtp	1911	tcp/udp	Starlight Networks Multimedia Transport Protocol
rhp-iibp	1912	tcp/udp	rhp-iibp—M-Ware
armadp	1913	tcp/udp	armadp—Arm, ltd
elm-momentum	1914	tcp/udp	Elm-Momentum—MDS
facelink	1915	tcp/udp	FACELINK—Hiscom
persona	1916	tcp/udp	Persoft Persona
noagent	1917	tcp/udp	nOAgent—Datawatch
can-nds	1918	tcp/udp	Candle Directory Service - NDS
can-dch	1919	tcp/udp	Candle Directory Service - DCH
can-ferret	1920	tcp/udp	Candle Directory Service - FERRET
noadmin	1921	tcp/udp	NoAdmin—Datawatch
tapestry	1922	tcp/udp	Tapestry—Anritsu
spice	1923	tcp/udp	SPICE—Sendit
xiip	1924	tcp/udp	XIIP—HMRinc
#	1925-1929	tcp/udp	Unassigned
driveappserver	1930	tcp/udp	Drive AppServer—Demon
amdsched	1931	tcp/udp	AMD SCHED—Warwick
#	1932-1943	tcp/udp	Unassigned
close-combat	1944	tcp/udp	close-combat—Microsoft
dialogic-elmd	1945	tcp/udp	dialogic-elmd
tekpls	1946	tcp/udp	tekpls—Tektronix
hlserver	1947	tcp/udp	hlserver-Fast-ag
eye2eye	1948	tcp/udp	eye2eye—losoftware
ismaeasdaqlive	1949	tcp/udp	ISMA Easdaq Live
ismaeasdaqtest	1950	tcp/udp	ISMA Easdaq Test
bcs-lmserver	1951	tcp/udp	bcs-lmserver—Knoware BV
mpnjsc	1952	tcp/udp	mpnjsc—PFU Direct
rapidbase	1953	tcp/udp	Rapid Base—VTT
#	1954-1960	tcp/udp	Unassigned
bts-appserver	1961	tcp/udp	BTS APPSERVER—Sabre
solid-e-engine	1964	tcp/udp	SOLID E ENGINE—Solidtech
tivoli-npm	1965	tcp/udp	Tivoli NPM
slush	1966	tcp/udp	Slush—IBS
sns-quote	1967	tcp/udp	Quote—Calicotech
#	1968-1971	tcp/udp	Unassigned
intersys-cache	1972	tcp/udp	Cache
intersys-cache	1972	tcp/udp	Cache
dlsrap	1973	tcp/udp	Data Link Switching Remote Access Protocol—Cisco
drp	1974	tcp/udp	DRP—Cisco

Abbreviation	Port-Decimal	Protocol	Description/Owner
tcoflashagent	1975	tcp/udp	TCO Flash Agent—TCOsoft
tcoregagent	1976	tcp/udp	TCO Reg Agent—TCOsoft
tcoaddressbook	1977	tcp/udp	TCO Address Book—TCOsoft
unisql	1978	tcp/udp	UniSQL—Windtraveller
unisql-java	1979	tcp/udp	UniSQL Java—Windtraveller
#	1980-1983	tcp/udp	Unassigned
bb	1984	tcp/udp	BB—MacLawran
hsrp	1985	tcp/udp	Hot Standby Router Protocol—Cisco
licensedaemon	1986	tcp/udp	cisco license management
tr-rsrb-p1	1987	tcp/udp	cisco RSRB Priority 1 port
tr-rsrb-p2	1988	tcp/udp	cisco RSRB Priority 2 port
tr-rsrb-p3	1989	tcp/udp	cisco RSRB Priority 3 port
mshnet	1989	tcp/udp	MHSnet system (unassigned but in use)—Sidney U
stun-p1	1990	tcp/udp	cisco STUN Priority 1 port
stun-p2	1991	tcp/udp	cisco STUN Priority 2 port
stun-p3	1992	tcp/udp	cisco STUN Priority 3 port
ipsendmsg	1992	tcp/udp	IPsendmsg (unassigned but in use)—Sidney U
snmp-tcp-port	1993	tcp/udp	cisco SNMP TCP port
stun-port	1994	tcp/udp	cisco serial tunnel port
perf-port	1995	tcp/udp	cisco perf port
tr-rsrb-port	1996	tcp/udp	cisco Remote SRB port
gdp-port	1997	tcp/udp	cisco Gateway Discovery Protocol
x25-svc-port	1998	tcp/udp	cisco X.25 service (XOT)
tcp-id-port	1999	tcp/udp	cisco identification port
callbook	2000	tcp/udp	
dc	2001	tcp	
wizard	2001	udp	curry
globe	2002	tcp/udp	
mailbox	2004	tcp	
emce	2004	udp	CCWS mm conf
berknet	2005	tcp	
oracle	2005	udp	
invokator	2006	tcp	
raid-cc	2006	udp	raid
dectalk	2007	tcp	
raid-am	2007	udp	
conf	2008	tcp	
terminaldb	2008	udp	
news	2009	tcp	
whosockami	2009	udp	
search	2010	tcp	
pipe_server	2010	udp	
raid-cc	2011	tcp	raid
servserv	2011	udp	
ttyinfo	2012	tcp	
raid-ac	2012	udp	
raid-am	2013	tcp	
raid-cd	2013	udp	
troff	2014	tcp	
raid-sf	2014	udp	
cypress	2015	tcp	
raid-cs	2015	udp	
bootserver	2016	tcp/udp	
cypress-stat	2017	tcp	
bootclient	2017	udp	
terminaldb	2018	tcp	
rellpack	2018	udp	
whosockami	2019	tcp	
about	2019	udp	
xinupageserver	2020	tcp	
xinupageserver	2020	udp	
servexec	2021	tcp	
xinuexpansion1	2021	udp	
down	2022	tcp	
xinuexpansion2	2022	udp	
xinuexpansion3	2023	tcp/udp	
xinuexpansion4	2024	tcp/udp	
ellpack	2025	tcp	

Abbreviation	Port-Decimal	Protocol	Description/Owner
xribs	2025	udp	
scrabble	2026	tcp/udp	
shadowserver	2027	tcp/udp	
submitserver	2028	tcp/udp	
device2	2030	tcp/udp	
blackboard	2032	tcp/udp	
glogger	2033	tcp/udp	
scoremgr	2034	tcp/udp	
imsldoc	2035	tcp/udp	
objectmanager	2038	tcp/udp	
lam	2040	tcp/udp	
interbase	2041	tcp/udp	
isis	2042	tcp/udp	isis
isis-bcast	2043	tcp/udp	isis-bcast
rimsl	2044	tcp/udp	
cdfunc	2045	tcp/udp	
cdfunc	2045	tcp/udp	
sdfunc	2046	tcp/udp	
sdfunc	2046	tcp/udp	
dls	2047	tcp/udp	
dls	2047	tcp/udp	
dls-monitor	2048	tcp/udp	
dls-monitor	2048	tcp/udp	
*		tcp/udp	<== NOTE Conflict on 2049 !
shilp	2049	tcp/udp	
nfs	2049	tcp/udp	Network File System - Sun Microsystems
dlsrpn	2065	tcp/udp	Data Link Switch Read Port Number—Radware
dlswpn	2067	tcp/udp	Data Link Switch Write Port Number—Radware
lrp	2090	tcp/udp	Load Report Protocol—Radware
prp	2091	tcp/udp	PRP—Radware
descent3	2092	tcp/udp	Descent 3—Outrage
nbx-cc	2093	tcp/udp	NBX CC—NBXcorp
nbx-au	2094	tcp/udp	NBX AU—NBXcorp
nbx-ser	2095	tcp/udp	NBX SER—NBXcorp
nbx-dir	2096	tcp/udp	NBX DIR—NBXcorp
jetformpreview	2097	tcp/udp	Jet Form Preview
dialog-port	2098	tcp/udp	Dialog Port—Cisco
h2250-annex-g	2099	tcp/udp	H.225.0 Annex G—Vocaltec
amiganetfs	2100	tcp/udp	amiganetfs—U di Pisa
rtcm-sc104	2101	tcp/udp	rtcm-sc104—wsrcc
zephyr-srv	2102	tcp/udp	Zephyr server—MIT
zephyr-clt	2103	tcp/udp	Zephyr serv-hm connection—MIT
zephyr-hm	2104	tcp/udp	Zephyr hostmanager—MIT
minipay	2105	tcp/udp	MiniPay—IBM
mzap	2106	tcp/udp	MZAP—Microsoft
bintec-admin	2107	tcp/udp	BinTec Admin
comcam	2108	tcp/udp	Comcam
ergolight	2109	tcp/udp	Ergolight—Ledalite
#	2110-2164	tcp/udp	Unassigned
x-bone-api	2165	tcp/udp	X-Bone AP-ISI
#	2166-2179	tcp/udp	Unassigned
mc-gt-srv	2180	tcp/udp	Millicent Vendor Gateway Server—DEC
eforward	2181	tcp/udp	eforward—Flagship
ici	2200	tcp/udp	ICI—Unisys
ats	2201	tcp/udp	Advanced Training System Program
imtc-map	2202	tcp/udp	Multimedia Teleconferencing Cosortium—Databeam
kali	2213	tcp/udp	Kali—U of Georgia
ganymede	2220	tcp/udp	Ganymede
rockwell-csp1	2221	tcp/udp	Rockwell CSP1
rockwell-csp2	2222	tcp/udp	Rockwell CSP2
rockwell-csp3	2223	tcp/udp	Rockwell CSP3
ivs-video	2232	tcp/udp	IVS Video default—INRIA
infocrypt	2233	tcp/udp	INFOCRYPT—Isolation
directplay	2234	tcp/udp	DirectPlay—Microsoft
sercomm-wlink	2235	tcp/udp	Sercomm-Wlink—Sercomm
nani	2236	tcp/udp	Nani—EDS
optech-port1-lm	2237	tcp/udp	Optech Port1 License Manager—Opticaltech
aviva-sna	2238	tcp/udp	AVIVA SNA SER

Abbreviation	Port-Decimal	Protocol	Description/Owner
imagequery	2239	tcp/udp	Image Query—Numinous
recipe	2240	tcp/udp	RECIPe—GTE
ivsd	2241	tcp/udp	IVS Daemon—INRIA
foliocorp	2242	tcp/udp	Folio Remote Server
magicom	2243	tcp/udp	Magicom Protocol
nmsserver	2244	tcp/udp	NMS Server—NMSS
hao	2245	tcp/udp	HaO—HAO
#	2246-2278	tcp/udp	Unassigned
xmquery	2279	tcp/udp	xmquery—IBM
lnvpoller	2280	tcp/udp	LNVPOLLER—Lotus
lnvconsole	2281	tcp/udp	LNVCONSOLE—Lotus
lnvalarm	2282	tcp/udp	LNVALARM—Lotus
lnvstatus	2283	tcp/udp	LNVSTATUS—Lotus
lnvmaps	2284	tcp/udp	LNVMAPS—Lotus
lnvmailmon	2285	tcp/udp	LNVMAILMON—Lotus
nas-metering	2286	tcp/udp	NAS-Metering—Symantec
dna	2287	tcp/udp	DNA—Ricochet
netml	2288	tcp/udp	NETML—Krypton
#	2289-2293	tcp/udp	Unassigned
konshus-lm	2294	tcp/udp	Konshus License Manager (FLEX)
advant-lm	2295	tcp/udp	Advant License Manager—The ABB Group
theta-lm	2296	tcp/udp	Theta License Manager (Rainbow)
d2k-datamover1	2297	tcp/udp	D2K DataMover 1
d2k-datamover2	2298	tcp/udp	D2K DataMover 2
pc-telecommute	2299	tcp/udp	PC Telecommute—Symantec
cvmmon	2300	tcp/udp	CVMMON—HP
cpq-wbem	2301	tcp/udp	Compaq HTTP
binderysupport	2302	tcp/udp	Bindery Support—Novell
proxy-gateway	2303	tcp/udp	Proxy Gateway—Funk Software
attachmate-uts	2304	tcp/udp	Attachmate UTS
mt-scaleserver	2305	tcp/udp	MT ScaleServer
tappi-boxnet	2306	tcp/udp	TAPPI BoxNet—Alliancetechnical
pehelp	2307	tcp/udp	pehelp—HP
sdhelp	2308	tcp/udp	sdhelp—HP
sdserver	2309	tcp/udp	SD Server—HP
sdclient	2310	tcp/udp	SD Client—HP
messageservice	2311	tcp/udp	Message Service
iapp	2313	tcp/udp	IAPP (Inter Access Point Protocol)—Lucent
cr-websystems	2314	tcp/udp	CR WebSystems—Peddie
precise-sft	2315	tcp/udp	Precise Sft.
sent-lm	2316	tcp/udp	SENT License Manager—Axix Inc
attachmate-g32	2317	tcp/udp	Attachmate G32
cadencecontrol	2318	tcp/udp	Cadence Control—Polybon
infolibria	2319	tcp/udp	InfoLibria
siebel-ns	2320	tcp/udp	Siebel NS
rdlap	2321	tcp/udp	RDLAP over UDP—Motorola Inc.
ofsd	2322	tcp/udp	ofsd—Commvault
3d-nfsd	2323	tcp/udp	3d-nfsd—Commvault
cosmocall	2324	tcp/udp	Cosmocall
designspace-lm	2325	tcp/udp	Design Space License Management—Ansys
idcp	2326	tcp/udp	IDCP—KGC
xingcsm	2327	tcp/udp	xingcsm—Xingtech
netrix-sftm	2328	tcp/udp	Netrix SFTM
nvd	2329	tcp/udp	NVD—Noblenet
tscchat	2330	tcp/udp	TSCCHAT
agentview	2331	tcp/udp	AGENTVIEW—Aaccorp
rcc-host	2332	tcp/udp	RCC Host—Rapistan
snapp	2333	tcp/udp	SNAPP—Meridian-data
ace-client	2334	tcp/udp	ACE Client Auth—Securitydynamics
ace-proxy	2335	tcp/udp	ACE Proxy—Securitydynamics
appleugcontrol	2336	tcp/udp	Apple UG Control
ideesrv	2337	tcp/udp	ideesrv—Ideetique
norton-lambert	2338	tcp/udp	Norton Lambert
3com-webview	2339	tcp/udp	3Com WebView
wrs_registry	2340	tcp/udp	WRS Registry
xiostatus	2341	tcp/udp	XIO Status—Xiotech
manage-exec	2342	tcp/udp	Seagate Manage Exec
nati-logos	2343	tcp/udp	nati logos—Natinst

Abbreviation	Port-Decimal	Protocol	Description/Owner
fcmsys	2344	tcp/udp	fcmsys—Fujitsu
dbm	2345	tcp/udp	dbm—Fujitsu
redstorm_join	2346	tcp/udp	Game Connection Port—Redstorm
redstorm_find	2347	tcp/udp	Game Announcement and Location—Redstorm
redstorm_info	2348	tcp/udp	Information to query for game status—Redstorm
redstorm_diag	2349	tcp/udp	Diagnostics Port—Redstorm
psbserver	2350	tcp/udp	psbserver—Pharos
psrserver	2351	tcp/udp	psrserver—Pharos
pslserver	2352	tcp/udp	pslserver—Pharos
pspserver	2353	tcp/udp	pspserver—Pharos
psprserver	2354	tcp/udp	psprserver—Pharos
psdbserver	2355	tcp/udp	psdbserver—Pharos
gxtelmd	2356	tcp/udp	GXT License Managemant
unihub-server	2357	tcp/udp	UniHub Server—CSIcontrols
futrix	2358	tcp/udp	Futrix—IBM
flukeserver	2359	tcp/udp	FlukeServer—Forte Networks
nexstorindltd	2360	tcp/udp	NexstorIndLtd
tl1	2361	tcp/udp	TL1—Cerent
digiman	2362	tcp/udp	digiman
mediacntrlnfsd	2363	tcp/udp	Media Central NFSD—Sun
oi-2000	2364	tcp/udp	OI-2000—Software Horizons Inc.
dbref	2365	tcp/udp	dbref—Fujitsu
qip-login	2366	tcp/udp	qip-login—Lucent
service-ctrl	2367	tcp/udp	Service Control—HP
opentable	2368	tcp/udp	OpenTable
acs2000-dsp	2369	tcp/udp	ACS2000 DSP
l3-hbmon	2370	tcp/udp	L3-HBMon—L-3security
#	2371-2380	tcp/udp	Unassigned
compaq-https	2381	tcp/udp	Compaq HTTPS
ms-olap3	2382	tcp/udp	Microsoft OLAP
ms-olap4	2383	tcp/udp	Microsoft OLAP
#	2384-2388	tcp/udp	Unassigned
ovsessionmgr	2389	tcp/udp	OpenView Session Mgr—HP
rsmtp	2390	tcp/udp	RSMTP—Bell Labs
3com-net-mgmt	2391	tcp/udp	3COM Net Management
tacticalauth	2392	tcp/udp	Tactical Auth
ms-olap1	2393	tcp/udp	MS OLAP 1—Microsoft
ms-olap2	2394	tcp/udp	MS OLAP 2—Microsoft
lan900_remote	2395	tcp/udp	LAN900 Remote—DigiTechINC
wusage	2396	tcp/udp	Wusage—Thomas Boutell
ncl	2397	tcp/udp	NCL—Motorola Inc.
orbiter	2398	tcp/udp	Orbiter—Callisto
fmpro-fdal	2399	tcp/udp	"FileMaker, Inc. - Data Access Layer"
opequus-server	2400	tcp/udp	OpEquus Server—tesl
cvspserver	2401	tcp/udp	cvspserver—Cyclic
taskmaster2000	2402	tcp/udp	TaskMaster 2000 Server—Datacap
taskmaster2000	2403	tcp/udp	TaskMaster 2000 Web—Datacap
iec870-5-104	2404	tcp/udp	IEC870-5-104—Sat-Automation
trc-netpoll	2405	tcp/udp	TRC Netpoll—Telcores
jediserver	2406	tcp/udp	JediServer—NCR
orion	2407	tcp/udp	Orion—PC Users Group
optimanet	2408	tcp/udp	OptimaNet—Optimal
sns-protocol	2409	tcp/udp	SNS Protocol—NetManage
vrts-registry	2410	tcp/udp	VRTS Registry—VERITAS Software
netwave-ap-mgmt	2411	tcp/udp	Netwave AP Management
cdn	2412	tcp/udp	CDN—Netmind
orion-rmi-reg	2413	tcp/udp	orion-rmi-reg —IBM
interlingua	2414	tcp/udp	Interlingua—Virtual-Unlimited
comtest	2415	tcp/udp	COMTEST
rmtserver	2416	tcp/udp	RMT Server—SimWare
composit-server	2417	tcp/udp	Composit Server—PFU Direct
cas	2418	tcp/udp	cas—Fujitsu
attachmate-s2s	2419	tcp/udp	Attachmate S2S
dslremote-mgmt	2420	tcp/udp	DSL Remote Management—Westell, Inc
g-talk	2421	tcp/udp	G-Talk—Four-Sight
crmsbits	2422	tcp/udp	CRMSBITS—Telstra
rnrp	2423	tcp/udp	RNRP—ABB Sweden
kofax-svr	2424	tcp/udp	KOFAX-SVR

Abbreviation	Port-Decimal	Protocol	Description/Owner
fjitsuappmgr	2425	tcp/udp	Fujitsu App Manager
applianttcp	2426	tcp/udp	Appliant TCP
mgcp-gateway	2427	tcp/udp	Media Gateway Control Protocol Gateway-Telcordia
ott	2428	tcp/udp	One Way Trip Time—BBN
ft-role	2429	tcp/udp	FT-ROLE—CLR
venus	2430	tcp/udp	venus—Robert Watson
venus-se	2431	tcp/udp	venus-se—Robert Watson
codasrv	2432	tcp/udp	codasrv—Robert Watson
codasrv-se	2433	tcp/udp	codasrv-se—Robert Watson
pxc-epmap	2434	tcp/udp	pxc-epmap—Xerox
optilogic	2435	tcp/udp	OptiLogic—Wirespeed
topx	2436	tcp/udp	TOP/X—ubbcluj
unicontrol	2437	tcp/udp	UniControl—Hardware Software Development GmBh
msp	2438	tcp/udp	MSP—ACC
sybasedbsynch	2439	tcp/udp	SybaseDBSynch
spearway	2440	tcp/udp	Spearway Lockers
pvsw-inet	2441	tcp/udp	pvsw-inet—Pervasive
netangel	2442	tcp/udp	Netangel—Identcode
powerclientcsf	2443	tcp/udp	PowerClient Central Storage Facility—Unisys
btpp2sectrans	2444	tcp/udp	BT PP2 Sectrans
dtn1	2445	tcp/udp	DTN1
bues_service	2446	tcp/udp	bues_service—telekom400
ovwdb	2447	tcp/udp	OpenView NNM daemon—HP
hpppssvr	2448	tcp/udp	hpppsvr—HP
ratl	2449	tcp/udp	RATL—Unisys
netadmin	2450	tcp/udp	netadmin—CU-muc
netchat	2451	tcp/udp	netchat—CU-muc
snifferclient	2452	tcp/udp	SnifferClient—NAI
madge-om	2453	tcp/udp	madge-om
indx-dds	2454	tcp/udp	IndX-DDS
wago-io-system	2455	tcp/udp	WAGO-IO-SYSTEM—IEEE
altav-remmgt	2456	tcp/udp	altav-remmgt—DEC
rapido-ip	2457	tcp/udp	Rapido_IP—Wamnet
griffin	2458	tcp/udp	griffin—Unisys
community	2459	tcp/udp	Community—Webmaster
ms-theater	2460	tcp/udp	ms-theater—Microsoft
qadmifoper	2461	tcp/udp	qadmifoper—Tellabs
qadmifevent	2462	tcp/udp	qadmifevent—Tellabs
symbios-raid	2463	tcp/udp	Symbios Raid
direcpc-si	2464	tcp/udp	DirecPC SI—Hughes Network Systems
lbm	2465	tcp/udp	Load Balance Management—PFU Direct
lbf	2466	tcp/udp	Load Balance Forwarding—PFU Direct
high-criteria	2467	tcp/udp	High Criteria
qip-msgd	2468	tcp/udp	qip_msgd—Lucent
mti-tcs-comm	2469	tcp/udp	MTI-TCS-COMM—Microtempus
taskman-port	2470	tcp/udp	taskman port—Himel
seaodbc	2471	tcp/udp	SeaODBC—Aran
c3	2472	tcp/udp	C3—com-on
aker-cdp	2473	tcp/udp	Aker-cdp
vitalanalysis	2474	tcp/udp	Vital Analysis—VitalSigns
ace-server	2475	tcp/udp	ACE Server
ace-svr-prop	2476	tcp/udp	ACE Server Propagation
ssm-cvs	2477	tcp/udp	SecurSight Certificate Valifation Service—SecurityDynamics
ssm-cssps	2478	tcp/udp	SecurSight Authentication Server (SLL)—SecurityDynamics
ssm-els	2479	tcp/udp	SecurSight Event Logging Server (SSL)—SecurityDynamics
lingwood	2480	tcp/udp	Lingwood's Detail—Btinternet
giop	2481	tcp/udp	Oracle GIOP
giop-ssl	2482	tcp/udp	Oracle GIOP SSL
ttc	2483	tcp/udp	Oracle TTC
ttc-ssl	2484	tcp/udp	Oracle TTC SSL
netobjects1	2485	tcp/udp	Net Objects1
netobjects2	2486	tcp/udp	Net Objects2
pns	2487	tcp/udp	Policy Notice Service—Fujitsu
moy-corp	2488	tcp/udp	Moy Corporation
tsilb	2489	tcp/udp	TSILB—Travsoft
qip-qdhcp	2490	tcp/udp	qip_qdhcp—Lucent
conclave-cpp	2491	tcp/udp	Conclave CPP—Interdyn
groove	2492	tcp/udp	GROOVE

Abbreviation	Port-Decimal	Protocol	Description/Owner
talarian-mqs	2493	tcp/udp	Talarian MQS
bmc-ar	2494	tcp/udp	BMC AR
fast-rem-serv	2495	tcp/udp	Fast Remote Services—Adaptec
dirgis	2496	tcp/udp	DIRGIS—DIRGE
quaddb	2497	tcp/udp	Quad DB—Quad-Sys
odn-castraq	2498	tcp/udp	ODN-CasTraq
unicontrol	2499	tcp/udp	UniControl
rtsserv	2500	tcp/udp	Resource Tracking system server—ETSUADMN.ETSU.EDU
rtsclient	2501	tcp/udp	Resource Tracking system client—ETSUADMN.ETSU.EDU
kentrox-prot	2502	tcp/udp	Kentrox Protocol
nms-dpnss	2503	tcp/udp	NMS-DPNSS—NMS-Europe
wlbs	2504	tcp/udp	WLBS—Microsoft
torque-traffic	2505	tcp/udp	torque-traffic
jbroker	2506	tcp/udp	jbroker—objectScape
spock	2507	tcp/udp	spock—Texas A&M U
jdatastore	2508	tcp/udp	JDataStore—Inprise
fjmpss	2509	tcp/udp	fjmpss—Fujitsu
fjappmgrbulk	2510	tcp/udp	fjappmgrbulk—Fujitsu
metastorm	2511	tcp/udp	Metastorm
citrixima	2512	tcp/udp	Citrix IMA
citrixadmin	2513	tcp/udp	Citrix ADMIN
facsys-ntp	2514	tcp/udp	Facsys NTP
facsys-router	2515	tcp/udp	Facsys Router
maincontrol	2516	tcp/udp	Main Control—Vocaltec
call-sig-trans	2517	tcp/udp	H.323 Annex E call signaling transport—Vocaltec
willy	2518	tcp/udp	Willy—Hudsmoar
globmsgsvc	2519	tcp/udp	globmsgsvc—Intel
pvsw	2520	tcp/udp	pvsw—Pervasive
adaptecmgr	2521	tcp/udp	Adaptec Manager
windb	2522	tcp/udp	WinDb—Arium
qke-llc-v3	2523	tcp/udp	Qke LLC V.3—Kecam-Han
optiwave-lm	2524	tcp/udp	Optiwave License Management
ms-v-worlds	2525	tcp/udp	MS V-Worlds—Microsoft
ema-sent-lm	2526	tcp/udp	EMA License Manager—EMA Denmark
iqserver	2527	tcp/udp	IQ Server—AKBS
ncr_ccl	2528	tcp/udp	NCR CCL
utsftp	2529	tcp/udp	UTS FTP
vrcommerce	2530	tcp/udp	VR Commerce—IBM
ito-e-gui	2531	tcp/udp	ITO-E GUI—HP
ovtopmd	2532	tcp/udp	OVTOPMD—HP
snifferserver	2533	tcp/udp	SnifferServer—NAI
combox-web-acc	2534	tcp/udp	Combox Web Access
mdhcp	2535	tcp/udp	MDHCP—Sun
btpp2audctr1	2536	tcp/udp	btpp2audctr1—British Telcom
upgrade	2537	tcp/udp	Upgrade Protocol—DST-Inc
vnwk-prapi	2538	tcp/udp	vnwk-prapi—VisualNetworks
vsiadmin	2539	tcp/udp	VSI Admin
lonworks	2540	tcp/udp	LonWorks—Echelon
lonworks2	2541	tcp/udp	LonWorks2—Echelon
davinci	2542	tcp/udp	daVinci—TZI
reftek	2543	tcp/udp	REFTEK
reftek	2543	udp	REFTEK
novell-zen	2544	tcp	Novell ZEN
sis-emt	2545	tcp/udp	sis-emt—Securicor
vytalvaultbrtp	2546	tcp/udp	vytalvaultbrtp—Vytalnet
vytalvaultvsmp	2547	tcp/udp	vytalvaultvsmp—Vytalnet
vytalvaultpipe	2548	tcp/udp	vytalvaultpipe—Vytalnet
ipass	2549	tcp/udp	IPASS—Pass
ads	2550	tcp/udp	ADS—Adobe
isg-uda-server	2551	tcp/udp	ISG UDA Server
call-logging	2552	tcp/udp	Call Logging—Ascend
efidiningport	2553	tcp/udp	efidiningport—ExecPC
vcnet-link-v10	2554	tcp/udp	VCnet-Link v10—Csaba Mate
compaq-wcp	2555	tcp/udp	Compaq WCP
nicetec-nmsvc	2556	tcp/udp	nicetec-nmsvc
nicetec-mgmt	2557	tcp/udp	nicetec-mgmt
pclemultimedia	2558	tcp/udp	PCLE Multi Media—PinnacleSys
lstp	2559	tcp/udp	LSTP—IBM

Abbreviation	Port-Decimal	Protocol	Description/Owner
labrat	2560	tcp/udp	labrat—IBM
mosaixcc	2561	tcp/udp	MosaixCC
delibo	2562	tcp/udp	Delibo—NovaWiz LTD
cti-redwood	2563	tcp/udp	CTI Redwood—Daou
hp-3000-telnet	2564	tcp	HP 3000 NS/VT block mode telnet
coord-svr	2565	tcp/udp	Coordinator Server—EnsembleSoft
pcs-pcw	2566	tcp/udp	pcs-pcw—Pcare
clp	2567	tcp/udp	Cisco Line Protocol
spamtrap	2568	tcp/udp	SPAM TRAP—BenaTong Software
sonuscallsig	2569	tcp/udp	Sonus Call Signal
hs-port	2570	tcp/udp	HS Port—NETect
cecsvc	2571	tcp/udp	CECSVC—Corder-Eng
ibp	2572	tcp/udp	IBP—Activ
trustestablish	2573	tcp/udp	Trust Establish—IBM
blockade-bpsp	2574	tcp/udp	Blockade BPSP
hl7	2575	tcp/udp	HL7
tclprodebugger	2576	tcp/udp	TCL Pro Debugger
tclprodebugger	2576	udp	TCL Pro Debugger
scipticslsrvr	2577	tcp/udp	Scriptics Lsrvr
rvs-isdn-dcp	2578	tcp/udp	RVS ISDN DCP
mpfoncl	2579	tcp/udp	mpfoncl—Fujitsu
tributary	2580	tcp/udp	Tributary—Bristol
argis-te	2581	tcp/udp	ARGIS TE
argis-ds	2582	tcp/udp	ARGIS DS
mon	2583	tcp/udp	MON—Transmeta
cyaserv	2584	tcp/udp	cyaserv—CYASolutions
netx-server	2585	tcp/udp	NETX Server—IPMetrics
netx-agent	2586	tcp/udp	NETX Agent—IPMetrics
masc	2587	tcp/udp	MASC—USC
privilege	2588	tcp/udp	Privilege—AKS
quartus-tcl	2589	tcp/udp	quartus tcl
idotdist	2590	tcp/udp	idotdist—Invino
maytagshuffle	2591	tcp/udp	Maytag Shuffle
netrek	2592	tcp/udp	netrek—TeamQuest
mns-mail	2593	tcp/udp	MNS Mail Notice Service—Fujitsu
dts	2594	tcp/udp	Data Base Server—Home-Online
worldfusion1	2595	tcp/udp	World Fusion 1
worldfusion2	2596	tcp/udp	World Fusion 2
homesteadglory	2597	tcp/udp	Homestead Glory
citriximaclient	2598	tcp/udp	Citrix MA Client
meridiandata	2599	tcp/udp	Meridian Data
hpstgmgr	2600	tcp/udp	HPSTGMGR—HP
discp-client	2601	tcp/udp	discp client—3Com
discp-server	2602	tcp/udp	discp server—3com
servicemeter	2603	tcp/udp	Service Meter—Synoia
nsc-ccs	2604	tcp/udp	NSC CCS—NetworkSciences
nsc-posa	2605	tcp/udp	NSC POSA—NetworkSciences
netmon	2606	tcp/udp	Dell Netmon
connection	2607	tcp/udp	Dell Connection
wag-service	2608	tcp/udp	Wag Service
system-monitor	2609	tcp/udp	System Monitor—AlphaLink
versa-tek	2610	tcp/udp	VersaTek
lionhead	2611	tcp/udp	LIONHEAD
qpasa-agent	2612	tcp/udp	Qpasa Agent—MQSoftware
smntubootstrap	2613	tcp/udp	SMNTUBootstrap—Metrics
neveroffline	2614	tcp/udp	Never Offline—AMO
firepower	2615	tcp/udp	firepower
appswitch-emp	2616	tcp/udp	appswitch-emp—TopLayer
cmadmin	2617	tcp/udp	Clinical Context Managers—Sentillion
priority-e-com	2618	tcp/udp	Priority E-Com—Eshbel
bruce	2619	tcp/udp	bruce—Sun
lpsrecommender	2620	tcp/udp	LPSRecommender—Andromedia
miles-apart	2621	tcp/udp	Miles Apart Jukebox Server—MilesInfo
metricadbc	2622	tcp/udp	MetricaDBC
lmdp	2623	tcp/udp	LMDP—RocketTalk
aria	2624	tcp/udp	Aria—Andromedia
blwnkl-port	2625	tcp/udp	Blwnkl Port—3Com
gbjd816	2626	tcp/udp	gbjd816—George Balesta

Abbreviation	Port-Decimal	Protocol	Description/Owner
moshebeeri	2627	tcp/udp	Moshe Beeri
dict	2628	tcp/udp	DICT—U North Carolina
sitaraserver	2629	tcp/udp	Sitara Server
sitaramgmt	2630	tcp/udp	Sitara Management
sitaradir	2631	tcp/udp	Sitara Dir
irdg-post	2632	tcp/udp	IRdg Post
interintelli	2633	tcp/udp	InterIntelli
pk-electronics	2634	tcp/udp	PK Electronics—PKWorld
backburner	2635	tcp/udp	Back Burner—MetaCreations
solve	2636	tcp/udp	Solve—Sterling
imdocsvc	2637	tcp/udp	Import Document Service—NetRight
sybaseanywhere	2638	tcp/udp	Sybase Anywhere
aminet	2639	tcp/udp	AMInet—Alcorn McBride Inc
sai_sentlm	2640	tcp/udp	Sabbagh Associates Licence Manager
hdl-srv	2641	tcp/udp	HDL Server—CNRI
tragic	2642	tcp/udp	Tragic—j51
gte-samp	2643	tcp/udp	GTE-SAMP
travsoft-ipx-t	2644	tcp/udp	Travsoft IPX Tunnel
novell-ipx-cmd	2645	tcp/udp	Novell IPX CMD
and-lm	2646	tcp/udp	AND Licence Manager
syncserver	2647	tcp/udp	SyncServer—SyncInc
upsnotifyprot	2648	tcp/udp	Upsnotifyprot—Via-RS
vpsipport	2649	tcp/udp	VPSIPPORT—CSIR
eristwoguns	2650	tcp/udp	eristwoguns—NetPro
ebinsite	2651	tcp/udp	EBInSite
interpathpanel	2652	tcp/udp	InterPathPanel
sonus	2653	tcp/udp	Sonus—SonusNet
corel_vncadmin	2654	tcp/udp	Corel VNC Admin—CorelComputer
unglue	2655	tcp/udp	UNIX Nt Glue—Pscomp
kana	2656	tcp/udp	Kana
sns-dispatcher	2657	tcp/udp	SNS Dispatcher—FirstFloor
sns-admin	2658	tcp/udp	SNS Admin—FirstFloor
sns-query	2659	tcp/udp	SNS Query—FirstFloor
gcmonitor	2660	tcp/udp	GC Monitor—Geodesic
olhost	2661	tcp/udp	OLHOST—LAN-Aces
bintec-capi	2662	tcp/udp	BinTec-CAPI
bintec-tapi	2663	tcp/udp	BinTec-TAPI
command-mq-gm	2664	tcp/udp	Command MQ GM—Boole&Babbage
command-mq-pm	2665	tcp/udp	Command MQ PM—Boole&Babbage
extensis	2666	tcp/udp	extensis
alarm-clock-s	2667	tcp/udp	Alarm Clock Server—Tatanka
alarm-clock-c	2668	tcp/udp	Alarm Clock Client—Tatanka
toad	2669	tcp/udp	TOAD—Tatanka
tve-announce	2670	tcp/udp	TVE Announce—WebTV
newlixreg	2671	tcp/udp	newlixreg—GAAJ
nhserver	2672	tcp/udp	nhserver—Aran
firstcall42	2673	tcp/udp	First Call 42—TFN
ewnn	2674	tcp/udp	ewnn—Omron Software
ttc-etap	2675	tcp/udp	TTC ETAP
simslink	2676	tcp/udp	SIMSLink—Simswsare
gadgetgate1way	2677	tcp/udp	Gadget Gate 1 Way—InternetGadgets
gadgetgate2way	2678	tcp/udp	Gadget Gate 2 Way—InternetGadgets
syncserverssl	2679	tcp/udp	Sync Server SSL—SyncInc
pxc-sapxom	2680	tcp/udp	pxc-sapxom—Xerox
mpnjsomb	2681	tcp/udp	mpnjsomb—PFU Direct
srsp	2682	tcp/udp	SRSP—umu
ncdloadbalance	2683	tcp/udp	NCDLoadBalance
mpnjsosv	2684	tcp/udp	mpnjsosv—PFU Direct
mpnjsocl	2685	tcp/udp	mpnjsocl—PFU Direct
mpnjsomg	2686	tcp/udp	mpnjsomg—PFU Direct
pq-lic-mgmt	2687	tcp/udp	pq-lic-mgmt—PQSystems
md-cg-http	2688	tcp/udp	md-cf-http—ExecMail
fastlynx	2689	tcp/udp	FastLynx—Dave Sewell
hp-nnm-data	2690	tcp/udp	HP NNM Embedded Database
itinternet	2691	tcp/udp	IT Internet
admins-lms	2692	tcp/udp	Admins LMS
belarc-http	2693	tcp/udp	belarc-http
pwrsevent	2694	tcp/udp	pwrsevent—Fujitsu

Abbreviation	Port-Decimal	Protocol	Description/Owner
vspread	2695	tcp/udp	VSPREAD—Fujitsu
unifyadmin	2696	tcp/udp	Unify Admin
oce-snmp-trap	2697	tcp/udp	Oce SNMP Trap Port
mck-ivpip	2698	tcp/udp	MCK-IVPIP
csoft-plusclnt	2699	tcp/udp	Csoft Plus Client
tqdata	2700	tcp/udp	tqdata—TeamQuest
sms-rcinfo	2701	tcp/udp	SMS RCINFO—Microsoft
sms-xfer	2702	tcp/udp	SMS XFER—Microsoft
sms-chat	2703	tcp/udp	SMS CHAT—Microsoft
sms-remctrl	2704	tcp/udp	SMS REMCTRL—Microsoft
sds-admin	2705	tcp/udp	SDS Admin—Sun
ncdmirroring	2706	tcp/udp	NCD Mirroring
emcsymapiport	2707	tcp/udp	EMCSYMAPIPORT
banyan-net	2708	tcp/udp	Banyan-Net
supermon	2709	tcp/udp	Supermon—LANL.gov
sso-service	2710	tcp/udp	SSO Service—Okiok
sso-control	2711	tcp/udp	SSO Control—Okiok
aocp	2712	tcp/udp	Axapta Object Communication Protocol—Damgaard
raven1	2713	tcp/udp	Raven1—Unified-Technologies
raven2	2714	tcp	Raven2—Unified-Technologies
hpstgmgr2	2715	tcp/udp	HPSTGMGR2—HP
inova-ip-disco	2716	tcp/udp	Inova IP Disco
pn-requester	2717	tcp/udp	PN REQUESTER—BMC Software, Inc.
pn-requester2	2718	tcp/udp	PN REQUESTER 2—BMC Software, Inc.
scan-change	2719	tcp/udp	Scan & Change—Lucent
wkars	2720	tcp/udp	wkars—WirelessKnowledge
smart-diagnose	2721	tcp/udp	Smart Diagnose—Geoffry Meek
proactivesrvr	2722	tcp/udp	Proactive Server—IBM
watchdognt	2723	tcp/udp	WatchDog NT
qotps	2724	tcp/udp	qotps—QueryObject
msolap-ptp2	2725	tcp/udp	MSOLAP PTP2—Microsoft
tams	2726	tcp/udp	TAMS—JTI
mgcp-callagent	2727	tcp/udp	Media Gateway Control Protocol Call Agent—Telcordia
sqdr	2728	tcp/udp	SQDR—Starquest
tcim-control	2729	tcp/udp	TCIM Control—Raytheon
nec-raidplus	2730	tcp/udp	NEC RaidPlus
netdragon-msngr	2731	tcp/udp	NetDragon Messanger—GlobalNet
g5m	2732	tcp/udp	G5M—ACM
signet-ctf	2733	tcp/udp	Signet CTF—SAC
ccs-software	2734	tcp/udp	CCS Software
monitorconsole	2735	tcp/udp	Monitor Console—Ganymede
radwiz-nms-srv	2736	tcp/udp	RADWIZ NMS SRV
srp-feedback	2737	tcp/udp	SRP Feedback—Swiss Federal Institute of Technology
ndl-tcp-ois-gw	2738	tcp/udp	NDL TCP-OSI Gateway
tn-timing	2739	tcp/udp	TN Timing
alarm	2740	tcp/udp	Alarm—BFPG
tsb	2741	tcp/udp	TSB—Lucent
tsb2	2742	tcp/udp	TSB2—Lucent
murx	2743	tcp/udp	murx—Dachbu
honyaku	2744	tcp/udp	honyaku—Omron Software
urbisnet	2745	tcp/udp	URBISNET
cpudpencap	2746	tcp/udp	CPUDPENCAP—Checkpoint
fjippol-swrly	2747	tcp/udp	Fujitsu
fjippol-polsvr	2748	tcp/udp	Fujitsu
fjippol-cnsl	2749	tcp/udp	Fujitsu
fjippol-port1	2750	tcp/udp	Fujitsu
fjippol-port2	2751	tcp/udp	Fujitsu
rsisysaccess	2752	tcp/udp	RSISYS ACCESS
de-spot	2753	tcp/udp	de-spot—DigitalEnvoy
apollo-cc	2754	tcp/udp	APOLLO CC—Brand Communications
expresspay	2755	tcp/udp	Express Pay—Netcom
simplement-tie	2756	tcp/udp	simplement-tie—Netvision
cnrp	2757	tcp/udp	CNRP—The ABB Group
apollo-status	2758	tcp/udp	APOLLO Status—Brand Communications
apollo-gms	2759	tcp/udp	APOLLO GMS—Brand Communications
sabams	2760	tcp/udp	Saba MS
dicom-iscl	2761	tcp/udp	DICOM ISCL—Siemens
dicom-tls	2762	tcp/udp	DICOM TLS—Siemens

Abbreviation	Port-Decimal	Protocol	Description/Owner
desktop-dna	2763	tcp/udp	Desktop DNA—MiramarSys
data-insurance	2764	tcp/udp	Data Insurance—Standard
qip-audup	2765	tcp/udp	qip-audup—Lucent
compaq-scp	2766	tcp/udp	Compaq SCP
uadtc	2767	tcp/udp	UADTC—AppliedIS
uacs	2768	tcp/udp	UACS—AppliedIS
singlept-mvs	2769	tcp/udp	Single Point MVS
veronica	2770	tcp/udp	Veronica—Coyote
vergencecm	2771	tcp/udp	Vergence CM—Sentillion
auris	2772	tcp/udp	auris—Telefónica Investigación y Desarrollo, S.A.
pcbakcup1	2773	tcp/udp	PC Backup—Remote Backup Systems
pcbakcup2	2774	tcp/udp	PC Backup—Remote Backup Systems
smpp	2775	tcp/udp	SMMP—Logica Aldiscon
ridgeway1	2776	tcp/udp	Ridgeway Systems & Software
ridgeway2	2777	tcp/udp	Ridgeway Systems & Software
gwen-sonya	2778	tcp/udp	Gwen-Sonya—Internet Connect Inc.
lbc-sync	2779	tcp/udp	LBC Sync—Fujitsu
lbc-control	2780	tcp/udp	LBC Control—Fujitsu
whosells	2781	tcp/udp	whosells—GNSS
everydayrc	2782	tcp/udp	everydayrc—Bluemoon Interactive
aises	2783	tcp/udp	AISES—Pgaero
www-dev	2784	tcp/udp	world wide web - development
aic-np	2785	tcp/udp	aic-np—Cisco
aic-oncrpc	2786	tcp/udp	aic-oncrpc - Destiny MCD database—Cisco
piccolo	2787	tcp/udp	piccolo - Cornerstone Software
fryeserv	2788	tcp/udp	NetWare Loadable Module - Seagate Software
media-agent	2789	tcp/udp	Media Agent—The BRM Group
plgproxy	2790	tcp/udp	PLG Proxy—Aladdin Inc.
mtport-regist	2791	tcp/udp	MT Port Registrator—Institute for Advanced Medical Education
f5-globalsite	2792	tcp/udp	f5-globalsite—F5 Networks, Inc.
initlsmsad	2793	tcp/udp	initlsmsad—Compaq
aaftp	2794	tcp/udp	aaftp—E. Jay Berkenbilt
livestats	2795	tcp/udp	LiveStats—Avid Sports
ac-tech	2796	tcp/udp	ac-tech
esp-encap	2797	tcp/udp	esp-encap—F-Secure Corp.
tmesis-upshot	2798	tcp/udp	TMESIS-UPShot
icon-discover	2799	tcp/udp	ICON Discover
acc-raid	2800	tcp/udp	ACC RAID—Adaptec
igcp	2801	tcp/udp	IGCP—Codemasters
veritas-tcp1	2802	tcp/udp	Veritas TCP1
btprjctrl	2803	tcp/udp	btprjctrl—British Telcom
telexis-vtu	2804	tcp/udp	Telexis VTU
wta-wsp-s	2805	tcp/udp	WTA WSP-S—Alcatel
cspuni	2806	tcp/udp	cspuni—PFU
cspmulti	2807	tcp/udp	cspmulti—PFU
j-lan-p	2808	tcp/udp	J-LAN-P—JDL Internet Services
corbaloc	2809	tcp/udp	CORBA LOC—Distributed Systems Tech Center
netsteward	2810	tcp/udp	NetSteward
gsiftp	2811	tcp/udp	GSI FTP—U Illinois Urbana-Champaign
atmtcp	2812	tcp/udp	atmtcp—Swiss Federal Institute of Technology
llm-pass	2813	tcp/udp	llm-pass
llm-csv	2814	tcp/udp	llm-csv
lbc-measure	2815	tcp/udp	LBC Measurement—Fujitsu
lbc-watchdog	2816	tcp/udp	LBC Watchdog—Fujitsu
nmsigport	2817	tcp/udp	NMSig Port—Inalp Networks
rmlnk	2818	tcp/udp	rmlnk
fc-faultnotify	2819	tcp/udp	FC Fault Notification—HP
univision	2820	tcp/udp	UniVision—FastFreeNet ISP
vml-dms	2821	tcp/udp	vml_dms—VERITAS Software
ka0wuc	2822	tcp/udp	ka0wuc
cqg-netlan	2823	tcp/udp	CQG Net/LAN—CQG Inc.
slc-systemlog	2826	tcp/udp	slc systemlog—Demon Internet ISP
slc-ctrlrloops	2827	tcp/udp	slc ctrlrloops—Demon Internet ISP
itm-lm	2828	tcp/udp	ITM License Manager
silkp1	2829	tcp/udp	silkp1—Silknet Inc.
silkp2	2830	tcp/udp	silkp2—Silknet Inc.
silkp3	2831	tcp/udp	silkp3—Silknet Inc.
silkp4	2832	tcp/udp	silkp4—Silknet Inc.

Abbreviation	Port-Decimal	Protocol	Description/Owner
glishd	2833	tcp/udp	glishd—cv.nrao.edu
evtp	2834	tcp/udp	EVTP—Solution Soft
evtp-data	2835	tcp/udp	EVTP-DATA—Solution Soft
catalyst	2836	tcp/udp	catalyst—JSB Software Technologies, PLC
repliweb	2837	tcp/udp	Repliweb—SoftLink
starbot	2838	tcp/udp	Starbot
nmsigport	2839	tcp/udp	NMSigPort—Inalp Networks
l3-exprt	2840	tcp/udp	l3-exprt—L3 Security
l3-ranger	2841	tcp/udp	l3-ranger—L3 Security
l3-hawk	2842	tcp/udp	l3-hawk—L3 Security
pdnet	2843	tcp/udp	Pdnet—MTS Internet ISP
bpcp-poll	2844	tcp/udp	BPCP POLL—SPX Corp.
bpcp-trap	2845	tcp/udp	BPCP TRAP—SPX Corp.
aimpp-hello	2846	tcp/udp	AIMPP Hello—Automation Intelligence
aimpp-port-req	2847	tcp/udp	AIMPP Port Req—Automation Intelligence
amt-blc-port	2848	tcp/udp	AMT-BLC-PORT—Telecom Italia
fxp	2849	tcp/udp	FXP—Oanda Inc.
metaconsole	2850	tcp/udp	MetaConsole—Netaphore Software
webemshttp	2851	tcp/udp	webemshttp—Jetstream Communications
bears-01	2852	tcp/udp	bears-01
ispipes	2853	tcp/udp	ISPipes—EMC2
infomover	2854	tcp/udp	InfoMover—EMC2
cesdinv	2856	tcp/udp	cesdinv—PFU Direct
simctlp	2857	tcp/udp	SimCtIP—GMX Inc.
ecnp	2858	tcp/udp	ECNP—Certified Software
activememory	2859	tcp/udp	Active Memory—EDA Consulting
dialpad-voice1	2860	tcp/udp	Dialpad Voice 1—dialpad.com, Inc.
dialpad-voice2	2861	tcp/udp	Dialpad Voice 2—dialpad.com, Inc.
ttg-protocol	2862	tcp/udp	TTG Protocol—Travel Technologies Group
sonardata	2863	tcp/udp	Sonar Data
astromed-main	2864	tcp/udp	main 5001 cmd—Astro-Med Inc.
pit-vpn	2865	tcp/udp	pit-vpn—Linux Networks
lwlistener	2866	tcp/udp	lwlistener—Quest
esps-portal	2867	tcp/udp	esps-portal
npep-messaging	2868	tcp/udp	NPEP Messaging—Norman
icslap	2869	tcp/udp	ICSLAP—Microsoft
daishi	2870	tcp/udp	daishi—U Memphis
msi-selectplay	2871	tcp/udp	MSI Select Play—MediaStation
contract	2872	tcp/udp	CONTRACT—Telnor
paspar2-zoomin	2873	tcp/udp	PASPAR2 ZoomIn
dxmessagebase1	2874	tcp/udp	dxmessagebase1—Delphix
dxmessagebase2	2875	tcp/udp	dxmessagebase2—Delphix
sps-tunnel	2876	tcp/udp	SPS Tunnel—FortressTech
bluelance	2877	tcp/udp	BLUELANCE
aap	2878	tcp/udp	AAP—Sun
ucentric-ds	2879	tcp/udp	ucentric-ds
synapse	2880	tcp/udp	synapse—BIX
ndsp	2881	tcp/udp	NDSP—GMX Inc.
ndtp	2882	tcp/udp	NDTP—GMX Inc.
ndnp	2883	tcp/udp	NDNP—GMX Inc.
flashmsg	2884	tcp/udp	Flash Msg—ZinkNet
topflow	2885	tcp/udp	TopFlow—TopLayer
#	2886-2896	tcp/udp	Unassigned
citrix-rtmp	2897	tcp/udp	Citrix RTMP
#	2898-2907	tcp/udp	Unassigned
mao	2908	tcp/udp	mao—ENSTA
funk-dialout	2909	tcp/udp	Funk Dialout
tdaccess	2910	tcp/udp	TDAccess—Metrics
blockade	2911	tcp/udp	Blockade—Interlog
epicon	2912	tcp/udp	Epicon
boosterware	2913	tcp/udp	Booster Ware—NetVision
gamelobby	2914	tcp/udp	Game Lobby—Pipex
tksocket	2915	tcp/udp	TK Socket—Protix
elvin_server	2916	tcp/udp	Elvin Server—Pobox
elvin_client	2917	tcp/udp	Elvin Client—Pobox
kastenchasepad	2918	tcp/udp	Kasten Chase Pad—Kasten Chase
#	2918-2970	tcp/udp	Unassigned
netclip	2971	tcp/udp	Net Clip—U di Pisa

Abbreviation	Port-Decimal	Protocol	Description/Owner
pmsm-webrctl	2972	tcp/udp	PMSM Webrctl
svnetworks	2973	tcp/udp	SV Networks
signal	2974	tcp/udp	Signal—Icall
fjmpcm	2975	tcp/udp	Fujitsu Configuration Management Service
cns-srv-port	2976	tcp/udp	CNS Server Port—Cisco
ttc-etap-ns	2977	tcp/udp	TTCs Enterprise Test Access Protocol - NS
ttc-etap-ds	2978	tcp/udp	TTCs Enterprise Test Access Protocol - DS
h263-video	2979	tcp/udp	H.263 Video Streaming —ACM
wimd	2980	tcp/udp	Instant Messaging Service—Pobox
mylxamport	2981	tcp/udp	MYLXAMPORT
#	2982-2995	tcp/udp	Unassigned
vsixml	2996	tcp/udp	vsixml
#	2997	tcp/udp	Unassigned
realsecure	2998	tcp/udp	Real Secure—ISS
remoteware-un	2999	tcp/udp	RemoteWare Unassigned—Xcellenet
hbci	3000	tcp/udp	HBCI—IBM
remoteware-cl	3000	tcp/udp	RemoteWare Client—Xcellenet
redwood-broker	3001	tcp/udp	Redwood Broker—PowerFrame
exlm-agent	3002	tcp/udp	EXLM Agent—Clemson U
remoteware-srv	3002	tcp/udp	RemoteWare Server—Xcellenet
cgms	3003	tcp/udp	CGMS—USCG Military
csoftragent	3004	tcp/udp	Csoft Agent
geniuslm	3005	tcp/udp	Genius License Manager
ii-admin	3006	tcp/udp	Instant Internet Admin—Bay Networks
lotusmtap	3007	tcp/udp	Lotus Mail Tracking Agent Protocol
midnight-tech	3008	tcp/udp	Midnight Technologies
pxc-ntfy	3009	tcp/udp	PXC-NTFY—Xerox
gw	3010	tcp/udp	Telerate Workstation—Dow Jones
ping-pong	3010	udp	Telerate Workstation—Dow Jones
trusted-web	3011	tcp/udp	Trusted Web—SSE
twsdss	3012	tcp/udp	Trusted Web Client—SSE
gilatskysurfer	3013	tcp/udp	Gilat Sky Surfer
broker_service	3014	tcp/udp	Broker Service—Novell
nati-dstp	3015	tcp/udp	NATI DSTP—Natinst
notify_srvr	3016	tcp/udp	Notify Server—Novell
event_listener	3017	tcp/udp	Event Listener—Novell
srvc_registry	3018	tcp/udp	Service Registry—Novell
resource_mgr	3019	tcp/udp	Resource Manager—Novell
cifs	3020	tcp/udp	CIFS—Microsoft
agriserver	3021	tcp/udp	AGRI Server—Agri-Datalog
csregagent	3022	tcp/udp	CSREGAGENT—Csoft
magicnotes	3023	tcp/udp	magicnotes
nds_sso	3024	tcp/udp	NDS_SSO—Novell
arepa-raft	3025	tcp/udp	Arepa Raft—IEEE
agri-gateway	3026	tcp/udp	AGRI Gateway—Agri-Datalog
LiebDevMgmt_C	3027	tcp/udp	LiebDevMgmt_C—Libert
LiebDevMgmt_DM	3028	tcp/udp	LiebDevMgmt_DM—Libert
LiebDevMgmt_A	3029	tcp/udp	LiebDevMgmt_A—Libert
arepa-cas	3030	tcp/udp	Arepa Cas
agentvu	3031	tcp/udp	AgentVU—Apple
redwood-chat	3032	tcp/udp	Redwood Chat—Daou
pdb	3033	tcp/udp	PDB—PixStream
osmosis-aeea	3034	tcp/udp	Osmosis AEEA—InterAccess
fjsv-gssagt	3035	tcp/udp	FJSV gssagt—Fujitsu
hagel-dump	3036	tcp/udp	Hagel DUMP
hp-san-mgmt	3037	tcp/udp	HP SAN Mgmt
santak-ups	3038	tcp/udp	Santak UPS
cogitate	3039	tcp/udp	Cogitate, Inc.
tomato-springs	3040	tcp/udp	Tomato Springs
di-traceware	3041	tcp/udp	di-traceware—Digisle
journee	3042	tcp/udp	journee
brp	3043	tcp/udp	BRP—Hughes Network Systems
responsenet	3045	tcp/udp	ResponseNet
di-ase	3046	tcp/udp	di-ase—Digisle
hlserver	3047	tcp/udp	Fast Security HL Server
pctrader	3048	tcp/udp	Sierra Net PC Trader
nsws	3049	tcp/udp	NSWS—PSLink
gds_db	3050	tcp/udp	gds_db—InterBase

Abbreviation	Port-Decimal	Protocol	Description/Owner
galaxy-server	3051	tcp/udp	Galaxy Server—Galaxy Verticals
apcpcns	3052	tcp/udp	APCPCNS—APPC
dsom-server	3053	tcp/udp	dsom-server—PNL Gov.
amt-cnf-prot	3054	tcp/udp	AMT CNF PROT—Telecom Italia
policyserver	3055	tcp/udp	Policy Server—SonusNet
cdl-server	3056	tcp/udp	CDL Server—Engineer
goahead-fldup	3057	tcp/udp	GoAhead FldUp
videobeans	3058	tcp/udp	videobeans—Sun
qsoft	3059	tcp/udp	qsoft—EarlHaig
interserver	3060	tcp/udp	interserver—InterBase
cautcpd	3061	tcp/udp	cautcpd—Computer Associates
ncacn-ip-tcp	3062	tcp/udp	ncacn-ip-tcp—Computer Associates
ncadg-ip-udp	3063	tcp/udp	ncadg-ip-udp—Computer Associates
slinterbase	3065	tcp/udp	slinterbase—InterBase
#	3066-3067	tcp/udp	Unassigned
ls3bcast	3068	tcp/udp	ls3 Broadcast—PowerWare
ls3	3069	tcp/udp	ls3—PowerWare
#	3070-3079	tcp/udp	Unassigned
stm_pproc	3080	tcp/udp	stm_pproc—STMI
tl1-lv	3081	tcp/udp	TL1-LV—SONET Internetworking Forum (SIF)
tl1-raw	3082	tcp/udp	TL1-RAW—SONET Internetworking Forum (SIF)
tl1-telnet	3083	tcp/udp	TL1-TELNET—SONET Internetworking Forum (SIF)
#	3084-3104	tcp/udp	Unassigned
cardbox	3105	tcp/udp	Cardbox
cardbox-http	3106	tcp/udp	Cardbox HTTP
#	3107-3129	tcp/udp	Unassigned
icpv2	3130	tcp/udp	ICPv2—NLANr
netbookmark	3131	tcp/udp	Net Book Mark—IBM
#	3132-3140	tcp/udp	Unassigned
vmodem	3141	tcp/udp	VMODEM—PSLink
rdc-wh-eos	3142	tcp/udp	RDC WH EOS—NCC
seaview	3143	tcp/udp	Sea View—Seagate
tarantella	3144	tcp/udp	Tarantella—SCO
csi-lfap	3145	tcp/udp	CSI-LFAP—Cabletron
#	3146	tcp/udp	Unassigned
rfio	3147	tcp/udp	RFIO—CERN
nm-game-admin	3148	tcp/udp	NetMike Game Administrator—VisionPlus
nm-game-server	3149	tcp/udp	NetMike Game Server—VisionPlus
nm-asses-admin	3150	tcp/udp	NetMike Assessor Administrator—VisionPlus
nm-assessor	3151	tcp/udp	NetMike Assessor—VisionPlus
#	3152-3179	tcp/udp	Unassigned
mc-brk-srv	3180	tcp/udp	Millicent Broker Server—DEC
bmcpatrolagent	3181	tcp/udp	BMC Patrol Agent—BMC Software, Inc.
bmcpatrolrnvu	3182	tcp/udp	BMC Patrol Rendezvous—BMC Software, Inc.
#	3183-3261	tcp/udp	Unassigned
necp	3262	tcp/udp	NECP—ISI
#	3263	tcp/udp	Unassigned
ccmail	3264	tcp/udp	cc:mail/lotus
altav-tunnel	3265	tcp/udp	Altav Tunnel—DEC
ns-cfg-server	3266	tcp/udp	NS CFG Server—NetSoft
ibm-dial-out	3267	tcp/udp	IBM Dial Out
msft-gc	3268	tcp/udp	Microsoft Global Catalog
msft-gc-ssl	3269	tcp/udp	Microsoft Global Catalog with LDAP/SSL
verismart	3270	tcp/udp	Verismart
csoft-prev	3271	tcp/udp	CSoft Prev Port—Iname
user-manager	3272	tcp/udp	Fujitsu User Manager
sxmp	3273	tcp/udp	Simple Extensible Multiplexed Protocol—WEBDash
ordinox-server	3274	tcp/udp	Ordinox Server
samd	3275	tcp/udp	SAMD—HP
maxim-asics	3276	tcp/udp	Maxim ASICs
awg-proxy	3277	tcp/udp	AWG Proxy—HP
lkcmserver	3278	tcp/udp	LKCM Server
admind	3279	tcp/udp	admind—ChisTech
vs-server	3280	tcp/udp	VS Server—Godlew
sysopt	3281	tcp/udp	SYSOPT—DEC
datusorb	3282	tcp/udp	Datusorb
net-assistant	3283	tcp/udp	Net Assistant—Apple
4talk	3284	tcp/udp	4Talk—Four-Sight

Abbreviation	Port-Decimal	Protocol	Description/Owner
plato	3285	tcp/udp	Plato—TRO
e-net	3286	tcp/udp	E-Net—Eent
directvdata	3287	tcp/udp	DIRECTVDATA
cops	3288	tcp/udp	COPS—IPHighway
enpc	3289	tcp/udp	ENPC—Epson
caps-lm	3290	tcp/udp	CAPS LOGISTICS TOOLKIT - LM
sah-lm	3291	tcp/udp	S A Holditch & Associates - LM
cart-o-rama	3292	tcp/udp	Cart O Rama—SeaLabs
fg-fps	3293	tcp/udp	fg-fps—Freegate
fg-gip	3294	tcp/udp	fg-gip—Freegate
dyniplookup	3295	tcp/udp	Dynamic IP Lookup—GeoCities
rib-slm	3296	tcp/udp	Rib License Manager
cytel-lm	3297	tcp/udp	Cytel License Manager
transview	3298	tcp/udp	Transview—SNI
pdrncs	3299	tcp/udp	pdrncs—Tektronix
###########	3300-3301	tcp/udp	Unauthorized Use by SAP R/3
mcs-fastmail	3302	tcp/udp	MCS Fastmail
opsession-clnt	3303	tcp/udp	OP Session Client—Net Manage
opsession-srvr	3304	tcp/udp	OP Session Server—Net Manage
odette-ftp	3305	tcp/udp	ODETTE-FTP—Ford Mtrs.
mysql	3306	tcp/udp	MySQL—Analytikerna
opsession-prxy	3307	tcp/udp	OP Session Proxy—Net Manage
tns-server	3308	tcp/udp	TNS Server—DEC
tns-adv	3309	tcp/udp	TNS ADV—DEC
dyna-access	3310	tcp/udp	Dyna Access—CornerstoneSoftware
mcns-tel-ret	3311	tcp/udp	MCNS Tel Ret
appman-server	3312	tcp/udp	Application Management Server
uorb	3313	tcp/udp	Unify Object Broker
uohost	3314	tcp/udp	Unify Object Host
cdid	3315	tcp/udp	CDID—Chat
aicc-cmi	3316	tcp/udp	AICC/CMI—Boeing
vsaiport	3317	tcp/udp	VSAI PORT—Mei
ssrip	3318	tcp/udp	Swith to Swith Routing Information Protocol—Fujitsu
sdt-lmd	3319	tcp/udp	SDT License Manager—Tin
officelink2000	3320	tcp/udp	Office Link 2000—Teltone
vnsstr	3321	tcp/udp	VNSSTR—Mei
active-net	3322-3325	tcp/udp	ISI
sftu	3326	tcp/udp	SFTU—Spacenet
bbars	3327	tcp/udp	BBARS—Bandl
egptlm	3328	tcp/udp	Eaglepoint License Manager
hp-device-disc	3329	tcp/udp	HP Device Disc
mcs-calypsoicf	3330	tcp/udp	MCS Calypso ICF
mcs-messaging	3331	tcp/udp	MCS Messaging
mcs-mailsvr	3332	tcp/udp	MCS Mail Server
dec-notes	3333	tcp/udp	DEC Notes
directv-web	3334	tcp/udp	Direct TV Webcasting
directv-soft	3335	tcp/udp	Direct TV Software Updates
directv-tick	3336	tcp/udp	Direct TV Tickers
directv-catlg	3337	tcp/udp	Direct TV Data Catalog
anet-b	3338	tcp/udp	OMF data b—ABB Sweden
anet-l	3339	tcp/udp	OMF data l—ABB Sweden
anet-m	3340	tcp/udp	OMF data m—ABB Sweden
anet-h	3341	tcp/udp	OMF data h—ABB Sweden
webtie	3342	tcp/udp	WebTIE—NOAA.gov
ms-cluster-net	3343	tcp/udp	MS Cluster Net
bnt-manager	3344	tcp/udp	BNT Manager
influence	3345	tcp/udp	Influence—Topia
trnsprntproxy	3346	tcp/udp	Trnsprnt Proxy—Intel
phoenix-rpc	3347	tcp/udp	Phoenix RPC
pangolin-laser	3348	tcp/udp	Pangolin Laser
chevinservices	3349	tcp/udp	Chevin Services
findviatv	3350	tcp/udp	FINDVIATV—8x8
btrieve	3351	tcp/udp	BTRIEVE—Pervasive-SW
ssql	3352	tcp/udp	SSQL—Pervasive-SW
fatpipe	3353	tcp/udp	FATPIPE—Ragula
suitjd	3354	tcp/udp	SUITJD—DEC
ordinox-dbase	3355	tcp/udp	Ordinox Dbase
upnotifyps	3356	tcp/udp	UPNOTIFYPS—UpLANe

Abbreviation	Port-Decimal	Protocol	Description/Owner
adtech-test	3357	tcp/udp	Adtech Test IP
mpsysrmsvr	3358	tcp/udp	Mp Sys Rmsvr—Fujitsu
wg-netforce	3359	tcp/udp	WG NetForce
kv-server	3360	tcp/udp	KV Server—CIPartners
kv-agent	3361	tcp/udp	KV Agent—CIPartners
dj-ilm	3362	tcp/udp	DJ ILM
nati-vi-server	3363	tcp/udp	NATI Vi Server—Natinst
creativeserver	3364	tcp/udp	Creative Server—Emotion
contentserver	3365	tcp/udp	Content Server—Emotion
creativepartnr	3366	tcp/udp	Creative Partner—Emotion
satvid-datalnk	3367-3371	tcp/udp	Satellite Video Data Link—Intel
tip2	3372	tcp/udp	TIP 2—Tandem
lavenir-lm	3373	tcp/udp	Lavenir License Manager
cluster-disc	3374	tcp/udp	Cluster Disc—NCR
vsnm-agent	3375	tcp/udp	VSNM Agent—VitalSigns
cdborker	3376	tcp/udp	CD Broker—Esps
cogsys-lm	3377	tcp/udp	Cogsys Network License Manager
wsicopy	3378	tcp/udp	WSICOPY
socorfs	3379	tcp/udp	SOCORFS
sns-channels	3380	tcp/udp	SNS Channels—FirstFloor
geneous	3381	tcp/udp	Geneous—HisTech
fujitsu-neat	3382	tcp/udp	Fujitsu Network Enhanced Antitheft function
esp-lm	3383	tcp/udp	Enterprise Software Products License Manager
hp-clic	3384	tcp/udp	Cluster Management Services—HP
hp-clic	3384	udp	Hardware Management—HP
qnxnetman	3385	tcp/udp	qnxnetman
gprs-data	3386	tcp/udp	GPRS Data—ETSI
backroomnet	3387	tcp/udp	Back Room Net—Carreker
cbserver	3388	tcp/udp	CB Server—Hyperion Inc.
ms-wbt-server	3389	tcp/udp	MS WBT Server
dsc	3390	tcp/udp	Distributed Service Coordinator—Secant
savant	3391	tcp/udp	SAVANT—TheProgrammers
efi-lm	3392	tcp/udp	EFI License Management—ExecPC
d2k-tapestry1	3393	tcp/udp	D2K Tapestry Client to Server
d2k-tapestry2	3394	tcp/udp	D2K Tapestry Server to Server
dyna-lm	3395	tcp/udp	Dyna License Manager (Elam)
printer_agent	3396	tcp/udp	Printer Agent—Novell
cloanto-lm	3397	tcp/udp	Cloanto License Manager
mercantile	3398	tcp/udp	Mercantile—UNI2
csms	3399	tcp/udp	CSMS—Cedros
csms2	3400	tcp/udp	CSMS2—Cedros
filecast	3401	tcp/udp	filecast—Pair
#	3402-3420	tcp/udp	Unassigned
bmap	3421	tcp/udp	Bull Apprise portmapper
mira	3454	tcp	Apple Remote Access Protocol—U Michigan
prsvp	3455	tcp/udp	RSVP Port—ISI
vat	3456	tcp/udp	VAT default data—LBL Gov.
vat-control	3457	tcp/udp	VAT default control—LBL Gov
d3winosfi	3458	tcp/udp	D3WinOsfi—PickSys
integral	3459	tcp/udp	Integral
edm-manager	3460	tcp/udp	EDM Manger—Novadigm
edm-stager	3461	tcp/udp	EDM Stager—Novadigm
edm-std-notify	3462	tcp/udp	EDM STD Notify—Novadigm
edm-adm-notify	3463	tcp/udp	EDM ADM Notify—Novadigm
edm-mgr-sync	3464	tcp/udp	EDM MGR Sync—Novadigm
edm-mgr-cntrl	3465	tcp/udp	EDM MGR Cntrl—Novadigm
workflow	3466	tcp/udp	WORKFLOW—Csesys
rcst	3467	tcp/udp	RCST—RemoteControlSexToys
ttcmremotectrl	3468	tcp/udp	TTCM Remote Controll—Proxy
pluribus	3469	tcp/udp	Pluribus—Caplet
jt400	3470	tcp/udp	jt400-IBM
jt400-ssl	3471	tcp/udp	jt400-ssl—IBM
#	3472-3562	tcp/udp	Unassigned
watcomdebug	3563	tcp/udp	Watcom Debug
#	3564-3671	tcp/udp	Unassigned
harlequinorb	3672	tcp/udp	harlequinorb
#	3673-3844	tcp/udp	Unassigned
v-one-spp	3845	tcp/udp	V-ONE Single Port Proxy

Abbreviation	Port-Decimal	Protocol	Description/Owner
#	3846-3899	tcp/udp	Unassigned
udt_os	3900	tcp/udp	Unidata UDT OS
mapper-nodemgr	3984	tcp/udp	MAPPER network node manager—Unisys
mapper-mapethd	3985	tcp/udp	MAPPER TCP/IP server—Unisys
mapper-ws_ethd	3986	tcp/udp	MAPPER workstation server—Unisys
centerline	3987	tcp/udp	Centerline
#	3988-3999	tcp/udp	Unassigned
terabase	4000	tcp/udp	Terabase
newoak	4001	tcp/udp	NewOak
pxc-spvr-ft	4002	tcp/udp	pxc-spvr-ft—Xerox
pxc-splr-ft	4003	tcp/udp	pxc-splr-ft—Xerox
pxc-roid	4004	tcp/udp	pxc-roid—Xerox
pxc-pin	4005	tcp/udp	pxc-pin—Xerox
pxc-spvr	4006	tcp/udp	pxc-spvr—Xerox
pxc-splr	4007	tcp/udp	pxc-splr—Xerox
netcheque	4008	tcp/udp	NetCheque accounting
chimera-hwm	4009	tcp/udp	Chimera HWM—UC Irvine
samsung-unidex	4010	tcp/udp	Samsung Unidex
altserviceboot	4011	tcp/udp	Alternate Service Boot—Intel
pda-gate	4012	tcp/udp	PDA Gate—IBM
acl-manager	4013	tcp/udp	ACL Manager—Fujitsu
taiclock	4014	tcp/udp	TAICLOCK—UIC
talarian-mcast1	4015	tcp/udp	Talarian Mcast
talarian-mcast2	4016	tcp/udp	Talarian Mcast
talarian-mcast3	4017	tcp/udp	Talarian Mcast
talarian-mcast4	4018	tcp/udp	Talarian Mcast
talarian-mcast5	4019	tcp/udp	Talarian Mcast
#	4020-4095	tcp/udp	Unassigned
bre	4096	tcp/udp	BRE (Bridge Relay Element)—Timeplex
patrolview	4097	tcp/udp	Patrol View—BMC Software, Inc.
drmsfsd	4098	tcp/udp	drmsfsd—Fujitsu
dpcp	4099	tcp/udp	DPCP—Demon
#	4100-4131	tcp/udp	Unassigned
nuts_dem	4132	tcp/udp	NUTS Daemon—SNI
nuts_bootp	4133	tcp/udp	NUTS Bootp Server—SNI
nifty-hmi	4134	tcp/udp	NIFTY-Serve HMI protocol
oirtgsvc	4141	tcp/udp	Workflow Server
oidocsvc	4142	tcp/udp	Document Server—SNS
oidsr	4143	tcp/udp	Document Replication—SNS
#########Compuserve (unoffically) is using port 4144 #########			
#	4144-4159	tcp/udp	Unassigned
jini-discovery	4160	tcp/udp	Jini Discovery—Sun
#	4161-4199	tcp/udp	Unassigned
vrml-multi-use	4200-4299	tcp/udp	VRML Multi User Systems
corelccam	4300	tcp/udp	Corel CCam
#	4301-4320	tcp/udp	Unassigned
rwhois	4321	tcp/udp	Remote Who Is—InterNIC
unicall	4343	tcp/udp	UNICALL—UniData
vinainstall	4344	tcp/udp	Vinalnstall—Vina-Tech
m4-network-as	4345	tcp/udp	Macro 4 Network AS
elanlm	4346	tcp/udp	ELAN LM—ProjTech
lansurveyor	4347	tcp/udp	LAN Surveyor—Neon
itose	4348	tcp/udp	ITOSE—HP
fsportmap	4349	tcp/udp	File System Port Map—Sarnoff
net-device	4350	tcp/udp	Net Device—Microsoft
plcy-net-svcs	4351	tcp/udp	PLCY Net Services
#	4352	tcp/udp	Unassigned
f5-iquery	4353	tcp/udp	F5 iQuery
#	4354-4443	tcp/udp	Unassigned
saris	4442	tcp/udp	Saris—TeleConsult GmbH
pharos	4443	tcp/udp	Pharos—TeleConsult GmbH
krb524	4444	tcp/udp	KRB524—ISI
nv-video	4444	tcp/udp	NV Video default—Xerox
upnotifyp	4445	tcp/udp	UPNOTIFYP—UpLANet
n1-fwp	4446	tcp/udp	N1-FWP—Network-1
n1-rmgmt	4447	tcp/udp	N1-RMGMT—Network-1
asc-slmd	4448	tcp/udp	ASC Licence Manager
privatewire	4449	tcp/udp	PrivateWire—ARX

Abbreviation	Port-Decimal	Protocol	Description/Owner
camp	4450	tcp/udp	Camp
ctisystemmsg	4451	tcp/udp	CTI System Msg—Control Technology
nssalertmgr	4453	tcp/udp	NSS Alert Manager—Symantec
nssagentmgr	4454	tcp/udp	NSS Agent Manager—Symantec
prchat-user	4455	tcp/udp	PR Chat User—IBM
prchat-server	4456	tcp/udp	PR Chat Server—IBM
prRegister	4457	tcp/udp	PR Register—IBM
#	4458-4499	tcp/udp	Unassigned
sae-urn	4500	tcp/udp	sae-urn—Proper
urn-x-cdchoice	4501	tcp/udp	urn-x-cdchoice—Proper
worldscores	4545	tcp/udp	WorldScores
sf-lm	4546	tcp/udp	SF License Manager (Sentinel)
lanner-lm	4547	tcp/udp	Lanner License Manager
#	4548-4566	tcp/udp	Unassigned
tram	4567	tcp/udp	TRAM—SUN
bmc-reporting	4568	tcp/udp	BMC Reporting—BMC Software, Inc.
#	4569-4599	tcp/udp	Unassigned
piranha1	4600	tcp/udp	Piranha1—Primark Corporation
piranha2	4601	tcp/udp	Piranha2—Primark Corporation
#	4602-4671	tcp/udp	Unassigned
rfa	4672	tcp/udp	remote file access server
#	4673-4799	tcp/udp	Unassigned
iims	4800	tcp/udp	Icona Instant Messenging System
iwec	4801	tcp/udp	Icona Web Embedded Chat
ilss	4802	tcp/udp	Icona License System Server
#	4803-4826	tcp/udp	Unassigned
htcp	4827	tcp/udp	HTCP—VIX
#	4828-4867	tcp/udp	Unassigned
phrelay	4868	tcp/udp	Photon Relay—QNX
phrelaydbg	4869	tcp/udp	Photon Relay Debug—QNX
abbs	4885	tcp/udp	ABBS—ML
att-intercom	4983	tcp/udp	AT&T Intercom
commplex-main	5000	tcp/udp	
commplex-link	5001	tcp/udp	
rfe	5002	tcp/udp	radio free ethernet
fmpro-internal	5003	tcp/udp	FileMaker, Inc. - Proprietary transport
avt-profile-1	5004	tcp/udp	avt-profile-1—GMD
avt-profile-2	5005	tcp/udp	avt-profile-2—GMD
wsm-server	5006	tcp/udp	wsm server—WRQ
wsm-server-ssl	5007	tcp/udp	wsm server ssl—WRQ
telelpathstart	5010	tcp/udp	TelepathStart—IMB
telelpathattack	5011	tcp/udp	TelepathAttack—IMB
zenginkyo-1	5020	tcp/udp	zenginkyo-1—NttData
zenginkyo-2	5021	tcp/udp	zenginkyo-2—NttData
asnaacceler8db	5042	tcp/udp	asnaacceler8db—Asna
mmcc	5050	tcp/udp	multimedia conference control tool—ISI
ita-agent	5051	tcp/udp	ITA Agent—Axent
ita-manager	5052	tcp/udp	ITA Manager—Axent
unot	5055	tcp/udp	UNOT—CmgiSolutions
sip	5060	tcp/udp	SIP—Columbia U
i-net-2000-npr	5069	tcp/udp	I/Net 2000-NPR—CSIControls
powerschool	5071	tcp/udp	PowerSchool
rmonitor_secure	5145	tcp/udp	
rmonitor_secure	5145	udp	
atmp	5150	tcp/udp	Ascend Tunnel Management Protocol
esri_sde	5151	tcp/udp	ESRI SDE Instance
sde-discovery	5152	tcp/udp	ESRI SDE Instance Discovery
ife_icorp	5165	tcp/udp	ife_1corp—Bull
aol	5190	tcp/udp	America-Online
aol-1	5191	tcp/udp	AmericaOnline1
aol-2	5192	tcp/udp	AmericaOnline2
aol-3	5193	tcp/udp	AmericaOnline3
targus-aib1	5200	tcp/udp	Targus AIB 1
targus-aib2	5201	tcp/udp	Targus AIB 2
targus-tnts1	5202	tcp/udp	Targus TNTS 1
targus-tnts2	5203	tcp/udp	Targus TNTS 2
padl2sim	5236	tcp/udp	
pk	5272	tcp/udp	PK—EBA

Abbreviation	Port-Decimal	Protocol	Description/Owner
hacl-hb	5300	tcp/udp	# HA cluster heartbeat—HP
hacl-gs	5301	tcp/udp	# HA cluster general services—HP
hacl-cfg	5302	tcp/udp	# HA cluster configuration—HP
hacl-probe	5303	tcp/udp	# HA cluster probing—HP
hacl-local	5304	tcp/udp	# HA Cluster Commands—HP
hacl-test	5305	tcp/udp	# HA Cluster Test
sun-mc-grp	5306	tcp/udp	Sun MC Group
sco-aip	5307	tcp/udp	SCO AIP
cfengine	5308	tcp/udp	Cfengine—HiOslo
jprinter	5309	tcp/udp	J Printer—Bristol
outlaws	5310	tcp/udp	Outlaws—Lucasarts
tmlogin	5311	tcp/udp	TM Login—Sun
#	5312-5399	tcp/udp	Unassigned
excerpt	5400	tcp/udp	Excerpt Search—Alma
excerpts	5401	tcp/udp	Excerpt Search Secure—Alma
mftp	5402	tcp/udp	MFTP—StarburstCom
hpoms-ci-lstn	5403	tcp/udp	HPOMS-CI-LSTN
hpoms-dps-lstn	5404	tcp/udp	HPOMS-DPS-LSTN
netsupport	5405	tcp/udp	NetSupport—Pipex
systemics-sox	5406	tcp/udp	Systemics Sox
foresyte-clear	5407	tcp/udp	Foresyte-Clear
foresyte-sec	5408	tcp/udp	Foresyte-Sec
salient-dtasrv	5409	tcp/udp	Salient Data Server
salient-usrmgr	5410	tcp/udp	Salient User Manager
salient-mux	5411	tcp/udp	Salient MUX
actnet	5411	tcp/udp	ActNet—ACT Research
continuus	5412	tcp/udp	Continuus
wwiotalk	5413	tcp/udp	WWIOTALK—Wonder Ware
statusd	5414	tcp/udp	StatusD—SatelNet
ns-server	5415	tcp/udp	NS Server—NetSof
sns-gateway	5416	tcp/udp	SNS Gateway—FirstFloor
sns-agent	5417	tcp/udp	SNS Agent—FirstFloor
mcntp	5418	tcp/udp	MCNTP—Pilhuhn
dj-ice	5419	tcp/udp	DJ-ICE—Dow Jones
cylink-c	5420	tcp/udp	Cylink-C
netsupport2	5421	tcp/udp	Net Support 2—Pipex
#	5422-5453	tcp/udp	Unassigned
apc-tcp-udp-4	5454	tcp/udp	apc-tcp-udp-4—APPC
apc-tcp-udp-5	5455	tcp/udp	apc-tcp-udp-5—APPC
apc-tcp-udp-6	5456	tcp/udp	apc-tcp-udp-6—APPC
#	5457-5499	tcp/udp	Unassigned
fcp-addr-srvr1	5500	tcp/udp	fcp-addr-srvr1—AC
fcp-addr-srvr2	5501	tcp/udp	fcp-addr-srvr2—AC
fcp-srvr-inst1	5502	tcp/udp	fcp-srvr-inst1—AC
fcp-srvr-inst2	5503	tcp/udp	fcp-srvr-inst2—AC
fcp-cics-gw1	5504	tcp/udp	fcp-cics-gw1—AC
#	5504-5553	tcp/udp	Unassigned
sgi-esphttp	5554	tcp/udp	SGI ESP HTTP
###########Port 5555 also used by HP Omniback###################			
personal-agent	5555	tcp/udp	Personal Agent—Infoseek
#	5556-5598	tcp/udp	Unassigned
esinstall	5599	tcp/udp	Enterprise Security Remote Install—Axent
esmmanager	5600	tcp/udp	Enterprise Security Manager—Axent
esmagent	5601	tcp/udp	Enterprise Security Agent—Axent
a1-msc	5602	tcp/udp	A1-MSC—Lucent
a1-bs	5603	tcp/udp	A1-BS—Lucent
a3-sdunode	5604	tcp/udp	A3-SDUNode—Lucent
a4-sdunode	5605	tcp/udp	A4-SDUNode—Lucent
pcanywheredata	5631	tcp/udp	pcANYWHEREdata—Symantec
pcanywherestat	5632	tcp/udp	pcANYWHEREstat—Symantec
rrac	5678	tcp/udp	Remote Replication Agent Connection—Microsoft
dccm	5679	tcp/udp	Direct Cable Connect Manager—Microsoft
proshareaudio	5713	tcp/udp	proshare conf audio—Intel
prosharevideo	5714	tcp/udp	proshare conf video—Intel
prosharedata	5715	tcp/udp	proshare conf data—Intel
prosharerequest	5716	tcp/udp	proshare conf request—Intel
openmail	5729	tcp/udp	Openmail User Agent Layer—HP
ida-discover1	5741	tcp/udp	IDA Discover Port 1—I-Data

Abbreviation	Port-Decimal	Protocol	Description/Owner
ida-discover2	5742	tcp/udp	IDA Discover Port 2—I-Data
fcopy-server	5745	tcp/udp	fcopy-server—SoftLink
fcopys-server	5746	tcp/udp	fcopys-server—SoftLink
openmailg	5755	tcp/udp	OpenMail Desk Gateway server—HP
x500ms	5757	tcp/udp	OpenMail X.500 Directory Server—HP
openmailns	5766	tcp/udp	OpenMail NewMail Server—HP
s-openmail	5767	tcp/udp	OpenMail Suer Agent Layer (Secure)—HP
openmailpxy	5768	tcp/udp	OpenMail CMTS Server—HP
netagent	5771	tcp/udp	NetAgent—Eshare
mppolicy-v5	5968	tcp/udp	mppolicy-v5—Fujitsu
mppolicy-mgr	5969	tcp/udp	mppolicy-mgr—Fujitsu
x11	6000-6063	tcp/udp	X Window System—MIT
softcm	6110	tcp/udp	HP SoftBench CM
softcm	6110	udp	HP SoftBench CM
spc	6111	tcp/udp	HP SoftBench Sub-Process Control
dtspcd	6112	tcp/udp	dtspcd—Sun
backup-express	6123	tcp/udp	Backup Express—SyncSort
meta-corp	6141	tcp/udp	Meta Corporation License Manager
aspentec-lm	6142	tcp/udp	Aspen Technology License Manager
watershed-lm	6143	tcp/udp	Watershed License Manager—Zion
statsci1-lm	6144	tcp/udp	StatSci License Manager - 1
statsci2-lm	6145	tcp/udp	StatSci License Manager - 2
lonewolf-lm	6146	tcp/udp	Lone Wolf Systems License Manager
montage-lm	6147	tcp/udp	Montage License Manager
ricardo-lm	6148	tcp/udp	Ricardo North America License Manager
tal-pod	6149	tcp/udp	tal-pod—Taligent
crip	6253	tcp/udp	CRIP—Ciena
clariion-evr01	6389	tcp/udp	clariion-evr01
info-aps	6400	tcp/udp	Seagate Software
info-was	6401	tcp/udp	Seagate Software
info-eventsvr	6402	tcp/udp	Seagate Software
info-cachesvr	6403	tcp/udp	Seagate Software
info-filesvr	6404	tcp/udp	Seagate Software
info-pagesvr	6405	tcp/udp	Seagate Software
info-processvr	6406	tcp/udp	Seagate Software
reserved1	6407	tcp/udp	Seagate Software
reserved2	6408	tcp/udp	Seagate Software
reserved3	6409	tcp/udp	Seagate Software
reserved4	6410	tcp/udp	Seagate Software
skip-cert-recv	6455	tcp	SKIP Certificate Receive—SUN
skip-cert-send	6456	tcp	SKIP Certificate Send—Sun
lvision-lm	6471	tcp/udp	LVision License Manager
#	6472-6499	tcp/udp	Unassigned
boks	6500	tcp/udp	BoKS Master
boks_servc	6501	tcp/udp	BoKS Servc
boks_servm	6502	tcp/udp	BoKS Servm
boks_clntd	6503	tcp/udp	BoKS Clntd
#	6504	tcp/udp	Unassigned
badm_priv	6505	tcp/udp	BoKS Admin Private Port—Dynas
badm_pub	6506	tcp/udp	BoKS Admin Public Port—Dynas
bdir_priv	6507	tcp/udp	"BoKS Dir Server, Private Port—Dynas"
bdir_pub	6508	tcp/udp	"BoKS Dir Server, Public Port—Dynas"
#	6509-6546	tcp/udp	Unassigned
apc-tcp-udp-1	6547	tcp/udp	apc-tcp-udp-1—APPC
apc-tcp-udp-2	6548	tcp/udp	apc-tcp-udp-2—APPC
apc-tcp-udp-3	6549	tcp/udp	apc-tcp-udp-3—APPC
fg-sysupdate	6550	tcp/udp	fg-sysupdate—Freegate
#	6551-6557	tcp/udp	Unassigned
xdsxdm	6558	tcp/udp	
ircu	6665-6669	tcp/udp	IRCU—Acrux
vocaltec-gold	6670	tcp/udp	Vocaltec Global Online Directory
vision_server	6672	tcp/udp	vision_server—SHL
vision_elmd	6673	tcp/udp	vision_elmd—SHL
kti-icad-srvr	6701	tcp/udp	KTI/ICAD Nameserver—KTIworld
hnmp	6790	tcp/udp	HNMP—NASA
ambit-lm	6831	tcp/udp	ambit-lm
netmo-default	6841	tcp/udp	Netmo Default—Netmosphere
netmo-http	6842	tcp/udp	Netmo HTTP—Netmosphere

Abbreviation	Port-Decimal	Protocol	Description/Owner
jmact3	6961	udp	JMACT3
jmevt2	6962	tcp/udp	jmevt2—Fujitsu
swismgr1	6963	tcp/udp	swismgr1—Fujitsu
swismgr2	6964	tcp/udp	swismgr2—Fujitsu
swistrap	6965	tcp/udp	swistrap—Fujitsu
swispol	6966	tcp/udp	swispol—Fujitsu
acmsoda	6969	tcp/udp	acmsoda—ACM
iatp-highpri	6998	tcp/udp	IATP-highPri—MCI
iatp-normalpri	6999	tcp/udp	IATP-normalPri—MCI
afs3-fileserver	7000	tcp/udp	file server itself
afs3-callback	7001	tcp/udp	callbacks to cache managers
afs3-prserver	7002	tcp/udp	users & groups database
afs3-vlserver	7003	tcp/udp	volume location database
afs3-kaserver	7004	tcp/udp	AFS/Kerberos authentication service
afs3-volser	7005	tcp/udp	volume managment server
afs3-errors	7006	tcp/udp	error interpretation service
afs3-bos	7007	tcp/udp	basic overseer process
afs3-update	7008	tcp/udp	server-to-server updater
afs3-rmtsys	7009	tcp/udp	remote cache manager service
ups-onlinet	7010	tcp/udp	onlinet uninterruptable power supplies—Exide
talon-disc	7011	tcp/udp	Talon Discovery Port—PowerWare
talon-engine	7012	tcp/udp	Talon Engine—PowerWare
dpserve	7020	tcp/udp	DP Serve—Hummingbird
dpserveadmin	7021	tcp/udp	DP Serve Admin—Hummingbird
arcp	7070	tcp/udp	ARCP—NASA
lazy-ptop	7099	tcp/udp	lazy-ptop—Scitex
font-service	7100	tcp/udp	X Font Service—MIT
virprot-lm	7121	tcp/udp	Virtual Prototypes License Manager
clutild	7174	tcp/udp	Clutild—IBM
fodms	7200	tcp/udp	FODMS FLIP—Rockwell
dlip	7201	tcp/udp	DLIP—Rockwell
swx	7300-7390	tcp/udp	The Swiss Exchange
#	7391-7394	tcp/udp	Unassigned
winqedit	7395	tcp/udp	winqedit—Robelle
#	7396-7425	tcp/udp	Unassigned
pmdmgr	7426	tcp/udp	OpenView DM Postmaster Manager
oveadmgr	7427	tcp/udp	OpenView DM Event Agent Manager—HP
ovladmgr	7428	tcp/udp	OpenView DM Log Agent Manager—HP
opi-sock	7429	tcp/udp	OpenView DM rqt communication—HP
xmpv7	7430	tcp/udp	OpenView DM xmpv7 api pipe—HP
pmd	7431	tcp/udp	OpenView DM ovc/xmpv3 api pipe—HP
faximum	7437	tcp/udp	Faximum
telops-lmd	7491	tcp/udp	telops-lmd
pafec-lm	7511	tcp/udp	pafec-lm
nta-ds	7544	tcp/udp	FlowAnalyzer DisplayServer—Cisco
nta-us	7545	tcp/udp	FlowAnalyzer UtilityServer—Cisco
vsi-omega	7566	tcp/udp	VSI Omega
aries-kfinder	7570	tcp/udp	Aries Kfinder
sun-lm	7588	tcp/udp	Sun License Manager
pmdfmgt	7633	tcp/udp	PMDF Management—InnoSoft
cbt	7777	tcp/udp	cbt—Univesity College London
accu-lmgr	7781	tcp/udp	accu-lmgr—Accugraph
t2-drm	7932	tcp/udp	Tier 2 Data Resource Manager
t2-brm	7933	tcp/udp	Tier 2 Business Rules Manager
supercell	7967	tcp/udp	Supercell—SigSci
quest-vista	7980	tcp/udp	Quest Vista
irdmi2	7999	tcp/udp	iRDMI2—Intel
irdmi	8000	tcp/udp	iRDMI—Intel
vcom-tunnel	8001	tcp/udp	VCOM Tunnel—IBM
teradataordbms	8002	tcp/udp	Teradata ORDBMS—NCR
http-alt	8008	tcp/udp	HTTP Alternate—World Wide Web Consortium
pro-ed	8032	tcp/udp	ProEd—Gold Rush
mindprint	8033	tcp/udp	MindPrint—Gold Rush
http-alt	8080	tcp/udp	HTTP Alternate (see port 80)—Percpt
patrol	8160	tcp/udp	Patrol—BMC Software, Inc
patrol-snmp	8161	tcp/udp	Patrol SNMP—BMC Software, Inc
trivnet1	8200	tcp/udp	TRIVNET
trivnet2	8201	tcp/udp	TRIVNET

Abbreviation	Port-Decimal	Protocol	Description/Owner
lm-perfworks	8204	tcp/udp	LM Perfworks—Landmark
lm-instmgr	8205	tcp/udp	LM Instmgr—Landmark
lm-dta	8206	tcp/udp	LM Dta—Landmark
lm-sserver	8207	tcp/udp	LM Sserver—Landmark
lm-webwatcher	8208	tcp/udp	LM Webwatcher—Landmark
server-find	8351	tcp/udp	Server Find—Chancery
cruise-enum	8376	tcp/udp	Cruise ENUMC—Cruise Technology
cruise-swroute	8377	tcp/udp	Cruise SWROUTE—Cruise Technology
cruise-config	8378	tcp/udp	Cruise CONFIG—Cruise Technology
cruise-diags	8379	tcp/udp	Cruise DIAGS—Cruise Technology
cruise-update	8380	tcp/udp	Cruise UPDATE—Cruise Technology
cvd	8400	tcp/udp	cvd—Commvault
sabarsd	8401	tcp/udp	sabarsd—Commvault
abarsd	8402	tcp/udp	abarsd—Commvault
admind	8403	tcp/udp	admind—Commvault
npmp	8450	tcp/udp	npmp—Demon
vp2p	8473	tcp/udp	Virtual Point to Point—Atos-Group
rtsp-alt	8554	tcp/udp	RTSP Alternate (see port 554)—Precept
ibus	8733	tcp/udp	iBus—SoftWired
ultraseek-http	8765	tcp/udp	Ultraseek HTTP—Infoseek
cddbp-alt	8880	tcp/udp	CDDBP—Moonsoft
truecm	8804	tcp/udp	truecm—True Blue Software
ddi-tcp-1	8888	tcp/udp	NewsEDGE server TCP (TCP 1)—DeskTop Data
ddi-tcp-2	8889	tcp/udp	Desktop Data TCP 1—DeskTop Data
ddi-tcp-3	8890	tcp/udp	Desktop Data TCP 2—DeskTop Data
ddi-tcp-4	8891	tcp/udp	Desktop Data TCP 3: NESS application—DeskTop Data
ddi-tcp-5	8892	tcp/udp	Desktop Data TCP 4: FARM product—DeskTop Data
ddi-tcp-6	8893	tcp/udp	Desktop Data TCP 5: NewsEDGE/Web application—DeskTop Data
ddi-tcp-7	8894	tcp/udp	Desktop Data TCP 6: COAL application—DeskTop Data
jmb-cds1	8900	tcp/udp	JMB-CDS 1—Creative Design
jmb-cds2	8901	tcp/udp	JMB-CDS 2—Creative Design
cslistener	9000	tcp/udp	Cslistener—Cincom
sctp	9006	tcp/udp	SCTP—Tagnet
websm	9090	tcp/udp	WebSM—IBM
netlock1	9160	tcp/udp	NetLOCK1
netlock2	9161	tcp/udp	NetLOCK2
netlock3	9162	tcp/udp	NetLOCK3
netlock4	9163	tcp/udp	NetLOCK4
netlock5	9164	tcp/udp	NetLOCK5
wap-wsp	9200	tcp/udp	WAP connectionless session service
wap-wsp-wtp	9201	tcp/udp	WAP session service
wap-wsp-s	9202	tcp/udp	WAP secure connectionless session service
wap-wsp-wtp-s	9203	tcp/udp	WAP secure session service
wap-vcard	9204	tcp/udp	WAP vCard
wap-vcal	9205	tcp/udp	WAP vCal
wap-vcard-s	9206	tcp/udp	WAP vCard Secure
wap-vcal-s	9207	tcp/udp	WAP vCal Secure
guibase	9321	tcp/udp	guibase—Fujitsu
mpidcmgr	9343	tcp/udp	MpIdcMgr—Fujitsu
mphlpdmc	9344	tcp/udp	Mphlpdmc—Fujitsu
fjdmimgr	9374	tcp/udp	fjdmimgr—Fujitsu
fjinvmgr	9396	tcp/udp	fjinvmgr—Fujitsu
mpidcagt	9397	tcp/udp	MpIdcAgt—Fujitsu
ismserver	9500	tcp/udp	ismserver—Micro Muse
man	9535	tcp/udp	Intel
msgsys	9594	tcp/udp	Message System—Intel
pds	9595	tcp/udp	Ping Discovery Service—Intel
rasadv	9753	tcp/udp	rasadv—Microsoft
sd	9876	tcp/udp	Session Director—LBL.GOV
cyborg-systems	9888	tcp/udp	CYBORG Systems
monkeycom	9898	tcp/udp	MonkeyCom
#	9899-9908	tcp/udp	Unassigned
domaintime	9909	tcp/udp	domaintime—GreyWare
#	9910-9991	tcp/udp	Unassigned
palace	9992	tcp/udp	The Palace
palace	9993	tcp/udp	The Palace
palace	9994	tcp/udp	The Palace
palace	9995	tcp/udp	The Palace

Abbreviation	Port-Decimal	Protocol	Description/Owner
palace	9996	tcp/udp	The Palace
palace	9997	tcp/udp	The Palace
distinct32	9998	tcp/udp	Distinct32
distinct	9999	tcp/udp	distinct
#	10001-10006	tcp/udp	Unassigned
mvs-capacity	10007	tcp/udp	MVS Capacity—IBM
#	10008-10079	tcp/udp	Unassigned
amanda	10080	tcp/udp	Amanda —Purdue U
#	10081-10114	tcp/udp	Unassigned
ganymede-endpt	10115	tcp/udp	Ganymede Endpoint
#	10116-10287	tcp/udp	Unassigned
blocks	10288	tcp/udp	Blocks
#	10289-10999	tcp/udp	Unassigned
irisa	11000	tcp/udp	IRISA—DataMedia
metasys	11001	tcp/udp	Metasys—JCI
vce	11111	tcp/udp	Viral Computing Environment (VCE)—Fred Cohen & Assoc.
atm-uhas	11367	tcp/udp	ATM UHAS—Attachmate
h323callsigalt	11720	tcp/udp	h323 Call Signal Alternate—Cisco
entextxid	12000	tcp/udp	IBM Enterprise Extender SNA XID Exchange
entextnetwk	12001	tcp/udp	IBM Enterprise Extender SNA COS Network Priority
entexthigh	12002	tcp/udp	IBM Enterprise Extender SNA COS High Priority
entextmed	12003	tcp/udp	IBM Enterprise Extender SNA COS Medium Priority
entextlow	12004	tcp/udp	IBM Enterprise Extender SNA COS Low Priority
hivep	12172	tcp/udp	HiveP—Skenbe
tsaf	12753	tcp/udp	tsaf port
i-zipqd	13160	tcp/udp	I-ZIPQD—QuarterDeck
bprd	13720	tcp/udp	BPRD Protocol (VERITAS NetBackup)
bpbrm	13721	tcp/udp	BPBRM Protocol (VERITAS NetBackup)
bpjava-msvc	13722	tcp/udp	BP Java MSVC Protocol—VERITAS Software
bpcd	13782	tcp/udp	VERITAS NetBackup
vopied	13783	tcp/udp	VOPIED Protocol—VERITAS Software
dsmcc-config	13818	tcp/udp	DSMCC Config—ISO/IEC13818-6 MPEG-2 DSM-CC
dsmcc-session	13819	tcp/udp	DSMCC Session Messages—ISO/IEC13818-6 MPEG-2 DSM-CC
dsmcc-passthru	13820	tcp/udp	DSMCC Pass-Thru Messages—ISO/IEC13818-6 MPEG-2 DSM-CC
dsmcc-download	13821	tcp/udp	DSMCC Download Protocol—ISO/IEC13818-6 MPEG-2 DSM-CC
dsmcc-ccp	13822	tcp/udp	DSMCC Channel Change Protocol—ISO/IEC13818-6 MPEG-2 DSM-CC
itu-sccp-ss7	14001	tcp/udp	ITU SCCP (SS7)—Ericsson
netserialext1	16360	tcp/udp	netserialext1—Aquila Group
netserialext2	16361	tcp/udp	netserialext2—Aquila Group
netserialext3	16367	tcp/udp	netserialext3—Aquila Group
netserialext4	16368	tcp/udp	netserialext4—Aquila Group
isode-dua	17007	tcp/udp	
chipper	17219	tcp/udp	Chipper
biimenu	18000	tcp/udp	Beckman Instruments, Inc.
opsec-cvp	18181	tcp/udp	OPSEC CVP—CheckPoint
opsec-ufp	18182	tcp/udp	OPSEC UFP-CheckPoint
ac-cluster	18463	tcp/udp	AC Cluster—HP
apc-necmp	18888	tcp/udp	APCNECMP—Astralpoint
keysrvr	19283	tcp/udp	Key Server for SASSAFRAS
keyshadow	19315	tcp/udp	Key Shadow for SASSAFRAS
hp-sco	19410	tcp/udp	hp-sco
hp-sca	19411	tcp/udp	hp-sca
jcp	19541	tcp/udp	JCP Client—JCI
dnp	20000	tcp/udp	DNP—ACSAtlanta
track	20670	tcp/udp	Track—US Naval Acadamy
vofr-gateway	21590	tcp/udp	VoFR Gateway—Tollbridge Technology
webphone	21845	tcp/udp	webphone
netspeak-is	21846	tcp/udp	NetSpeak Corp. Directory Services
netspeak-cs	21847	tcp/udp	NetSpeak Corp. Connection Services
netspeak-acd	21848	tcp/udp	NetSpeak Corp. Automatic Call Distribution
netspeak-cps	21849	tcp/udp	NetSpeak Corp. Credit Processing System
snapenetio	22000	tcp/udp	SNAPenetIO
optocontrol	22001	tcp/udp	OptoControl—Opto22
wnn6	22273	tcp/udp	wnn6—Omron Software
vocaltec-wconf	22555	tcp/udp	Vocaltec Web Conference
aws-brf	22800	tcp/udp	Telerate Information Platform LAN—Dow Jones
brf-gw	22951	tcp/udp	Telerate Information Platform WAN—Dow Jone
med-ltp	24000	tcp/udp	med-ltp—HP

Abbreviation	Port-Decimal	Protocol	Description/Owner
med-fsp-rx	24001	tcp/udp	med-fsp-rx—HP
med-fsp-tx	24002	tcp/udp	med-fsp-tx—HP
med-supp	24003	tcp/udp	med-supp—HP
med-ovw	24004	tcp/udp	med-ovw—HP
med-ci	24005	tcp/udp	med-ci—HP
med-net-svc	24006	tcp/udp	med-net-svc—HP
intel_rci	24386	tcp/udp	Intel RCI
icl-twobase1	25000	tcp/udp	icl-twobase1
icl-twobase2	25001	tcp/udp	icl-twobase2
icl-twobase3	25002	tcp/udp	icl-twobase3
icl-twobase4	25003	tcp/udp	icl-twobase4
icl-twobase5	25004	tcp/udp	icl-twobase5
icl-twobase6	25005	tcp/udp	icl-twobase6
icl-twobase7	25006	tcp/udp	icl-twobase7
icl-twobase8	25007	tcp/udp	icl-twobase8
icl-twobase9	25008	tcp/udp	icl-twobase9
icl-twobase10	25009	tcp/udp	icl-twobase10
vocaltec-hos	25793	tcp/udp	Vocaltec Address Server
quake	26000	tcp/udp	quake
wnn6-ds	26208	tcp/udp	wnn6-ds—Omron Software
flex-lm	27000-27009	tcp/udp	FLEX LM (1-10)—Globes
tw-auth-key	27999	tcp/udp	TW Authentication/Key Distribution and Attribute Certificate Services—SSE
filenet-tms	32768	tcp/udp	Filenet TMS
filenet-rpc	32769	tcp/udp	Filenet RPC
filenet-nch	32770	tcp/udp	Filenet NCH
traceroute	33434	tcp/udp	traceroute use—IANA
kastenxpipe	36865	tcp/udp	KastenX Pipe—Kasten Chase
rockwell-encap	44818	tcp/udp	Rockwell Encapsulation
eba	45678	tcp/udp	EBA PRISE
dbbrowse	47557	tcp/udp	Databeam Corporation
directplaysrvr	47624	tcp/udp	Direct Play Server—Microsoft
ap	47806	tcp/udp	ALC Protocol—Atuomated Logic
bacnet	47808	tcp/udp	Building Automation and Control Networks—Cornell U
nimcontroller	48000	tcp/udp	Nimbus Controller—Nimsoft
nimspooler	48001	tcp/udp	Nimbus Spooler—Nimsoft
nimhub	48002	tcp/udp	Nimbus Hub—Nimsoft
nimgtw	48003	tcp/udp	Nimbus Gateway—Nimsoft

Port Assignments (sorted by abbreviation):

Abbreviation	Port-Decimal	Protocol	Description/Owner
#	4	tcp/udp	Unassigned
#	6	tcp/udp	Unassigned
#	8	tcp/udp	Unassigned
#	10	tcp/udp	Unassigned
#	12	tcp/udp	Unassigned
#	14	tcp/udp	Unassigned
#	15	tcp/udp	Unassigned [was netstat]
#	16	tcp/udp	Unassigned
#	26	tcp/udp	Unassigned
#	28	tcp/udp	Unassigned
#	30	tcp/udp	Unassigned
#	32	tcp/udp	Unassigned
#	34	tcp/udp	Unassigned
#	36	tcp/udp	Unassigned
#	40	tcp/udp	Unassigned
#	60	tcp/udp	Unassigned
#	706	tcp/udp	Unassigned
#	708	tcp/udp	Unassigned
#	2997	tcp/udp	Unassigned
#	3146	tcp/udp	Unassigned
#	3263	tcp/udp	Unassigned
#	4352	tcp/udp	Unassigned
#	6504	tcp/udp	Unassigned
#	10001-10006	tcp/udp	Unassigned
#	10008-10079	tcp/udp	Unassigned
#	10081-10114	tcp/udp	Unassigned
#	10116-10287	tcp/udp	Unassigned
#	10289-10999	tcp/udp	Unassigned
#	1086-1096	tcp/udp	Unassigned
#	1203-1211	tcp/udp	Unassigned
#	1223-1233	tcp/udp	Unassigned
#	1235-1238	tcp/udp	Unassigned
#	1240-1247	tcp/udp	Unassigned
#	1249-1299	tcp/udp	Unassigned
#	1301-1309	tcp/udp	Unassigned
#	1315-1319	tcp/udp	Unassigned
#	1322-1334	tcp/udp	Unassigned
#	1336-1344	tcp/udp	Unassigned
#	1845-1849	tcp/udp	Unassigned
#	1852-1859	tcp/udp	Unassigned
#	1866-1869	tcp/udp	Unassigned
#	1875-1880	tcp/udp	Unassigned
#	1882-1894	tcp/udp	Unassigned
#	1896-1898	tcp/udp	Unassigned
#	1925-1929	tcp/udp	Unassigned
#	1932-1943	tcp/udp	Unassigned
#	1954-1960	tcp/udp	Unassigned
#	1968-1971	tcp/udp	Unassigned
#	1980-1983	tcp/udp	Unassigned
#	2110-2164	tcp/udp	Unassigned
#	2166-2179	tcp/udp	Unassigned
#	2246-2278	tcp/udp	Unassigned
#	225-241	tcp/udp	Reserved—ISI
#	2289-2293	tcp/udp	Unassigned
#	2371-2380	tcp/udp	Unassigned
#	2384-2388	tcp/udp	Unassigned
#	249-255	tcp/udp	Reserved—ISI
#	266-279	tcp/udp	Unassigned
#	285-307	tcp/udp	Unassigned
#	2886-2896	tcp/udp	Unassigned
#	2898-2907	tcp/udp	Unassigned
#	2918-2970	tcp/udp	Unassigned
#	2982-2995	tcp/udp	Unassigned
#	3066-3067	tcp/udp	Unassigned
#	3070-3079	tcp/udp	Unassigned

Abbreviation	Port-Decimal	Protocol	Description/Owner
#	3084-3104	tcp/udp	Unassigned
#	3107-3129	tcp/udp	Unassigned
#	3132-3140	tcp/udp	Unassigned
#	3152-3179	tcp/udp	Unassigned
#	3183-3261	tcp/udp	Unassigned
#	323-332	tcp/udp	Unassigned
#	334-343	tcp/udp	Unassigned
#	3402-3420	tcp/udp	Unassigned
#	3472-3562	tcp/udp	Unassigned
#	3564-3671	tcp/udp	Unassigned
#	3673-3844	tcp/udp	Unassigned
#	3846-3899	tcp/udp	Unassigned
#	3988-3999	tcp/udp	Unassigned
#	4020-4095	tcp/udp	Unassigned
#	4100-4131	tcp/udp	Unassigned
#	4144-4159	tcp/udp	Unassigned
#	4161-4199	tcp/udp	Unassigned
#	4301-4320	tcp/udp	Unassigned
#	4354-4443	tcp/udp	Unassigned
#	4458-4499	tcp/udp	Unassigned
#	4548-4566	tcp/udp	Unassigned
#	4569-4599	tcp/udp	Unassigned
#	4602-4671	tcp/udp	Unassigned
#	4673-4799	tcp/udp	Unassigned
#	4803-4826	tcp/udp	Unassigned
#	4828-4867	tcp/udp	Unassigned
#	5312-5399	tcp/udp	Unassigned
#	5422-5453	tcp/udp	Unassigned
#	5457-5499	tcp/udp	Unassigned
#	5504-5553	tcp/udp	Unassigned
#	5556-5598	tcp/udp	Unassigned
#	6472-6499	tcp/udp	Unassigned
#	6509-6546	tcp/udp	Unassigned
#	6551-6557	tcp/udp	Unassigned
#	695-703	tcp	Unassigned
#	712-728	tcp/udp	Unassigned
#	7391-7394	tcp/udp	Unassigned
#	7396-7425	tcp/udp	Unassigned
#	778-779	tcp/udp	Unassigned
#	788-799	tcp/udp	Unassigned
#	802-809	tcp/udp	Unassigned
#	811-827	tcp/udp	Unassigned
#	830-872	tcp/udp	Unassigned
#	875-885	tcp/udp	Unassigned
#	889-899	tcp/udp	Unassigned
#	901-910	tcp/udp	Unassigned
#	912-988	tcp/udp	Unassigned
#	9899-9908	tcp/udp	Unassigned
#	9910-9991	tcp/udp	Unassigned
#	1008	udp	Possibly used by Sun Solaris
#	1001-1009	tcp/udp	Unassigned
#	1011-1022	tcp/udp	Reserved
###########	3300-3301	tcp/udp	Unauthorized Use by SAP R/3
###########Port 5555 also used by HP Omniback####################			
#########Compuserve (unoffically) is using port 4144 #########			
#unitary	126	tcp/udp	Unisys Unitary Login
*	90	tcp	also being used unofficially by Pointcast
*	126	tcp	Previously assigned to application below
*	222	tcp	Possible Conflict of Port 222 with "Masqdialer" below
*		tcp/udp	<== NOTE Conflict on 2049 !
3com-amp3	629	tcp/udp	3Com AMP3
3com-net-mgmt	2391	tcp/udp	3COM Net Management
3Com-nsd	1742	tcp/udp	3Com-nsd
3com-tsmux	106	tcp/udp	3COM-TSMUX
3com-webview	2339	tcp/udp	3Com WebView
3d-nfsd	2323	tcp/udp	3d-nfsd—Commvault
3ds-lm	1538	tcp/udp	3ds-lm
3L-L1	1511	tcp/udp	3L-L1

Abbreviation	Port-Decimal	Protocol	Description/Owner
3m-image-lm	1550	tcp/udp	Image Storage license manager 3M Company
4talk	3284	tcp/udp	4Talk—Four-Sight
914c/g	211	tcp/udp	Texas Instruments 914C/G Terminal
9pfs	564	tcp/udp	plan 9 file service
a1-bs	5603	tcp/udp	A1-BS—Lucent
a1-msc	5602	tcp/udp	A1-MSC—Lucent
a3-sdunode	5604	tcp/udp	A3-SDUNode—Lucent
a4-sdunode	5605	tcp/udp	A4-SDUNode—Lucent
aaftp	2794	tcp/udp	aaftp—E. Jay Berkenbilt
aal-lm	1469	tcp/udp	Active Analysis Limited License Manager
aap	2878	tcp/udp	AAP—Sun
aas	1601	tcp/udp	aas—Lachman
abarsd	8402	tcp/udp	abarsd—Commvault
abbaccuray	1546	tcp/udp	abbaccuray
abbs	4885	tcp/udp	ABBS—ML
about	2019	udp	
acap	674	tcp/udp	ACAP—Innosoft
acas	62	tcp/udp	ACA Services—DEC
accessbuilder	888	tcp/udp	AccessBuilder-3Com
ac-cluster	18463	tcp/udp	AC Cluster—HP
acc-raid	2800	tcp/udp	ACC RAID—Adaptec
accu-lmgr	7781	tcp/udp	accu-lmgr—Accugraph
ace-client	2334	tcp/udp	ACE Client Auth—Securitydynamics
ace-proxy	2335	tcp/udp	ACE Proxy—Securitydynamics
ace-server	2475	tcp/udp	ACE Server
ace-svr-prop	2476	tcp/udp	ACE Server Propagation
aci	187	tcp/udp	Application Communication Interface—TI
acl-manager	4013	tcp/udp	ACL Manager—Fujitsu
acmaint_dbd	774	udp	
acmaint_transd	775	udp	
acmsoda	6969	tcp/udp	acmsoda—ACM
acp	599	tcp/udp	Aeolon Core Protocol
acr-nema	104	tcp/udp	ACR-NEMA Digital Imag. & Comm. 300
acs2000-dsp	2369	tcp/udp	ACS2000 DSP
ac-tech	2796	tcp/udp	ac-tech
activememory	2859	tcp/udp	Active Memory—EDA Consulting
active-net	3322-3325	tcp/udp	ISI
actnet	5411	tcp/udp	ActNet—ACT Research
adaptecmgr	2521	tcp/udp	Adaptec Manager
adapt-sna	1365	tcp/udp	Network Software Associates
admind	3279	tcp/udp	admind—ChisTech
admind	8403	tcp/udp	admind—Commvault
admins-lms	2692	tcp/udp	Admins LMS
ads	2550	tcp/udp	ADS—Adobe
adtech-test	3357	tcp/udp	Adtech Test IP
advant-lm	2295	tcp/udp	Advant License Manager—The ABB Group
aed-512	149	tcp/udp	AED 512 Emulation Service—UPENN
af	1411	tcp/udp	AudioFile—DEC
afpovertcp	548	tcp/udp	AFP over TCP—Apple
afs	1483	tcp/udp	AFS License Manager
afs3-bos	7007	tcp/udp	basic overseer process
afs3-callback	7001	tcp/udp	callbacks to cache managers
afs3-errors	7006	tcp/udp	error interpretation service
afs3-fileserver	7000	tcp/udp	file server itself
afs3-kaserver	7004	tcp/udp	AFS/Kerberos authentication service
afs3-prserver	7002	tcp/udp	users & groups database
afs3-rmtsys	7009	tcp/udp	remote cache manager service
afs3-update	7008	tcp/udp	server-to-server updater
afs3-vlserver	7003	tcp/udp	volume location database
afs3-volser	7005	tcp/udp	volume managment server
agentview	2331	tcp/udp	AGENTVIEW—Aaccorp
agentvu	3031	tcp/udp	AgentVU—Apple
agentx	705	tcp/udp	AgentX—acec
agri-gateway	3026	tcp/udp	AGRI Gateway—Agri-Datalog
agriserver	3021	tcp/udp	AGRI Server—Agri-Datalog
aicc-cmi	3316	tcp/udp	AICC/CMI—Boeing
aic-np	2785	tcp/udp	aic-np—Cisco
aic-oncrpc	2786	tcp/udp	aic-oncrpc - Destiny MCD database—Cisco
aimpp-hello	2846	tcp/udp	AIMPP Hello—Automation Intelligence

Abbreviation	Port-Decimal	Protocol	Description/Owner
aimpp-port-req	2847	tcp/udp	AIMPP Port Req—Automation Intelligence
airs	1481	tcp/udp	AIRS
aises	2783	tcp/udp	AISES—Pgaero
aker-cdp	2473	tcp/udp	Aker-cdp
alarm	2740	tcp/udp	Alarm—BFPG
alarm-clock-c	2668	tcp/udp	Alarm Clock Client—Tatanka
alarm-clock-s	2667	tcp/udp	Alarm Clock Server—Tatanka
alpes	463	tcp/udp	alpes—IMAGE
alphatech-lm	1653	tcp/udp	alphatech-lm
alta-ana-lm	1346	tcp/udp	Alta Analytics License Manager
altav-remmgt	2456	tcp/udp	altav-remmgt—DEC
altav-tunnel	3265	tcp/udp	Altav Tunnel—DEC
altserviceboot	4011	tcp/udp	Alternate Service Boot—Intel
amanda	10080	tcp/udp	Amanda —Purdue U
ambit-lm	6831	tcp/udp	ambit-lm
amdsched	1931	tcp/udp	AMD SCHED—Warwick
amiganetfs	2100	tcp/udp	amiganetfs—U di Pisa
aminet	649	tcp/udp	Aminet—Adaptivemedia
aminet	2639	tcp/udp	AMInet—Alcorn McBride Inc
ampr-info	1535	tcp/udp	ampr-info
ampr-inter	1536	tcp/udp	ampr-inter
ampr-rcmd	459	tcp/udp	ampr-rcmd
amt-blc-port	2848	tcp/udp	AMT-BLC-PORT—Telecom Italia
amt-cnf-prot	3054	tcp/udp	AMT CNF PROT—Telecom Italia
and-lm	2646	tcp/udp	AND Licence Manager
anet	212	tcp/udp	ATEXSSTR—Kodak
anet-b	3338	tcp/udp	OMF data b—ABB Sweden
anet-h	3341	tcp/udp	OMF data h—ABB Sweden
anet-l	3339	tcp/udp	OMF data l—ABB Sweden
anet-m	3340	tcp/udp	OMF data m—ABB Sweden
ansanotify	116	tcp/udp	ANSA REX Notify—ANSA
ansatrader	124	tcp/udp	ANSA REX Trader—ANSA
ansoft-lm-1	1083	tcp/udp	Anasoft License Manager
ansoft-lm-2	1084	tcp/udp	Anasoft License Manager
answersoft-lm	1781	tcp/udp	answersoft-lm
ansys-lm	1800	tcp/udp	ANSYS-License manager
anynetgateway	1491	tcp/udp	anynetgateway
aocp	2712	tcp/udp	Axapta Object Communication Protocol—Damgaard
aodv	654	tcp/udp	AODV—Sun
aol	5190	tcp/udp	America-Online
aol-1	5191	tcp/udp	AmericaOnline1
aol-2	5192	tcp/udp	AmericaOnline2
aol-3	5193	tcp/udp	AmericaOnline3
ap	47806	tcp/udp	ALC Protocol—Atuomated Logic
apc-necmp	18888	tcp/udp	APCNECMP—Astralpoint
apcpcns	3052	tcp/udp	APCPCNS—APPC
apc-tcp-udp-1	6547	tcp/udp	apc-tcp-udp-1—APPC
apc-tcp-udp-2	6548	tcp/udp	apc-tcp-udp-2—APPC
apc-tcp-udp-3	6549	tcp/udp	apc-tcp-udp-3—APPC
apc-tcp-udp-4	5454	tcp/udp	apc-tcp-udp-4—APPC
apc-tcp-udp-5	5455	tcp/udp	apc-tcp-udp-5—APPC
apc-tcp-udp-6	5456	tcp/udp	apc-tcp-udp-6—APPC
apertus-ldp	539	tcp/udp	Apertus Technologies Load Determination
apollo-cc	2754	tcp/udp	APOLLO CC—Brand Communications
apollo-gms	2759	tcp/udp	APOLLO GMS—Brand Communications
apollo-status	2758	tcp/udp	APOLLO Status—Brand Communications
apple-licman	1381	tcp/udp	Apple Network License Manager
appleqtc	458	tcp/udp	apple quick time
appleqtcsrvr	545	tcp/udp	appleqtcsrvr
appleugcontrol	2336	tcp/udp	Apple UG Control
applianttcp	2426	tcp/udp	Appliant TCP
applix	999	udp	Applix ac
appman-server	3312	tcp/udp	Application Management Server
appswitch-emp	2616	tcp/udp	appswitch-emp—TopLayer
apri-lm	1447	tcp/udp	Applied Parallel Research LM
arbortext-lm	1557	tcp/udp	ArborText License Manager
arcisdms	262	tcp/udp	Arcisdms
arcp	7070	tcp/udp	ARCP—NASA
ardt	1826	tcp/udp	ARDT—Ardent

Abbreviation	Port-Decimal	Protocol	Description/Owner
ardusmul	1835	tcp/udp	ARDUS Multicast
ardusuni	1834	tcp/udp	ARDUS Unicast
arepa-cas	3030	tcp/udp	Arepa Cas
arepa-raft	3025	tcp/udp	Arepa Raft—IEEE
argis-ds	2582	tcp/udp	ARGIS DS
argis-te	2581	tcp/udp	ARGIS TE
aria	2624	tcp/udp	Aria—Andromedia
ariel1	419	tcp/udp	Ariel—Research Libraries Group
ariel2	421	tcp/udp	Ariel—Research Libraries Group
ariel3	422	tcp/udp	Ariel—Research Libraries Group
aries-kfinder	7570	tcp/udp	Aries Kfinder
armadp	1913	tcp/udp	armadp—Arm, ltd
arns	384	tcp/udp	A Remote Network Server System
asa	386	tcp/udp	ASA Message Router Object Def—Allen-Bradley
asa-appl-proto	502	tcp/udp	asa-appl-proto—Modicon
asci-val	1560	tcp/udp	asci-val—Advanced Systems
asc-slmd	4448	tcp/udp	ASC Licence Manager
ashwin	1556	tcp/udp	AshWin CI Tecnologies
asi	1827	tcp/udp	ASI—The ABB Group
asia	626	tcp/udp	ASIA—Apple
asipregistry	687	tcp/udp	asipregistry—Apple
asip-webadmin	311	tcp/udp	AppleShare IP WebAdmin
asnaacceler8db	5042	tcp/udp	asnaacceler8db—Asna
aspeclmd	1544	tcp/udp	aspeclmd
aspen-services	1749	tcp/udp	aspen-services—Aspen Graphics Technical Analysis
aspentec-lm	6142	tcp/udp	Aspen Technology License Manager
as-servermap	449	tcp/udp	AS Server Mapper—IBM
astromed-main	2864	tcp/udp	main 5001 cmd—Astro-Med Inc.
at-3	203	tcp/udp	AppleTalk Unused—CMU
at-5	205	tcp/udp	AppleTalk Unused—CMU
at-7	207	tcp/udp	AppleTalk Unused—CMU
at-8	208	tcp/udp	AppleTalk Unused—CMU
at-echo	204	tcp/udp	AppleTalk Echo—CMU
atex_elmd	1385	tcp/udp	Atex Publishing License Manager
atmp	5150	tcp/udp	Ascend Tunnel Management Protocol
atmtcp	2812	tcp/udp	atmtcp—Swiss Federal Institute of Technology
atm-uhas	11367	tcp/udp	ATM UHAS—Attachmate
atm-zip-office	1520	tcp/udp	atm zip office
at-nbp	202	tcp/udp	AppleTalk Name Binding—CMU
at-rtmp	201	tcp/udp	AppleTalk Routing Maintenance—CMU
ats	2201	tcp/udp	Advanced Training System Program
attachmate-g32	2317	tcp/udp	Attachmate G32
attachmate-s2s	2419	tcp/udp	Attachmate S2S
attachmate-uts	2304	tcp/udp	Attachmate UTS
att-intercom	4983	tcp/udp	AT&T Intercom
at-zis	206	tcp/udp	AppleTalk Zone Information—CMU
audio-activmail	1397	tcp/udp	Audio Active Mail
audionews	114	tcp/udp	Audio News Multicast—Chalmers U of Technology
audit	182	tcp/udp	Unisys Audit SITP
auditd	48	tcp/udp	Digital Audit Daemon
auris	2772	tcp/udp	auris—Telefónica Investigación y Desarrollo, S.A.
aurora-cmgr	364	tcp/udp	Aurora CMGR—Aurora Technology
aurp	387	tcp/udp	Appletalk Update-Based Routing Pro—Novell
auth	113	tcp/udp	Authentication Service—ARPA
autodesk-lm	1422	tcp/udp	Autodesk License Manager
aux_bus_shunt	623	tcp/udp	Aux Bus Shunt—Intel
avian	486	tcp/udp	avian—Lotus
aviva-sna	2238	tcp/udp	AVIVA SNA SER
avt-profile-1	5004	tcp/udp	avt-profile-1—GMD
avt-profile-2	5005	tcp/udp	avt-profile-2—GMD
awg-proxy	3277	tcp/udp	AWG Proxy—HP
aws-brf	22800	tcp/udp	Telerate Information Platform LAN—Dow Jones
axon-lm	1548	tcp/udp	Axon License Manager
backburner	2635	tcp/udp	Back Burner—MetaCreations
backroomnet	3387	tcp/udp	Back Room Net—Carreker
backup-express	6123	tcp/udp	Backup Express—SyncSort
bacnet	47808	tcp/udp	Building Automation and Control Networks—Cornell U
badm_priv	6505	tcp/udp	BoKS Admin Private Port—Dynas
badm_pub	6506	tcp/udp	BoKS Admin Public Port—Dynas

Abbreviation	Port-Decimal	Protocol	Description/Owner
banyan-net	2708	tcp/udp	Banyan-Net
banyan-rpc	567	tcp/udp	banyan-rpc
banyan-vip	573	tcp/udp	banyan-vip
bb	1984	tcp/udp	BB—MacLawran
bbars	3327	tcp/udp	BBARS—Bandl
bbn-mmc	1347	tcp/udp	multi media conferencing
bbn-mmx	1348	tcp/udp	multi media conferencing
bcs-broker	1704	tcp/udp	bcs-broker—Knoware BV
bcs-lmserver	1951	tcp/udp	bcs-lmserver—Knoware BV
bdir_priv	6507	tcp/udp	"BoKS Dir Server, Private Port—Dynas"
bdir_pub	6508	tcp/udp	"BoKS Dir Server, Public Port—Dynas"
bdp	581	tcp/udp	Bundle Discovery Protocol—Xylogics
bears-01	2852	tcp/udp	bears-01
belarc-http	2693	tcp/udp	belarc-http
berknet	2005	tcp	
bftp	152	tcp/udp	Background File Transfer Program—ISI
bgmp	264	tcp/udp	BGMP—U Michigan
bgp	179	tcp/udp	Border Gateway Protocol—Cisco
bgs-nsi	482	tcp/udp	bgs-nsi
bh611	354	tcp/udp	bh611—Bell&Howell
bhevent	357	tcp/udp	bhevent—Bell&Howell
bhfhs	248	tcp/udp	bhfhs—Bell&Howell
bhmds	310	tcp/udp	bhmds—Bell&Howell
bhoedap4	352	tcp/udp	bhoedap4 (added 5/21/97-not really assigned)—Bell&Howell
bhoetty	351	tcp/udp	bhoetty (added 5/21/97-not really assigned)—Bell&Howell
biff	512	udp	notify users of new mail received
biimenu	18000	tcp/udp	Beckman Instruments, Inc.
binderysupport	2302	tcp/udp	Bindery Support—Novell
bintec-admin	2107	tcp/udp	BinTec Admin
bintec-capi	2662	tcp/udp	BinTec-CAPI
bintec-tapi	2663	tcp/udp	BinTec-TAPI
blackboard	2032	tcp/udp	
blackjack	1025	tcp/udp	network blackjack
bl-idm	142	tcp/udp	Britton-Lee IDM
blockade	2911	tcp/udp	Blockade—Interlog
blockade-bpsp	2574	tcp/udp	Blockade BPSP
blocks	10288	tcp/udp	Blocks
blueberry-lm	1432	tcp/udp	Blueberry Software License Manager
bluelance	2877	tcp/udp	BLUELANCE
blwnkl-port	2625	tcp/udp	Blwnkl Port—3Com
bmap	3421	tcp/udp	Bull Apprise portmapper
bmc_patroldb	1313	tcp/udp	BMC_PATROLDB
bmc-ar	2494	tcp/udp	BMC AR
bmc-net-adm	1769	tcp/udp	bmc-net-adm—BMC Software, Inc.
bmc-net-svc	1770	tcp/udp	bmc-net-svc—BMC Software, Inc.
bmcpatrolagent	3181	tcp/udp	BMC Patrol Agent—BMC Software, Inc.
bmcpatrolrnvu	3182	tcp/udp	BMC Patrol Rendezvous—BMC Software, Inc.
bmc-reporting	4568	tcp/udp	BMC Reporting—BMC Software, Inc.
bmpp	632	tcp/udp	bmpp—Corvu
bnet	415	tcp/udp	Bnet—Unisys
bnt-manager	3344	tcp/udp	BNT Manager
boks	6500	tcp/udp	BoKS Master
boks_clntd	6503	tcp/udp	BoKS Clntd
boks_servc	6501	tcp/udp	BoKS Servc
boks_servm	6502	tcp/udp	BoKS Servm
boosterware	2913	tcp/udp	Booster Ware—NetVision
bootclient	2017	udp	
bootpc	68	tcp/udp	Bootstrap Protocol Client—Stanford U
bootps	67	tcp/udp	Bootstrap Protocol Server—Stanford U
bootserver	2016	tcp/udp	
borland-dsj	707	tcp/udp	Borland DSJ
bpbrm	13721	tcp/udp	BPBRM Protocol (VERITAS NetBackup)
bpcd	13782	tcp/udp	VERITAS NetBackup
bpcp-poll	2844	tcp/udp	BPCP POLL—SPX Corp.
bpcp-trap	2845	tcp/udp	BPCP TRAP—SPX Corp.
bpjava-msvc	13722	tcp/udp	BP Java MSVC Protocol—VERITAS Software
bprd	13720	tcp/udp	BPRD Protocol (VERITAS NetBackup)
bre	4096	tcp/udp	BRE (Bridge Relay Element)—Timeplex
brf-gw	22951	tcp/udp	Telerate Information Platform WAN—Dow Jone

Abbreviation	Port-Decimal	Protocol	Description/Owner
broker_service	3014	tcp/udp	Broker Service—Novell
brp	3043	tcp/udp	BRP—Hughes Network Systems
bruce	2619	tcp/udp	bruce—Sun
btpp2audctr1	2536	tcp/udp	btpp2audctr1—British Telcom
btpp2sectrans	2444	tcp/udp	BT PP2 Sectrans
btprjctrl	2803	tcp/udp	btprjctrl—British Telcom
btrieve	3351	tcp/udp	BTRIEVE—Pervasive-SW
bts-appserver	1961	tcp/udp	BTS APPSERVER—Sabre
bues_service	2446	tcp/udp	bues_service—telekom400
busboy	998	tcp	
bytex	1375	tcp/udp	Bytex
c3	2472	tcp/udp	C3—com-on
cableport-ax	282	tcp/udp	Cable Port A/X—Intel
cab-protocol	595	tcp/udp	CAB Protocol
caci-lm	1554	tcp/udp	CACI Products Company License Manager
cadabra-lm	1563	tcp/udp	Cadabra License Manager
cadencecontrol	2318	tcp/udp	Cadence Control—Polybon
cadis-1	1441	tcp/udp	Cadis License Management
cadis-2	1442	tcp/udp	Cadis License Management
cadkey-licman	1399	tcp/udp	Cadkey License Manager
cadkey-tablet	1400	tcp/udp	Cadkey Tablet Daemon
cadlock	770	tcp/udp	
cadlock2	1000	tcp/udp	
cadsi-lm	1387	tcp/udp	Computer Aided Design Software Inc LM
caicci	1721	tcp/udp	caicci—Computer Associates
caiccipc	1202	tcp/udp	caiccipc—CAI
CAllic	216	tcp/udp	Computer Associates Int'l License Server
cal	588	tcp/udp	CAL—Intel
callbook	2000	tcp/udp	
call-logging	2552	tcp/udp	Call Logging—Ascend
call-sig-trans	2517	tcp/udp	H.323 Annex E call signaling transport—Vocaltec
cambertx-lm	1734	tcp/udp	Camber Corporation License Management
camp	4450	tcp/udp	Camp
can-dch	1919	tcp/udp	Candle Directory Service - DCH
can-ferret	1920	tcp/udp	Candle Directory Service - FERRET
can-nds	1918	tcp/udp	Candle Directory Service - NDS
canocentral0	1871	tcp/udp	Cano Central 0—Canon
canocentral1	1872	tcp/udp	Cano Central 1—Canon
capfast-lmd	1756	tcp/udp	capfast-lmd—Phase Three Logic Inc.
caps-lm	3290	tcp/udp	CAPS LOGISTICS TOOLKIT - LM
cardbox	3105	tcp/udp	Cardbox
cardbox-http	3106	tcp/udp	Cardbox HTTP
cart-o-rama	3292	tcp/udp	Cart O Rama—SeaLabs
cas	2418	tcp/udp	cas—Fujitsu
catalyst	2836	tcp/udp	catalyst—JSB Software Technologies, PLC
cautcpd	3061	tcp/udp	cautcpd—Computer Associates
cbserver	3388	tcp/udp	CB Server—Hyperion Inc.
cbt	7777	tcp/udp	cbt—Univesity College London
ccmail	3264	tcp/udp	cc:mail/lotus
ccs-software	2734	tcp/udp	CCS Software
cdborker	3376	tcp/udp	CD Broker—Esps
cdc	223	tcp/udp	Certificate Distribution Center—DEC
cddbp	888	tcp/udp	CD Database Protocol-unassigned—Moonsoft
cddbp-alt	8880	tcp/udp	CDDBP—Moonsoft
cdfunc	2045	tcp/udp	
cdfunc	2045	tcp/udp	
cdid	3315	tcp/udp	CDID—Chat
cdl-server	3056	tcp/udp	CDL Server—Engineer
cdn	2412	tcp/udp	CDN—Netmind
cecsvc	2571	tcp/udp	CECSVC—Corder-Eng
centerline	3987	tcp/udp	Centerline
centra	1709	tcp/udp	centra—Centra Inc.
cera-bcm	1794	tcp/udp	cera-bcm—IBM
cert-initiator	1639	tcp/udp	cert-initiator—Sun
cert-responder	1640	tcp/udp	cert-responder—Sun
cesdinv	2856	tcp/udp	cesdinv—PFU Direct
cfdptkt	120	tcp/udp	CFDPTKT—UCOLUMBIA
cfengine	5308	tcp/udp	Cfengine—HiOslo
cft-0	1761	tcp/udp	cft-0

Abbreviation	Port-Decimal	Protocol	Description/Owner
cft-1	1762	tcp/udp	cft-1
cft-2	1763	tcp/udp	cft-2
cft-3	1764	tcp/udp	cft-3
cft-4	1765	tcp/udp	cft-4
cft-5	1766	tcp/udp	cft-5
cft-6	1767	tcp/udp	cft-6
cft-7	1768	tcp/udp	cft-7
cgms	3003	tcp/udp	CGMS—USCG Military
chargen	19	tcp/udp	Character Generator
checksum	1386	tcp/udp	CheckSum License Manager
chevinservices	3349	tcp/udp	Chevin Services
chimera-hwm	4009	tcp/udp	Chimera HWM—UC Irvine
chip-lm	1572	tcp/udp	Chipcom License Manager
chipper	17219	tcp/udp	Chipper
chromagrafx	1373	tcp/udp	Chromagrafx
chshell	562	tcp/udp	chcmd
cichild-lm	1523	tcp/udp	cichild
cichlid	1377	tcp/udp	Cichlid License Manager
cifs	3020	tcp/udp	CIFS—Microsoft
cimplex	673	tcp/udp	CIMPLEX—Cesi
cinegrfx-lm	1743	tcp/udp	Cinema Graphics License Manager—RFX Systems Integration
cisco-fna	130	tcp/udp	cisco FNATIVE
cisco-net-mgmt	1741	tcp/udp	cisco-net-mgmt
cisco-sys	132	tcp/udp	cisco SYSMAINT
cisco-tdp	711	tcp/udp	Cisco TDP
cisco-tna	131	tcp/udp	cisco TNATIVE
citadel	504	tcp/udp	citadel
citrixadmin	2513	tcp/udp	Citrix ADMIN
citrixima	2512	tcp/udp	Citrix IMA
citriximaclient	2598	tcp/udp	Citrix MA Client
citrix-rtmp	2897	tcp/udp	Citrix RTMP
citynl	1729	tcp/udp	CityNL License Management
cl/1	172	tcp/udp	Network Innovations CL/1
clariion-evr01	6389	tcp/udp	clariion-evr01
clearcase	371	tcp/udp	Clearcase—Atria
cloanto-lm	3397	tcp/udp	Cloanto License Manager
cloanto-net-1	356	tcp/udp	Cloanto Net 1
close-combat	1944	tcp/udp	close-combat—Microsoft
clp	2567	tcp/udp	Cisco Line Protocol
cluster-disc	3374	tcp/udp	Cluster Disc—NCR
clutild	7174	tcp/udp	Clutild—IBM
clvm-cfg	1476	tcp/udp	clvm-cfg
cmadmin	2617	tcp/udp	Clinical Context Managers—Sentillion
cmip-agent	164	tcp/udp	CMIP/TCP Agent
cmip-man	163	tcp/udp	CMIP/TCP Manager
cnap	1637	tcp/udp	CableNet Admin Protocol
cncp	1636	tcp/udp	CableNet Control Protocol
cnhrp	1757	tcp/udp	cnhrp—Virata Semiconductor
cnip	1638	tcp/udp	CableNet Info Protocol
cnrp	2757	tcp/udp	CNRP—The ABB Group
cns-srv-port	2976	tcp/udp	CNS Server Port—Cisco
coauthor	1529	tcp/udp	oracle
codaauth2	370	tcp/udp	codaauth2—FreeBSD Project
codasrv	2432	tcp/udp	codasrv—Robert Watson
codasrv-se	2433	tcp/udp	codasrv-se—Robert Watson
cogitate	3039	tcp/udp	Cogitate, Inc.
cogsys-lm	3377	tcp/udp	Cogsys Network License Manager
collaborator	622	tcp/udp	Collaborator—Opteamasoft
combox-web-acc	2534	tcp/udp	Combox Web Access
comcam	2108	tcp/udp	Comcam
command-mq-gm	2664	tcp/udp	Command MQ GM—Boole&Babbage
command-mq-pm	2665	tcp/udp	Command MQ PM—Boole&Babbage
commerce	542	tcp/udp	commerce
commonspace	1592	tcp/udp	commonspace—Within Technology
commplex-link	5001	tcp/udp	
commplex-main	5000	tcp/udp	
community	2459	tcp/udp	Community—Webmaster
compaq-https	2381	tcp/udp	Compaq HTTPS
compaq-scp	2766	tcp/udp	Compaq SCP

Abbreviation	Port-Decimal	Protocol	Description/Owner
compaq-wcp	2555	tcp/udp	Compaq WCP
composit-server	2417	tcp/udp	Composit Server—PFU Direct
compressnet	2	tcp/udp	Management Utility—Process
compressnet	3	tcp/udp	Compression—Process
comscm	437	tcp/udp	comscm—DEC
comtest	2415	tcp/udp	COMTEST
con	759	tcp/udp	
concert	786	tcp/udp	Concert—IBM
conclave-cpp	2491	tcp/udp	Conclave CPP—Interdyn
concomp1	1802	tcp/udp	ConComp1
concurrent-lm	1648	tcp/udp	concurrent-lm
conf	2008	tcp	
conference	531	tcp/udp	chat
conferencetalk	1713	tcp/udp	ConferenceTalk—Videoserver
confluent	1484	tcp/udp	Confluent License Manager
connection	2607	tcp/udp	Dell Connection
connendp	693	tcp/udp	connendp—Future-Gate
connlcli	1358	tcp/udp	CONNLCLI
contentserver	454	tcp/udp	ContentServer—Emotion
contentserver	3365	tcp/udp	Content Server—Emotion
continuus	5412	tcp/udp	Continuus
contract	2872	tcp/udp	CONTRACT—Telnor
coord-svr	2565	tcp/udp	Coordinator Server—EnsembleSoft
cops	3288	tcp/udp	COPS—IPHighway
corba-iiop	683	tcp/udp	CORBA IIOP—Object Management Group
corba-iiop-ssl	684	tcp/udp	CORBA IIOP SSL—Object Management Group
corbaloc	2809	tcp/udp	CORBA LOC—Distributed Systems Tech Center
corel_vncadmin	2654	tcp/udp	Corel VNC Admin—CorelComputer
corelccam	4300	tcp/udp	Corel CCam
corelvideo	1566	tcp/udp	CORELVIDEO
corerjd	284	tcp/udp	corerjd—Core Networks
cosmocall	2324	tcp/udp	Cosmocall
courier	530	tcp/udp	rpc
covia	64	tcp/udp	Communications Integrator (CI)—The Clout Project (Chicago)
cpq-wbem	2301	tcp/udp	Compaq HTTP
cpudpencap	2746	tcp/udp	CPUDPENCAP—Checkpoint
cqg-netlan	2823	tcp/udp	CQG Net/LAN—CQG Inc.
creativepartnr	455	tcp/udp	CreativePartnr—Emotion
creativepartnr	3366	tcp/udp	Creative Partner—Emotion
creativeserver	453	tcp/udp	CreativeServer—Emotion
creativeserver	3364	tcp/udp	Creative Server—Emotion
crip	6253	tcp/udp	CRIP—Ciena
crmsbits	2422	tcp/udp	CRMSBITS—Telstra
crs	507	tcp/udp	crs—Microsoft
cruise-config	8378	tcp/udp	Cruise CONFIG—Cruise Technology
cruise-diags	8379	tcp/udp	Cruise DIAGS—Cruise Technology
cruise-enum	8376	tcp/udp	Cruise ENUMC—Cruise Technology
cruise-swroute	8377	tcp/udp	Cruise SWROUTE—Cruise Technology
cruise-update	8380	tcp/udp	Cruise UPDATE—Cruise Technology
cr-websystems	2314	tcp/udp	CR WebSystems—Peddie
cryptoadmin	624	tcp/udp	Crypto Admin—Cyberus
csbphonemaster	1724	tcp/udp	csbphonemaster
csdm	1468	tcp/udp	CSDM
csdm	1472	tcp/udp	csdm
csdmbase	1467	tcp/udp	CSDMBASE
csdmbase	1471	tcp/udp	csdmbase
csi-lfap	3145	tcp/udp	CSI-LFAP—Cabletron
csi-sgwp	348	tcp/udp	Cabletron Management Protocol
cslistener	9000	tcp/udp	Cslistener—Cincom
csms	3399	tcp/udp	CSMS—Cedros
csms2	3400	tcp/udp	CSMS2—Cedros
csnet-ns	105	tcp/udp	Mailbox Name Nameserver—UWISC
cso	105	tcp/udp	CCSO name server protocol—eSpecials
csoft1	1837	tcp/udp	csoft1
csoft-plusclnt	2699	tcp/udp	Csoft Plus Client
csoft-prev	3271	tcp/udp	CSoft Prev Port—Iname
csoftragent	3004	tcp/udp	Csoft Agent
cspmulti	2807	tcp/udp	cspmulti—PFU
cspuni	2806	tcp/udp	cspuni—PFU

Abbreviation	Port-Decimal	Protocol	Description/Owner
csregagent	3022	tcp/udp	CSREGAGENT—Csoft
ctcd	1851	tcp/udp	ctcd—Cybertrace
ctf	84	tcp/udp	Common Trace Facility—DEC
cti-redwood	2563	tcp/udp	CTI Redwood—Daou
ctisystemmsg	4451	tcp/udp	CTI System Msg—Control Technology
cuillamartin	1356	tcp/udp	CuillaMartin Company
custix	528	tcp/udp	Customer—IXChange
cvc	1495	tcp/udp	cvc
cvc_hostd	442	tcp/udp	cvc_hostd—Cray
cvd	8400	tcp/udp	cvd—Commvault
cvmmon	2300	tcp/udp	CVMMON—HP
cvmon	1686	tcp/udp	cvmon—HP
cvspserver	2401	tcp/udp	cvspserver—Cyclic
cyaserv	2584	tcp/udp	cyaserv—CYASolutions
cybercash	551	tcp/udp	cybercash
cyborg-systems	9888	tcp/udp	CYBORG Systems
cycleserv	763	tcp/udp	
cycleserv2	772	tcp/udp	
cylink-c	5420	tcp/udp	Cylink-C
cypress	2015	tcp	
cypress-stat	2017	tcp	
cytel-lm	3297	tcp/udp	Cytel License Manager
d2k-datamover1	2297	tcp/udp	D2K DataMover 1
d2k-datamover2	2298	tcp/udp	D2K DataMover 2
d2k-tapestry1	3393	tcp/udp	D2K Tapestry Client to Server
d2k-tapestry2	3394	tcp/udp	D2K Tapestry Server to Server
d3winosfi	3458	tcp/udp	D3WinOsfi—PickSys
daishi	2870	tcp/udp	daishi—U Memphis
dantz	497	tcp/udp	dantz
darcorp-lm	1679	tcp/udp	darcorp-lm
dasp	439	tcp/udp	dasp
data-insurance	2764	tcp/udp	Data Insurance—Standard
datametrics	1645	tcp/udp	datametrics
datasurfsrv	461	tcp/udp	DataRampSrv—MV Communications Inc.
datasurfsrvsec	462	tcp/udp	DataRampSrvSec—MV Communications Inc.
datex-asn	355	tcp/udp	DATEX-ASN
datusorb	3282	tcp/udp	Datusorb
davinci	2542	tcp/udp	daVinci—TZI
dawn	1908	tcp/udp	Dawn—Travsoft
daytime	13	tcp/udp	Daytime (RFC 867)
dbase	217	tcp/udp	dBASE Unix—Ashton-Tate
dbbrowse	47557	tcp/udp	Databeam Corporation
dberegister	1479	tcp/udp	dberegister
dbm	2345	tcp/udp	dbm—Fujitsu
dbref	2365	tcp/udp	dbref—Fujitsu
dbreporter	1379	tcp/udp	Integrity Solutions
dbsa-lm	1407	tcp/udp	DBSA License Manager
dbstar	1415	tcp/udp	DBStar
dc	2001	tcp	
dca	1456	tcp/udp	DCA
dccm	5679	tcp/udp	Direct Cable Connect Manager—Microsoft
dcp	93	tcp/udp	Device Control Protocol—BBN
dcs	1367	tcp/udp	DCS
dctp	675	tcp/udp	DCTP—Ansa
ddi-tcp-1	8888	tcp/udp	NewsEDGE server TCP (TCP 1)—DeskTop Data
ddi-tcp-2	8889	tcp/udp	Desktop Data TCP 1—DeskTop Data
ddi-tcp-3	8890	tcp/udp	Desktop Data TCP 2—DeskTop Data
ddi-tcp-4	8891	tcp/udp	Desktop Data TCP 3: NESS application—DeskTop Data
ddi-tcp-5	8892	tcp/udp	Desktop Data TCP 4: FARM product—DeskTop Data
ddi-tcp-6	8893	tcp/udp	Desktop Data TCP 5: NewsEDGE/Web application—DeskTop Data
ddi-tcp-7	8894	tcp/udp	Desktop Data TCP 6: COAL application—DeskTop Data
ddm-dfm	447	tcp/udp	DDM-RFM—IBM
ddm-rdb	446	tcp/udp	DDM-RDB—IBM
ddm-ssl	448	tcp/udp	DDM-SSL—IBM
dec_dlm	625	tcp/udp	DEC DLM
decap	403	tcp/udp	decap
decauth	316	tcp/udp	decAuth
decbsrv	579	tcp/udp	decbsrv
decladebug	410	tcp/udp	DECLadebug Remote Debug Protocol

Abbreviation	Port-Decimal	Protocol	Description/Owner
dec-mbadmin	1655	tcp/udp	dec-mbadmin
dec-mbadmin-h	1656	tcp/udp	dec-mbadmin-h
dec-notes	3333	tcp/udp	DEC Notes
dectalk	2007	tcp	
decvms-sysmgt	441	tcp/udp	decvms-sysmgt
dei-icda	618	tcp/udp	DEI-ICDA—Tbaytel
delibo	2562	tcp/udp	Delibo—NovaWiz LTD
deos	76	tcp/udp	Distributed External Object Store-The World Kiosk
descent3	2092	tcp/udp	Descent 3—Outrage
designspace-lm	2325	tcp/udp	Design Space License Management—Ansys
deskshare	1702	tcp/udp	deskshare
deskshare	1702	tcp/udp	deskshare
desktop-dna	2763	tcp/udp	Desktop DNA—MiramarSys
de-spot	2753	tcp/udp	de-spot—DigitalEnvoy
device	801	tcp/udp	
device2	2030	tcp/udp	
deviceshare	552	tcp/udp	deviceshare—Advsyscon
dhcp-failover	647	tcp/udp	DHCP Failover—Process
dhcpv6-client	546	tcp/udp	DHCPv6 Client—DEC
dhcpv6-server	547	tcp/udp	DHCPv6 Server—DEC
diagmond	1508	tcp/udp	diagmond
dialogic-elmd	1945	tcp/udp	dialogic-elmd
dialog-port	2098	tcp/udp	Dialog Port—Cisco
dialpad-voice1	2860	tcp/udp	Dialpad Voice 1—dialpad.com, Inc.
dialpad-voice2	2861	tcp/udp	Dialpad Voice 2—dialpad.com, Inc.
di-ase	3046	tcp/udp	di-ase—Digisle
dicom-iscl	2761	tcp/udp	DICOM ISCL—Siemens
dicom-tls	2762	tcp/udp	DICOM TLS—Siemens
dict	2628	tcp/udp	DICT—U North Carolina
digiman	2362	tcp/udp	digiman
digital-evm	619	tcp/udp	Digital EVM
digital-notary	1335	tcp/udp	Digital Notary Protocol
digital-vrc	466	tcp/udp	digital-vrc
direcpc-dll	1844	tcp/udp	DirecPC-DLL—Hughes Network Systems
direcpc-si	2464	tcp/udp	DirecPC SI—Hughes Network Systems
direcpc-video	1825	tcp/udp	DirecPC Video—Hughes Network Systems
direct	242	tcp/udp	Direct—Peer Direct Inc.
directplay	2234	tcp/udp	DirectPlay—Microsoft
directplaysrvr	47624	tcp/udp	Direct Play Server—Microsoft
directv-catlg	3337	tcp/udp	Direct TV Data Catalog
directvdata	3287	tcp/udp	DIRECTVDATA
directv-soft	3335	tcp/udp	Direct TV Software Updates
directv-tick	3336	tcp/udp	Direct TV Tickers
directv-web	3334	tcp/udp	Direct TV Webcasting
dirgis	2496	tcp/udp	DIRGIS—DIRGE
dis	393	tcp/udp	Data Interpretation System—Metaphor
discard	9	tcp/udp	Discard—ISI
disclose	667	tcp/udp	campaign contribution disclosures - SDR Technologies
discp-client	2601	tcp/udp	discp client—3Com
discp-server	2602	tcp/udp	discp server—3com
distinct	9999	tcp/udp	distinct
distinct32	9998	tcp/udp	Distinct32
di-traceware	3041	tcp/udp	di-traceware—Digisle
dixie	96	tcp/udp	DIXIE Protocol Specification—Umich
dj-ice	5419	tcp/udp	DJ-ICE—Dow Jones
dj-ilm	3362	tcp/udp	DJ ILM
dlip	7201	tcp/udp	DLIP—Rockwell
dls	197	tcp/udp	Directory Location Service—Purdue U
dls	2047	tcp/udp	
dls	2047	tcp/udp	
dls-mon	198	tcp/udp	Directory Location Service Monitor—Purdue U
dls-monitor	2048	tcp/udp	
dls-monitor	2048	tcp/udp	
dlsrap	1973	tcp/udp	Data Link Switching Remote Access Protocol—Cisco
dlsrpn	2065	tcp/udp	Data Link Switch Read Port Number—Radware
dlswpn	2067	tcp/udp	Data Link Switch Write Port Number—Radware
dmdocbroker	1489	tcp/udp	dmdocbroker
dn6-nlm-aud	195	tcp/udp	DNSIX Network Level Module Audit—NAVY.MIL
dn6-smm-red	196	tcp/udp	DNSIX Session Mgt Module Audit Redir—NAVY.MIL

Abbreviation	Port-Decimal	Protocol	Description/Owner
dna	2287	tcp/udp	DNA—Ricochet
dna-cml	436	tcp/udp	DNA-CML—DEC
dnp	20000	tcp/udp	DNP—ACSAtlanta
dnsix	90	tcp/udp	DNSIX Securit Attribute Token Map
docstor	1488	tcp/udp	DocStor
domain	53	tcp/udp	Domain Name Server—ISI
domaintime	9909	tcp/udp	domaintime—GreyWare
donnyworld	1821	tcp/udp	donnyworld
doom	666	tcp/udp	doom Id Software
down	2022	tcp	
dpcp	4099	tcp/udp	DPCP—Demon
dpi-proxy	1795	tcp/udp	dpi-proxy—NETsilicon
dpkeyserv	1780	tcp/udp	dpkeyserv—Omron Software
dpserve	7020	tcp/udp	DP Serve—Hummingbird
dpserveadmin	7021	tcp/udp	DP Serve Admin—Hummingbird
dpsi	315	tcp/udp	DPSI—DeskTopPaging
driveappserver	1930	tcp/udp	Drive AppServer—Demon
drmsfsd	4098	tcp/udp	drmsfsd—Fujitsu
drp	1974	tcp/udp	DRP—Cisco
dsc	3390	tcp/udp	Distributed Service Coordinator—Secant
dsETOS	378	tcp/udp	NEC Corporation
dsf	555	tcp/udp	
dsfgw	438	tcp/udp	dsfgw
dslremote-mgmt	2420	tcp/udp	DSL Remote Management—Westell, Inc
dsmcc-ccp	13822	tcp/udp	DSMCC Channel Change Protocol—ISO/IEC13818-6 MPEG-2 DSM-CC
dsmcc-config	13818	tcp/udp	DSMCC Config—ISO/IEC13818-6 MPEG-2 DSM-CC
dsmcc-download	13821	tcp/udp	DSMCC Download Protocol—ISO/IEC13818-6 MPEG-2 DSM-CC
dsmcc-passthru	13820	tcp/udp	DSMCC Pass-Thru Messages—ISO/IEC13818-6 MPEG-2 DSM-CC
dsmcc-session	13819	tcp/udp	DSMCC Session Messages—ISO/IEC13818-6 MPEG-2 DSM-CC
dsom-server	3053	tcp/udp	dsom-server—PNL Gov.
dsp	33	tcp/udp	Display Support Protocol—dca.mil
dsp3270	246	tcp/udp	Display Systems Protocol—dca.mil
dtag-ste-sb	352	tcp/udp	DTAG (assigned long ago)—Deutsche Telekom
dtk	365	tcp/udp	DTK—Fred Cohen & Assoc.
dtn1	2445	tcp/udp	DTN1
dts	2594	tcp/udp	Data Base Server—Home-Online
dtspcd	6112	tcp/udp	dtspcd—Sun
dvl-activemail	1396	tcp/udp	DVL Active Mail
dwf	1450	tcp/udp	Tandem Distributed Workbench Facility
dwr	644	tcp/udp	dwr—Xerox
dxmessagebase1	2874	tcp/udp	dxmessagebase1—Delphix
dxmessagebase2	2875	tcp/udp	dxmessagebase2—Delphix
dyna-access	3310	tcp/udp	Dyna Access—CornerstoneSoftware
dyna-lm	3395	tcp/udp	Dyna License Manager (Elam)
dyniplookup	3295	tcp/udp	Dynamic IP Lookup—GeoCities
ea1	1791	tcp/udp	EA1
eba	45678	tcp/udp	EBA PRISE
ebinsite	2651	tcp/udp	EBInSite
echo	7	tcp/udp	Echo—ISI
ecnp	2858	tcp/udp	ECNP—Certified Software
edb-server1	1635	tcp/udp	EDB Server 1—Simple Software Solutions
editbench	1350	tcp/udp	Registration Network Protocol—NEXT
edm-adm-notify	3463	tcp/udp	EDM ADM Notify—Novadigm
edm-manager	3460	tcp/udp	EDM Manger—Novadigm
edm-mgr-cntrl	3465	tcp/udp	EDM MGR Cntrl—Novadigm
edm-mgr-sync	3464	tcp/udp	EDM MGR Sync—Novadigm
edm-stager	3461	tcp/udp	EDM Stager—Novadigm
edm-std-notify	3462	tcp/udp	EDM STD Notify—Novadigm
efidiningport	2553	tcp/udp	efidiningport—ExecPC
efi-lm	3392	tcp/udp	EFI License Management—ExecPC
eforward	2181	tcp/udp	eforward—Flagship
efs	520	tcp	extended file name server
egptlm	3328	tcp/udp	Eaglepoint License Manager
eicon-server	1438	tcp/udp	Eicon Security Agent Server
eicon-slp	1440	tcp/udp	Eicon Service Location Protocol
eicon-x25	1439	tcp/udp	Eicon X25 SNA Gateway
elan	1378	tcp/udp	Elan License Manager
elanlm	4346	tcp/udp	ELAN LM—ProjTech
elcsd	704	tcp/udp	errlog copy/server daemon

Abbreviation	Port-Decimal	Protocol	Description/Owner
ellpack	2025	tcp	
elm-momentum	1914	tcp/udp	Elm-Momentum—MDS
elvin_client	2917	tcp/udp	Elvin Client—Pobox
elvin_server	2916	tcp/udp	Elvin Server—Pobox
ema-sent-lm	2526	tcp/udp	EMA License Manager—EMA Denmark
embl-ndt	394	tcp/udp	EMBL Nucleic Data Transfer—Upsala U
emce	2004	udp	CCWS mm conf
emcsymapiport	2707	tcp/udp	EMCSYMAPIPORT
emfis-cntl	141	tcp/udp	EMFIS Control Service—ISI
emfis-data	140	tcp/udp	EMFIS Data Service—ISI
empire-empuma	1691	tcp/udp	empire-empuma—Empire Technology
encore	1740	tcp/udp	encore—PRI Automation
e-net	3286	tcp/udp	E-Net—Eent
enl	1804	tcp/udp	ENL—Seagate Software
enl-name	1805	tcp/udp	ENL-Name—Seagate Software
enpc	3289	tcp/udp	ENPC—Epson
entexthigh	12002	tcp/udp	IBM Enterprise Extender SNA COS High Priority
entextlow	12004	tcp/udp	IBM Enterprise Extender SNA COS Low Priority
entextmed	12003	tcp/udp	IBM Enterprise Extender SNA COS Medium Priority
entextnetwk	12001	tcp/udp	IBM Enterprise Extender SNA COS Network Priority
entextxid	12000	tcp/udp	IBM Enterprise Extender SNA XID Exchange
entomb	775	tcp	
entp	1865	tcp/udp	ENTP—Epson
entrust-aaas	680	tcp/udp	entrust-aaas
entrust-aams	681	tcp/udp	entrust-aams
entrust-ash	710	tcp/udp	Entrust Administration Service Handler
entrust-kmsh	709	tcp/udp	Entrust Key Management Service Handler
entrust-sps	640	tcp/udp	entrust-sps
entrusttime	309	tcp/udp	EntrustTime
epicon	2912	tcp/udp	Epicon
epmap	135	tcp/udp	DCE endpoint resolution—HP
equationbuilder	1351	tcp/udp	Digital Tool Works (MIT)
ergolight	2109	tcp/udp	Ergolight—Ledalite
eristwoguns	2650	tcp/udp	eristwoguns—NetPro
erpc	121	tcp/udp	Encore Expedited Remote Pro.Call
escp-ip	621	tcp/udp	ESCP—Pobox
es-elmd	1822	tcp/udp	es-elmd
esinstall	5599	tcp/udp	Enterprise Security Remote Install—Axent
esl-lm	1455	tcp/udp	ESL License Manager
esmagent	5601	tcp/udp	Enterprise Security Agent—Axent
esmmanager	5600	tcp/udp	Enterprise Security Manager—Axent
esp-encap	2797	tcp/udp	esp-encap—F-Secure Corp.
esp-lm	3383	tcp/udp	Enterprise Software Products License Manager
esps-portal	2867	tcp/udp	esps-portal
esri_sde	5151	tcp/udp	ESRI SDE Instance
esro-emsdp	642	tcp/udp	ESRO-EMSDP V1.3—Neda
esro-gen	259	tcp/udp	Efficient Short Remote Operations—Neda Communications, Inc.
essbase	1423	tcp/udp	Essbase Arbor Software
essweb-gw	1772	tcp/udp	EssWeb Gateway—Hyperion Inc.
etftp	1818	tcp/udp	Enhanced Trivial File Transfer Protocol—Mitre
etp	1798	tcp/udp	Event Transfer Protocol—HP
ets	1569	tcp/udp	ets
eudora-set	592	tcp/udp	Eudora Set—Qualcomm
evb-elm	1504	tcp/udp	EVB Software Engineering License Manager
event_listener	3017	tcp/udp	Event Listener—Novell
everydayrc	2782	tcp/udp	everydayrc—Bluemoon Interactive
evtp	2834	tcp/udp	EVTP—Solution Soft
evtp-data	2835	tcp/udp	EVTP-DATA—Solution Soft
ewnn	2674	tcp/udp	ewnn—Omron Software
excerpt	5400	tcp/udp	Excerpt Search—Alma
excerpts	5401	tcp/udp	Excerpt Search Secure—Alma
exec	512	tcp	remote process execution
exlm-agent	3002	tcp/udp	EXLM Agent—Clemson U
expresspay	2755	tcp/udp	Express Pay—Netcom
extensis	2666	tcp/udp	extensis
eye2eye	1948	tcp/udp	eye2eye—Iosoftware
eyelink	589	tcp/udp	EyeLink—U Tornonto
f5-globalsite	2792	tcp/udp	f5-globalsite—F5 Networks, Inc.
f5-iquery	4353	tcp/udp	F5 iQuery

Abbreviation	Port-Decimal	Protocol	Description/Owner
facelink	1915	tcp/udp	FACELINK—Hiscom
facilityview	1561	tcp/udp	facilityview—Particle Measuring Systems Inc.
facsys-ntp	2514	tcp/udp	Facsys NTP
facsys-router	2515	tcp/udp	Facsys Router
fastlynx	2689	tcp/udp	FastLynx—Dave Sewell
fast-rem-serv	2495	tcp/udp	Fast Remote Services—Adaptec
fatpipe	3353	tcp/udp	FATPIPE—Ragula
fatserv	347	tcp/udp	Fatmen Server
faximum	7437	tcp/udp	Faximum
faxportwinport	1620	tcp/udp	faxportwinport—LanSource
fc-cli	1371	tcp/udp	Fujitsu Config Protocol
fc-faultnotify	2819	tcp/udp	FC Fault Notification—HP
fcmsys	2344	tcp/udp	fcmsys—Fujitsu
fcopy-server	5745	tcp/udp	fcopy-server—SoftLink
fcopys-server	5746	tcp/udp	fcopys-server—SoftLink
fcp	510	tcp/udp	FirstClass Protocol—Softarc
fcp-addr-srvr1	5500	tcp/udp	fcp-addr-srvr1—AC
fcp-addr-srvr2	5501	tcp/udp	fcp-addr-srvr2—AC
fcp-cics-gw1	5504	tcp/udp	fcp-cics-gw1—AC
fcp-srvr-inst1	5502	tcp/udp	fcp-srvr-inst1—AC
fcp-srvr-inst2	5503	tcp/udp	fcp-srvr-inst2—AC
fcp-udp	810	tcp/udp	FCP—Softarc
fc-ser	1372	tcp/udp	Fujitsu Config Protocol
femis	1776	tcp/udp	Federal Emergency Management Information System—pl.gov
fg-fps	3293	tcp/udp	fg-fps—Freegate
fg-gip	3294	tcp/udp	fg-gip—Freegate
fg-sysupdate	6550	tcp/udp	fg-sysupdate—Freegate
fhc	1499	tcp/udp	Federico Heinz Consultora
fhsp	1807	tcp/udp	Fujitsu Hot Standby Protocol
filecast	3401	tcp/udp	filecast—Pair
filenet-nch	32770	tcp/udp	Filenet NCH
filenet-rpc	32769	tcp/udp	Filenet RPC
filenet-tms	32768	tcp/udp	Filenet TMS
findviatv	3350	tcp/udp	FINDVIATV—8x8
finger	79	tcp/udp	Finger—Rutgers U
finle-lm	1784	tcp/udp	Finle License Manager
firefox	1689	tcp/udp	firefox
firepower	2615	tcp/udp	firepower
firstcall42	2673	tcp/udp	First Call 42—TFN
fjappmgrbulk	2510	tcp/udp	fjappmgrbulk—Fujitsu
fjdmimgr	9374	tcp/udp	fjdmimgr—Fujitsu
fj-hdnet	1717	tcp/udp	fj-hdnet—Fujitsu
fjicl-tep-a	1901	tcp/udp	Fujitsu ICL Terminal Emulator Program A
fjicl-tep-b	1902	tcp/udp	Fujitsu ICL Terminal Emulator Program B
fjicl-tep-c	1904	tcp/udp	Fujitsu ICL Terminal Emulator Program C
fjinvmgr	9396	tcp/udp	fjinvmgr—Fujitsu
fjippol-cnsl	2749	tcp/udp	Fujitsu
fjippol-polsvr	2748	tcp/udp	Fujitsu
fjippol-port1	2750	tcp/udp	Fujitsu
fjippol-port2	2751	tcp/udp	Fujitsu
fjippol-swrly	2747	tcp/udp	Fujitsu
fjitsuappmgr	2425	tcp/udp	Fujitsu App Manager
fjmpcm	2975	tcp/udp	Fujitsu Configuration Management Service
fjmpjps	1873	tcp/udp	jps—Fujitsu
fjmpss	2509	tcp/udp	fjmpss—Fujitsu
fjris	1783	tcp/udp	Fujitsu Remote Install Service
fjsv-gssagt	3035	tcp/udp	FJSV gssagt—Fujitsu
fjswapsnp	1874	tcp/udp	swapsnp—Fujitsu
flashmsg	2884	tcp/udp	Flash Msg—ZinkNet
flexlm	744	tcp/udp	Flexible License Manager—Olivetti
flex-lm	27000-27009	tcp/udp	FLEX LM (1-10)—Globes
fln-spx	221	tcp/udp	Berkeley rlogind with SPX auth
flukeserver	2359	tcp/udp	FlukeServer—Forte Networks
fmpro-fdal	2399	tcp/udp	"FileMaker, Inc. - Data Access Layer"
fmpro-internal	5003	tcp/udp	FileMaker, Inc. - Proprietary transport
fodms	7200	tcp/udp	FODMS FLIP—Rockwell
foliocorp	2242	tcp/udp	Folio Remote Server
font-service	7100	tcp/udp	X Font Service—MIT
foresyte-clear	5407	tcp/udp	Foresyte-Clear

Abbreviation	Port-Decimal	Protocol	Description/Owner
foresyte-sec	5408	tcp/udp	Foresyte-Sec
fryeserv	2788	tcp/udp	NetWare Loadable Module - Seagate Software
fsportmap	4349	tcp/udp	File System Port Map—Sarnoff
ftp	21	tcp/udp	File Transfer [Control]—ISI
ftp-agent	574	tcp/udp	FTP Software Agent System
ftp-data	20	tcp/udp	File Transfer [Default Data]—ISI
ftps	990	tcp/udp	ftp protocol, control, over TLS/SSL—Consensus
ftps-data	989	tcp/udp	ftp protocol, data, over TLS/SSL—Consensus
ftrapid-1	1746	tcp/udp	ftrapid-1—Unisys
ftrapid-2	1747	tcp/udp	ftrapid-2—Unisys
ft-role	2429	tcp/udp	FT-ROLE—CLR
ftsrv	1359	tcp/udp	FTSRV—BITNET
fujitsu-dev	747	tcp/udp	Fujitsu Device Control
fujitsu-dtc	1513	tcp/udp	Fujitsu DTC
fujitsu-dtcns	1514	tcp/udp	Fujitsu DTCNS
fujitsu-mmpdc	1657	tcp/udp	fujitsu-mmpdc
fujitsu-neat	3382	tcp/udp	Fujitsu Network Enhanced Antitheft function
funk-dialout	2909	tcp/udp	Funk Dialout
funk-license	1787	tcp/udp	funk-license
funk-logger	1786	tcp/udp	funk-logger
funkproxy	1505	tcp/udp	Funk Software, Inc.
futrix	2358	tcp/udp	Futrix—IBM
fxp	2849	tcp/udp	FXP—Oanda Inc.
g5m	2732	tcp/udp	G5M—ACM
gacp	190	tcp/udp	Gateway Access Control Protocol-LANL.GOV
gadgetgate1way	2677	tcp/udp	Gadget Gate 1 Way—InternetGadgets
gadgetgate2way	2678	tcp/udp	Gadget Gate 2 Way—InternetGadgets
galaxy-server	3051	tcp/udp	Galaxy Server—Galaxy Verticals
gamegen1	1738	tcp/udp	GameGen1—MultiGen-Paradigm Inc.
gamelobby	2914	tcp/udp	Game Lobby—Pipex
gandalf-lm	1421	tcp/udp	Gandalf License Manager
ganymede	2220	tcp/udp	Ganymede
ganymede-endpt	10115	tcp/udp	Ganymede Endpoint
garcon	999	tcp	
gat-lmd	1708	tcp/udp	gat-lmd—global-tech.com
gbjd816	2626	tcp/udp	gbjd816—George Balesta
gcmonitor	2660	tcp/udp	GC Monitor—Geodesic
gdomap	538	tcp/udp	gdomap—Brainstorm
gdp-port	1997	tcp/udp	cisco Gateway Discovery Protocol
gds_db	3050	tcp/udp	gds_db—InterBase
gemini-lm	1590	tcp/udp	gemini-lm—OpenGVS
geneous	3381	tcp/udp	Geneous—HisTech
genie	402	tcp/udp	Genie Protocol
genie-lm	1453	tcp/udp	Genie License Manager
geniuslm	3005	tcp/udp	Genius License Manager
genrad-mux	176	tcp/udp	GENRAD-MUX
ggf-ncp	678	tcp/udp	GNU Gereration Foundation NCP
gilatskysurfer	3013	tcp/udp	Gilat Sky Surfer
ginad	634	tcp/udp	ginad—Cal State U
giop	2481	tcp/udp	Oracle GIOP
giop-ssl	2482	tcp/udp	Oracle GIOP SSL
glishd	2833	tcp/udp	glishd—cv.nrao.edu
global-dtserv	1774	tcp/udp	global-dtserv—Global Communications
global-wlink	1909	tcp/udp	Global World Link—Global Communications
globe	2002	tcp/udp	
globmsgsvc	2519	tcp/udp	globmsgsvc—Intel
glogger	2033	tcp/udp	
goahead-fldup	3057	tcp/udp	GoAhead FldUp
goldleaf-licman	1401	tcp/udp	Goldleaf License Manager
go-login	491	tcp/udp	go-login—Graphn
gopher	70	tcp/udp	Gopher—UMN
gppitnp	103	tcp/udp	Genesis Point-to-Point Trans Net
gprs-data	3386	tcp/udp	GPRS Data—ETSI
graphics	41	tcp/udp	Graphics
gridgen-elmd	1542	tcp/udp	gridgen-elmd
griffin	2458	tcp/udp	griffin—Unisys
groove	2492	tcp/udp	GROOVE
groupwise	1677	tcp/udp	groupwise—Novell

Abbreviation	Port-Decimal	Protocol	Description/Owner
gsi	1850	tcp/udp	GSI
gsiftp	2811	tcp/udp	GSI FTP—U Illinois Urbana-Champaign
gss-http	488	tcp/udp	gss-http—Galaxy Verticals
gss-xlicen	128	tcp/udp	GSS X License Verification
g-talk	2421	tcp/udp	G-Talk—Four-Sight
gtegsc-lm	1452	tcp/udp	GTE Government Systems License Man
gte-samp	2643	tcp/udp	GTE-SAMP
guibase	9321	tcp/udp	guibase—Fujitsu
gv-us	1369	tcp/udp	GlobalView to Unix Shell—Fuji-Xerox
gw	3010	tcp/udp	Telerate Workstation—Dow Jones
gwen-sonya	2778	tcp/udp	Gwen-Sonya—Internet Connect Inc.
gwha	1383	tcp/udp	GW Hannaway Network License Manager
gxtelmd	2356	tcp/udp	GXT License Managemant
h2250-annex-g	2099	tcp/udp	H.225.0 Annex G—Vocaltec
h263-video	2979	tcp/udp	H.263 Video Streaming —ACM
h323callsigalt	11720	tcp/udp	h323 Call Signal Alternate—Cisco
h323gatedisc	1718	tcp/udp	h323gatedisc—Intel
h323gatestat	1719	tcp/udp	h323gatestat—Intel
h323hostcall	1720	tcp/udp	h323hostcall—Intel
h323hostcallsc	1300	tcp/udp	H323 Host Call Secure—Intel
hacl-cfg	5302	tcp/udp	# HA cluster configuration—HP
hacl-gs	5301	tcp/udp	# HA cluster general services—HP
hacl-hb	5300	tcp/udp	# HA cluster heartbeat—HP
hacl-local	5304	tcp/udp	# HA Cluster Commands—HP
hacl-probe	5303	tcp/udp	# HA cluster probing—HP
hacl-test	5305	tcp/udp	# HA Cluster Test
ha-cluster	694	tcp/udp	ha-cluster—Bell Labs
hagel-dump	3036	tcp/udp	Hagel DUMP
hao	2245	tcp/udp	HaO—HAO
hap	661	tcp/udp	HAP—Uroam
harlequinorb	3672	tcp/udp	harlequinorb
harp	1816	tcp/udp	HARP—Portland State U
hassle	375	tcp/udp	Hassle—U Basel
hbci	3000	tcp/udp	HBCI—IBM
hb-engine	1703	tcp/udp	hb-engine—Zoom Inc.
hcp-wismar	686	tcp/udp	Hardware Control Protocol Wismar—Liverpool U
hdap	263	tcp/udp	HDAP—Zyxel
hdl-srv	2641	tcp/udp	HDL Server—CNRI
health-polling	1161	tcp/udp	Health Polling—Triosoftware
health-trap	1162	tcp/udp	Health Trap—Triosoftware
hecmtl-db	1551	tcp/udp	HECMTL-DB
hello	1789	tcp/udp	hello—U Illinois Chicago
hems	151	tcp/udp	HEMS
here-lm	1409	tcp/udp	Here License Manager
hermes	1248	tcp/udp	
high-criteria	2467	tcp/udp	High Criteria
hiq	1410	tcp/udp	HiQ License Manager
hivep	12172	tcp/udp	HiveP—Skenbe
hks-lm	1722	tcp/udp	HKS License Manager
hl7	2575	tcp/udp	HL7
hlserver	1947	tcp/udp	hlserver-Fast-ag
hlserver	3047	tcp/udp	Fast Security HL Server
hmmp-ind	612	tcp/udp	HMMP Indication—Microsoft
hmmp-op	613	tcp/udp	HMMP Operation—Microsoft
hnmp	6790	tcp/udp	HNMP—NASA
homesteadglory	2597	tcp/udp	Homestead Glory
honyaku	2744	tcp/udp	honyaku—Omron Software
hostname	101	tcp/udp	NIC Host Name Server—ISI
hosts2-ns	81	tcp/udp	HOSTS2 Name Server-S1.gov
houdini-lm	1715	tcp/udp	houdini-lm—Side Effects SW
hp-3000-telnet	2564	tcp	HP 3000 NS/VT block mode telnet
hp-alarm-mgr	383	tcp/udp	hp performance data alarm manager
hp-clic	3384	tcp/udp	Cluster Management Services—HP
hp-clic	3384	udp	Hardware Management—HP
hp-collector	381	tcp/udp	hp performance data collector
hp-device-disc	3329	tcp/udp	HP Device Disc
hp-hcip	1782	tcp/udp	hp-hcip
hp-hcip-gwy	1803	tcp/udp	HP-HCIP-GWY

Abbreviation	Port-Decimal	Protocol	Description/Owner
hp-managed-node	382	tcp/udp	hp performance data managed node
hp-nnm-data	2690	tcp/udp	HP NNM Embedded Database
hpoms-ci-lstn	5403	tcp/udp	HPOMS-CI-LSTN
hpoms-dps-lstn	5404	tcp/udp	HPOMS-DPS-LSTN
hpppssvr	2448	tcp/udp	hpppsvr—HP
hp-san-mgmt	3037	tcp/udp	HP SAN Mgmt
hp-sca	19411	tcp/udp	hp-sca
hp-sco	19410	tcp/udp	hp-sco
hpstgmgr	2600	tcp/udp	HPSTGMGR—HP
hpstgmgr2	2715	tcp/udp	HPSTGMGR2—HP
hp-webadmin	1188	tcp/udp	HP Web Admin
hs-port	2570	tcp/udp	HS Port—NETect
hsrp	1985	tcp/udp	Hot Standby Router Protocol—Cisco
htcp	4827	tcp/udp	HTCP—VIX
http	80	tcp/udp	World Wide Web HTTP-World Wide Web Consortium
http-alt	591	tcp/udp	FileMaker, Inc. - HTTP Alternate (see Port 80)
http-alt	8008	tcp/udp	HTTP Alternate—World Wide Web Consortium
http-alt	8080	tcp/udp	HTTP Alternate (see port 80)—Percpt
http-mgmt	280	tcp/udp	http-mgmt—HP-UnitedKingdom
http-rpc-epmap	593	tcp/udp	HTTP RPC Ep Map—Microsoft
https	443	tcp/udp	http protocol over TLS/SSL—Netscape
husky	1310	tcp/udp	Husky—Zang
hybrid	1424	tcp/udp	Hybrid Encryption Protocol
hybrid-pop	473	tcp/udp	hybrid-pop
hypercube-lm	1577	tcp/udp	hypercube-lm
hyper-g	418	tcp/udp	Hyper-G—IICM
hyperwave-isp	692	tcp/udp	Hyperwave-ISP
iad1	1030	tcp/udp	BBN IAD
iad2	1031	tcp/udp	BBN IAD
iad3	1032	tcp/udp	BBN IAD
iafdbase	480	tcp/udp	iafdbase
iafserver	479	tcp/udp	iafserver—Solect
iapp	2313	tcp/udp	IAPP (Inter Access Point Protocol)—Lucent
iasd	432	tcp/udp	IASD—Encore
iatp-highpri	6998	tcp/udp	IATP-highPri—MCI
iatp-normalpri	6999	tcp/udp	IATP-normalPri—MCI
iberiagames	1726	tcp/udp	IBERIAGAMES
ibm_wrless_lan	1461	tcp/udp	IBM Wireless LAN
ibm-abtact	1586	tcp/udp	ibm-abtact—Stanford U
ibm-app	385	tcp/udp	IBM Application
ibm-cics	1435	tcp/udp	IBM CISC
ibm-db2	523	tcp/udp	IBM-DB2
ibm-dial-out	3267	tcp/udp	IBM Dial Out
ibm-dt-2	1792	tcp/udp	ibm-dt-2
ibm-mqseries	1414	tcp/udp	IBM MQSeries
ibm-mqseries2	1881	tcp/udp	IBM MQSeries
ibm-pps	1376	tcp/udp	IBM Person to Person Software
ibm-res	1405	tcp/udp	IBM Remote Execution Starter
ibp	2572	tcp/udp	IBP—Activ
ibus	8733	tcp/udp	iBus—SoftWired
ica	1494	tcp/udp	ica
icabrowser	1604	tcp/udp	icabrowser—Citrix
icad-el	425	tcp/udp	ICAD
ici	2200	tcp/udp	ICI—Unisys
iclcnet_svinfo	887	tcp/udp	ICL coNETion server info
iclcnet-locate	886	tcp/udp	ICL coNETion locate server
iclpv-dm	1389	tcp/udp	Document Manager—ICL
iclpv-nlc	1394	tcp/udp	Network Log Client—ICL
iclpv-nls	1393	tcp/udp	Network Log Server—ICL
iclpv-pm	1392	tcp/udp	Print Manager—ICL
iclpv-sas	1391	tcp/udp	Storage Access Server—ICL
iclpv-sc	1390	tcp/udp	Storage Controller—ICL
iclpv-wsm	1395	tcp/udp	PC Workstation Manager software—ICL
icl-twobase1	25000	tcp/udp	icl-twobase1
icl-twobase10	25009	tcp/udp	icl-twobase10
icl-twobase2	25001	tcp/udp	icl-twobase2
icl-twobase3	25002	tcp/udp	icl-twobase3
icl-twobase4	25003	tcp/udp	icl-twobase4

Abbreviation	Port-Decimal	Protocol	Description/Owner
icl-twobase5	25004	tcp/udp	icl-twobase5
icl-twobase6	25005	tcp/udp	icl-twobase6
icl-twobase7	25006	tcp/udp	icl-twobase7
icl-twobase8	25007	tcp/udp	icl-twobase8
icl-twobase9	25008	tcp/udp	icl-twobase9
icon-discover	2799	tcp/udp	ICON Discover
icpv2	3130	tcp/udp	ICPv2—NLANr
icslap	2869	tcp/udp	ICSLAP—Microsoft
ida-discover1	5741	tcp/udp	IDA Discover Port 1—I-Data
ida-discover2	5742	tcp/udp	IDA Discover Port 2—I-Data
idcp	2326	tcp/udp	IDCP—KGC
ideesrv	2337	tcp/udp	ideesrv—Ideetique
iden-ralp	1725	tcp/udp	iden-ralp—Motorola Inc.
ident	113	tcp	
idfp	549	tcp/udp	IDFP—Kovi
idotdist	2590	tcp/udp	idotdist—Invino
iec870-5-104	2404	tcp/udp	IEC870-5-104—Sat-Automation
ieee-mms	651	tcp/udp	IEEE MMS—SGI
ies-lm	1443	tcp/udp	Integrated Engineering Software
ife_icorp	5165	tcp/udp	ife_1corp—Bull
ifor-protocol	1515	tcp/udp	ifor-protocol
igcp	2801	tcp/udp	IGCP—Codemasters
igi-lm	1404	tcp/udp	Infinite Graphics License Manager
ii-admin	3006	tcp/udp	Instant Internet Admin—Bay Networks
iims	4800	tcp/udp	Icona Instant Messenging System
iiop	535	tcp/udp	iiop—DEC
ill	1611	tcp/udp	Inter Library Loan—University College Dublin
ilss	4802	tcp/udp	Icona License System Server
imagequery	2239	tcp/udp	Image Query—Numinous
imap	143	tcp/udp	Internet Message Access Protocol—U Washington
imap3	220	tcp/udp	Interactive Mail Access Protocol v3—Stanford U
imap4-ssl	585	tcp/udp	IMAP4+SSL (use 993 instead)—U Washington
imaps	993	tcp/udp	imap4 protocol over TLS/SSL—Consensus
imdocsvc	2637	tcp/udp	Import Document Service—NetRight
impera	1710	tcp/udp	impera
imsldoc	2035	tcp/udp	
imsp	406	tcp/udp	Interactive Mail Support Protocol—CMU
imtc-map	2202	tcp/udp	Multimedia Teleconferencing Cosortium—Databeam
imtc-mcs	1503	tcp/udp	Databeam
inbusiness	244	tcp/udp	inbusiness—Intel
indx-dds	2454	tcp/udp	IndX-DDS
i-net-2000-npr	5069	tcp/udp	I/Net 2000-NPR—CSIControls
influence	3345	tcp/udp	Influence—Topia
info-aps	6400	tcp/udp	Seagate Software
info-cachesvr	6403	tcp/udp	Seagate Software
infocrypt	2233	tcp/udp	INFOCRYPT—Isolation
info-eventsvr	6402	tcp/udp	Seagate Software
info-filesvr	6404	tcp/udp	Seagate Software
infolibria	2319	tcp/udp	InfoLibria
infoman	1451	tcp/udp	IBM Information Management
infomover	2854	tcp/udp	InfoMover—EMC2
info-pagesvr	6405	tcp/udp	Seagate Software
info-processvr	6406	tcp/udp	Seagate Software
informatik-lm	1428	tcp/udp	Informatik License Manager
infoseek	414	tcp/udp	InfoSeek
info-was	6401	tcp/udp	Seagate Software
ingreslock	1524	tcp/udp	ingres
ingres-net	134	tcp/udp	INGRES-NET Service
initlsmsad	2793	tcp/udp	initlsmsad—Compaq
innosys	1412	tcp/udp	InnoSys
innosys-acl	1413	tcp/udp	Innosys-ACL
inova-ip-disco	2716	tcp/udp	Inova IP Disco
insitu-conf	1490	tcp/udp	insitu-conf
inspect	1602	tcp/udp	inspect—DEC
instl_bootc	1068	tcp/udp	Installation Bootstrap Proto. Cli.—HP
instl_boots	1067	tcp/udp	Installation Bootstrap Proto. Serv.—HP
intecourier	495	tcp/udp	intecourier—Intecom
integral	3459	tcp/udp	Integral

Abbreviation	Port-Decimal	Protocol	Description/Owner
integra-sme	484	tcp/udp	Integra Software Management Environment
intel_rci	24386	tcp/udp	Intel RCI
intellistor-lm	1539	tcp/udp	Intellistor License Manager
interbase	2041	tcp/udp	
interhdl_elmd	1454	tcp/udp	interHDL License Manager
interintelli	2633	tcp/udp	InterIntelli
interlingua	2414	tcp/udp	Interlingua—Virtual-Unlimited
interpathpanel	2652	tcp/udp	InterPathPanel
interserver	3060	tcp/udp	interserver—InterBase
intersys-cache	1972	tcp/udp	Cache
intersys-cache	1972	tcp/udp	Cache
intrastar	1907	tcp/udp	IntraSTAR—IntraSTAR
intrinsa	503	tcp/udp	Intrinsa
intuitive edge	1355	tcp/udp	Intuitive Edge—Ohio State U
intv	1585	tcp/udp	intv—Claddagh Films
invision	1641	tcp/udp	InVision—CSK Software
invokator	2006	tcp	
ioc-sea-lm	1579	tcp/udp	ioc-sea-lm
ipass	2549	tcp/udp	IPASS—Pass
ipcd	576	tcp/udp	ipcd—Ipass
ipcserver	600	tcp/udp	Sun IPC server
ipdd	578	tcp/udp	ipdd—Ipass
ipp	631	tcp/udp	IPP (Internet Printing Protocol)—Xerox
ipsendmsg	1992	tcp/udp	IPsendmsg (unassigned but in use)—Sidney U
ipx	213	tcp/udp	IPX—Novell
iqserver	2527	tcp/udp	IQ Server—AKBS
irc	194	tcp/udp	Internet Relay Chat Protocol—Oulun Tliopisto
ircs	994	tcp/udp	irc protocol over TLS/SSL—Consensus
irc-serv	529	tcp/udp	IRC-SERV—Acrux
ircu	6665-6669	tcp/udp	IRCU—Acrux
irdg-post	2632	tcp/udp	IRdg Post
irdmi	8000	tcp/udp	iRDMI—Intel
irdmi2	7999	tcp/udp	iRDMI2—Intel
irisa	11000	tcp/udp	IRISA—DataMedia
is99c	379	tcp/udp	TIA/EIA/IS-99 modem client—Qualcomm
is99s	380	tcp/udp	TIA/EIA/IS-99 modem server—Qualcomm
isakmp	500	tcp/udp	isakmp—NSCS-Military
isg-uda-server	2551	tcp/udp	ISG UDA Server
isi-gl	55	tcp/udp	ISI Graphics Language
isis	2042	tcp/udp	isis
isis-am	1642	tcp/udp	isis-am
isis-ambc	1643	tcp/udp	isis-ambc
isis-ambc	1643	tcp/udp	isis-ambc
isis-bcast	2043	tcp/udp	isis-bcast
ismaeasdaqlive	1949	tcp/udp	ISMA Easdaq Live
ismaeasdaqtest	1950	tcp/udp	ISMA Easdaq Test
ismserver	9500	tcp/udp	ismserver—Micro Muse
isode-dua	17007	tcp/udp	
iso-ill	499	tcp/udp	ISO ILL Protocol—U Copenhagen
iso-ip	147	tcp/udp	ISO-IP—Data Broadcasting Corporation
iso-tp0	146	tcp/udp	ISO-IP0—Data Broadcasting Corporation
iso-tsap	102	tcp/udp	ISO-TSAP Class 0—Data Broadcasting Corporation
iso-tsap-c2	399	tcp/udp	ISO Transport Class 2 Non-Control over TCP—DEC
ispipes	2853	tcp/udp	ISPipes—EMC2
issd	1600	tcp/udp	
isysg-lm	1609	tcp/udp	isysg-lm
ita-agent	5051	tcp/udp	ITA Agent—Axent
ita-manager	5052	tcp/udp	ITA Manager—Axent
itinternet	2691	tcp/udp	IT Internet
itm-lm	2828	tcp/udp	ITM License Manager
itm-mcell-s	828	tcp/udp	itm-mcell-s—Itmasters
itm-mcell-u	1828	tcp/udp	itm-mcell-u—ITmasters
ito-e-gui	2531	tcp/udp	ITO-E GUI—HP
itose	4348	tcp/udp	ITOSE—HP
itscomm-ns	1573	tcp/udp	itscomm-ns—IBM
itu-sccp-ss7	14001	tcp/udp	ITU SCCP (SS7)—Ericsson
ivsd	2241	tcp/udp	IVS Daemon—INRIA
ivs-video	2232	tcp/udp	IVS Video default—INRIA

Abbreviation	Port-Decimal	Protocol	Description/Owner
iwec	4801	tcp/udp	Icona Web Embedded Chat
i-zipqd	13160	tcp/udp	I-ZIPQD—QuarterDeck
jacobus-lm	1578	tcp/udp	Jacobus License Manager
jaleosnd	1623	tcp/udp	jaleosnd
jargon	148	tcp/udp	Jargon—Bearnet
jbroker	2506	tcp/udp	jbroker—objectScape
jcp	19541	tcp/udp	JCP Client—JCI
jdatastore	2508	tcp/udp	JDataStore—Inprise
jediserver	2406	tcp/udp	JediServer—NCR
jerand-lm	1810	tcp/udp	Jerand License Manager
jetform	1706	tcp/udp	jetform
jetformpreview	2097	tcp/udp	Jet Form Preview
jini-discovery	4160	tcp/udp	Jini Discovery—Sun
j-lan-p	2808	tcp/udp	J-LAN-P—JDL Internet Services
jlicelmd	1567	tcp/udp	jlicelmd
jmact3	6961	udp	JMACT3
jmb-cds1	8900	tcp/udp	JMB-CDS 1—Creative Design
jmb-cds2	8901	tcp/udp	JMB-CDS 2—Creative Design
jmevt2	6962	tcp/udp	jmevt2—Fujitsu
journee	3042	tcp/udp	journee
jprinter	5309	tcp/udp	J Printer—Bristol
jt400	3470	tcp/udp	jt400-IBM
jt400-ssl	3471	tcp/udp	jt400-ssl—IBM
ka0wuc	2822	tcp/udp	ka0wuc
kali	2213	tcp/udp	Kali—U of Georgia
kana	2656	tcp/udp	Kana
kastenchasepad	2918	tcp/udp	Kasten Chase Pad—Kasten Chase
kastenxpipe	36865	tcp/udp	KastenX Pipe—Kasten Chase
kentrox-prot	2502	tcp/udp	Kentrox Protocol
kerberos	88	tcp/udp	Kerberos—ISI
kerberos-adm	749	tcp/udp	kerberos administration—eSpecials
kerberos-iv	750	udp	kerberos version iv—eSpecials
kermit	1649	tcp/udp	kermit—Columbia U
keyserver	584	tcp/udp	Key Server—Systemics
keyshadow	19315	tcp/udp	Key Shadow for SASSAFRAS
keysrvr	19283	tcp/udp	Key Server for SASSAFRAS
kis	186	tcp/udp	KIS Protocol
klogin	543	tcp/udp	
kmscontrol	1773	tcp/udp	KMSControl—KMSystems Inc.
knet-cmp	157	tcp/udp	KNET/VM Command/Message Protocol—XYLOGICS
kofax-svr	2424	tcp/udp	KOFAX-SVR
konshus-lm	2294	tcp/udp	Konshus License Manager (FLEX)
kpasswd	464	tcp/udp	kpasswd—MIT
krb524	4444	tcp/udp	KRB524—ISI
kryptolan	398	tcp/udp	Kryptolan—Sectra
kshell	544	tcp/udp	krcmd
kti-icad-srvr	6701	tcp/udp	KTI/ICAD Nameserver—KTIworld
kv-agent	3361	tcp/udp	KV Agent—CIPartners
kv-server	3360	tcp/udp	KV Server—CIPartners
l2f	1701	tcp/udp	l2f—Cisco
l2tp	1701	tcp/udp	l2tp—Cisco
l3-exprt	2840	tcp/udp	l3-exprt—L3 Security
l3-hawk	2842	tcp/udp	l3-hawk—L3 Security
l3-hbmon	2370	tcp/udp	L3-HBMon—L-3security
l3-ranger	2841	tcp/udp	l3-ranger—L3 Security
labrat	2560	tcp/udp	labrat—IBM
lam	2040	tcp/udp	
la-maint	51	tcp/udp	IMP Logical Address Maintenance—Timeplex
lan900_remote	2395	tcp/udp	LAN900 Remote—DigiTechINC
lanner-lm	4547	tcp/udp	Lanner License Manager
lanserver	637	tcp/udp	lanserver—IBM
lansource	1485	tcp/udp	LANSource
lansurveyor	4347	tcp/udp	LAN Surveyor—Neon
lanyon-lantern	1682	tcp/udp	lanyon-lantern
laplink	1547	tcp/udp	laplink—Traveling Software
lavenir-lm	3373	tcp/udp	Lavenir License Manager
lazy-ptop	7099	tcp/udp	lazy-ptop—Scitex
lbc-control	2780	tcp/udp	LBC Control—Fujitsu

Abbreviation	Port-Decimal	Protocol	Description/Owner
lbc-measure	2815	tcp/udp	LBC Measurement—Fujitsu
lbc-sync	2779	tcp/udp	LBC Sync—Fujitsu
lbc-watchdog	2816	tcp/udp	LBC Watchdog—Fujitsu
lbf	2466	tcp/udp	Load Balance Forwarding—PFU Direct
lbm	2465	tcp/udp	Load Balance Management—PFU Direct
ldap	389	tcp/udp	Lightweight Directory Access Protocol—U Michigan
ldaps	636	tcp/udp	ldap protocol over TLS/SSL (was sldap)—Xcert
ldp	646	tcp/udp	LDP—Cisco
lecroy-vicp	1861	tcp/udp	LeCroy VICP
legent-1	373	tcp/udp	Legent Corporation
legent-2	374	tcp/udp	Legent Corporation
liberty-lm	1496	tcp/udp	liberty-lm
licensedaemon	1986	tcp/udp	cisco license management
LiebDevMgmt_A	3029	tcp/udp	LiebDevMgmt_A—Libert
LiebDevMgmt_C	3027	tcp/udp	LiebDevMgmt_C—Libert
LiebDevMgmt_DM	3028	tcp/udp	LiebDevMgmt_DM—Libert
lingwood	2480	tcp/udp	Lingwood's Detail—Btinternet
link	245	tcp/udp	LINK
linkname	1903	tcp/udp	Local Link Name Resolution—Lucent
linx	1361	tcp/udp	LinX
lionhead	2611	tcp/udp	LIONHEAD
livelan	1555	tcp/udp	livelan—Pictel
livestats	2795	tcp/udp	LiveStats—Avid Sports
ljk-login	472	tcp/udp	ljk-login—LJK Software
lkcmserver	3278	tcp/udp	LKCM Server
llm-csv	2814	tcp/udp	llm-csv
llm-pass	2813	tcp/udp	llm-pass
lmdp	2623	tcp/udp	LMDP—RocketTalk
lm-dta	8206	tcp/udp	LM Dta—Landmark
lm-instmgr	8205	tcp/udp	LM Instmgr—Landmark
lm-perfworks	8204	tcp/udp	LM Perfworks—Landmark
lm-sserver	8207	tcp/udp	LM Sserver—Landmark
lm-webwatcher	8208	tcp/udp	LM Webwatcher—Landmark
lnvalarm	2282	tcp/udp	LNVALARM—Lotus
lnvconsole	2281	tcp/udp	LNVCONSOLE—Lotus
lnvmailmon	2285	tcp/udp	LNVMAILMON—Lotus
lnvmaps	2284	tcp/udp	LNVMAPS—Lotus
lnvpoller	2280	tcp/udp	LNVPOLLER—Lotus
lnvstatus	2283	tcp/udp	LNVSTATUS—Lotus
loadav	750	udp	
loaprobe	1634	tcp/udp	Log On America Probe
localinfosrvr	1487	tcp/udp	LocalInfoSrvr
locus-con	127	tcp/udp	Locus PC-Interface Conn Server—UCLA
locus-map	125	tcp/udp	Locus PC-Interface Net Map Ser—UCLA
lofr-lm	1752	tcp/udp	Leap of Faith Research License Manager
login	513	tcp	remote login a la telnet;
lonewolf-lm	6146	tcp/udp	Lone Wolf Systems License Manager
lontalk-norm	1628	tcp/udp	Lontalk normal—Cisco
lontalk-urgnt	1629	tcp/udp	Lontalk urgent—Cisco
lonworks	2540	tcp/udp	LonWorks—Echelon
lonworks2	2541	tcp/udp	LonWorks2—Echelon
lotusmtap	3007	tcp/udp	Lotus Mail Tracking Agent Protocol
lotusnote	1352	tcp/udp	Lotus Note—Iris
lpsrecommender	2620	tcp/udp	LPSRecommender—Andromedia
lrp	2090	tcp/udp	Load Report Protocol—Radware
ls3	3069	tcp/udp	ls3—PowerWare
ls3bcast	3068	tcp/udp	ls3 Broadcast—PowerWare
lstp	2559	tcp/udp	LSTP—IBM
lupa	1212	tcp/udp	lupa—Databus
lvision-lm	6471	tcp/udp	LVision License Manager
lwlistener	2866	tcp/udp	lwlistener—Quest
m4-network-as	4345	tcp/udp	Macro 4 Network AS
macon-tcp	456	tcp/udp	macon-tcp—Fujitsu
mac-srvr-admin	660	tcp/udp	MacOS Server Admin—Apple
madge-om	2453	tcp/udp	madge-om
magenta-logic	313	tcp/udp	Magenta Logic-NetFusion
magicnotes	3023	tcp/udp	magicnotes
magicom	2243	tcp/udp	Magicom Protocol
mailbox	2004	tcp	

Abbreviation	Port-Decimal	Protocol	Description/Owner
mailbox-lm	505	tcp/udp	mailbox-lm—Stercomm
mailq	174	tcp/udp	MAILQ—U Toronto
maincontrol	2516	tcp/udp	Main Control—Vocaltec
mainsoft-lm	1593	tcp/udp	mainsoft-lm
maitrd	997	tcp/udp	
man	9535	tcp/udp	Intel
manage-exec	2342	tcp/udp	Seagate Manage Exec
mao	2908	tcp/udp	mao—ENSTA
mapper-mapethd	3985	tcp/udp	MAPPER TCP/IP server—Unisys
mapper-nodemgr	3984	tcp/udp	MAPPER network node manager—Unisys
mapper-ws_ethd	3986	tcp/udp	MAPPER workstation server—Unisys
marcam-lm	1444	tcp/udp	Marcam License Management
masc	2587	tcp/udp	MASC—USC
masqdialer	224	tcp/udp	masqdialer—VillageNet
matip-type-a	350	tcp/udp	MATIP Type A—SITA Inc.
matip-type-b	351	tcp/udp	MATIP Type B—SITA Inc.
maxim-asics	3276	tcp/udp	Maxim ASICs
maytagshuffle	2591	tcp/udp	Maytag Shuffle
mc2studios	1899	tcp/udp	MC2Studios—TheCube
mcagent	1820	tcp/udp	mcagent—IBM
mc-brk-srv	3180	tcp/udp	Millicent Broker Server—DEC
mc-client	1180	tcp/udp	Millicent Client Proxy—DEC
mc-gt-srv	2180	tcp/udp	Millicent Vendor Gateway Server—DEC
mciautoreg	1528	tcp/udp	micautoreg
mcidas	112	tcp/udp	McIDAS Data Transmission Protocol—UCAR
mck-ivpip	2698	tcp/udp	MCK-IVPIP
mcns-sec	638	tcp/udp	mcns-sec—Cablelabs
mcns-tel-ret	3311	tcp/udp	MCNS Tel Ret
mcntp	5418	tcp/udp	MCNTP—Pilhuhn
mcs-calypsoicf	3330	tcp/udp	MCS Calypso ICF
mcs-fastmail	3302	tcp/udp	MCS Fastmail
mcs-mailsvr	3332	tcp/udp	MCS Mail Server
mcs-messaging	3331	tcp/udp	MCS Messaging
mdbs_daemon	800	tcp/udp	
md-cg-http	2688	tcp/udp	md-cf-http—ExecMail
mdc-portmapper	685	tcp/udp	MDC Port Mapper—Altavista
mdhcp	2535	tcp/udp	MDHCP—Sun
mdqs	666	tcp/udp	
mecomm	668	tcp/udp	MeComm—Me Technology
med-ci	24005	tcp/udp	med-ci—HP
med-fsp-rx	24001	tcp/udp	med-fsp-rx—HP
med-fsp-tx	24002	tcp/udp	med-fsp-tx—HP
media-agent	2789	tcp/udp	Media Agent—The BRM Group
mediacntrlnfsd	2363	tcp/udp	Media Central NFSD—Sun
med-ltp	24000	tcp/udp	med-ltp—HP
med-net-svc	24006	tcp/udp	med-net-svc—HP
med-ovw	24004	tcp/udp	med-ovw—HP
med-supp	24003	tcp/udp	med-supp—HP
mercantile	3398	tcp/udp	Mercantile—UNI2
meregister	669	tcp/udp	MeRegister—Me Technology
meridiandata	2599	tcp/udp	Meridian Data
messageservice	2311	tcp/udp	Message Service
metaconsole	2850	tcp/udp	MetaConsole—Netaphore Software
meta-corp	6141	tcp/udp	Meta Corporation License Manager
metagram	99	tcp/udp	Metagram Relay
metastorm	2511	tcp/udp	Metastorm
metasys	11001	tcp/udp	Metasys—JCI
meter	570	tcp/udp	demon
meter	571	tcp/udp	udemon
metricadbc	2622	tcp/udp	MetricaDBC
metrics-pas	1824	tcp/udp	metrics-pas
mfcobol	86	tcp/udp	Micro Focus Cobol
mftp	349	tcp/udp	mftp—Microsoft
mftp	5402	tcp/udp	MFTP—StarburstCom
mgcp-callagent	2727	tcp/udp	Media Gateway Control Protocol Call Agent—Telcordia
mgcp-gateway	2427	tcp/udp	Media Gateway Control Protocol Gateway-Telcordia
micom-pfs	490	tcp/udp	micom-pfs
microcom-sbp	1680	tcp/udp	microcom-sbp
micromuse-lm	1534	tcp/udp	micromuse-lm

Abbreviation	Port-Decimal	Protocol	Description/Owner
microsoft-ds	445	tcp/udp	Microsoft-DS
midnight-tech	3008	tcp/udp	Midnight Technologies
mil-2045-47001	1581	tcp/udp	MIL-2045-47001—ITT Industries
miles-apart	2621	tcp/udp	Miles Apart Jukebox Server—MilesInfo
mimer	1360	tcp/udp	MIMER
mindprint	8033	tcp/udp	MindPrint—Gold Rush
minipay	2105	tcp/udp	MiniPay—IBM
mini-sql	1114	tcp/udp	Mini SQL—Hughes Australia
mira	3454	tcp	Apple Remote Access Protocol—U Michigan
miroconnect	1532	tcp/udp	miroconnect
mit-dov	91	tcp/udp	MIT Dover Spooler
miteksys-lm	1482	tcp/udp	Miteksys License Manager
mit-ml-dev	83	tcp/udp	MIT ML Device
mit-ml-dev	85	tcp/udp	MIT ML Device
mloadd	1427	tcp/udp	mloadd monitoring tool
mm-admin	534	tcp/udp	MegaMedia Admin—AIS-gmbh
mmcc	5050	tcp/udp	multimedia conference control tool—ISI
mmpft	1815	tcp/udp	MMPFT
mns-mail	2593	tcp/udp	MNS Mail Notice Service—Fujitsu
mobileip-agent	434	tcp/udp	MobileIP-Agent—DEC
mobilip-mn	435	tcp/udp	MobilIP-MN—DEC
moldflow-lm	1576	tcp/udp	moldflow-lm
molly	1374	tcp/udp	EPI Software Systems
mon	2583	tcp/udp	MON—Transmeta
mondex	471	tcp/udp	Mondex—NatWest Offshore Ltd.
monitor	561	tcp/udp	
monitorconsole	2735	tcp/udp	Monitor Console—Ganymede
monkeycom	9898	tcp/udp	MonkeyCom
montage-lm	6147	tcp/udp	Montage License Manager
mortgageware	367	tcp/udp	MortgageWare—InterLinq
mosaixcc	2561	tcp/udp	MosaixCC
moshebeeri	2627	tcp/udp	Moshe Beeri
moy-corp	2488	tcp/udp	Moy Corporation
mpfoncl	2579	tcp/udp	mpfoncl—Fujitsu
mphlpdmc	9344	tcp/udp	Mphlpdmc—Fujitsu
mpidcagt	9397	tcp/udp	MpldcAgt—Fujitsu
mpidcmgr	9343	tcp/udp	MpldcMgr—Fujitsu
mpm	45	tcp/udp	Message Processing Module [recv]
mpm-flags	44	tcp/udp	MPM FLAGS Protocol
mpm-snd	46	tcp/udp	MPM [default send]—ISI
mpnjsc	1952	tcp/udp	mpnjsc—PFU Direct
mpnjsocl	2685	tcp/udp	mpnjsocl—PFU Direct
mpnjsomb	2681	tcp/udp	mpnjsomb—PFU Direct
mpnjsomg	2686	tcp/udp	mpnjsomg—PFU Direct
mpnjsosv	2684	tcp/udp	mpnjsosv—PFU Direct
mpp	218	tcp/udp	Netix Message Posting Protocol
mppolicy-mgr	5969	tcp/udp	mppolicy-mgr—Fujitsu
mppolicy-v5	5968	tcp/udp	mppolicy-v5—Fujitsu
mps-raft	1700	tcp/udp	mps-raft—AOL
mpsysrmsvr	3358	tcp/udp	Mp Sys Rmsvr—Fujitsu
mptn	397	tcp/udp	Multi Protocol Trans. Net—IBM
mrm	679	tcp/udp	MRM—Cisco
ms-cluster-net	3343	tcp/udp	MS Cluster Net
msdp	639	tcp/udp	MSDP—Cisco
msexch-routing	691	tcp/udp	MS Exchange Routing—Microsoft
msft-gc	3268	tcp/udp	Microsoft Global Catalog
msft-gc-ssl	3269	tcp/udp	Microsoft Global Catalog with LDAP/SSL
msg-auth	31	tcp/udp	MSG Authentication—BBN
msg-icp	29	tcp/udp	MSG ICP—BBN
msgsys	9594	tcp/udp	Message System—Intel
mshnet	1989	tcp/udp	MHSnet system (unassigned but in use)—Sidney U
msiccp	1731	tcp/udp	MSICCP—Microsoft
msims	1582	tcp/udp	MSIMS
msi-selectplay	2871	tcp/udp	MSI Select Play—MediaStation
msl_lmd	1464	tcp/udp	MSL License Manager
msmq	1801	tcp/udp	Microsoft Message Que
msnp	1863	tcp/udp	MSNP—Microsoft
ms-olap1	2393	tcp/udp	MS OLAP 1—Microsoft

Abbreviation	Port-Decimal	Protocol	Description/Owner
ms-olap2	2394	tcp/udp	MS OLAP 2—Microsoft
ms-olap3	2382	tcp/udp	Microsoft OLAP
ms-olap4	2383	tcp/udp	Microsoft OLAP
msolap-ptp2	2725	tcp/udp	MSOLAP PTP2—Microsoft
msp	18	tcp/udp	Message Send Protocol
msp	2438	tcp/udp	MSP—ACC
ms-rome	569	tcp/udp	microsoft rome
ms-shuttle	568	tcp/udp	microsoft shuttle
ms-sna-base	1478	tcp/udp	ms-sna-base
ms-sna-server	1477	tcp/udp	ms-sna-server
ms-sql-m	1434	tcp/udp	Microsoft-SQL-Monitor
ms-sql-s	1433	tcp/udp	Microsoft-SQL-Server
ms-streaming	1755	tcp/udp	ms-streaming
ms-theater	2460	tcp/udp	ms-theater—Microsoft
ms-v-worlds	2525	tcp/udp	MS V-Worlds—Microsoft
ms-wbt-server	3389	tcp/udp	MS WBT Server
mti-tcs-comm	2469	tcp/udp	MTI-TCS-COMM—Microtempus
mtp	1911	tcp/udp	Starlight Networks Multimedia Transport Protocol
mtport-regist	2791	tcp/udp	MT Port Registrator—Institute for Advanced Medical Education
mt-scaleserver	2305	tcp/udp	MT ScaleServer
multiling-http	777	tcp/udp	Multiling HTTP—CTV
multiplex	171	tcp/udp	Network Innovations Multiplex
mumps	188	tcp/udp	Plus Five's MUMPS
murray	1123	tcp/udp	Murray
murx	2743	tcp/udp	murx—Dachbu
musiconline	1806	tcp/udp	Musiconline—SysSpace
mvel-lm	1574	tcp/udp	mvel-lm—Midland Valley
mvs-capacity	10007	tcp/udp	MVS Capacity—IBM
mvx-lm	1510	tcp/udp	Midland Valley Exploration Ltd. Lic. Man.
mylex-mapd	467	tcp/udp	mylex-mapd
mylxamport	2981	tcp/udp	MYLXAMPORT
myrtle	1831	tcp/udp	Myrtle—Genscan
mysql	3306	tcp/udp	MySQL—Analytikerna
mzap	2106	tcp/udp	MZAP—Microsoft
n1-fwp	4446	tcp/udp	N1-FWP—Network-1
n1-rmgmt	4447	tcp/udp	N1-RMGMT—Network-1
n2nremote	1685	tcp/udp	n2nremote—Net2Net
name	42	tcp/udp	Host Name Server
nameserver	42	tcp/udp	Host Name Server
namp	167	tcp/udp	NAMP—NY State Ed Net (NYSER.NET)
nani	2236	tcp/udp	Nani—EDS
nas	991	tcp/udp	Netnews Administration System—Fu-Berlin
nas-metering	2286	tcp/udp	NAS-Metering—Symantec
nati-dstp	3015	tcp/udp	NATI DSTP—Natinst
nati-logos	2343	tcp/udp	nati logos—Natinst
nati-vi-server	3363	tcp/udp	NATI Vi Server—Natinst
nbx-au	2094	tcp/udp	NBX AU—NBXcorp
nbx-cc	2093	tcp/udp	NBX CC—NBXcorp
nbx-dir	2096	tcp/udp	NBX DIR—NBXcorp
nbx-ser	2095	tcp/udp	NBX SER—NBXcorp
ncacn-ip-tcp	3062	tcp/udp	ncacn-ip-tcp—Computer Associates
ncadg-ip-udp	3063	tcp/udp	ncadg-ip-udp—Computer Associates
ncdloadbalance	2683	tcp/udp	NCDLoadBalance
ncdmirroring	2706	tcp/udp	NCD Mirroring
nced	404	tcp/udp	nced
ncl	2397	tcp/udp	NCL—Motorola Inc.
ncld	405	tcp/udp	ncld
ncp	524	tcp/udp	NCP—Novell
ncpm-ft	1744	tcp/udp	ncpm-ft—HP
ncpm-hip	1683	tcp/udp	ncpm-hip—HP
ncpm-pm	1591	tcp/udp	ncpm-pm—HP
ncr_ccl	2528	tcp/udp	NCR CCL
ncube-lm	1521	tcp/udp	nCube License Manager
ndl-tcp-ois-gw	2738	tcp/udp	NDL TCP-OSI Gateway
ndm-requester	1363	tcp/udp	Network DataMover Requester—Ricoh
ndm-server	1364	tcp/udp	Network DataMover Server——Ricoh
ndnp	2883	tcp/udp	NDNP—GMX Inc.
nds_sso	3024	tcp/udp	NDS_SSO—Novell

Abbreviation	Port-Decimal	Protocol	Description/Owner
ndsauth	353	tcp/udp	NDSAUTH—Novell
ndsp	2881	tcp/udp	NDSP—GMX Inc.
ndtp	2882	tcp/udp	NDTP—GMX Inc.
necp	3262	tcp/udp	NECP—ISI
nec-raidplus	2730	tcp/udp	NEC RaidPlus
neod1	1047	tcp/udp	Sun's NEO Object Request Broker
neod2	1048	tcp/udp	Sun's NEO Object Request Broker
nerv	1222	tcp/udp	SNI R&D network
nest-protocol	489	tcp/udp	nest-protocol—Novell
net8-cman	1830	tcp/udp	Oracle Net8 CMan Admin
netadmin	2450	tcp/udp	netadmin—CU-muc
netagent	5771	tcp/udp	NetAgent—Eshare
netangel	2442	tcp/udp	Netangel—Identcode
net-assistant	3283	tcp/udp	Net Assistant—Apple
netbill-auth	1615	tcp/udp	NetBill Authorization Server—CMU
netbill-cred	1614	tcp/udp	NetBill Credential Server—CMU
netbill-keyrep	1613	tcp/udp	NetBill Key Repository—CMU
netbill-prod	1616	tcp/udp	NetBill Product Server—CMU
netbill-trans	1612	tcp/udp	NetBill Transaction Server—CMU
netbios-dgm	138	tcp/udp	NETBIOS Datagram Service—ISI
netbios-ns	137	tcp/udp	NETBIOS Name Service—ISI
netbios-ssn	139	tcp/udp	NETBIOS Session Service—ISI
netbookmark	3131	tcp/udp	Net Book Mark—IBM
netchat	2451	tcp/udp	netchat—CU-muc
netcheque	4008	tcp/udp	NetCheque accounting
netclip	2971	tcp/udp	Net Clip—U di Pisa
netcomm1	1676	tcp	netcomm1—Symantec
netcomm2	1676	udp	netcomm2—Symantec
netcp	395	tcp/udp	NETscout Control Protocol
net-device	4350	tcp/udp	Net Device—Microsoft
netdragon-msngr	2731	tcp/udp	NetDragon Messanger—GlobalNet
netgw	741	tcp/udp	netGW—netcs
netlabs-lm	1406	tcp/udp	NetLabs License Manager
netlock1	9160	tcp/udp	NetLOCK1
netlock2	9161	tcp/udp	NetLOCK2
netlock3	9162	tcp/udp	NetLOCK3
netlock4	9163	tcp/udp	NetLOCK4
netlock5	9164	tcp/udp	NetLOCK5
netmap_lm	1493	tcp/udp	netmap_lm
netml	2288	tcp/udp	NETML—Krypton
netmo-default	6841	tcp/udp	Netmo Default—Netmosphere
netmo-http	6842	tcp/udp	Netmo HTTP—Netmosphere
netmon	2606	tcp/udp	Dell Netmon
netnews	532	tcp/udp	readnews
netobjects1	2485	tcp/udp	Net Objects1
netobjects2	2486	tcp/udp	Net Objects2
netopia-vo1	1839	tcp/udp	netopia-vo1
netopia-vo2	1840	tcp/udp	netopia-vo2
netopia-vo3	1841	tcp/udp	netopia-vo3
netopia-vo4	1842	tcp/udp	netopia-vo4
netopia-vo5	1843	tcp/udp	netopia-vo5
netrcs	742	tcp/udp	Network based Rev. Cont. Sys
netrek	2592	tcp/udp	netrek—TeamQuest
netrisk	1799	tcp/udp	NETRISK—Trident Data Systems
netrix-sftm	2328	tcp/udp	Netrix SFTM
netrjs-1	71	tcp/udp	Remote Job Service—ISI
netrjs-2	72	tcp/udp	Remote Job Service—ISI
netrjs-3	73	tcp/udp	Remote Job Service—ISI
netrjs-4	74	tcp/udp	Remote Job Service—ISI
netsc-dev	155	tcp/udp	NETSC—CSC
netsc-prod	154	tcp/udp	NETSC—CSC
netserialext1	16360	tcp/udp	netserialext1—Aquila Group
netserialext2	16361	tcp/udp	netserialext2—Aquila Group
netserialext3	16367	tcp/udp	netserialext3—Aquila Group
netserialext4	16368	tcp/udp	netserialext4—Aquila Group
netspeak-acd	21848	tcp/udp	NetSpeak Corp. Automatic Call Distribution
netspeak-cps	21849	tcp/udp	NetSpeak Corp. Credit Processing System
netspeak-cs	21847	tcp/udp	NetSpeak Corp. Connection Services
netspeak-is	21846	tcp/udp	NetSpeak Corp. Directory Services

Abbreviation	Port-Decimal	Protocol	Description/Owner
netsteward	2810	tcp/udp	NetSteward
netsupport	5405	tcp/udp	NetSupport—Pipex
netsupport2	5421	tcp/udp	Net Support 2—Pipex
netview-aix-1	1661	tcp/udp	netview-aix-1—IBM
netview-aix-10	1670	tcp/udp	netview-aix-10—IBM
netview-aix-11	1671	tcp/udp	netview-aix-11—IBM
netview-aix-12	1672	tcp/udp	netview-aix-12—IBM
netview-aix-2	1662	tcp/udp	netview-aix-2—IBM
netview-aix-3	1663	tcp/udp	netview-aix-3—IBM
netview-aix-4	1664	tcp/udp	netview-aix-4—IBM
netview-aix-5	1665	tcp/udp	netview-aix-5—IBM
netview-aix-6	1666	tcp/udp	netview-aix-6—IBM
netview-aix-7	1667	tcp/udp	netview-aix-7—IBM
netview-aix-8	1668	tcp/udp	netview-aix-8—IBM
netview-aix-9	1669	tcp/udp	netview-aix-9—IBM
netviewdm1	729	tcp/udp	IBM NetView DM/6000 Server/Client
netviewdm2	730	tcp/udp	IBM NetView DM/6000 send/tcp
netviewdm3	731	tcp/udp	IBM NetView DM/6000 receive/tcp
netwall	533	tcp/udp	for emergency broadcasts
netware-csp	1366	tcp/udp	Novell NetWare Comm Service Platform
netware-ip	396	tcp/udp	Novell Netware over IP
netwave-ap-mgmt	2411	tcp/udp	Netwave AP Management
netx-agent	2586	tcp/udp	NETX Agent—IPMetrics
netx-server	2585	tcp/udp	NETX Server—IPMetrics
neveroffline	2614	tcp/udp	Never Offline—AMO
newacct	100	tcp	[unauthorized use]
newlixreg	2671	tcp/udp	newlixreg—GAAJ
newoak	4001	tcp/udp	NewOak
new-rwho	550	tcp/udp	new-who
news	2009	tcp	
nexstorindltd	2360	tcp/udp	NexstorIndLtd
nextstep	178	tcp/udp	NextStep Window Server—NEXT
nfa	1155	tcp/udp	Network File Access
nfs	2049	tcp/udp	Network File System - Sun Microsystems
nfsd-status	1110	tcp/udp	Cluster status info—HP
ng-umds	1690	tcp/udp	ng-umds
nhserver	2672	tcp/udp	nhserver—Aran
nicetec-mgmt	2557	tcp/udp	nicetec-mgmt
nicetec-nmsvc	2556	tcp/udp	nicetec-nmsvc
nicname	43	tcp/udp	Who Is
ni-ftp	47	tcp/udp	NI FTP—Isode Internet/X.400 Messaging
nifty-hmi	4134	tcp/udp	NIFTY-Serve HMI protocol
nim	1058	tcp/udp	nim—IBM
ni-mail	61	tcp/udp	NI MAIL—Isode Internet/X.400 Messaging
nimcontroller	48000	tcp/udp	Nimbus Controller—Nimsoft
nimgtw	48003	tcp/udp	Nimbus Gateway—Nimsoft
nimhub	48002	tcp/udp	Nimbus Hub—Nimsoft
nimreg	1059	tcp/udp	nimreg—IBM
nimrod-agent	1617	tcp/udp	nimrod Inter-Agent Communication—BBN
nimspooler	48001	tcp/udp	Nimbus Spooler—Nimsoft
nip	376	tcp/udp	Amiga Envoy Network Inquiry Proto—I-AM Inc.
nkd	1650	tcp/udp	
nlogin	758	tcp/udp	
nmap	689	tcp/udp	NMAP—Novonyx
nm-asses-admin	3150	tcp/udp	NetMike Assessor Administrator—VisionPlus
nm-assessor	3151	tcp/udp	NetMike Assessor—VisionPlus
nm-game-admin	3148	tcp/udp	NetMike Game Administrator—VisionPlus
nm-game-server	3149	tcp/udp	NetMike Game Server—VisionPlus
nms	1429	tcp/udp	Hypercom NMS
nms_topo_serv	1486	tcp/udp	nms_topo_serv
nmsd	1239	tcp/udp	NMSD—PTC
nms-dpnss	2503	tcp/udp	NMS-DPNSS—NMS-Europe
nmsigport	2817	tcp/udp	NMSig Port—Inalp Networks
nmsigport	2839	tcp/udp	NMSigPort—Inalp Networks
nmsp	537	tcp/udp	Networked Media Streaming Protocol—Narrative
nmsp	1790	tcp/udp	Narrative Media Streaming Protocol
nmsserver	2244	tcp/udp	NMS Server—NMSS
nnsp	433	tcp/udp	NNSP—UC Berkeley

Abbreviation	Port-Decimal	Protocol	Description/Owner
nntp	119	tcp/udp	Network News Transfer Protocol-UC BERKLEY
nntps	563	tcp/udp	nntp protocol over TLS/SSL (was snntp)—Netscape
noadmin	1921	tcp/udp	NoAdmin—Datawatch
noagent	1917	tcp/udp	nOAgent—Datawatch
norton-lambert	2338	tcp/udp	Norton Lambert
notify_srvr	3016	tcp/udp	Notify Server—Novell
novastorbakcup	308	tcp/udp	Novastor Backup
novell-ipx-cmd	2645	tcp/udp	Novell IPX CMD
novell-lu6.2	1416	tcp/udp	Novell LU6.2
novell-zen	2544	tcp	Novell ZEN
npep-messaging	2868	tcp/udp	NPEP Messaging—Norman
npmp	8450	tcp/udp	npmp—Demon
npmp-gui	611	tcp/udp	npmp-gui—CRL
npmp-local	610	tcp/udp	npmp-local—CRL
npmp-trap	609	tcp/udp	npmp-trap—CRL
npp	92	tcp/udp	Network Printing Protocol—UMD
nqs	607	tcp/udp	nqs—Cray
nrcabq-lm	1458	tcp/udp	Nichols Research Corp.
ns	760	tcp/udp	
nsc-ccs	2604	tcp/udp	NSC CCS—NetworkSciences
ns-cfg-server	3266	tcp/udp	NS CFG Server—NetSoft
nsc-posa	2605	tcp/udp	NSC POSA—NetworkSciences
nsiiops	261	tcp/udp	IIOP Name Service over TLS/SSL—Netscape
nsjtp-ctrl	1687	tcp/udp	nsjtp-ctrl—HP
nsjtp-data	1688	tcp/udp	nsjtp-data—HP
nssagentmgr	4454	tcp/udp	NSS Agent Manager—Symantec
nssalertmgr	4453	tcp/udp	NSS Alert Manager—Symantec
ns-server	5415	tcp/udp	NS Server—NetSof
nss-routing	159	tcp/udp	NSS-Routing—IBM
nsw-fe	27	tcp/udp	NSW User System FE—BBN
nsws	3049	tcp/udp	NSWS—PSLink
nta-ds	7544	tcp/udp	FlowAnalyzer DisplayServer—Cisco
ntalk	518	tcp/udp	
nta-us	7545	tcp/udp	FlowAnalyzer UtilityServer—Cisco
ntp	123	tcp/udp	Network Time Protocol—UDEL
nucleus	1463	tcp/udp	Nucleus
nuts_bootp	4133	tcp/udp	NUTS Bootp Server—SNI
nuts_dem	4132	tcp/udp	NUTS Daemon—SNI
nvd	2329	tcp/udp	NVD—Noblenet
nv-video	4444	tcp/udp	NV Video default—Xerox
nxedit	126	tcp/udp	NXEdit—Unisys
obex	650	tcp/udp	OBEX—Motorola
objcall	94	tcp/udp	Tivoli Object Dispatcher
objective-dbc	1388	tcp/udp	Objective Solutions DataBase Cache
objectmanager	2038	tcp/udp	
ocbinder	183	tcp/udp	OCBinder
oceansoft-lm	1466	tcp/udp	Ocean Software License Manager
oce-snmp-trap	2697	tcp/udp	Oce SNMP Trap Port
oc-lm	1448	tcp/udp	OpenConnect License Manager
ocs_amu	429	tcp/udp	OCS_AMU—US Air Force
ocs_cmu	428	tcp/udp	OCS_CMU—US Air Force
ocserver	184	tcp/udp	OCServer
odette-ftp	3305	tcp/udp	ODETTE-FTP—Ford Mtrs.
odmr	366	tcp/udp	ODMR—Qualcomm
odn-castraq	2498	tcp/udp	ODN-CasTraq
officelink2000	3320	tcp/udp	Office Link 2000—Teltone
ofsd	2322	tcp/udp	ofsd—Commvault
ohimsrv	506	tcp/udp	ohimsrv—Openhorizon
oi-2000	2364	tcp/udp	OI-2000—Software Horizons Inc.
oidocsvc	4142	tcp/udp	Document Server—SNS
oidsr	4143	tcp/udp	Document Replication—SNS
oirtgsvc	4141	tcp/udp	Workflow Server
olhost	2661	tcp/udp	OLHOST—LAN-Aces
omginitialrefs	900	tcp/udp	OMG Initial Refs—Sun
omserv	764	tcp/udp	
onmux	417	tcp/udp	Onmux-The World Kiosk
ontime	1622	tcp/udp	ontime
opalis-rdv	536	tcp/udp	opalis-rdv

Abbreviation	Port-Decimal	Protocol	Description/Owner
opalis-robot	314	tcp/udp	Opalis Robot
opc-job-start	423	tcp/udp	IBM Operations Planning and Control Start
opc-job-track	424	tcp/udp	IBM Operations Planning and Control Track
openmail	5729	tcp/udp	Openmail User Agent Layer—HP
openmailg	5755	tcp/udp	OpenMail Desk Gateway server—HP
openmailns	5766	tcp/udp	OpenMail NewMail Server—HP
openmailpxy	5768	tcp/udp	OpenMail CMTS Server—HP
openmath	1473	tcp/udp	OpenMath
openport	260	tcp/udp	Openport
opentable	2368	tcp/udp	OpenTable
openvms-sysipc	557	tcp/udp	openvms-sysipc—DEC
opequus-server	2400	tcp/udp	OpEquus Server—tesl
opi-sock	7429	tcp/udp	OpenView DM rqt communication—HP
opsec-cvp	18181	tcp/udp	OPSEC CVP—CheckPoint
opsec-ufp	18182	tcp/udp	OPSEC UFP-CheckPoint
opsession-clnt	3303	tcp/udp	OP Session Client—Net Manage
opsession-prxy	3307	tcp/udp	OP Session Proxy—Net Manage
opsession-srvr	3304	tcp/udp	OP Session Server—Net Manage
optech-port1-lm	2237	tcp/udp	Optech Port1 License Manager—Opticaltech
optika-emedia	1829	tcp/udp	Optika eMedia
optilogic	2435	tcp/udp	OptiLogic—Wirespeed
optimanet	2408	tcp/udp	OptimaNet—Optimal
optiwave-lm	2524	tcp/udp	Optiwave License Management
optocontrol	22001	tcp/udp	OptoControl—Opto22
oracle	2005	udp	
oracle-em1	1748	tcp/udp	oracle-em1
oracle-em2	1754	tcp/udp	oracle-em2
oraclenames	1575	tcp/udp	oraclenames
oraclenet8cman	1630	tcp/udp	Oracle Net8 Cman
oracle-vp1	1809	tcp/udp	Oracle-VP1
oracle-vp2	1808	tcp/udp	Oracle-VP2
ora-lm	1446	tcp/udp	Optical Research Associates License Manager
orasrv	1525	tcp/udp	oracle
orbiter	2398	tcp/udp	Orbiter—Callisto
orbixd	1570	tcp/udp	orbixd—Iona Technologies
orbplus-iiop	1597	tcp/udp	orbplus-iiop—HP (Apollo)
ordinox-dbase	3355	tcp/udp	Ordinox Dbase
ordinox-server	3274	tcp/udp	Ordinox Server
orion	2407	tcp/udp	Orion—PC Users Group
orion-rmi-reg	2413	tcp/udp	orion-rmi-reg —IBM
os-licman	1384	tcp/udp	Objective Solutions License Manager
osmosis-aeea	3034	tcp/udp	Osmosis AEEA—InterAccess
osu-nms	192	tcp/udp	OSU Network Monitoring System—OHIO-STATE U
ott	2428	tcp/udp	One Way Trip Time—BBN
outlaws	5310	tcp/udp	Outlaws—Lucasarts
oveadmgr	7427	tcp/udp	OpenView DM Event Agent Manager—HP
ovladmgr	7428	tcp/udp	OpenView DM Log Agent Manager—HP
ovsessionmgr	2389	tcp/udp	OpenView Session Mgr—HP
ovtopmd	2532	tcp/udp	OVTOPMD—HP
ovwdb	2447	tcp/udp	OpenView NNM daemon—HP
pacerforum	1480	tcp/udp	PacerForum
padl2sim	5236	tcp/udp	
pafec-lm	7511	tcp/udp	pafec-lm
palace	9992	tcp/udp	The Palace
palace	9993	tcp/udp	The Palace
palace	9994	tcp/udp	The Palace
palace	9995	tcp/udp	The Palace
palace	9996	tcp/udp	The Palace
palace	9997	tcp/udp	The Palace
pammratc	1632	tcp/udp	PAMMRATC—Peak Audio
pammrpc	1633	tcp/udp	PAMMRPC—Peak Audio
pangolin-laser	3348	tcp/udp	Pangolin Laser
panja-axbnet	1320	tcp/udp	Panja-AXBNET
paradym-31port	1864	tcp/udp	Paradym 31 Port—Wizdom
paspar2-zoomin	2873	tcp/udp	PASPAR2 ZoomIn
passgo	511	tcp/udp	PassGo
passgo-tivoli	627	tcp/udp	PassGo Tivoli
password-chg	586	tcp/udp	Password Change

Abbreviation	Port-Decimal	Protocol	Description/Owner
patrol	8160	tcp/udp	Patrol—BMC Software, Inc
patrol-snmp	8161	tcp/udp	Patrol SNMP—BMC Software, Inc
patrolview	4097	tcp/udp	Patrol View—BMC Software, Inc.
pawserv	345	tcp/udp	Perf Analysis Workbench
pay-per-view	1564	tcp/udp	Pay-Per-View—ISI
pcanywheredata	5631	tcp/udp	pcANYWHEREdata—Symantec
pcanywherestat	5632	tcp/udp	pcANYWHEREstat—Symantec
pcbakcup1	2773	tcp/udp	PC Backup—Remote Backup Systems
pcbakcup2	2774	tcp/udp	PC Backup—Remote Backup Systems
pciarray	1552	tcp/udp	pciarray—AVL
pclemultimedia	2558	tcp/udp	PCLE Multi Media—PinnacleSys
pcmail-srv	158	tcp/udp	PCMail Server—MIT
pconnectmgr	1562	tcp/udp	pconnectmgr—Veritas
pcs-pcw	2566	tcp/udp	pcs-pcw—Pcare
pc-telecommute	2299	tcp/udp	PC Telecommute—Symantec
pctrader	3048	tcp/udp	Sierra Net PC Trader
pda-gate	4012	tcp/udp	PDA Gate—IBM
pdap	344	tcp/udp	Prospero Data Access Protocol—ISI
pdap-np	1526	tcp/udp	Prospero Data Access Prot non-priv
pdb	3033	tcp/udp	PDB—PixStream
pdnet	2843	tcp/udp	Pdnet—MTS Internet ISP
pdp	1675	tcp/udp	Pacific Data Products
pdps	1314	tcp/udp	Photoscript Distributed Printing System—Compulink
pdrncs	3299	tcp/udp	pdrncs—Tektronix
pds	9595	tcp/udp	Ping Discovery Service—Intel
pegboard	1357	tcp/udp	Electronic PegBoard
pehelp	2307	tcp/udp	pehelp—HP
peport	1449	tcp/udp	PEport
perf-port	1995	tcp/udp	cisco perf port
persona	1916	tcp/udp	Persoft Persona
personal-agent	5555	tcp/udp	Personal Agent—Infoseek
personal-link	281	tcp/udp	Personal Link—Computer Network Resources, Inc.
pftp	662	tcp/udp	PFTP
ph	481	tcp/udp	Ph service—UMDAC
pharmasoft	1779	tcp/udp	pharmasoft
pharos	4443	tcp/udp	Pharos—TeleConsult GmbH
philips-vc	583	tcp/udp	Philips Video-Conferencing
phoenix-rpc	3347	tcp/udp	Phoenix RPC
phonebook	767	tcp/udp	phone
photuris	468	tcp/udp	proturis—U Michigan
phrelay	4868	tcp/udp	Photon Relay—QNX
phrelaydbg	4869	tcp/udp	Photon Relay Debug—QNX
piccolo	2787	tcp/udp	piccolo - Cornerstone Software
picknfs	1598	tcp/udp	picknfs—Pick Systems
picodbc	1603	tcp/udp	pickodbc—picknfs
pim-rp-disc	496	tcp/udp	PIM-RP-DISC—Cisco
ping-pong	3010	udp	Telerate Workstation—Dow Jones
pip	321	tcp/udp	PIP—Activerse Inc.
pip	1321	tcp/udp	PIP—Activerse
pipe_server	2010	udp	
pipes	1465	tcp/udp	Pipes Platform
piranha1	4600	tcp/udp	Piranha1—Primark Corporation
piranha2	4601	tcp/udp	Piranha2—Primark Corporation
pirp	553	tcp/udp	pirp—UC Berkeley
pit-vpn	2865	tcp/udp	pit-vpn—Linux Networks
pk	5272	tcp/udp	PK—EBA
pk-electronics	2634	tcp/udp	PK Electronics—PKWorld
pkix-3-ca-ra	829	tcp/udp	PKIX-3 CA/RA—Entrust
pkix-timestamp	318	tcp/udp	PKIX TimeStamp—Entrust
plato	3285	tcp/udp	Plato—TRO
plato-lm	1819	tcp/udp	Plato License Manager—Ermuk
plcy-net-svcs	4351	tcp/udp	PLCY Net Services
plgproxy	2790	tcp/udp	PLG Proxy—Aladdin Inc.
pluribus	3469	tcp/udp	Pluribus—Caplet
pmd	7431	tcp/udp	OpenView DM ovc/xmpv3 api pipe—HP
pmdfmgt	7633	tcp/udp	PMDF Management—InnoSoft
pmdmgr	7426	tcp/udp	OpenView DM Postmaster Manager
pmsm-webrctl	2972	tcp/udp	PMSM Webrctl

Abbreviation	Port-Decimal	Protocol	Description/Owner
pn-requester	2717	tcp/udp	PN REQUESTER—BMC Software, Inc.
pn-requester2	2718	tcp/udp	PN REQUESTER 2—BMC Software, Inc.
pns	2487	tcp/udp	Policy Notice Service—Fujitsu
policyserver	3055	tcp/udp	Policy Server—SonusNet
pop2	109	tcp/udp	Post Office Protocol - Version 2—ISI
pop3	110	tcp/udp	Post Office Protocol - Version 3—Data Broadcasting Corporation
pop3s	995	tcp/udp	pop3 protocol over TLS/SSL (was spop3—Microsoft)
pov-ray	494	tcp/udp	POV-Ray
powerburst	485	tcp/udp	Air Soft Power Burst
powerclientcsf	2443	tcp/udp	PowerClient Central Storage Facility—Unisys
powerguardian	1777	tcp/udp	powerguardian—BenaTong Software
powerschool	5071	tcp/udp	PowerSchool
pptconference	1711	tcp/udp	pptconference—Microsoft
pptp	1723	tcp/udp	pptp—Microsoft
pq-lic-mgmt	2687	tcp/udp	pq-lic-mgmt—PQSystems
pra_elmd	1587	tcp/udp	pra_elmd—Platte River Associates, Inc.
prchat-server	4456	tcp/udp	PR Chat Server—IBM
prchat-user	4455	tcp/udp	PR Chat User—IBM
precise-sft	2315	tcp/udp	Precise Sft.
printer	515	tcp/udp	spooler
printer_agent	3396	tcp/udp	Printer Agent—Novell
print-srv	170	tcp/udp	Network PostScript—DEC
priority-e-com	2618	tcp/udp	Priority E-Com—Eshbel
privatechat	1735	tcp/udp	PrivateChat
privatewire	4449	tcp/udp	PrivateWire—ARX
privilege	2588	tcp/udp	Privilege—AKS
prm-nm	409	tcp/udp	Prospero Resource Manager Node Man—ISI
prm-nm-np	1403	tcp	Prospero Resource Manager—ISI
prm-sm	408	tcp/udp	Prospero Resource Manager Sys. Man—ISI
prm-sm-np	1402	tcp	Prospero Resource Manager-ISI
proactivesrvr	2722	tcp/udp	Proactive Server—IBM
prodigy-intrnet	1778	tcp/udp	prodigy-internet
pro-ed	8032	tcp/udp	ProEd—Gold Rush
profile	136	tcp/udp	PROFILE Naming System—U ARIZONA
prolink	1678	tcp/udp	prolink—Tektronix
proshare1	1459	tcp/udp	Proshare Notebook Application
proshare2	1460	tcp/udp	Proshare Notebook Application
proshareaudio	5713	tcp/udp	proshare conf audio—Intel
prosharedata	5715	tcp/udp	proshare conf data—Intel
proshare-mc-1	1673	tcp/udp	Intel Proshare Multicast
proshare-mc-2	1674	tcp/udp	Intel Proshare Multicast
prosharerequest	5716	tcp/udp	proshare conf request—Intel
prosharevideo	5714	tcp/udp	proshare conf video—Intel
prospero	191	tcp/udp	Prospero Directory Service—ISI
prospero-np	1525	tcp/udp	Prospero Directory Service non-priv
proxim	1732	tcp/udp	proxim
proxima-lm	1445	tcp/udp	Proxima License Manager
proxy-gateway	2303	tcp/udp	Proxy Gateway—Funk Software
prp	2091	tcp/udp	PRP—Radware
prRegister	4457	tcp/udp	PR Register—IBM
prsvp	3455	tcp/udp	RSVP Port—ISI
psbserver	2350	tcp/udp	psbserver—Pharos
psdbserver	2355	tcp/udp	psdbserver—Pharos
pslserver	2352	tcp/udp	pslserver—Pharos
psmond	1788	tcp/udp	psmond—HP
psprserver	2354	tcp/udp	psprserver—Pharos
pspserver	2353	tcp/udp	pspserver—Pharos
psrserver	2351	tcp/udp	psrserver—Pharos
pssc	645	tcp/udp	PSSC—DLR
ptcnameservice	597	tcp/udp	PTC Name Service
ptp-event	319	tcp/udp	PTP Event—HP
ptp-general	320	tcp/udp	PTP General—HP
pump	751	tcp/udp	
puparp	998	udp	
puprouter	999	tcp/udp	
purenoise	663	tcp/udp	PureNoise—MailCity
pvsw	2520	tcp/udp	pvsw—Pervasive
pvsw-inet	2441	tcp/udp	pvsw-inet—Pervasive

Abbreviation	Port-Decimal	Protocol	Description/Owner
pwdgen	129	tcp/udp	Password Generator Protocol—ARMY.MIL
pwrsevent	2694	tcp/udp	pwrsevent—Fujitsu
pxc-epmap	2434	tcp/udp	pxc-epmap—Xerox
pxc-ntfy	3009	tcp/udp	PXC-NTFY—Xerox
pxc-pin	4005	tcp/udp	pxc-pin—Xerox
pxc-roid	4004	tcp/udp	pxc-roid—Xerox
pxc-sapxom	2680	tcp/udp	pxc-sapxom—Xerox
pxc-splr	4007	tcp/udp	pxc-splr—Xerox
pxc-splr-ft	4003	tcp/udp	pxc-splr-ft—Xerox
pxc-spvr	4006	tcp/udp	pxc-spvr—Xerox
pxc-spvr-ft	4002	tcp/udp	pxc-spvr-ft—Xerox
qadmifevent	2462	tcp/udp	qadmifevent—Tellabs
qadmifoper	2461	tcp/udp	qadmifoper—Tellabs
qbikgdp	368	tcp/udp	QbikGDP
qft	189	tcp/udp	Queued File Transport—San Diego State U
qip-audup	2765	tcp/udp	qip-audup—Lucent
qip-login	2366	tcp/udp	qip-login—Lucent
qip-msgd	2468	tcp/udp	qip_msgd—Lucent
qip-qdhcp	2490	tcp/udp	qip_qdhcp—Lucent
qke-llc-v3	2523	tcp/udp	Qke LLC V.3—Kecam-Han
qmqp	628	tcp/udp	QMQP
qmtp	209	tcp/udp	The Quick Mail Transfer Protocol—UC Berkeley
qnxnetman	3385	tcp/udp	qnxnetman
qotd	17	tcp/udp	Quote of the Day—ISI
qotps	2724	tcp/udp	qotps—QueryObject
qpasa-agent	2612	tcp/udp	Qpasa Agent—MQSoftware
qrh	752	tcp/udp	
qsc	787	tcp/udp	QSC—Stomped
qsoft	3059	tcp/udp	qsoft—EarlHaig
quaddb	2497	tcp/udp	Quad DB—Quad-Sys
quake	26000	tcp/udp	quake
quartus-tcl	2589	tcp/udp	quartus tcl
quest-vista	7980	tcp/udp	Quest Vista
quotad	762	tcp/udp	
radio	1595	tcp/udp	radio—ISI
radio-bc	1596	udp	radio-bc—ISI
radio-sm	1596	tcp	radio-sm—ISI
radius	1812	tcp/udp	RADIUS—Livingston
radius-acct	1813	tcp/udp	RADIUS Accounting—Livingston
radwiz-nms-srv	2736	tcp/udp	RADWIZ NMS SRV
raid-ac	2012	udp	
raid-am	2007	udp	
raid-am	2013	tcp	
raid-cc	2006	udp	raid
raid-cc	2011	tcp	raid
raid-cd	2013	udp	
raid-cs	2015	udp	
raid-sf	2014	udp	
rap	38	tcp/udp	Route Access Protocol—The World Kiosk
rap	256	tcp/udp	RAP—IBM
rapidbase	1953	tcp/udp	Rapid Base—VTT
rapido-ip	2457	tcp/udp	Rapido_IP—Wamnet
rap-listen	1531	tcp/udp	rap-listen
rap-service	1530	tcp/udp	rap-service
rasadv	9753	tcp/udp	rasadv—Microsoft
ratl	2449	tcp/udp	RATL—Unisys
raven1	2713	tcp/udp	Raven1—Unified-Technologies
raven2	2714	tcp	Raven2—Unified-Technologies
rcc-host	2332	tcp/udp	RCC Host—Rapistan
rcp	469	tcp/udp	Radio Control Protocol
rcst	3467	tcp/udp	RCST—RemoteControlSexToys
rda	630	tcp/udp	RDA—Minster
rdb-dbs-disp	1571	tcp/udp	Oracle Remote Data Base
rdc-wh-eos	3142	tcp/udp	RDC WH EOS—NCC
rdlap	2321	tcp/udp	RDLAP over UDP—Motorola Inc.
rds	1540	tcp/udp	rds
rds2	1541	tcp/udp	rds2
realm-rusd	688	tcp/udp	REALM-RUSD—Realminfo

Abbreviation	Port-Decimal	Protocol	Description/Owner
realsecure	2998	tcp/udp	Real Secure—ISS
recipe	2240	tcp/udp	RECIPe—GTE
redstorm_diag	2349	tcp/udp	Diagnostics Port—Redstorm
redstorm_find	2347	tcp/udp	Game Announcement and Location—Redstorm
redstorm_info	2348	tcp/udp	Information to query for game status—Redstorm
redstorm_join	2346	tcp/udp	Game Connection Port—Redstorm
redwood-broker	3001	tcp/udp	Redwood Broker—PowerFrame
redwood-chat	3032	tcp/udp	Redwood Chat—Daou
reftek	2543	tcp/udp	REFTEK
reftek	2543	udp	REFTEK
registrar	1712	tcp/udp	resource monitoring service—HP
relief	1353	tcp/udp	Relief Consulting—PSI
rellpack	2018	udp	
re-mail-ck	50	tcp/udp	Remote Mail Checking Protocol—UIUC
remotefs	556	tcp/udp	rfs server
remote-kis	185	tcp/udp	Remote-KIS
remoteware-cl	3000	tcp/udp	RemoteWare Client—Xcellenet
remoteware-srv	3002	tcp/udp	RemoteWare Server—Xcellenet
remoteware-un	2999	tcp/udp	RemoteWare Unassigned—Xcellenet
remote-winsock	1745	tcp/udp	remote-winsock—Microsoft
repcmd	641	tcp/udp	repcmd—Replicase
repliweb	2837	tcp/udp	Repliweb—SoftLink
repscmd	653	tcp/udp	RepCmd—Tioga
rescap	283	tcp/udp	rescap—Internet Mail Consortium
Reserved	1024	tcp/udp	Reserved
reserved1	6407	tcp/udp	Seagate Software
reserved2	6408	tcp/udp	Seagate Software
reserved3	6409	tcp/udp	Seagate Software
reserved4	6410	tcp/udp	Seagate Software
resource_mgr	3019	tcp/udp	Resource Manager—Novell
responsenet	3045	tcp/udp	ResponseNet
rfa	4672	tcp/udp	remote file access server
rfe	5002	tcp/udp	radio free ethernet
rfile	750	tcp	
rfio	3147	tcp/udp	RFIO—CERN
rfx-lm	1497	tcp/udp	rfx-lm
rgtp	1431	tcp/udp	Reverse Gossip Transport
rhp-iibp	1912	tcp/udp	rhp-iibp—M-Ware
rib-slm	3296	tcp/udp	Rib License Manager
ricardo-lm	6148	tcp/udp	Ricardo North America License Manager
ridgeway1	2776	tcp/udp	Ridgeway Systems & Software
ridgeway2	2777	tcp/udp	Ridgeway Systems & Software
rightbrain	1354	tcp/udp	RightBrain Software
rimsl	2044	tcp/udp	
ripng	521	tcp/udp	ripng—Ipsilon
ris	180	tcp/udp	Intergraph
ris-cm	748	tcp/udp	Russell Info Sci Calendar Manager
rje	5	tcp/udp	Remote Job Entry—ISI
rkb-oscs	1817	tcp/udp	RKB-OSCS
rlp	39	tcp/udp	Resource Location Protocol—CMU
rlzdbase	635	tcp/udp	RLZ Dbase—Tyxar
rmc	657	tcp/udp	RMC—IBM
rmiactivation	1098	tcp/udp	RMI Activation—Sun
rmiregistry	1099	tcp/udp	RMI Registry—Sun
rmlnk	2818	tcp/udp	rmlnk
rmonitor	560	tcp/udp	rmonitord
rmonitor_secure	5145	tcp/udp	
rmonitor_secure	5145	udp	
rmt	411	tcp/udp	Remote MT Protocol—Lysator
rmtserver	2416	tcp/udp	RMT Server—SimWare
rna-lm	1522	tcp/udp	Ricardo North America License Manager
rnrp	2423	tcp/udp	RNRP—ABB Sweden
robcad-lm	1509	tcp/udp	Robcad, Ltd. License Manager
rockwell-csp1	2221	tcp/udp	Rockwell CSP1
rockwell-csp2	2222	tcp/udp	Rockwell CSP2
rockwell-csp3	2223	tcp/udp	Rockwell CSP3
rockwell-encap	44818	tcp/udp	Rockwell Encapsulation
roketz	1730	tcp/udp	roketz—Bluemoon Interactive

Abbreviation	Port-Decimal	Protocol	Description/Owner
router	520	udp	local routing process (on site)
rpasswd	774	tcp	
rpc2portmap	369	tcp/udp	rpc2portmap—FreeBSD Project
rrac	5678	tcp/udp	Remote Replication Agent Connection—Microsoft
rrh	753	tcp/udp	
rrifmm	1696	tcp/udp	rrifmm—RRI Inc.
rrilwm	1695	tcp/udp	rrilwm—RRI Inc.
rrimwm	1694	tcp/udp	rrimwm—RRI Inc.
rrirtr	1693	tcp/udp	rrirtr—RRI Inc.
rrisat	1697	tcp/udp	rrisat—RRI Inc.
rrp	648	tcp/udp	Registry Registrar Protocol (RRP)—Netsol
rsap	1647	tcp/udp	rsap—TU Illmenau (Tech University)
rsc-robot	1793	tcp/udp	rsc-robot
rsh-spx	222	tcp/udp	Berkeley rshd with SPX auth
rsisysaccess	2752	tcp/udp	RSISYS ACCESS
rsmtp	2390	tcp/udp	RSMTP—Bell Labs
rsvd	168	tcp/udp	RSVD—Imperial Software
rsvp_tunnel	363	tcp/udp	RSVP Tunnel—UCLA
rsvp-encap-1	1698	tcp/udp	RSVP-ENCAPSULATION-1—ISI
rsvp-encap-2	1699	tcp/udp	RSVP-ENCAPSULATION-2—ISI
rsync	873	tcp/udp	rsync—Australian National U
rtcm-sc104	2101	tcp/udp	rtcm-sc104—wsrcc
rtelnet	107	tcp/udp	Remote Telnet Service—ISI
rtip	771	tcp/udp	
rtsclient	2501	tcp/udp	Resource Tracking system client—ETSUADMN.ETSU.EDU
rtsp	554	tcp/udp	Real Time Stream Control Protocol—Prognet
rtsp-alt	8554	tcp/udp	RTSP Alternate (see port 554)—Precept
rtsps	322	tcp/udp	RTSPS—Microsoft
rtsserv	2500	tcp/udp	Resource Tracking system server—ETSUADMN.ETSU.EDU
rvs-isdn-dcp	2578	tcp/udp	RVS ISDN DCP
rwhois	4321	tcp/udp	Remote Who Is—InterNIC
rxe	761	tcp/udp	
rxmon	1311	tcp/udp	RxMon—Dell
sabams	2760	tcp/udp	Saba MS
sabarsd	8401	tcp/udp	sabarsd—Commvault
sae-urn	4500	tcp/udp	sae-urn—Proper
saft	487	tcp/udp	saft Simple Asynchronous File Transfer—U Stuttgart
sah-lm	3291	tcp/udp	S A Holditch & Associates - LM
sai_sentlm	2640	tcp/udp	Sabbagh Associates Licence Manager
saiseh	1644	tcp	Satellite-data Acquisition System 4—U Wisconsin
salient-dtasrv	5409	tcp/udp	Salient Data Server
salient-mux	5411	tcp/udp	Salient MUX
salient-usrmgr	5410	tcp/udp	Salient User Manager
samd	3275	tcp/udp	SAMD—HP
sa-msg-port	1646	tcp/udp	sa-msg-port—ITT Industries
samsung-unidex	4010	tcp/udp	Samsung Unidex
sanity	643	tcp/udp	SANity—OSR
santak-ups	3038	tcp/udp	Santak UPS
saris	4442	tcp/udp	Saris—TeleConsult GmbH
sas-1	1426	tcp/udp	Satellite-data Acquisition System 1
sas-2	1436	tcp/udp	Satellite-data Acquisition System 2
sas-3	1501	tcp/udp	Satellite-data Acquisition System 3
satvid-datalnk	3367-3371	tcp/udp	Satellite Video Data Link—Intel
savant	3391	tcp/udp	SAVANT—TheProgrammers
sbook	1349	tcp/udp	Registration Network Protocol—NEXT
scan-change	2719	tcp/udp	Scan & Change—Lucent
scanstat-1	1215	tcp/udp	scanSTAT 1.0—ISI
scc-security	582	tcp/udp	SCC Security—Semaphorecom
scientia-sdb	1811	tcp/udp	Scientia-SDB
scipticslsrvr	2577	tcp/udp	Scriptics Lsrvr
sco-aip	5307	tcp/udp	SCO AIP
sco-dtmgr	617	tcp/udp	SCO Desktop Administration Server
scohelp	457	tcp/udp	scohelp
scoi2odialog	360	tcp/udp	scoi2odialog
sco-inetmgr	615	tcp/udp	Internet Configuration Manager
scol	1200	tcp/udp	SCOL—Cryo-Networks
scoremgr	2034	tcp/udp	
sco-sysmgr	616	tcp/udp	SCO System Administration Server
sco-websrvrmg3	598	tcp/udp	SCO Web Server Manager 3
sco-websrvrmgr	620	tcp/udp	SCO WebServer Manager

Abbreviation	Port-Decimal	Protocol	Description/Owner
scrabble	2026	tcp/udp	
screencast	1368	tcp/udp	ScreenCast
sctp	9006	tcp/udp	SCTP—Tagnet
scx-proxy	470	tcp/udp	scx-proxy—Cray
sd	9876	tcp/udp	Session Director—LBL.GOV
sdclient	2310	tcp/udp	SD Client—HP
sde-discovery	5152	tcp/udp	ESRI SDE Instance Discovery
sd-elmd	1681	tcp/udp	sd-elmd—SoftDesk
sdfunc	2046	tcp/udp	
sdfunc	2046	tcp/udp	
sdhelp	2308	tcp/udp	sdhelp—HP
sdnskmp	558	tcp/udp	SDNSKMP
sds-admin	2705	tcp/udp	SDS Admin—Sun
sdsc-lm	1537	tcp/udp	isi-lm
sdserver	2309	tcp/udp	SD Server—HP
sdt-lmd	3319	tcp/udp	SDT License Manager—Tin
seaodbc	2471	tcp/udp	SeaODBC—Aran
search	2010	tcp	
search-agent	1234	tcp/udp	Infoseek Search Agent
seaview	3143	tcp/udp	Sea View—Seagate
secure-aux-bus	664	tcp/udp	Secure Aux Bus—Intel
semantix	361	tcp/udp	Semantix
send	169	tcp/udp	SEND—U Alaska
sent-lm	2316	tcp/udp	SENT License Manager—Axix Inc
sercomm-wlink	2235	tcp/udp	Sercomm-Wlink—Sercomm
server-find	8351	tcp/udp	Server Find—Chancery
servexec	2021	tcp	
service-ctrl	2367	tcp/udp	Service Control—HP
servicemeter	2603	tcp/udp	Service Meter—Synoia
servserv	2011	udp	
servstat	633	tcp/udp	Service Status update (Sterling Software)
sesi-lm	1714	tcp/udp	sesi-lm—Side Effects SW
set	257	tcp/udp	Secure Electronic Transaction—Pothole Industries
sf-lm	4546	tcp/udp	SF License Manager (Sentinel)
sfs-config	452	tcp/udp	Cray SFS config server
sfs-smp-net	451	tcp/udp	Cray Network Semaphore server
sftp	115	tcp/udp	Simple File Transfer Protocol—SRI
sftu	3326	tcp/udp	SFTU—Spacenet
sgcp	440	tcp/udp	sgcp—Data Broadcasting Corporation
sgi-esphttp	5554	tcp/udp	SGI ESP HTTP
sg-lm	1659	tcp/udp	Silicon Grail License Manager
sgmp	153	tcp/udp	SGMP—NY State Ed Net (NYSER.NET)
sgmp-traps	160	tcp/udp	SGMP-TRAPS—NY State Ed Net (NYSER.NET)
shadowserver	2027	tcp/udp	
shell	514	tcp	"cmd like exec, but automatic authentication"
shilp	2049	tcp/udp	
shiva_confsrvr	1651	tcp/udp	shiva_confsrvr
shivadiscovery	1502	tcp/udp	Shiva
shivahose	1549	tcp	Shiva Hose
shivasound	1549	udp	Shiva Sound
shockwave	1626	tcp/udp	Shockwave—Macromedia
shrinkwrap	358	tcp/udp	Shrinkwrap—Green Dragon
siam	498	tcp/udp	siam—Computer Answer Line
siebel-ns	2320	tcp/udp	Siebel NS
sift-uft	608	tcp/udp	Sender-Initiated/Unsolicited File Transfer—Rice U
signal	2974	tcp/udp	Signal—Icall
signet-ctf	2733	tcp/udp	Signet CTF—SAC
siipat	1733	tcp/udp	SIMS - SIIPAT Protocol for Alarm Transmission—SimWare
silkp1	2829	tcp/udp	silkp1—Silknet Inc.
silkp2	2830	tcp/udp	silkp2—Silknet Inc.
silkp3	2831	tcp/udp	silkp3—Silknet Inc.
silkp4	2832	tcp/udp	silkp4—Silknet Inc.
silverplatter	416	tcp/udp	Silverplatter
simba-cs	1543	tcp/udp	simba-cs
simbaexpress	1583	tcp/udp	simbaexpress
simbaservices	1599	tcp/udp	simbaservices
simctlp	2857	tcp/udp	SimCtIP—GMX Inc.
simplement-tie	2756	tcp/udp	simplement-tie—Netvision
simslink	2676	tcp/udp	SIMSLink—Simswsare
singlept-mvs	2769	tcp/udp	Single Point MVS

Abbreviation	Port-Decimal	Protocol	Description/Owner
sip	5060	tcp/udp	SIP—Columbia U
sis-emt	2545	tcp/udp	sis-emt—Securicor
sitaradir	2631	tcp/udp	Sitara Dir
sitaramgmt	2630	tcp/udp	Sitara Management
sitaraserver	2629	tcp/udp	Sitara Server
sixnetudr	1658	tcp/udp	sixnetudr—Wizvax Communications
sixtrak	1594	tcp/udp	sixtrak—Wizvax Communications
skip-cert-recv	6455	tcp	SKIP Certificate Receive—SUN
skip-cert-send	6456	tcp	SKIP Certificate Send—Sun
skip-mc-gikreq	1660	tcp/udp	skip-mc-gikreq—Sun
skronk	460	tcp/udp	skronk
skytelnet	1618	tcp/udp	skytelnet—Bell South
slc-ctrlrloops	2827	tcp/udp	slc ctrlrloops—Demon Internet ISP
slc-systemlog	2826	tcp/udp	slc systemlog—Demon Internet ISP
slingshot	1705	tcp/udp	slingshot—CSK Software
slinterbase	3065	tcp/udp	slinterbase—InterBase
slm-api	1606	tcp/udp	Salutation Manager (SLM-API)—IBM
slp	1605	tcp/udp	Salutation Manager (Salutation Protocol)—IBM
slush	1966	tcp/udp	Slush—IBS
smakynet	122	tcp/udp	SMAKYNET—Swiss Federal Institute of Technology
smart-diagnose	2721	tcp/udp	Smart Diagnose—Geoffry Meek
smart-lm	1608	tcp/udp	Smart Corp. License Manager
smartsdp	426	tcp/udp	smartsdp
smntubootstrap	2613	tcp/udp	SMNTUBootstrap—Metrics
smpp	2775	tcp/udp	SMMP—Logica Aldiscon
smpte	420	tcp/udp	SMPTE—CompuServe
sms-chat	2703	tcp/udp	SMS CHAT—Microsoft
smsd	596	tcp/udp	SMSD—DEC
smsp	413	tcp/udp	SMSP
sms-rcinfo	2701	tcp/udp	SMS RCINFO—Microsoft
sms-remctrl	2704	tcp/udp	SMS REMCTRL—Microsoft
sms-xfer	2702	tcp/udp	SMS XFER—Microsoft
smtp	25	tcp/udp	Simple Mail Transfer—ISI
smux	199	tcp/udp	SMUX—Data Broadcasting Corporation
sna-cs	1553	tcp/udp	sna-cs—Data Connection Ltd.
snagas	108	tcp/udp	SNA Gateway Access Server—DEC
snapenetio	22000	tcp/udp	SNAPenetIO
snapp	2333	tcp/udp	SNAPP—Meridian-data
snare	509	tcp/udp	snare—Capres
snaresecure	1684	tcp/udp	SnareSecure—Copenhagen Applied Research
s-net	166	tcp/udp	Sirius Systems
snifferclient	2452	tcp/udp	SnifferClient—NAI
snifferserver	2533	tcp/udp	SnifferServer—NAI
snmp	161	tcp/udp	SNMP—Data Broadcasting Corporation
snmp-tcp-port	1993	tcp/udp	cisco SNMP TCP port
snmptrap	162	tcp/udp	SNMPTRAP—Data Broadcasting Corporation
snpp	444	tcp/udp	Simple Network Paging Protocol [RFC1568]
sns-admin	2658	tcp/udp	SNS Admin—FirstFloor
sns-agent	5417	tcp/udp	SNS Agent—FirstFloor
sns-channels	3380	tcp/udp	SNS Channels—FirstFloor
sns-dispatcher	2657	tcp/udp	SNS Dispatcher—FirstFloor
sns-gateway	5416	tcp/udp	SNS Gateway—FirstFloor
sns-protocol	2409	tcp/udp	SNS Protocol—NetManage
sns-query	2659	tcp/udp	SNS Query—FirstFloor
sns-quote	1967	tcp/udp	Quote—Calicotech
sntp-heartbeat	580	tcp/udp	SNTP HEARTBEAT—UUNET (owned by MCI)
socks	1080	tcp/udp	Socks—NEC
socorfs	3379	tcp/udp	SOCORFS
softcm	6110	tcp/udp	HP SoftBench CM
softcm	6110	udp	HP SoftBench CM
softdataphone	1621	tcp/udp	softdataphone—IBM
softpc	215	tcp/udp	Insignia Solutions
solid-e-engine	1964	tcp/udp	SOLID E ENGINE—Solidtech
solve	2636	tcp/udp	Solve—Sterling
sonar	572	tcp/udp	sonar—U Tennessee Knoxville
sonardata	2863	tcp/udp	Sonar Data
sonus	2653	tcp/udp	Sonus—SonusNet
sonuscallsig	2569	tcp/udp	Sonus Call Signal
s-openmail	5767	tcp/udp	OpenMail Suer Agent Layer (Secure)—HP
sophia-lm	1408	tcp/udp	Sophia License Manager

Abbreviation	Port-Decimal	Protocol	Description/Owner
spamtrap	2568	tcp/udp	SPAM TRAP—BenaTong Software
spc	6111	tcp/udp	HP SoftBench Sub-Process Control
spearway	2440	tcp/udp	Spearway Lockers
spice	1923	tcp/udp	SPICE—Sendit
spmp	656	tcp/udp	SPMP—NoDomainName
spock	2507	tcp/udp	spock—Texas A&M U
spsc	478	tcp/udp	spsc—Software Partners
spss-lm	1759	tcp/udp	SPSS License Manager
sps-tunnel	2876	tcp/udp	SPS Tunnel—FortressTech
sqdr	2728	tcp/udp	SQDR—Starquest
sql*net	66	tcp/udp	Oracle SQL*NET
sql-net	150	tcp/udp	SQL-NET
sqlserv	118	tcp/udp	SQL Services—DEC
sqlsrv	156	tcp/udp	SQL Service—ISI
src	200	tcp/udp	IBM System Resource Controller
srmp	193	tcp/udp	Spider Remote Monitoring Protocol—SPIDER Systems UK
srp-feedback	2737	tcp/udp	SRP Feedback—Swiss Federal Institute of Technology
srsp	2682	tcp/udp	SRSP—umu
srssend	362	tcp/udp	SRS Send—Emergent
srvc_registry	3018	tcp/udp	Service Registry—Novell
ss7ns	477	tcp/udp	ss7ns—DEC
ssdp	1900	tcp/udp	SSDP—Microsoft
ssh	22	tcp/udp	SSH Remote Login Protocol—Helsinki U
sshell	614	tcp/udp	SSLshell—Quick
sslp	1750	tcp/udp	Simple Socket Library's PortMaster—NASA
ssm-cssps	2478	tcp/udp	SecurSight Authentication Server (SLL)—SecurityDynamics
ssm-cvs	2477	tcp/udp	SecurSight Certificate Valifation Service—SecurityDynamics
ssm-els	2479	tcp/udp	SecurSight Event Logging Server (SSL)—SecurityDynamics
sso-control	2711	tcp/udp	SSO Control—Okiok
sso-service	2710	tcp/udp	SSO Service—Okiok
ssql	3352	tcp/udp	SSQL—Pervasive-SW
ssrip	3318	tcp/udp	Swith to Swith Routing Information Protocol—Fujitsu
sstsys-lm	1692	tcp/udp	sstsys-lm—Netcom
starbot	2838	tcp/udp	Starbot
stargatealerts	1654	tcp/udp	stargatealerts—Intel
statsci1-lm	6144	tcp/udp	StatSci License Manager - 1
statsci2-lm	6145	tcp/udp	StatSci License Manager - 2
statsrv	133	tcp/udp	Statistics Service—U Delaware
statusd	5414	tcp/udp	StatusD—SatelNet
ste-smsc	1836	tcp/udp	ste-smsc—ST-Electronics
sti-envision	1312	tcp/udp	STI Envision—STIsystems
stm_pproc	3080	tcp/udp	stm_pproc—STMI
stmf	501	tcp/udp	STMF—Farradyne
stone-design-1	1492	tcp/udp	stone-design-1
street-stream	1736	tcp/udp	street-stream—Netcom
streettalk	566	tcp/udp	streettalk
stt	1607	tcp/udp	stt—Microsoft
stun-p1	1990	tcp/udp	cisco STUN Priority 1 port
stun-p2	1991	tcp/udp	cisco STUN Priority 2 port
stun-p3	1992	tcp/udp	cisco STUN Priority 3 port
stun-port	1994	tcp/udp	cisco serial tunnel port
stx	527	tcp/udp	Stock—IXChange
submission	587	tcp/udp	Submission—Qualcomm
submit	773	tcp	
submitserver	2028	tcp/udp	
subntbcst_tftp	247	tcp/udp	SUBNTBCST_TFTP—IBM
sugp	1905	tcp/udp	Secure UP.Link Gateway Protocol—UpLANet
suitjd	3354	tcp/udp	SUITJD—DEC
su-mit-tg	89	tcp/udp	SU/MIT Telnet Gateway
sunclustermgr	1097	tcp/udp	Sun Cluster Manager
sun-dr	665	tcp/udp	Sun DR
sun-lm	7588	tcp/udp	Sun License Manager
sun-mc-grp	5306	tcp/udp	Sun MC Group
sunrpc	111	tcp/udp	SUN Remote Procedure Call
sunscalar-dns	1870	tcp/udp	SunSCALAR DNS Service
sunscalar-svc	1860	tcp/udp	SunSCALAR Services
supdup	95	tcp/udp	SUPDUP—panda.com
supercell	7967	tcp/udp	Supercell—SigSci
supermon	2709	tcp/udp	Supermon—LANL.gov
surf	1010	tcp/udp	surf—Peapod

Abbreviation	Port-Decimal	Protocol	Description/Owner
sur-meas	243	tcp/udp	Survey Measurement—MIT
svnetworks	2973	tcp/udp	SV Networks
svrloc	427	tcp/udp	Server Location—FTP Software
svs-omagent	1625	tcp/udp	sys-omagent—IBM
swiftnet	1751	tcp/udp	"SwiftNet-Pentek, Inc."
swift-rvf	97	tcp/udp	Swift Remote Virtural File Protocol—U Florida
swismgr1	6963	tcp/udp	swismgr1—Fujitsu
swismgr2	6964	tcp/udp	swismgr2—Fujitsu
swispol	6966	tcp/udp	swispol—Fujitsu
swistrap	6965	tcp/udp	swistrap—Fujitsu
swx	7300-7390	tcp/udp	The Swiss Exchange
sxmp	3273	tcp/udp	Simple Extensible Multiplexed Protocol—WEBDash
sybaseanywhere	2638	tcp/udp	Sybase Anywhere
sybasedbsynch	2439	tcp/udp	SybaseDBSynch
symbios-raid	2463	tcp/udp	Symbios Raid
symplex	1507	tcp/udp	symplex
synapse	2880	tcp/udp	synapse—BIX
syncserver	2647	tcp/udp	SyncServer—SyncInc
syncserverssl	2679	tcp/udp	Sync Server SSL—SyncInc
synoptics-trap	412	tcp/udp	Trap Convention Port
synotics-broker	392	tcp/udp	SynOptics Port Broker Port
synotics-relay	391	tcp/udp	SynOptics SNMP Relay Port
syslog	514	udp	
sysopt	3281	tcp/udp	SYSOPT—DEC
systat	11	tcp/udp	Active Users-ISI
systemics-sox	5406	tcp/udp	Systemics Sox
system-monitor	2609	tcp/udp	System Monitor—AlphaLink
t128-gateway	1627	tcp/udp	T.128 Gateway—Data Connection Ltd.
t2-brm	7933	tcp/udp	Tier 2 Business Rules Manager
t2-drm	7932	tcp/udp	Tier 2 Data Resource Manager
tabula	1437	tcp/udp	Tabula
tacacs	49	tcp/udp	Login Host Protocol (TACACS)—BBN
tacacs-ds	65	tcp/udp	TACACS-Database Service—BBN
tacnews	98	tcp/udp	TAC News—ISI
tacticalauth	2392	tcp/udp	Tactical Auth
taiclock	4014	tcp/udp	TAICLOCK—UIC
talarian-mcast1	4015	tcp/udp	Talarian Mcast
talarian-mcast2	4016	tcp/udp	Talarian Mcast
talarian-mcast3	4017	tcp/udp	Talarian Mcast
talarian-mcast4	4018	tcp/udp	Talarian Mcast
talarian-mcast5	4019	tcp/udp	Talarian Mcast
talarian-mqs	2493	tcp/udp	Talarian MQS
taligent-lm	1475	tcp/udp	Taligent License Manager
talk	517	tcp/udp	"like tenex link, but across machines"
talnet	1838	tcp/udp	TALNET-Taltrade
talon-disc	7011	tcp/udp	Talon Discovery Port—PowerWare
talon-engine	7012	tcp/udp	Talon Engine—PowerWare
tal-pod	6149	tcp/udp	tal-pod—Taligent
tams	2726	tcp/udp	TAMS—JTI
tapestry	1922	tcp/udp	Tapestry—Anritsu
tappi-boxnet	2306	tcp/udp	TAPPI BoxNet—Alliancetechnical
tarantella	3144	tcp/udp	Tarantella—SCO
targus-aib1	5200	tcp/udp	Targus AIB 1
targus-aib2	5201	tcp/udp	Targus AIB 2
targus-tnts1	5202	tcp/udp	Targus TNTS 1
targus-tnts2	5203	tcp/udp	Targus TNTS 2
taskman-port	2470	tcp/udp	taskman port—Himel
taskmaster2000	2402	tcp/udp	TaskMaster 2000 Server—Datacap
taskmaster2000	2403	tcp/udp	TaskMaster 2000 Web—Datacap
taurus-wh	1610	tcp/udp	taurus-wh
tcim-control	2729	tcp/udp	TCIM Control—Raytheon
tclprodebugger	2576	tcp/udp	TCL Pro Debugger
tclprodebugger	2576	udp	TCL Pro Debugger
tcoaddressbook	1977	tcp/udp	TCO Address Book—TCOsoft
tcoflashagent	1975	tcp/udp	TCO Flash Agent—TCOsoft
tcoregagent	1976	tcp/udp	TCO Reg Agent—TCOsoft
tcp-id-port	1999	tcp/udp	cisco identification port
tcpmux	1	tcp/udp	TCP Port Service Multiplexer—SRI
tcpnethaspsrv	475	tcp/udp	tcpnethaspsrv—Aladdin
tdaccess	2910	tcp/udp	TDAccess—Metrics

Abbreviation	Port-Decimal	Protocol	Description/Owner
tdp-suite	1814	tcp/udp	TDP Suite—Motorola Inc.
techra-server	1862	tcp/udp	techra-server—Maxware
teedtap	559	tcp/udp	TEEDTAP—GTEgsc
tekpls	1946	tcp/udp	tekpls—Tektronix
telefinder	1474	tcp/udp	Telefinder
telelpathattack	5011	tcp/udp	TelepathAttack—IMB
telelpathstart	5010	tcp/udp	TelepathStart—IMB
telesis-licman	1380	tcp/udp	Telesis Network License Manager
telexis-vtu	2804	tcp/udp	Telexis VTU
telindus	1728	tcp/udp	TELINDUS
tell	754	tcp/udp	send
telnet	23	tcp/udp	Telnet—ISI
telnets	992	tcp/udp	telnet protocol over TLS/SSL—Consensus
telops-lmd	7491	tcp/udp	telops-lmd
tempo	526	tcp/udp	newdate
tenebris_nts	359	tcp/udp	Tenebris Network Trace Service
tenfold	658	tcp/udp	TenFold
terabase	4000	tcp/udp	Terabase
teradataordbms	8002	tcp/udp	Teradata ORDBMS—NCR
terminaldb	2008	udp	
terminaldb	2018	tcp	
texar	333	tcp/udp	Texar Security Port
tftp	69	tcp/udp	Trivial File Transfer—MIT
tftp-mcast	1758	tcp/udp	tftp-mcast—LanWorks
theta-lm	2296	tcp/udp	Theta License Manager (Rainbow)
tht-treasure	1832	tcp/udp	ThoughtTreasure—Signiform
ticf-1	492	tcp/udp	Transport Independent Convergence for FNA-PFU Direct
ticf-2	493	tcp/udp	Transport Independent Convergence for FNA
timbuktu	407	tcp/udp	Timbuktu—Netopia
timbuktu-srv1	1417	tcp/udp	Timbuktu Service 1 Port—Netopia
timbuktu-srv2	1418	tcp/udp	Timbuktu Service 2 Port—Netopia
timbuktu-srv3	1419	tcp/udp	Timbuktu Service 3 Port—Netopia
timbuktu-srv4	1420	tcp/udp	Timbuktu Service 4 Port—Netopia
time	37	tcp/udp	Time—ISI
timed	525	tcp/udp	timeserver
timeflies	1362	tcp/udp	TimeFlies—nwnexus
tinc	655	tcp/udp	TINC—Bigfoot
tip2	3372	tcp/udp	TIP 2—Tandem
tivoli-npm	1965	tcp/udp	Tivoli NPM
tksocket	2915	tcp/udp	TK Socket—Protix
tl1	2361	tcp/udp	TL1—Cerent
tl1-lv	3081	tcp/udp	TL1-LV—SONET Internetworking Forum (SIF)
tl1-raw	3082	tcp/udp	TL1-RAW—SONET Internetworking Forum (SIF)
tl1-telnet	3083	tcp/udp	TL1-TELNET—SONET Internetworking Forum (SIF)
tlisrv	1527	tcp/udp	oracle
tmesis-upshot	2798	tcp/udp	TMESIS-UPShot
tmlogin	5311	tcp/udp	TM Login—Sun
tnETOS	377	tcp/udp	NEC Corporation
tns-adv	3309	tcp/udp	TNS ADV—DEC
tns-cml	590	tcp/udp	TNS CML—DEC
tns-server	3308	tcp/udp	TNS Server—DEC
tn-timing	2739	tcp/udp	TN Timing
tn-tl-fd1	476	tcp/udp	tn-tl-fd1—ThinkNet
tn-tl-fd2	1584	tcp/udp	tn-tl-fd2—ThinkNet Inc.
tn-tl-r2	1580	tcp/udp	tn-tl-r2—Thinknet
tn-tl-w1	474	tcp/udp	tn-tl-w1—ThinkNet
toad	2669	tcp/udp	TOAD—Tatanka
tomato-springs	3040	tcp/udp	Tomato Springs
topflow	2885	tcp/udp	TopFlow—TopLayer
topx	2436	tcp/udp	TOP/X—ubbcluj
torque-traffic	2505	tcp/udp	torque-traffic
tpdu	1430	tcp/udp	Hypercom TPDU
tpip	594	tcp/udp	TPIP—Platinum
tpmd	1906	tcp/udp	TPortMapperReq—IBM
tqdata	2700	tcp/udp	tqdata—TeamQuest
traceroute	33434	tcp/udp	traceroute use—IANA
track	20670	tcp/udp	Track—US Naval Academy
tragic	2642	tcp/udp	Tragic—j51
tram	4567	tcp/udp	TRAM—SUN
translogic-lm	1753	tcp/udp	Translogic License Manager

Abbreviation	Port-Decimal	Protocol	Description/Owner
transview	3298	tcp/udp	Transview—SNI
travsoft-ipx-t	2644	tcp/udp	Travsoft IPX Tunnel
trc-netpoll	2405	tcp/udp	TRC Netpoll—Telcores
tributary	2580	tcp/udp	Tributary—Bristol
triquest-lm	1588	tcp/udp	triquest-lm
trivnet1	8200	tcp/udp	TRIVNET
trivnet2	8201	tcp/udp	TRIVNET
trnsprntproxy	3346	tcp/udp	Trnsprnt Proxy—Intel
troff	2014	tcp	
tr-rsrb-p1	1987	tcp/udp	cisco RSRB Priority 1 port
tr-rsrb-p2	1988	tcp/udp	cisco RSRB Priority 2 port
tr-rsrb-p3	1989	tcp/udp	cisco RSRB Priority 3 port
tr-rsrb-port	1996	tcp/udp	cisco Remote SRB port
truecm	8804	tcp/udp	truecm—True Blue Software
trusted-web	3011	tcp/udp	Trusted Web—SSE
trustestablish	2573	tcp/udp	Trust Establish—IBM
tsaf	12753	tcp/udp	tsaf port
tsb	2741	tcp/udp	TSB—Lucent
tsb2	2742	tcp/udp	TSB2—Lucent
tscchat	2330	tcp/udp	TSCCHAT
tserver	450	tcp/udp	Tserver—AT&T
tsilb	2489	tcp/udp	TSILB—Travsoft
tsspmap	1568	tcp/udp	tsspmap
ttc	2483	tcp/udp	Oracle TTC
ttc-etap	2675	tcp/udp	TTC ETAP
ttc-etap-ds	2978	tcp/udp	TTCs Enterprise Test Access Protocol - DS
ttc-etap-ns	2977	tcp/udp	TTCs Enterprise Test Access Protocol - NS
ttcmremotectrl	3468	tcp/udp	TTCM Remote Controll—Proxy
ttc-ssl	2484	tcp/udp	Oracle TTC SSL
ttg-protocol	2862	tcp/udp	TTG Protocol—Travel Technologies Group
ttyinfo	2012	tcp	
tve-announce	2670	tcp/udp	TVE Announce—WebTV
tw-auth-key	27999	tcp/udp	TW Authentication/Key Distribution and Attribute Certificate
twsdss	3012	tcp/udp	Trusted Web Client—SSE
uaac	145	tcp/udp	UAAC Protocol—Mitre
uacs	2768	tcp/udp	UACS—AppliedIS
uadtc	2767	tcp/udp	UADTC—AppliedIS
uaiact	1470	tcp/udp	Universal Analytics
uarps	219	tcp/udp	Unisys ARPs
ucentric-ds	2879	tcp/udp	ucentric-ds
udlr-dtcp	652	tcp/udp	UDLR_DTCP—INRIA
udpradio	1833	tcp/udp	udpradio—Warande
udp-sr-port	1624	tcp/udp	udp-sr-port—ITT Industries
udt_os	1382	tcp/udp	
udt_os	3900	tcp/udp	Unidata UDT OS
uis	390	tcp/udp	UIS
ulistproc	372	tcp/udp	ListProcessor—Boston U
ulp	522	tcp/udp	ULP—Microsoft
ulpnet	483	tcp/udp	ulpnet—Unibol
ultimad	1737	tcp/udp	ultimad—Ultima Interconnect Technology
ultrabac	1910	tcp/udp	ultrabac—CNW
ultraseek-http	8765	tcp/udp	Ultraseek HTTP—Infoseek
uma	144	tcp/udp	Universal Management Architecture—Power Center Software
uma	1797	tcp/udp	UMA—The Open Group
unglue	2655	tcp/udp	UNIX Nt Glue—Pscomp
unicall	4343	tcp/udp	UNICALL—UniData
unicontrol	2437	tcp/udp	UniControl—Hardware Software Development GmBh
unicontrol	2499	tcp/udp	UniControl
unidata-ldm	388	tcp/udp	Unidata LDM Version 4
unify	181	tcp/udp	Unify
unifyadmin	2696	tcp/udp	Unify Admin
unihub-server	2357	tcp/udp	UniHub Server—CSIcontrols
unisql	1978	tcp/udp	UniSQL—Windtraveller
unisql-java	1979	tcp/udp	UniSQL Java—Windtraveller
unisys-lm	1823	tcp/udp	Unisys Natural Language License Manager
univision	2820	tcp/udp	UniVision—FastFreeNet ISP
unot	5055	tcp/udp	UNOT—CmgiSolutions
uohost	3314	tcp/udp	Unify Object Host
uorb	3313	tcp/udp	Unify Object Broker
upgrade	2537	tcp/udp	Upgrade Protocol—DST-Inc

Abbreviation	Port-Decimal	Protocol	Description/Owner
upnotifyp	4445	tcp/udp	UPNOTIFYP—UpLANet
upnotifyps	3356	tcp/udp	UPNOTIFYPS—UpLANe
ups	401	tcp/udp	Uninterruptible Power Supply—BenaTong Softgware
upsnotifyprot	2648	tcp/udp	Upsnotifyprot—Via-RS
ups-onlinet	7010	tcp/udp	onlinet uninterruptable power supplies—Exide
urbisnet	2745	tcp/udp	URBISNET
url-rendezvous	659	tcp/udp	URL Rendezvous—Cisco
urm	606	tcp/udp	Cray Unified Resource Manager
urn-x-cdchoice	4501	tcp/udp	urn-x-cdchoice—Proper
user-manager	3272	tcp/udp	Fujitsu User Manager
us-gv	1370	tcp/udp	Unix Shell to GlobalView—Fuji-Xerox
utcd	1506	tcp/udp	Universal Time daemon (utcd)
utime	519	tcp/udp	unixtime
utmpcd	431	tcp/udp	UTMPCD
utmpsd	430	tcp/udp	UTMPSD
utsftp	2529	tcp/udp	UTS FTP
uucp	540	tcp/udp	uucpd
uucp-path	117	tcp/udp	UUCP Path Service
uucp-rlogin	541	tcp/udp	uucp-rlogin—Wimsey
vacdsm-app	671	tcp/udp	VACDSM-APP
vacdsm-sws	670	tcp/udp	VACDSM-SWS
valisys-lm	1457	tcp/udp	Valisys License Manager
vat	3456	tcp/udp	VAT default data—LBL Gov.
vat-control	3457	tcp/udp	VAT default control—LBL Gov
vatp	690	tcp/udp	VATP—Aticasoft
vaultbase	1771	tcp/udp	vaultbase
vce	11111	tcp/udp	Viral Computing Environment (VCE)—Fred Cohen & Assoc.
vcnet-link-v10	2554	tcp/udp	VCnet-Link v10—Csaba Mate
vcom-tunnel	8001	tcp/udp	VCOM Tunnel—IBM
vdmplay	1707	tcp/udp	vdmplay
vemmi	575	tcp/udp	VEMMI—Monaco Telematique
venus	2430	tcp/udp	venus—Robert Watson
venus-se	2431	tcp/udp	venus-se—Robert Watson
vergencecm	2771	tcp/udp	Vergence CM—Sentillion
verismart	3270	tcp/udp	Verismart
veritas-tcp1	2802	tcp/udp	Veritas TCP1
veronica	2770	tcp/udp	Veronica—Coyote
versa-tek	2610	tcp/udp	VersaTek
vettcp	78	tcp/udp	vettcp—DEC
vid	769	tcp/udp	
video-activmail	1398	tcp/udp	Video Active Mail
videobeans	3058	tcp/udp	videobeans—Sun
videotex	516	tcp/udp	videotex—Monaco Telamatique
vinainstall	4344	tcp/udp	VinaInstall—Vina-Tech
virprot-lm	7121	tcp/udp	Virtual Prototypes License Manager
virtual-places	1533	tcp/udp	Virtual Places Software
vision_elmd	6673	tcp/udp	vision_elmd—SHL
vision_server	6672	tcp/udp	vision_server—SHL
visitview	1631	tcp/udp	Visit view—U Wisconsin
vista-4gl	1895	tcp/udp	vista 4GL—Vistacomp
vistium-share	1545	tcp/udp	vistium-share
vitalanalysis	2474	tcp/udp	Vital Analysis—VitalSigns
vlsi-lm	1500	tcp/udp	VLSI License Manager
vml-dms	2821	tcp/udp	vml_dms—VERITAS Software
vmnet	175	tcp/udp	VMNET—Princeton U
vmodem	3141	tcp/udp	VMODEM—PSLink
vmpwscs	214	tcp/udp	VM PWSCS
vnas	577	tcp/udp	vnas—Ipass
vnsstr	3321	tcp/udp	VNSSTR—Mei
vnwk-prapi	2538	tcp/udp	vnwk-prapi—VisualNetworks
vocaltec-admin	1796	tcp/udp	Vocaltec Server Administration
vocaltec-gold	6670	tcp/udp	Vocaltec Global Online Directory
vocaltec-hos	25793	tcp/udp	Vocaltec Address Server
vocaltec-wconf	22555	tcp/udp	Vocaltec Web Conference
vofr-gateway	21590	tcp/udp	VoFR Gateway—Tollbridge Technology
v-one-spp	3845	tcp/udp	V-ONE Single Port Proxy
vopied	13783	tcp/udp	VOPIED Protocol—VERITAS Software
vp2p	8473	tcp/udp	Virtual Point to Point—Atos-Group
vpac	1517	tcp/udp	Virtual Places Audio control
vpad	1516	tcp/udp	Virtual Places Audio data

Abbreviation	Port-Decimal	Protocol	Description/Owner
vpjp	1345	tcp/udp	VPJP—America Online
vpp	677	tcp/udp	Virtual Presence Protocol—Cobrow
vpps-qua	672	tcp/udp	VPPS-QUA
vpps-via	676	tcp/udp	VPPS Via—Cesi
vpsipport	2649	tcp/udp	VPSIPPORT—CSIR
vpvc	1519	tcp/udp	Virtual Places Video control
vpvd	1518	tcp/udp	Virtual Places Video data
vqp	1589	tcp/udp	VQP—Cisco
vrcommerce	2530	tcp/udp	VR Commerce—IBM
vrml-multi-use	4200-4299	tcp/udp	VRML Multi User Systems
vrts-registry	2410	tcp/udp	VRTS Registry—VERITAS Software
vsaiport	3317	tcp/udp	VSAI PORT—Mei
vsiadmin	2539	tcp/udp	VSI Admin
vsinet	996	tcp/udp	vsinet—VSI
vsi-omega	7566	tcp/udp	VSI Omega
vsixml	2996	tcp/udp	vsixml
vslmp	312	tcp/udp	VSLMP
vsnm-agent	3375	tcp/udp	VSNM Agent—VitalSigns
vspread	2695	tcp/udp	VSPREAD—Fujitsu
vs-server	3280	tcp/udp	VS Server—Godlew
vytalvaultbrtp	2546	tcp/udp	vytalvaultbrtp—Vytalnet
vytalvaultpipe	2548	tcp/udp	vytalvaultpipe—Vytalnet
vytalvaultvsmp	2547	tcp/udp	vytalvaultvsmp—Vytalnet
wago-io-system	2455	tcp/udp	WAGO-IO-SYSTEM—IEEE
wag-service	2608	tcp/udp	Wag Service
wap-vcal	9205	tcp/udp	WAP vCal
wap-vcal-s	9207	tcp/udp	WAP vCal Secure
wap-vcard	9204	tcp/udp	WAP vCard
wap-vcard-s	9206	tcp/udp	WAP vCard Secure
wap-wsp	9200	tcp/udp	WAP connectionless session service
wap-wsp-s	9202	tcp/udp	WAP secure connectionless session service
wap-wsp-wtp	9201	tcp/udp	WAP session service
wap-wsp-wtp-s	9203	tcp/udp	WAP secure session service
watchdognt	2723	tcp/udp	WatchDog NT
watcomdebug	3563	tcp/udp	Watcom Debug
watcom-sql	1498	tcp/udp	Watcom-SQL
watershed-lm	6143	tcp/udp	Watershed License Manager—Zion
web2host	1559	tcp/udp	web2host—Mindspring
webaccess	1739	tcp/udp	webaccess—Asymetrix Learning Systems
webemshttp	2851	tcp/udp	webemshttp—Jetstream Communications
webobjects	1085	tcp/udp	Web Objects—Apple
webphone	21845	tcp/udp	webphone
websm	9090	tcp/udp	WebSM—IBM
webster	765	tcp/udp	
webtie	3342	tcp/udp	WebTIE—NOAA.gov
wg-netforce	3359	tcp/udp	WG NetForce
who	513	udp	maintains data bases showing who's logged in
whoami	565	tcp/udp	whoami
whois++	63	tcp/udp	whois++—Sunet
whosells	2781	tcp/udp	whosells—GNSS
whosockami	2009	udp	
whosockami	2019	tcp	
willy	2518	tcp/udp	Willy—Hudsmoar
wimd	2980	tcp/udp	Instant Messaging Service—Pobox
windb	2522	tcp/udp	WinDb—Arium
winddlb	1565	tcp/udp	WinDD—Tektronix
winddx	1727	tcp/udp	winddx—Tektronix
windlm	1785	tcp/udp	Wind River Systems License Manager
winqedit	7395	tcp/udp	winqedit—Robelle
wins	1512	tcp/udp	Microsoft's Windows Internet Name Service
wizard	2001	udp	curry
wkars	2720	tcp/udp	wkars—WirelessKnowledge
wlbs	2504	tcp/udp	WLBS—Microsoft
wnn6	22273	tcp/udp	wnn6—Omron Software
wnn6-ds	26208	tcp/udp	wnn6-ds—Omron Software
workflow	3466	tcp/udp	WORKFLOW—Csesys
work-sol	400	tcp/udp	Workstation Solutions
worldfusion1	2595	tcp/udp	World Fusion 1
worldfusion2	2596	tcp/udp	World Fusion 2

Abbreviation	Port-Decimal	Protocol	Description/Owner
world-lm	1462	tcp/udp	World License Manager
worldscores	4545	tcp/udp	WorldScores
wpages	776	tcp/udp	
wpgs	780	tcp/udp	
wrs_registry	2340	tcp/udp	WRS Registry
wsicopy	3378	tcp/udp	WSICOPY
wsm-server	5006	tcp/udp	wsm server—WRQ
wsm-server-ssl	5007	tcp/udp	wsm server ssl—WRQ
wta-wsp-s	2805	tcp/udp	WTA WSP-S—Alcatel
wusage	2396	tcp/udp	Wusage—Thomas Boutell
wwiotalk	5413	tcp/udp	WWIOTALK—Wonder Ware
www	80	tcp/udp	World Wide Web HTTP-World Wide Web Consortium
www-dev	2784	tcp/udp	world wide web - development
www-http	80	tcp/udp	World Wide Web HTTP-World Wide Web Consortium
www-ldap-gw	1760	tcp/udp	www-ldap-gw—DEC
x11	6000-6063	tcp/udp	X Window System—MIT
x25-svc-port	1998	tcp/udp	cisco X.25 service (XOT)
x500ms	5757	tcp/udp	OpenMail X.500 Directory Server—HP
xact-backup	911	tcp/udp	xact-backup—Xactlabs
x-bone-api	2165	tcp/udp	X-Bone AP-ISI
x-bone-ctl	265	tcp/udp	X-Bone CTL—ISI
xdmcp	177	tcp/udp	X Display Manager Control Protocol—MIT
xdsxdm	6558	tcp/udp	
xfer	82	tcp/udp	XFER Utility—Lockheed Martin
xfr	682	tcp/udp	XFR—Ultranet
xiip	1924	tcp/udp	XIIP—HMRinc
xingcsm	2327	tcp/udp	xingcsm—Xingtech
xingmpeg	1558	tcp/udp	xingmpeg—xingtech.com
xinuexpansion1	2021	udp	
xinuexpansion2	2022	udp	
xinuexpansion3	2023	tcp/udp	
xinuexpansion4	2024	tcp/udp	
xinupageserver	2020	tcp	
xinupageserver	2020	udp	
xiostatus	2341	tcp/udp	XIO Status—Xiotech
xmpv7	7430	tcp/udp	OpenView DM xmpv7 api pipe—HP
xmquery	2279	tcp/udp	xmquery—IBM
xmsg	1716	tcp/udp	xmsg—Xantel Connex
xnmp	1652	tcp/udp	xnmp—At&T Cerfnet
xns-auth	56	tcp/udp	XNS Authentication—Xerox
xns-ch	54	tcp/udp	XNS Clearinghouse—Xerox
xns-courier	165	tcp/udp	Xerox
xns-mail	58	tcp/udp	XNS Mail—Xerox
xns-time	52	tcp/udp	XNS Time Protocol—Xerox
xribs	2025	udp	
xs-openstorage	1619	tcp/udp	xs-openstorage—XuiSis Software Ltd.
xvttp	508	tcp/udp	xvttp
xyplex-mux	173	tcp/udp	Xyplex
yak-chat	258	tcp/udp	Yak Winsock Personal Chat
z39.50	210	tcp/udp	ANSI Z39.50
zannet	317	tcp/udp	Zannet-Access One
zenginkyo-1	5020	tcp/udp	zenginkyo-1—NttData
zenginkyo-2	5021	tcp/udp	zenginkyo-2—NttData
zephyr-clt	2103	tcp/udp	Zephyr serv-hm connection—MIT
zephyr-hm	2104	tcp/udp	Zephyr hostmanager—MIT
zephyr-srv	2102	tcp/udp	Zephyr server—MIT
zion-lm	1425	tcp/udp	Zion Software License Manager
zserv	346	tcp/udp	Zebra server
	0	tcp/udp	Reserved—ISI
	24	tcp/udp	any private mail system—UUNET
	35	tcp/udp	any private printer server—ISI
	57	tcp/udp	any private terminal access—ISI
	59	tcp/udp	any private file service—ISI
	75	tcp/udp	any private dial out service—ISI
	77	tcp/udp	any private RJE service—ISI
	87	tcp/udp	any private terminal link—ISI
	1023	tcp/udp	Reserved—IANA
			Services—SSE

IPX
Novell IPX

Ethernet LAN

Ethernet II

Dst	Src	Etype	Layers 3-7	FCS
6	6	8137		

802.3 w/802.2 SAP

Dst	Src	Len	SAP	CTL	Layers 3-7	FCS
			E0 | E0	03		

802.3 w/802.2 SNAP

Dst	Src	Len	SAP	CTL	OUI	Etype	Layers 3-7	FCS
			AA | AA	03	000000	8137		

802.3 w/out 802.2 (Novell Proprietary)

Dst	Src	Len	FFFF	Layers 3-7	FCS

Token Ring LAN

802.5 w/802.2 SAP

AC	FC	Dst	Src	SAP	CTL	Layers 3-7	FCS
				E0 | E0	03		

802.5 w/802.2 SNAP

AC	FC	Dst	Src	SAP	CTL	OUI	Etype	Layers 3-7	FCS
				AA | AA	03	000000	8137		

FDDI LAN

FDDI w/802.2 SAP

FC	Dst	Src	SAP	CTL	Layers 3-7	FCS
			E0 | E0	03		

FDDI w/802.2 SNAP

FC	Dst	Src	SAP	CTL	OUI	Etype	Layers 3-7	FCS
			AA | AA	03		8137		

CTL = Control Field, see IEEE 802.2 Reference

IPX Packet Structure

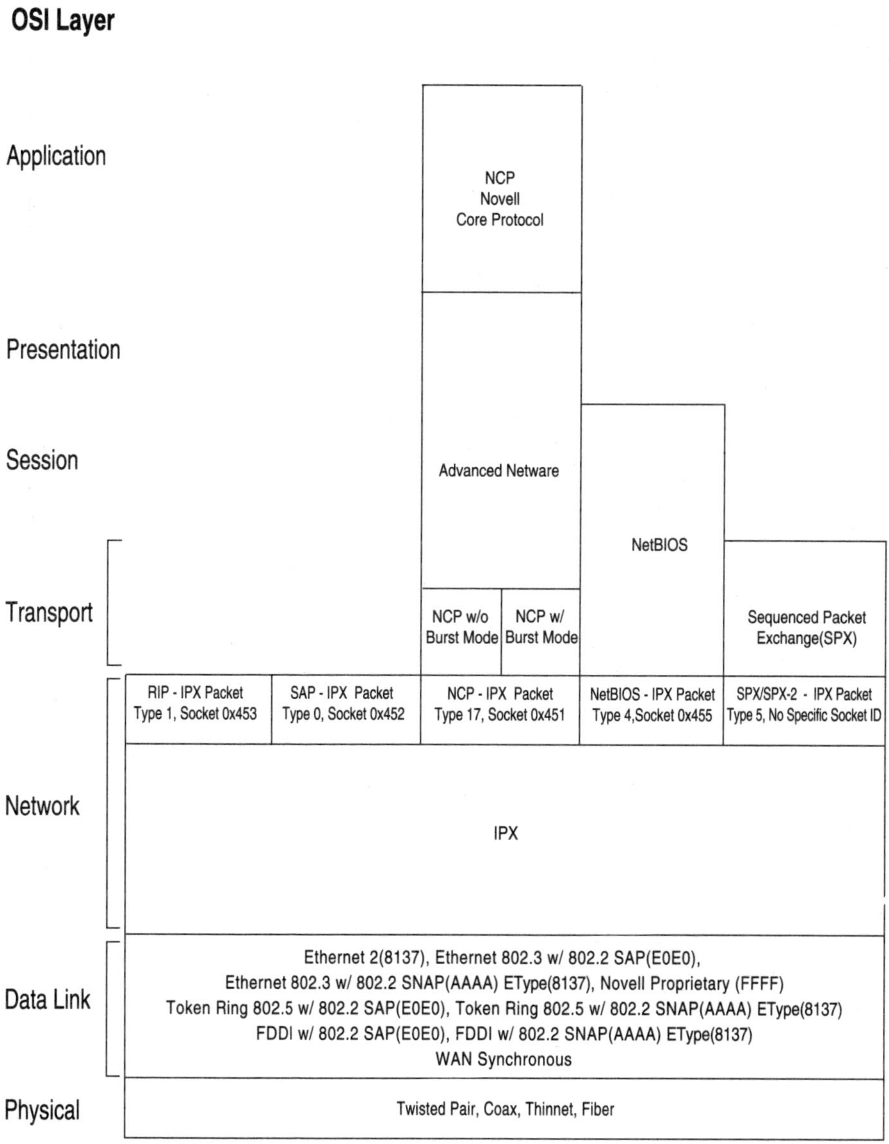

IPX Packet Details

IPX Header

<table>
<tr><td colspan="2">Checksum=FFFF</td><td colspan="2">Packet Length - IPX header plus additional layers after IPX header</td></tr>
<tr><td>Transport Control RRRRHHHH</td><td>Packet Type i.e. SPX,NCP</td><td colspan="2">Destination Network Address</td></tr>
<tr><td colspan="2">Destination Network Address</td><td colspan="2"></td></tr>
<tr><td colspan="4">Destination MAC Address</td></tr>
<tr><td colspan="2">Destination Socket Number</td><td colspan="2">Source Network Address</td></tr>
<tr><td colspan="2">Source Network Address</td><td colspan="2"></td></tr>
<tr><td colspan="4">Source MAC Address</td></tr>
<tr><td colspan="2">Source Socket Number</td><td colspan="2"></td></tr>
<tr><td colspan="4">RIP, SAP, SPX, NCP, Netbios Layer</td></tr>
</table>

IPX Header Details

Transport Control - R R R R H H H H

RRRR = Reserved

HHHH = Indicates the number of routers the packet has passed through. The maximum number of routers the packet will pass through is 15. The packet will be discarded by the 16th router.

Packet Type

00 – Novell
- SAP Reply to Get Nearest File Server (source socket 1106)

01 – Routing Information Protocol (RIP) or Netware Link State Protocol (NLSP)
- RIP Request-to find a file server network address (source & destination socket 1107, operation 1 = request), Destination network address = 00000000.FFFFFFFFFFFF
- RIP Reply-to the RIP Request Station (source and destination socket 1107, operation of 2 = response), Destination network address = Network.MAC address of RIP Request station
- RIP 60 second Broadcast – hop count to all network numbers (source and destination socket 1107, operation of 2 = response), Destination network address = Network.FFFFFFFFFFFF

02 – Echo Packet

03 – Error Packet

04 – Internetwork Packet Exchange (IPX)
- SAP 60 second broadcast (destination socket 1106), Destination network address = Network.FFFFFFFFFFFF
- NetBIOS functions (source and destination socket 1109)

05 – Sequenced Packet Exchange (SPX)

Continued on next page

17 – Netware Core Protocol (NCP)
– SAP Request to locate Nearest File Server (destination socket 1106)
– NCP functions (socket 1105)
20 – NetBIOS

Destination/Source Socket Number – Identifies the socket number of the internode process. In one direction, the socket number will be randomly assigned on a number between 0x4000 and 0x8000. (See Private Sockets in Apendix D)

0x0451(1105) NCP (Netware Core Protocol)
0x0452(1106) SAP (Service Advertising Protocol)
0x0453(1107) RIP (Routing Information Protocol)
0x0455(1109) NetBIOS
0x0456(1110) Diagnostic Packet
0x0457(1111) Serialization or Security Packet

SPX Header

<table>
<tr><td>Connection Control</td><td>Datastream Type</td><td>Source ID - connection number assigned by source station</td></tr>
<tr><td colspan="2">Destination ID - connection number assigned by destination station</td><td>Sequence Number - count of the packets transmitted from a station</td></tr>
<tr><td colspan="2">Acknowledgment Num.- count of the packets received from a station</td><td>Allocation Number - number of RX buffer available at a station</td></tr>
<tr><td colspan="3">Data(0-534 bytes)</td></tr>
</table>

Connection Control
0x10 - End of message
0x20 - Attention
0x40 - Acknowledgment required
0x80 - System packet

Data Stream Type
OxFE - End of connection
OxFF - End of Connection acknowledgement

SPX-2 Header

<table>
<tr><td>Connection Control</td><td>Datastream Type</td><td>Source ID - connection number assigned by source station</td></tr>
<tr><td colspan="2">Destination ID -connection number assigned by destination station</td><td>Sequence Number - count of the packets transmitted from a station</td></tr>
<tr><td colspan="2">Acknowledgment Num.</td><td>Allocation Number - number of RX buffer available at a station</td></tr>
<tr><td colspan="2">Extended Acknowledgment</td><td></td></tr>
<tr><td colspan="3">Data(0-1500 bytes)</td></tr>
</table>

Connection Control
0x10 - End of message
0x20 - Attention
0x40 - Acknowledgment required
0x80 - System packet
0x04 - Negotiate packet size
0x08 - SPX II type packet

Data Stream Type
OxFE - End of connection
OxFF - End of Connection acknowledgment
OxFC - Orderly release request
OxFD - Orderly release acknowledgement

NCP w/o Burst Mode Header

<table>
<tr><td colspan="2">Request Type</td><td>Sequence Number</td><td>Connection Number Low</td></tr>
<tr><td>Task Number</td><td>Connection Number High</td><td colspan="2"></td></tr>
<tr><td colspan="4">Request Type Details</td></tr>
</table>

NCP w/ Burst Mode Header

<table>
<tr><td>Request Type</td><td rowspan="2"></td></tr>
<tr></tr>
<tr><td colspan="2">Request Type Details</td></tr>
</table>

NCP Header

Request Type – Indicates the type of request sent from client to server

1111 - Create service connection

2222 - NCP Service Connection

3333 - Reply

5555 - End a Service Connection

7777 - Burst Mode Transfer

9999 - Server Busy

NetBIOS – Network Basic Input/Output System

Ethernet LAN

Ethernet II(bridging)

Destination	Source	SAP	CTL	NetBIOS Header	Layers 5-7	FCS
6	6	F0|F0				

Ethernet II over IP

Destination	Source	Etype	IP Layer	TCP Layer	NetBIOS Header	Layers 5-7	FCS
6	6	0800	20	20			

802.3 with 802.2 IPX SAP

Destination	Source	Length	SAP	CTL	IPX Layer	NetBIOS Header	Layers 5-7	FCS
6	6		E0|E0		30			

Token Ring LAN

802.5 with 802.2 SAP(bridging)

AC	FC	Destination	Source	SAP	CTL	NetBIOS Header	Layers 5-7	FCS
		6	6	F0|F0				

802.5 with 802.2 IP SNAP

AC	FC	Destination	Source	SAP	CTL	OUI	Etype	IP Layer	TCP Layer	NetBIOS Header	Layers 5-7	FCS
		6	6	AA|AA		6	0800	20	20			

802.5 with 802.2 IPX SAP

AC	FC	Destination	Source	SAP	CTL	IPX Layer	NetBIOS Header	Layers 5-7	FCS
		6	6	E0|E0		30			

FDDI LAN

FDDI with 802.2 SAP(bridging)

FC	Destination	Source	SAP	CTL	NetBIOS Header	Layers 5-7	FCS
	6	6	F0|F0				

FDDI with 802.2 IP SNAP

FC	Destination	Source	SAP	CTL	OUI	Etype	IP Layer	TCP Layer	NetBIOS Header	Layers 5-7	FCS
	6	6	AA|AA		6	0800	20	20			

FDDI with 802.2 IPX SAP

FC	Destination	Source	SAP	CTL	IPX Layer	NetBIOS Header	Layers 5-7	FCS
	6	6	E0|E0		30			

NetBIOS Packet Structure

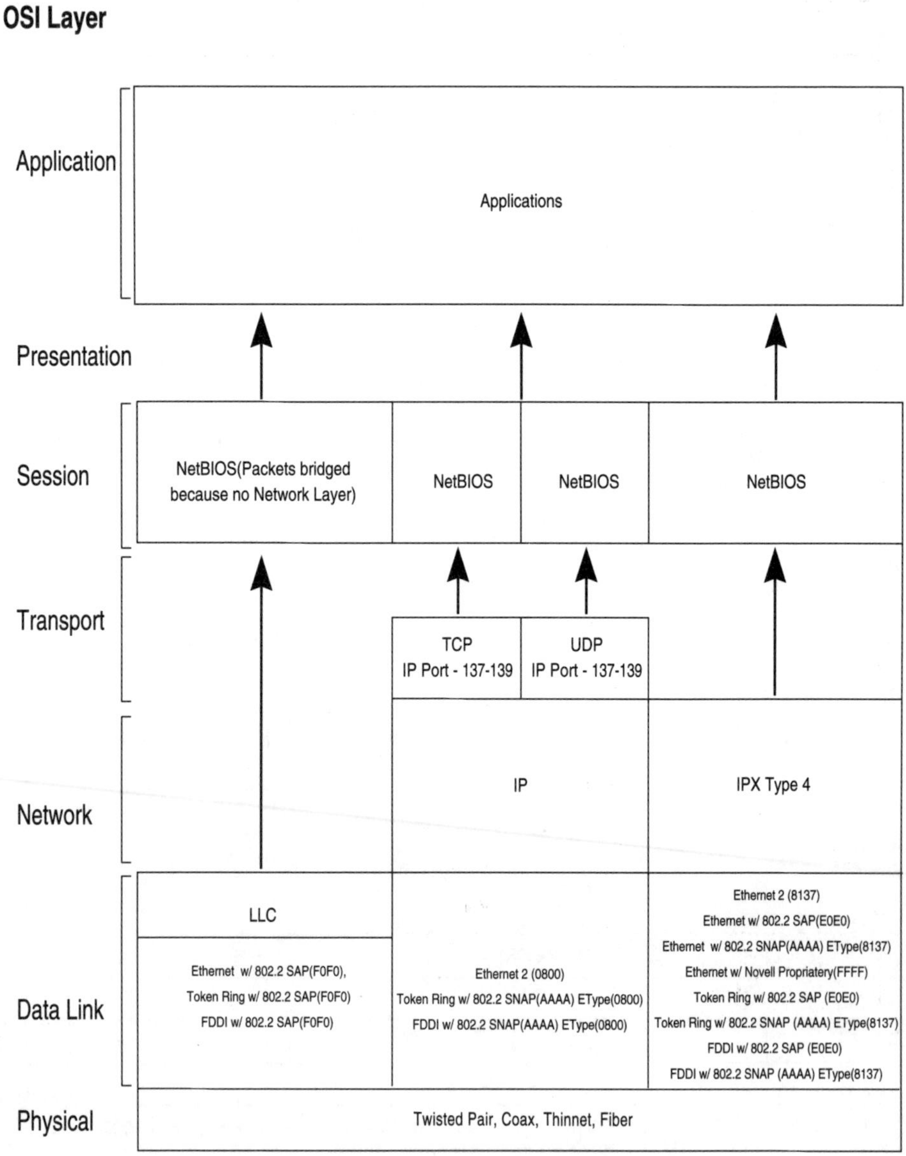

NetBIOS Packet Details

Type 137(x89) – NetBIOS Name Services

Transaction ID	Name Service Flags
Name Service Details	

Type 138(x8A) – NetBIOS Datagram Services

Datagram Service Type	Datagram Service Flag	Datagram ID
Datagram Service Packet Type Details		
Layers 5-7		

Type 139(x8B) – NetBIOS Session Services

Session Service Type	Session Services Flag	Length of Session Service Type Details
Session Service Type Details		

AppleTalk Protocol

Ethernet LAN

Ethernet II — Phase I

Dst 6	Src 6	Etype 809B	Layers 3-7	FCS

802.3 w/802.2 SNAP — Phase II

Dst	Src	Len	SAP AA | AA	CTL 03	OUI 080007	Etype 809B	Layers 3-7	FCS

Token Ring LAN

802.5 w/802.2 SNAP

AC	FC	Dst	Src	SAP AA | AA	CTL 03	OUI 080007	Etype 809B	Layers 3-7	FCS

FDDI LAN

FDDI w/802.2 SNAP

FC	Dst 6	Src 6	SAP AA | AA	CTL 03	OUI 080007	Etype 809B	Layers 3-7	FCS

AppleTalk Packet Structure

OSI Layer

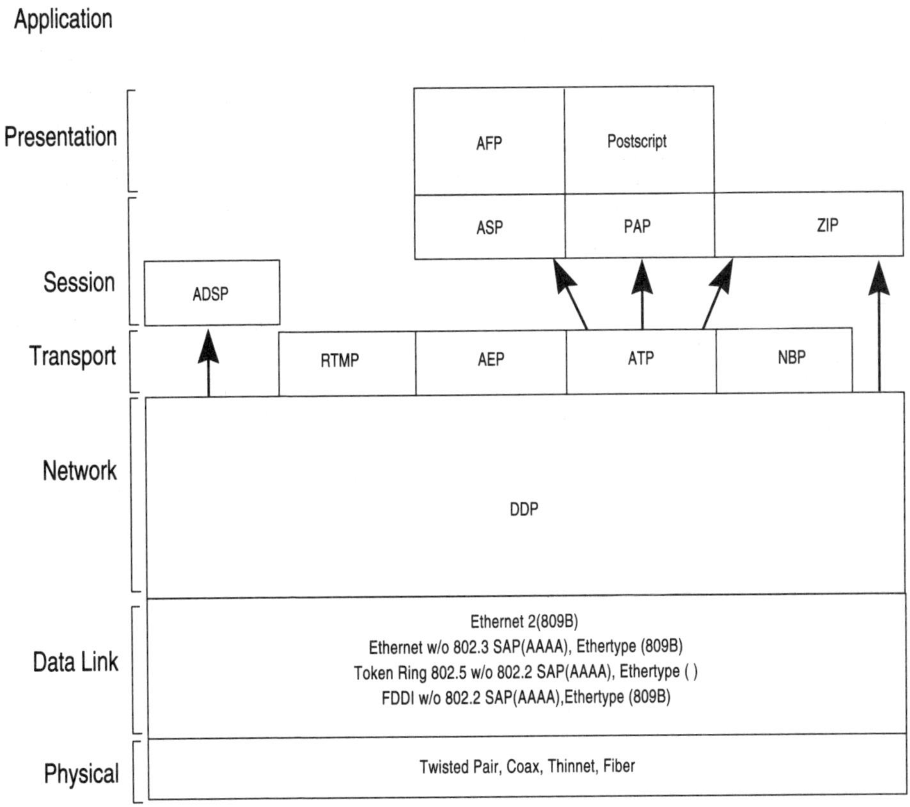

DDP - Datagram Delivery Protocol
RTMP - Routing Table Maintenance Protocol
AEP - Appletalk Echo Protocol
ATP - Appletalk Transaction Protocol
NBP - Name Binding Protocol
ADSP - Appletalk Data Stream Protocol
ASP - Appletalk Session Protocol
PAP - Printer Access Protocol
ZIP - Zone Information Protocol
AFP - Appletalk Filing Protocol

AppleTalk Packet Details

DDP Header

DDP Header Unused (2)	Hop Ct. (4)	Datagram Length (10)	DDP Checksum (8)
Destination Network # (16)		Source Network # (16)	
Destination Node (8)	Source Node (8)	Dest Socket (8)	Source Socket (8)
DDP Protocol Type (8)			

Hop Count - Number of routers a packet has passed through on a network. A packet cannot pass through more than 15 routers.

Datagram Length - Length of the DDP header.

DDP Checksum - CRC value for the DDP Layer.

Dest Socket - Unique number assigned to each individual session between two stations.

Source Socket - Unique number assigned to each individual session between two stations.

DDP Type - Identifies the type of protocol that the packet is intended for. Different DDP numbers can be used for data, messages or other routines.

DECNet Protocol

Ethernet LAN

Ethernet II

Dst 6	Src 6	Etype 6003	Layers 3-7	FCS

Token Ring LAN

802.5 w/802.2 SNAP

A C	F C	Dst 6	Src 6	SAP AA | AA	03	OUI 000000	Etype 6003	Layers 3-7	FCS

FDDI LAN

FDDI w/802.2 SNAP

F C	Dst 6	Src 6	SAP AA | AA	03	OUI 000000	Etype 6003	Layers 3-7	FCS

DECNet Packet Structure

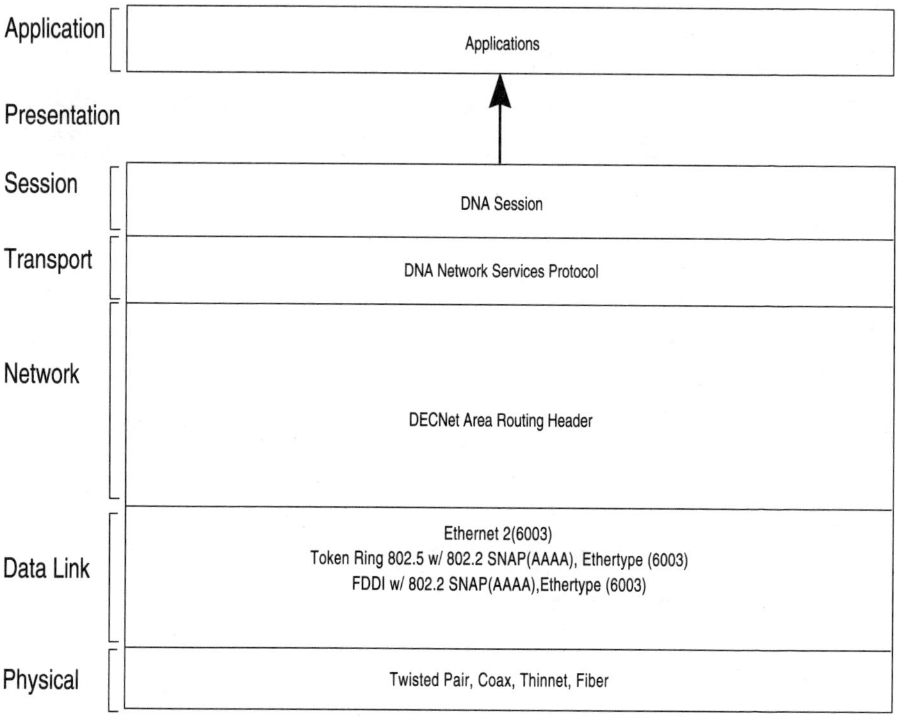

DECNet Packet Details

DECNet Area Routing Header

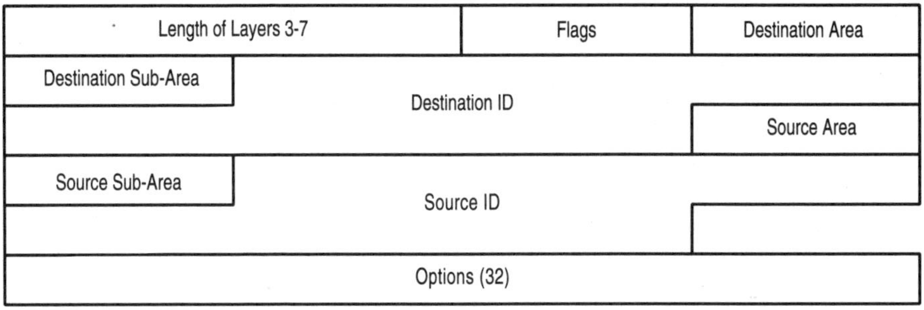

Converting Network and Node Number to Station MAC Address

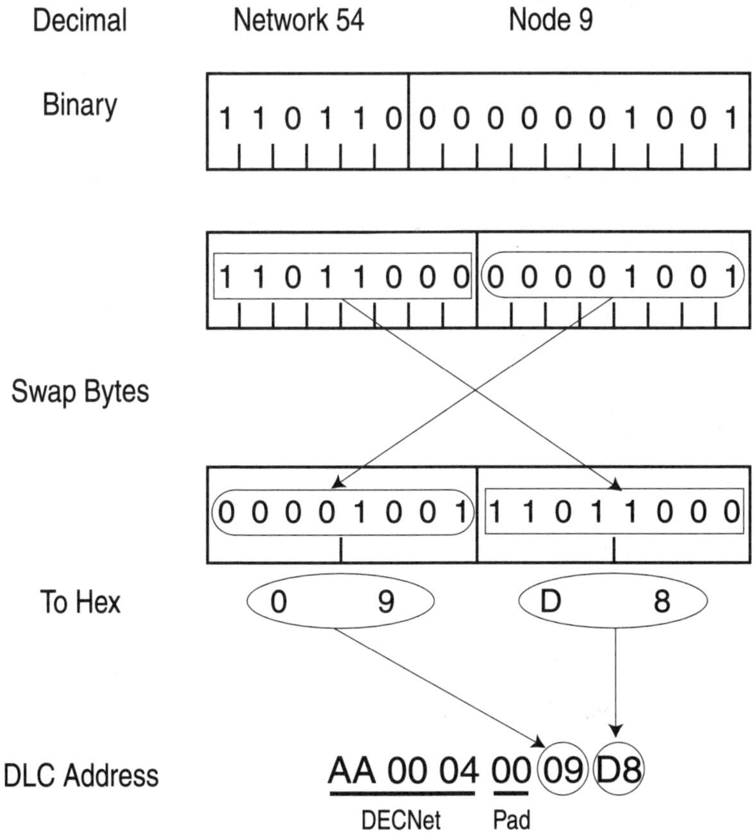

Network Documentation

Pine Mountain Group offers a method for collecting and displaying information critical in the design and maintenance of a Wide Area Network. The information is displayed in a fashion that allows one set of documentation to be used by all employees. WAN design, WAN maintenance, Transmission and Application personnel can review network design and applications with one common set of network documentation. It provides a method which allows all personnel to *"sing from the same song sheet."*

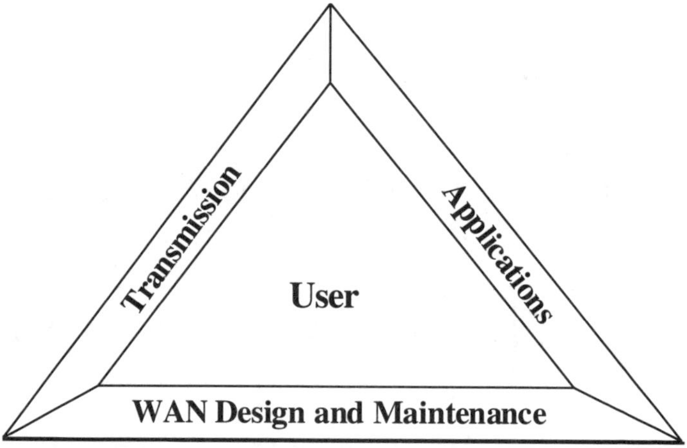

Network Documentation Goals:

(1) To display the connectivity and redundancy of a WAN equipment.

(2) Understand how traffic is passed between two network locations.

(3) Have all information available in one place to minimize the time it takes to resolve a network problem.

(4) Have updated information available for planning the addition of applications or equipment to the network.

Protocol Layers – one layer is generated for each protocol (IP, IPX, bridge, *etc.*). Each layer will show protocol network numbers, path cost, router port addresses, server addresses and LAN type.

Transmission Layer – provides information concerning the transmission systems connecting routers between buildings or cities. The information would include circuit identification number, circuit provider information and interface speed.

IP

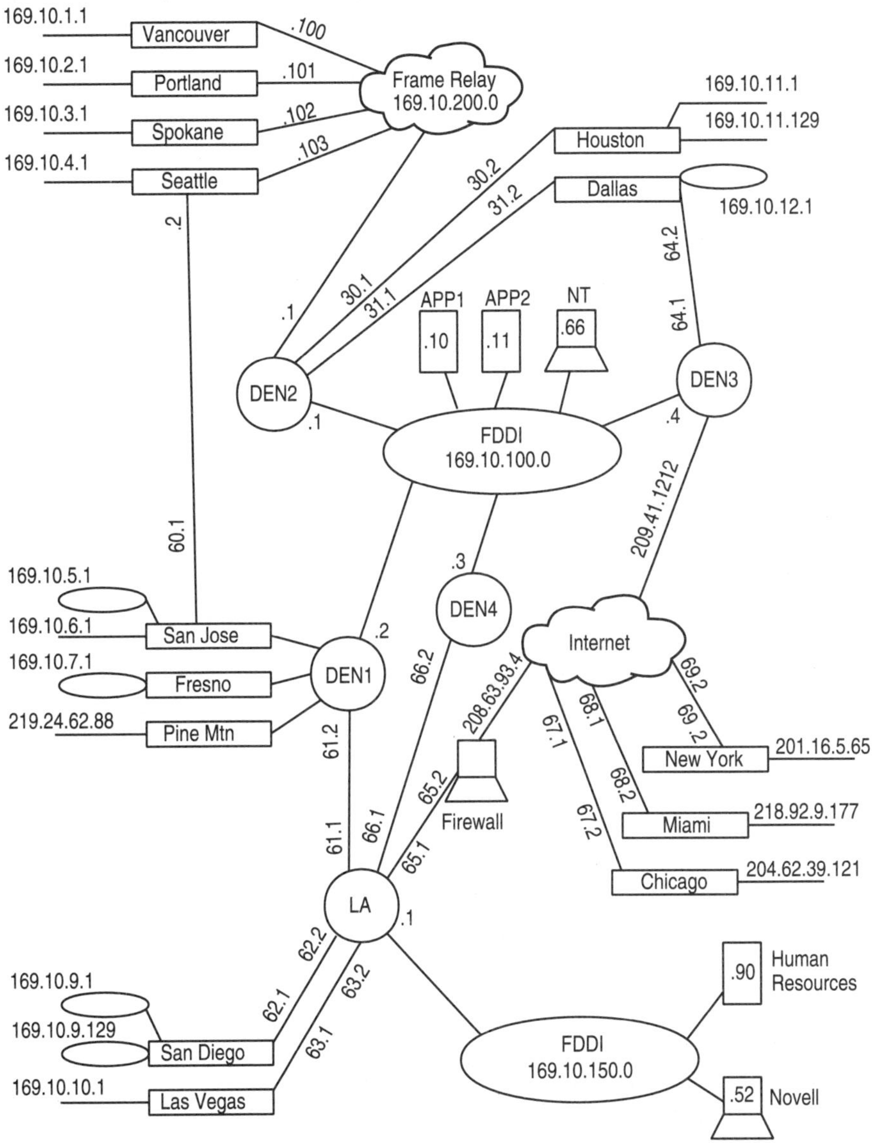

Internet Service Providers Denver (Company 1) 1-800-987-6789 Cust ID - 5555 Los Angeles (Company 2) 1-800-221-9898 Cust ID - 6666	**Company XYZ** **IP Protocol**

Note: Drawings would also contain IP addresses on the synchronous lines and port numbers used on the routers. Other packet types (bridging or routing) can be documented on separate drawing.

IPX

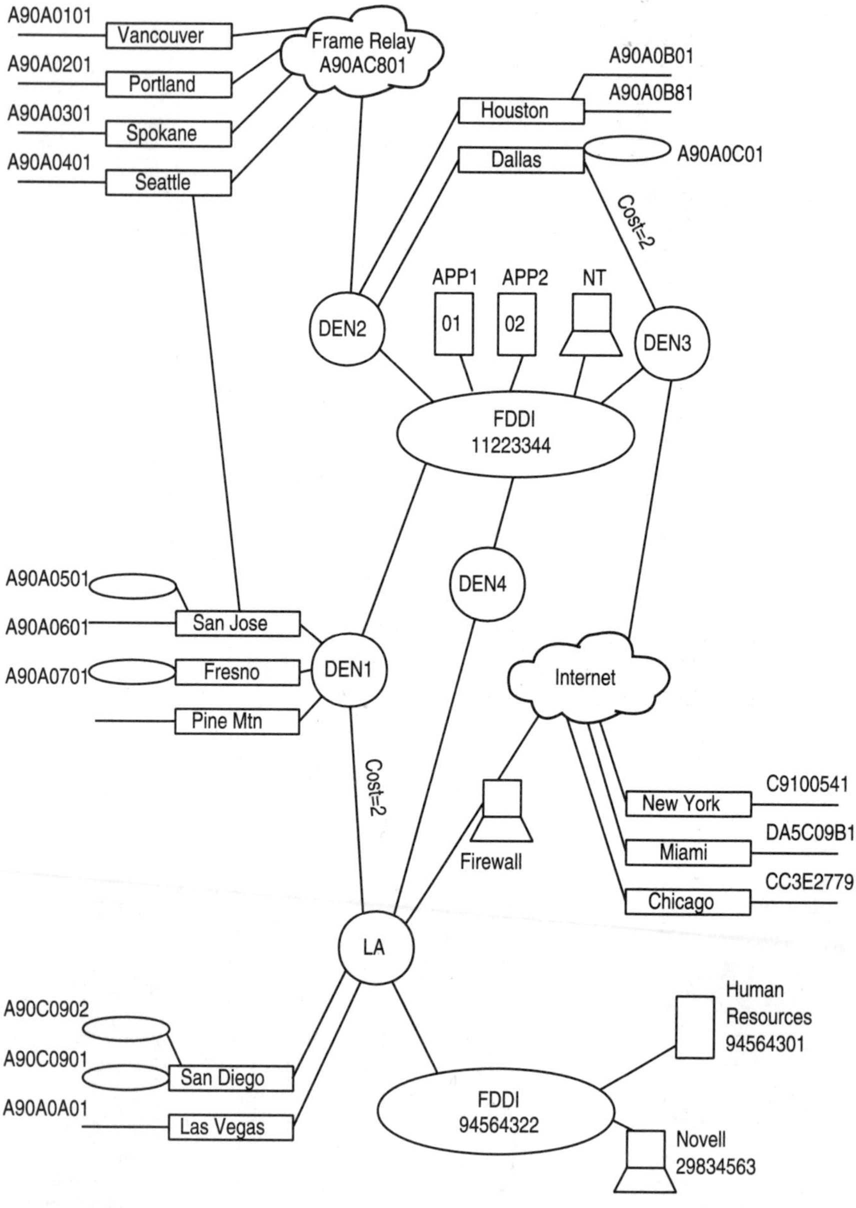

IPX Notes
1) Network numbers converted from IP address

Company XYZ

IPX Protocol

Note: Drawings would also contain IPX addresses on the synchronous lines

Transmission

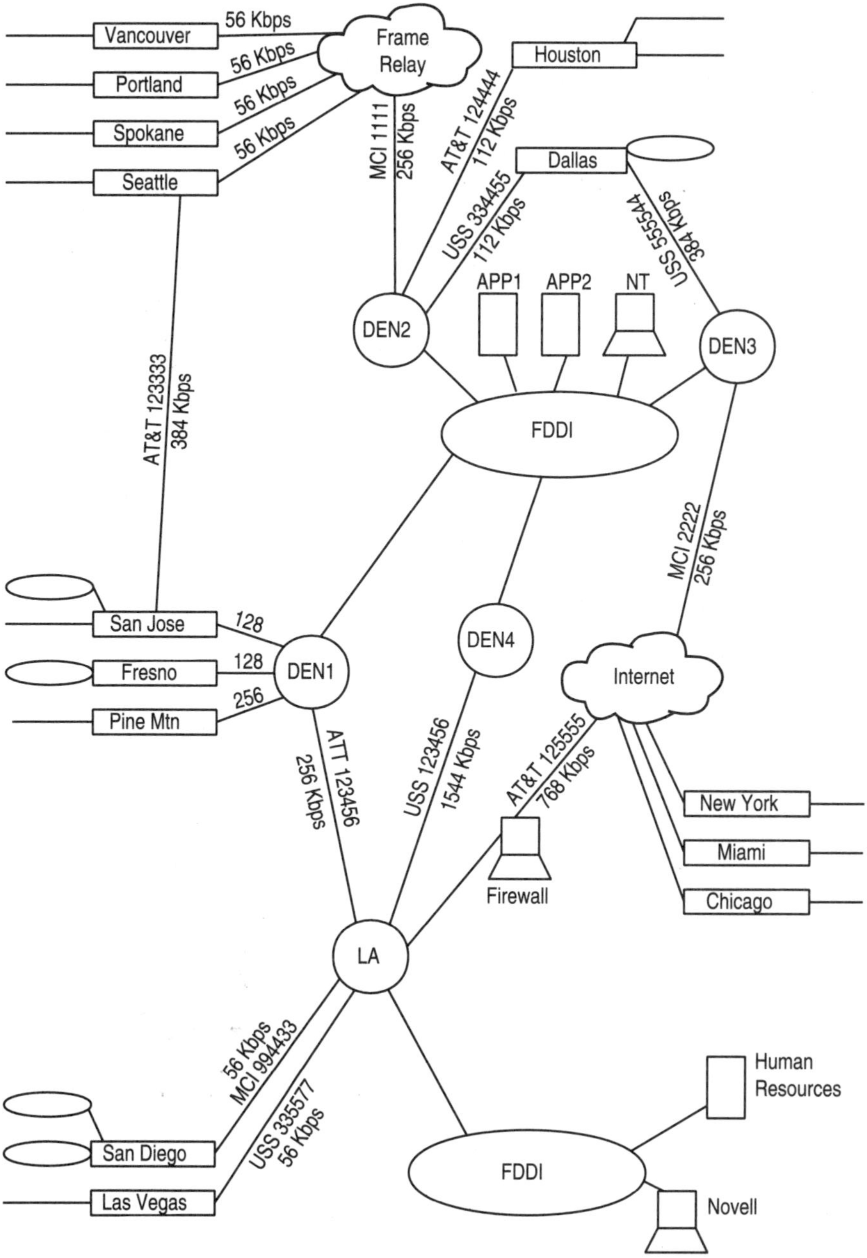

Leased Line Contacts	**Company XYZ**
AT&T 1-800-123-4567 Custtomer ID 1111 MCI 1-800-555-6666 Customer ID 2222 Sprint 1-800-222-6789 Customer ID 3333	**IP Protocol**

Note: Lease Line number and line speed can be shown on each Synchronous, Frame Relay or Internet connection

Multicast Addresses

00 00 69 02 XX XX	Concord Communications-DTQNA
01 00 0C CC CC CC	Cisco Systems-(SNAP) MDLP, MNRP, SME
01 00 10 00 00 20	Hughes Lan Systems-Terminal Server S/W download
01 00 10 FF FF 20	Hughes Lan Systems-Terminal Server S/W request
01 00 1D 00 00 00	Cabletron-PC-OV PC discover (on demand)
01 00 1D 42 00 00	Cabletron-PC-OV Bridge discover (on demand)
01 00 1D 52 00 00	Cabletron PC-OV MMAC discover (on demand)
01 00 5E 00 00 00	DoD Internet Multicast Reserved (RFC-1112)
01 00 5E 00 00 01	DoD Internet Multicast All Systems on this Subnet (RFC-1112)
01 00 5E 00 00 02	DoD Internet Multicast All Routers on this Subnet (RFC-1112)
01 00 5E 00 00 03	DoD Internet Multicast Unassigned (RFC-1112)
01 00 5E 00 00 04	DoD Internet Multicast DVMRP Routers (RFC-1075)
01 00 5E 00 00 05	OSPFIGP OSPFIGP All Routers (RFC-2328)
01 00 5E 00 00 06	OSPFIGP OSPFIGP Designated Routers (RFC-2328)
01 00 5E 00 00 07	ST Routers (RFC-1190)
01 00 5E 00 00 08	ST Hosts (RFC-1190)
01 00 5E 00 00 09	RIP2 Routers (RFC-1723)
01 00 5E 00 00 0A	Cisco Systems-IGRP & EIGRP
01 00 5E 00 00 0B	U.Michigan-Mobile-Agents
01 00 5E 00 00 0C	DHCP Server / Relay Agent (RFC-1884)
01 00 5E 00 00 0D	Cisco Systems-All PIM Routers
01 00 5E 00 00 0E	ISI-RSVP-ENCAPSULATION
01 00 5E 00 00 0F	cs.ucl.ac.uk-all-cbt-routers
01 00 5E 00 00 10	Cisco Systems-designated-sbm
01 00 5E 00 00 11	Cisco Systems-all-sbms
01 00 5E 00 00 12	Ipsilon-VRRP
01 00 5E 00 00 13	Siara-IP All L1 ISs
01 00 5E 00 00 14	Siara-IP All L2 ISs
01 00 5E 00 00 15	Siara-IP All Intermediate Systems
01 00 5E 00 00 16	Cisco Systems-IGMP
01 00 5E 00 00 17 through 01 00 5E 00 00 FF	Unassigned
01 00 5E 00 01 00	VMTP Managers Group (RFC-1045)
01 00 5E 00 01 01	NTP-Network Time Protocol (RFC-1119)
01 00 5E 00 01 02	SGI (Silicon Graphics)-Dogfight
01 00 5E 00 01 03	Xerox Parc-Rwhod
01 00 5E 00 01 04	Stanford University-VNP
01 00 5E 00 01 05	Artificial Horizons-Aviator
01 00 5E 00 01 06	Xerox Parc-NSS-Name Service Server
01 00 5E 00 01 07	Chalmers-AUDIONEWS-Audio News Multicast
01 00 5E 00 01 08	SUN NIS+ Information Service
01 00 5E 00 01 09	Xerox-MTP Multicast Transport Protocol
01 00 5E 00 01 0A	Precept-IETF-1-LOW-AUDIO
01 00 5E 00 01 0B	Precept-IETF-1-AUDIO
01 00 5E 00 01 0C	Precept-IETF-1-VIDEO
01 00 5E 00 01 0D	Precept-IETF-2-LOW-AUDIO
01 00 5E 00 01 0E	Precept-IETF-2-AUDIO
01 00 5E 00 01 0F	Precept-IETF-2-VIDEO
01 00 5E 00 01 10	Guido van Rossum-MUSIC-SERVICE
01 00 5E 00 01 11	SEANET-TELEMETRY
01 00 5E 00 01 12	SEANET-IMAGE

01 00 5E 00 01 13	ISI-MLOADD
01 00 5E 00 01 14	ISI-any private experiment
01 00 5E 00 01 15	Casc-DVMRP on MOSPF
01 00 5E 00 01 16	TGV-SVRLOC
01 00 5E 00 01 17	Xingtech-XINGTV
01 00 5E 00 01 18	Microsoft-ds
01 00 5E 00 01 19	birch.crd.ge.com-nbc-pro
01 00 5E 00 01 1A	birch.crd.ge.com-nbc-pfn
01 00 5E 00 01 1B	Lockheed-lmsc-calren-1
01 00 5E 00 01 1C	Lockheed-lmsc-calren-2
01 00 5E 00 01 1D	Lockheed-lmsc-calren-3
01 00 5E 00 01 1E	Lockheed-lmsc-calren-4
01 00 5E 00 01 1F	ampr-info
01 00 5E 00 01 20	ISI-mtrace
01 00 5E 80 00 00 through 01 00 5E FF FF FF	DoD Internet reserved by IANA
01 00 81 00 00 00	SynOptics-Network Management
01 00 81 00 00 02	SynOptics-Network Management
01 00 81 00 01 00	SynOptics-(now Nortel) (SNAP)Network Management Autotopology Packets
01 00 81 00 01 01	SynOptics-(now Nortel) (SNAP)Network Management Autotopology Packets
01 00 81 00 02 00	SynOptics-(now Nortel) (SNAP)Network Management Autotopology Packets
01 80 C2 00 00 00	IEEE 802.1d Spanning tree Bridge Group Address)
01 80 C2 00 00 01 through 01 80 C2 00 00 0F	IEEE 802.1d alternate Spanning tree multicast
01 80 C2 00 00 10	IEEE 802.1d All LANs Bridge Management Address
01 80 C2 00 00 11	IEEE 802.1e Load Server Group Address
01 80 C2 00 00 12	IEEE 802.1e Loadable Device Group Address
01 80 C2 00 00 14	ISO ISISDP 10589 OSI Route level 1 (within area) IS hello
01 80 C2 00 00 15	ISO ISISDP 10589 OSI Route level 2 (between area) IS hello
01 80 C2 00 00 16	ISO 10030 All CONS End Systems
01 80 C2 00 00 17	ISO 10030 All CONS SNAREs
01 80 C2 00 01 00	ANSI FDDI RMT Directed Beacon Group Address
01 80 C2 00 01 10	ANSI FDDI Status Report Frame Address
01 80 C2 00 01 20	ANSI FDDI SMT All Root Concentrator MAC
01 80 24 00 00 00	Kalpana Etherswitch every 60 seconds
01 DD 00 FF FF FF	Ungermann-Bass boot-me requests
01 DD 01 00 00 00	Ungermann-Bass Spanning Tree
03 00 00 00 00 01	NetBEUI
03 00 00 00 00 10	OS/2 1.3 EE + Communications Manager
03 00 00 00 00 40	OS/2 1.3 EE + Communications Manager
08 00 2B 0X XX XX through 08 00 2B 1X XX XX	DEC-23-365A1-00 PROM
08 00 2B 20 XX XX	DEC-23-00XEC-00 Multi-Address ROM
08 00 2B 22 00 00	DEC-Bridge Management
08 00 2B 23 XX XX	DEC-23-365A1-00 PROM
08 00 2B 23 XX XX through 08 00 2B 3X XX XX	DEC-PROM 23-365A1-00
08 00 2B 4X XX XX	DEC-23-365A1-00 Shadow PROM
08 00 2B 5X XX XX through 08 00 2B 7X XX XX	DEC-23-365A1-00 Shadow PROM
08 00 2B AX XX XX	DEC-23-00XEC-00 Multi-Address ROM

08 00 2B E0 XX XX	DEC-VAX 3000 Fault-Tolerant LAN
08 00 2B F0 XX XX	DEC-VAXft 3000 Fault-Tolerant LAN
08 00 2B XX XX XX	DEC-Servers/Lan Bridges
08 00 7C XX XX XX	Vitalink-Bridges
09 00 02 04 00 01	Vitalink-Printer Messages
09 00 02 04 00 02	Vitalink-Bridge Management
09 00 07 00 00 00 through 09 00 07 00 00 FC	Apple-AppleTalk Zone multicast addresses
09 00 07 FF FF FF	Apple-AppleTalk broadcast address
09 00 09 00 00 01	HP-Probe
09 00 09 00 00 01	HP-Probe
09 00 09 00 00 04	HP-DTC
09 00 0D XX XX XX	ICL-Oslan Multicast
09 00 0D 02 00 00	ICL-Oslan Service discover only on boot
09 00 0D 02 0A 38	ICL-Oslan Service discover only on boot
09 00 0D 02 0A 39	ICL-Oslan Service discover only on boot
09 00 0D 02 0A 3C	ICL-Oslan Service discover only on boot
09 00 0D 02 FF FF	ICL-Oslan Service discover only on boot
09 00 0D 09 00 00	ICL-Oslan Service discover as required
09 00 1E 00 00 00	Apollo (HP)-DOMAIN
09 00 26 01 00 01	Vitalink-TransLAN bridge management
09 00 2B 00 00 00	DEC-MUMPS
09 00 2B 00 00 01	DEC-DSM/DDP
09 00 2B 00 00 02	DEC-VAXELN
09 00 2B 00 00 03	DEC-Lanbridge Traffic Monitor (LTM)
09 00 2B 00 00 04	DEC-MAP End System Hello
09 00 2B 00 00 05	DEC-MAP Intermediate System Hello
09 00 2B 00 00 06	DEC-CSMA/CD Encryption
09 00 2B 00 00 07	DEC-NetBios Emulator (PCSG)
09 00 2B 00 00 0F	DEC-Local Area Transport (LAT)
09 00 2B 00 00 1x	DEC-Experimental
09 00 2B 01 00 00	DEC-LanBridge Copy packets (All bridges)
09 00 2B 01 00 01	DEC-LanBridge Hello packets (All local bridges) 1 packet per second, sent by the designated LanBridge
09 00 2B 02 00 00	DEC-DNA Level 2 Routing Layer routers
09 00 2B 02 01 00	DEC-DNA Naming Service Advertisement
09 00 2B 02 01 01	DEC-DNA Naming Service Solicitation
09 00 2B 02 01 09	DEC-Availability Manager for Distributed Systems DECamds
09 00 2B 02 01 02	DEC-Distributed Time Service
09 00 2B 03 xx xx	DEC-default filtering by bridges
09 00 2B 04 00 00	DEC-Local Area System Transport (LAST)
09 00 2B 23 00 00	DEC-Argonaut Console
09 00 39 00 70 00	Spider Systems-Bridge Hello packet
09 00 4C 00 00 00	BICC-802.1 management
09 00 4C 00 00 02	BICC-802.1 management
09 00 4C 00 00 06	BICC-Local bridge STA 802.1(D) Rev6
09 00 4C 00 00 0C	BICC-Remote bridge STA 802.1(D) Rev8
09 00 4C 00 00 0F	BICC-Remote bridge ADAPTIVE ROUTING (e.g. to Retix)
09 00 4E 00 00 02	Novell-IPX
09 00 56 00 00 00 through 09 00 56 FE FF FF	Stanford University-reserved
09 00 56 FF 00 00 through 09 00 56 FF FF FF	Stanford University-V Kernel, version 6.0

09 00 6A 00 01 00	TOP-NetBIOS
09 00 77 00 00 00	Retix-Bridge Local Management System
09 00 77 00 00 01	Retix-Spanning Tree Bridges
09 00 77 00 00 02	Retix-Bridge Adaptive routing
09 00 7C 01 00 01	Vitalink-DLS Multicast
09 00 7C 01 00 03	Vitalink-DLS Inlink
09 00 7C 01 00 04	Vitalink-DLS and non DLS Multicast
09 00 7C 02 00 05	Vitalink-Diagnostics
09 00 7C 05 00 01	Vitalink-Gateway
09 00 7C 05 00 02	Vitalink-Network Validation Message
09 00 87 80 FF FF	Xyplex-Terminal Servers
09 00 87 90 FF FF	Xyplex-Terminal Servers
0D 1E 15 BA DD 06	H.P.
AB 00 00 01 00 00	DEC-Maintenance Operation Protocol (MOP) Dump/Load Assistance
AB 00 00 02 00 00	DEC-Maintenance Operation Protocol (MOP) Remote Console. 1 System ID packet every 8-10 minutes, by every: DEC-LanBridge DEC-DEUNA interface DEC-DELUA interface DEC-DEQNA interface (in a certain mode)
AB 00 00 03 00 00	DEC-DECNET Phase IV end node Hello packets 1 packet every 15 seconds, sent by each DECNET host
AB 00 00 04 00 00	DEC-DECNET Phase IV Router Hello packets 1 packet every 15 seconds, sent by the DECNET router
AB 00 00 05 00 00 through AB 00 03 FF FF FF	DEC-Reserved
AB 00 03 00 00 00	DEC-Local Area Transport (LAT) - old
AB 00 04 00 xx xx	DEC-Reserved-Customer Private Use
AB 00 04 01 xx yy	DEC-Local Area VAX Cluster groups System Communication Architecture (SCA)
C0 00 00 00 00 01	IEEE 802.5 Active Monitor
C0 00 00 00 00 02	IEEE 802.5 Ring Parameter Server
C0 00 00 00 00 04	IEEE 802.5 Network Server Heartbeat
C0 00 00 00 00 08	IEEE 802.5 Ring Error Monitor
C0 00 00 00 00 10	IEEE 802.5 Configuration Report Server
C0 00 00 00 00 20	IEEE 802.5 Synchronous Bandwidth Manager
C0 00 00 00 00 40	IEEE 802.5 Locate - Directory Server
C0 00 00 00 00 80	IEEE 802.5 NETBIOS
C0 00 00 00 01 00	IEEE 802.5 Bridge
C0 00 00 00 02 00	IEEE 802.5 IMPL Server
C0 00 00 00 04 00	IEEE 802.5 Ring Authorization Server
C0 00 00 00 08 00	IEEE 802.5 LAN Gateway
C0 00 00 00 10 00	IEEE 802.5 Ring Wiring Concentrator
C0 00 00 00 20 00	IEEE 802.5 LAN Manager
C0 00 08 00 00 00	IEEE 802.5 Novell Netware
C0 00 FF FF FF FF	IEEE 802.5 Broadcast
CF 00 00 00 00 00	Ethernet Configuration Test protocol (Loopback)
FF FF 00 60 00 04	Lantastic
FF FF 00 40 00 01	Lantastic
FF FF 01 E0 00 04	Lantastic

IP Multicast Ranges

Host Extensions for IP Multicasting [RFC1112] specifies the extensions required of a host implementation of the Internet Protocol (IP) to support multicasting. The multicast addressess are in the range 224.0.0.0 through 239.255.255.255.

Multicast Address(s)	Usage	Responsible Person
224.0.0.0	Base Address (Reserved)	[RFC1112,JBP]
224.0.0.1	All Systems on this Subnet	[RFC1112,JBP]
224.0.0.2	All Routers on this Subnet	[JBP]
224.0.0.3	Unassigned	[JBP]
224.0.0.4	DVMRP Routers	[RFC1075,JBP]
224.0.0.5	OSPFIGP OSPFIGP All Routers	[RFC2328,JXM1]
224.0.0.6	OSPFIGP OSPFIGP Designated Routers	[RFC2328,JXM1]
224.0.0.7	ST Routers	[RFC1190,KS14]
224.0.0.8	ST Hosts	[RFC1190,KS14]
224.0.0.9	RIP2 Routers	[RFC1723,GSM11]
224.0.0.10	IGRP Routers	[Farinacci]
224.0.0.11	Mobile-Agents	[Bill Simpson]
224.0.0.12	DHCP Server / Relay Agent	[RFC1884]
224.0.0.13	All PIM Routers	[Farinacci]
224.0.0.14	RSVP-ENCAPSULATION	[Braden]
224.0.0.15	all-cbt-routers	[Ballardie]
224.0.0.16	designated-sbm	[Baker]
224.0.0.17	all-sbms	[Baker]
224.0.0.18	VRRP	[Hinden]
224.0.0.19	IP All L1 ISs	[Przygienda]
224.0.0.20	IP All L2 ISs	[Przygienda]
224.0.0.21	IP All Intermediate Systems	[Przygienda]
224.0.0.22	IGMP	[Deering]
224.0.0.23-224.0.0.255	Unassigned	[JBP]
224.0.1.0	VMTP Managers Group	[RFC1045,DRC3]
224.0.1.1	NTP-Network Time Protocol	[RFC1119,DLM1]
224.0.1.2	SGI-Dogfight	[AXC]
224.0.1.3	Rwhod	[SXD]
224.0.1.4	VNP	[DRC3]
224.0.1.5	Artificial Horizons - Aviator	[BXF]
224.0.1.6	NSS - Name Service Server	[BXS2]
224.0.1.7	AUDIONEWS - Audio News Multicast	[MXF2]
224.0.1.8	SUN NIS+ Information Service	[CXM3]
224.0.1.9	MTP Multicast Transport Protocol	[SXA]
224.0.1.10	IETF-1-LOW-AUDIO	[SC3]
224.0.1.11	IETF-1-AUDIO	[SC3]
224.0.1.12	IETF-1-VIDEO	[SC3]
224.0.1.13	IETF-2-LOW-AUDIO	[SC3]
224.0.1.14	IETF-2-AUDIO	[SC3]
224.0.1.15	IETF-2-VIDEO	[SC3]
224.0.1.16	MUSIC-SERVICE	[Guido van Rossum]
224.0.1.17	SEANET-TELEMETRY	[Andrew Maffei]
224.0.1.18	SEANET-IMAGE	[Andrew Maffei]

Multicast Address(s)	Usage	Responsible Person
224.0.1.19	MLOADD	[Braden]
224.0.1.20	any private experiment	[JBP]
224.0.1.21	DVMRP on MOSPF	[John Moy]
224.0.1.22	SVRLOC	[Veizades]
224.0.1.23	XINGTV	[Gordon]
224.0.1.24	microsoft-ds	<arnoldm@microsoft.com>
224.0.1.25	nbc-pro	<bloomer@birch.crd.ge.com>
224.0.1.26	nbc-pfn	<bloomer@birch.crd.ge.com>
224.0.1.27	lmsc-calren-1	[Uang]
224.0.1.28	lmsc-calren-2	[Uang]
224.0.1.29	lmsc-calren-3	[Uang]
224.0.1.30	lmsc-calren-4	[Uang]
224.0.1.31	ampr-info	[Janssen]
224.0.1.32	mtrace	[Casner]
224.0.1.33	RSVP-encap-1	[Braden]
224.0.1.34	RSVP-encap-2	[Braden]
224.0.1.35	SVRLOC-DA	[Veizades]
224.0.1.36	rln-server	[Kean]
224.0.1.37	proshare-mc	[Lewis]
224.0.1.38	dantz	[Yackle]
224.0.1.39	cisco-rp-announce	[Farinacci]
224.0.1.40	cisco-rp-discovery	[Farinacci]
224.0.1.41	gatekeeper	[Toga]
224.0.1.42	iberiagames	[Marocho]
224.0.1.43	nwn-discovery	[Zwemmer]
224.0.1.44	nwn-adaptor	[Zwemmer]
224.0.1.45	isma-1	[Dunne]
224.0.1.46	isma-2	[Dunne]
224.0.1.47	telerate	[Peng]
224.0.1.48	ciena	[Rodbell]
224.0.1.49	dcap-servers	[RFC2114]
224.0.1.50	dcap-clients	[RFC2114]
224.0.1.51	mcntp-directory	[Rupp]
224.0.1.52	mbone-vcr-directory	[Holfelder]
224.0.1.53	heartbeat	[Mamakos]
224.0.1.54	sun-mc-grp	[DeMoney]
224.0.1.55	extended-sys	[Poole]
224.0.1.56	pdrncs	[Wissenbach]
224.0.1.57	tns-adv-multi	[Albin]
224.0.1.58	vcals-dmu	[Shindoh]
224.0.1.59	zuba	[Jackson]
224.0.1.60	hp-device-disc	[Albright]
224.0.1.61	tms-production	[Gilani]
224.0.1.62	sunscalar	[Gibson]
224.0.1.63	mmtp-poll	[Costales]
224.0.1.64	compaq-peer	[Volpe]
224.0.1.65	iapp	[Meier]
224.0.1.66	multihasc-com	[Brockbank]
224.0.1.67	serv-discovery	[Honton]
224.0.1.68	mdhcpdisover	[Patel]
224.0.1.69	MMP-bundle-discovery1	[Malkin]
224.0.1.70	MMP-bundle-discovery2	[Malkin]

Multicast Address(s)	Usage	Responsible Person
224.0.1.71	XYPOINT DGPS Data Feed	[Green]
224.0.1.72	GilatSkySurfer	[Gal]
224.0.1.73	SharesLive	[Rowatt]
224.0.1.74	NorthernData	[Sheers]
224.0.1.75	SIP	[Schulzrinne]
224.0.1.76	IAPP	[Moelard]
224.0.1.77	AGENTVIEW	[Iyer]
224.0.1.78	Tibco Multicast1	[Shum]
224.0.1.79	Tibco Multicast2	[Shum]
224.0.1.80	MSP	[Caves]
224.0.1.81	OTT (One-way Trip Time)	[Schwartz]
224.0.1.82	TRACKTICKER	[Novick]
224.0.1.83	dtn-mc	[Gaddie]
224.0.1.84	jini-announcement	[Scheifler]
224.0.1.85	jini-request	[Scheifler]
224.0.1.86	sde-discovery	[Aronson]
224.0.1.87	DirecPC-SI	[Dillon]
224.0.1.88	B1RMonitor	[Purkiss]
224.0.1.89	3Com-AMP3 dRMON	[Banthia]
224.0.1.90	imFtmSvc	[Bhatti]
224.0.1.91	NQDS4	[Flynn]
224.0.1.92	NQDS5	[Flynn]
224.0.1.93	NQDS6	[Flynn]
224.0.1.94	NLVL12	[Flynn]
224.0.1.95	NTDS1	[Flynn]
224.0.1.96	NTDS2	[Flynn]
224.0.1.97	NODSA	[Flynn]
224.0.1.98	NODSB	[Flynn]
224.0.1.99	NODSC	[Flynn]
224.0.1.100	NODSD	[Flynn]
224.0.1.101	NQDS4R	[Flynn]
224.0.1.102	NQDS5R	[Flynn]
224.0.1.103	NQDS6R	[Flynn]
224.0.1.104	NLVL12R	[Flynn]
224.0.1.105	NTDS1R	[Flynn]
224.0.1.106	NTDS2R	[Flynn]
224.0.1.107	NODSAR	[Flynn]
224.0.1.108	NODSBR	[Flynn]
224.0.1.109	NODSCR	[Flynn]
224.0.1.110	NODSDR	[Flynn]
224.0.1.111	MRM	[Wei]
224.0.1.112	TVE-FILE	[Blackketter]
224.0.1.113	TVE-ANNOUNCE	[Blackketter]
224.0.1.114	Mac Srv Loc	[Woodcock]
224.0.1.115	Simple Multicast	[Crowcroft]
224.0.1.116	SpectraLinkGW	[Hamilton]
224.0.1.117	dieboldmcast	[Marsh]
224.0.1.118	Tivoli Systems	[Gabriel]
224.0.1.119	pq-lic-mcast	[Sledge]
224.0.1.120	HYPERFEED	[Kreutzjans]
224.0.1.121	Pipesplatform	[Dissett]

Multicast Address(s)	Usage	Responsible Person
224.0.1.122	LiebDevMgmg-DM	[Velten]
224.0.1.123	TRIBALVOICE	[Thompson]
224.0.1.124	UDLR-DTCP	[Cipiere]
224.0.1.125	PolyCom Relay1	[Coutiere]
224.0.1.126	Infront Multi1	[Lindeman]
224.0.1.127	XRX DEVICE DISC	[Wang]
224.0.1.128	CNN	[Lynch]
224.0.1.129	PTP-primary	[Eidson]
224.0.1.130	PTP-alternate1	[Eidson]
224.0.1.131	PTP-alternate2	[Eidson]
224.0.1.132	PTP-alternate3	[Eidson]
224.0.1.133	ProCast	[Revzen]
224.0.1.134	3Com Discp	[White]
224.0.1.135	CS-Multicasting	[Stanev]
224.0.1.136	TS-MC-1	[Sveistrup]
224.0.1.137	Make Source	[Daga]
224.0.1.138	Teleborsa	[Strazzera]
224.0.1.139	SUMAConfig	[Wallach]
224.0.1.140	Unassigned	
224.0.1.141	DHCP-SERVERS	[Hall]
224.0.1.142	CN Router-LL]	[Armitage
224.0.1.143	EMWIN	[Querubin]
224.0.1.144	Alchemy Cluster	[O'Rourke]
224.0.1.145	Satcast One	[Nevell]
224.0.1.146	Satcast Two	[Nevell]
224.0.1.147	Satcast Three	[Nevell]
224.0.1.148	Intline	[Sliwinski]
224.0.1.149	8x8 Multicast	[Roper]
224.0.1.150-224.0.1.255	Unassigned	[JBP]
224.0.2.1	"rwho" Group (BSD) (unofficial)	[JBP]
224.0.2.2	SUN RPC PMAPPROC_CALLIT	[BXE1]
224.0.2.064-224.0.2.095	SIAC MDD Service	[Tse]
224.0.2.096-224.0.2.127	CoolCast	[Ballister]
224.0.2.128-224.0.2.191	WOZ-Garage	[Marquardt]
224.0.2.192-224.0.2.255	SIAC MDD Market Service	[Lamberg]
224.0.3.000-224.0.3.255	RFE Generic Service	[DXS3]
224.0.4.000-224.0.4.255	RFE Individual Conferences	[DXS3]
224.0.5.000-224.0.5.127	CDPD Groups	[Bob Brenner]
224.0.5.128-224.0.5.191	SIAC Market Service	[Cho]
224.0.5.192-224.0.5.255	Unassigned	[IANA]
224.0.6.000-224.0.6.127	Cornell ISIS Project	[Tim Clark]
224.0.6.128-224.0.6.255	Unassigned	[IANA]
224.0.7.000-224.0.7.255	Where-Are-You	[Simpson]
224.0.8.000-224.0.8.255	INTV	[Tynan]
224.0.9.000-224.0.9.255	Invisible Worlds	[Malamud]
224.0.10.000-224.0.10.255	DLSw Groups	[Lee]
224.0.11.000-224.0.11.255	NCC.NET Audio	[Rubin]
224.0.12.000-224.0.12.063	Microsoft and MSNBC	[Blank]
224.0.13.000-224.0.13.255	UUNET PIPEX Net News	[Barber]
224.0.14.000-224.0.14.255	NLANR	[Wessels]
224.0.15.000-224.0.15.255	Hewlett Packard	[van der Meulen]

Multicast Address(s)	Usage	Responsible Person
224.0.16.000-224.0.16.255	XingNet	[Uusitalo]
224.0.17.000-224.0.17.031	Mercantile&Commodity Exchange	[Gilani]
224.0.17.032-224.0.17.063	NDQMD1	[Nelson]
224.0.17.064-224.0.17.127	ODN-DTVs	[Hodge]
224.0.18.000-224.0.18.255	Dow Jones	[Peng]
224.0.19.000-224.0.19.063	Walt Disney Company	[Watson]
224.0.19.064-224.0.19.095	Cal Multicast	[Moran]
224.0.19.096-224.0.19.127	SIAC Market Service	[Roy]
224.0.19.128-224.0.19.191	IIG Multicast	[Carr]
224.0.19.192-224.0.19.207	Metropol	[Crawford]
224.0.19.208-224.0.19.239	Xenoscience, Inc.	[Timm]
224.0.19.240-224.0.19.255	HYPERFEED	[Felix]
224.0.20.000-224.0.20.063	MS-IP/TV	[Wong]
224.0.20.064-224.0.20.127	Reliable Network Solutions	[Vogels]
224.0.20.128-224.0.20.143	TRACKTICKER Group	[Novick]
224.0.20.144-224.0.20.207	CNR Rebroadcast MCA	[Sautter]
224.0.21.000-224.0.21.127	Talarian MCAST	[Mendal]
224.0.22.000-224.0.22.255	WORLD MCAST	[Stewart]
224.0.252.000-224.0.252.255	Domain Scoped Group	[Fenner]
224.0.253.000-224.0.253.255	Report Group	[Fenner]
224.0.254.000-224.0.254.255	Query Group	[Fenner]
224.0.255.000-224.0.255.255	Border Routers	[Fenner]
224.1.0.0-224.1.255.255	ST Multicast Groups	[RFC1190,KS14]
224.2.0.0-224.2.127.253	Multimedia Conference Calls	[SC3]
224.2.127.254	SAPv1 Announcements	[SC3]
224.2.127.255	SAPv0 Announcements (deprecated)	[SC3]
224.2.128.0-224.2.255.255	SAP Dynamic Assignments	[SC3]
224.252.0.0-224.255.255.255	DIS transient groups	[Joel Snyder]
225.0.0.0-225.255.255.255	MALLOC (temp - renew 12/99)	[Handley]
232.0.0.0-232.255.255.255	VMTP transient groups see single-source-multicast file	[DRC3]
233.0.0.0-233.255.255.255	Static Allocations (temp - renew 6/00)	[Meyer2]
239.000.000.000-239.255.255.255	Administratively Scoped	[IANA,RFC2365]
239.000.000.000-239.063.255.255	Reserved	[IANA]
239.064.000.000-239.127.255.255	Reserved	[IANA]
239.128.000.000-239.191.255.255	Reserved	[IANA]
239.192.000.000-239.251.255.255	Organization-Local Scope	[Meyer,RFC2365]
239.252.000.000-239.252.255.255	Site-Local Scope (reserved)	[Meyer,RFC2365]
239.253.000.000-239.253.255.255	Site-Local Scope (reserved)	[Meyer,RFC2365]
239.254.000.000-239.254.255.255	Site-Local Scope (reserved)	[Meyer,RFC2365]
239.255.000.000-239.255.255.255	Site-Local Scope	[Meyer,RFC2365]

Note that when used on an Ethernet or IEEE 802 network, the 23 low-order bits of the IP Multicast address are placed in the low-order 23 bits of the Ethernet or IEEE 802 net multicast address 1.0.94.0.0.0.

Assigned OUI Addresses

IEEE Organizationally Unique Identifiers (OUI)

OUI	Vendor/Comment
000000	Xerox Corporation
000001	Xerox Corporation—ZEROX SYSTEMS INSTITUTE
000002	Xerox Corporation—ZEROX SYSTEMS INSTITUTE
000003	Xerox Corporation—ZEROX SYSTEMS INSTITUTE—SMC seen on Token Ring—SMC assigned OUIs 00800F, 00E029, 004027
000004	Xerox Corporation—OFFICE SYSTEMS DIVISION
000005	Xerox Corporation—OFFICE SYSTEMS DIVISION
000006	Xerox Corporation—OFFICE SYSTEMS DIVISION
000007	Xerox Corporation—OFFICE SYSTEMS DIVISION
000008	Xerox Corporation—OFFICE SYSTEMS DIVISION
000009	Xerox Corporation
00000A	Omron Tateisi Electronics Co.Section NFF, System R&D Labs. Research & Tech. Assessmnt Div.
00000B	Matrix Corporation
00000C	Cisco Systems, Inc.—Cisco sends bit reversed as 000030 on Token Ring
00000D	Fibronics Ltd.
00000E	Fujitsu Limited
00000F	NeXT, Inc.
000010	Hughes LAN Systems (formerly Sytek, Inc.)
000011	Normerel Systemes
000012	Information Technology Limited
000013	Camex
000014	Netronix
000015	Datapoint Corporation—Stratus Corp seen on Token Ring—bit reversed of assigned OUI 0000A8
000016	Du Pont Pixel Systems
000018	Webster Computer Corporation—Appletalk
000019	Applied Dynamics International
00001A	Advanced Micro Devices (AMD)—Racore Computer Products Inc. seen on Token Ring—bit reversed of assigned OUI 000058
00001B	Novell, Inc. (sold this business line to Eagle Technology)
00001C	Bell Technologies (formerly JDR Microdevices—generic, NE2000 drivers)
00001D	Cabletron Systems, Inc.—Cabletron sends bit reversed as 0000B8 on Token Ring
00001E	Telsist Industria Electronica
00001F	Cryptall LP
000020	DIAB (Data Intdustrier AB)
000021	SC&C—Sureman Comp. & Commun. Corp.
000022	Visual Technology Inc.—Castelle Corp seen on Token Ring—bit reversed of assigned OUI 000044
000023	ABB Industrial Systems AB
000024	Connect AS (now owned by Olicom)
000025	Ramtek Corp.
000026	Sha-Ken Co., Ltd.
000027	Japan Radio Company
000028	Prodigy Systems Corporation
000029	IMC—IMC Networks Corp.
00002A	TRW—SEDD/INP
00002B	Crisp Automation, Inc.
00002C	Autotote Limited—Network Resources Corporation (NRC) - MultiGate Hub1+, Hub2, etc. sends bit reversed of assigned OUI 000034)
00002D	Chromatics Inc.
00002E	Societe Evira
00002F	Timeplex Inc.
000030	VG Laboratory Systems Ltd.—Cisco Systems seen on Token Ring—bit reversed of assigned OUI 00000C
000031	Qpsx Communications Pty Ltd.
000032	GPT Limited (reassigned from GEC Computers Ltd)
000033	Egan Machinery Company
000034	Network Resources Corporation
000035	Spectragraphics Corporation (Lan Manufacturing Engineer)
000036	Atari Corporation
000037	Oxford Metrics Ltd
000038	CSS LABS
000039	Toshiba Corporation—Computer Division
00003A	Chyron Corporation
00003B	Hyundai Electronics
00003C	Auspex Systems Inc.
00003D	Unisys
00003E	Simpact
00003F	Syntrex Inc.
000040	Applicon, Inc.
000041	ICE Corporation
000042	Metier Management Systems Ltd.
000043	Micro Technology
000044	Castelle Corporation—Castelle sends bit reversed as 000022 on Token Ring
000045	Ford Aerospace & Comm. Corp.—Wellfleet seen on Token Ring—bit reversed of assigned OUI 0000A2
000046	ISC-Bunker Ramo, An Olivetti America Company
000047	Nicolet Instruments Corp.
000048	Seiko Epson Corporation
000049	Apricot Computers Ltd.
00004A	ADC Codenoll Technology Corp.
00004B	ICL Data OY— APT and Nokia Data Systems also seen or reported
00004C	NEC Corporation
00004D	DCI Corporation
00004E	Ampex Corporation
00004F	Logicraft 386-Ware P.C. Emulator
000050	Radisys Corporation
000051	Hob Electronic GmBh & Co. KG
000052	Optical Data Systems
000053	Compucorp
000054	Modicon, Inc.
000055	Commissariat a l'Energie Atom.—AT&T seen under this OUI
000056	Dr. B. Struck
000057	Scitex Corporation Ltd.
000058	Racore Computer Products Inc.
000059	Hellige GmBh E-SW
00005A	Syskonnect—A Business Unit of Schneider & Koch Co. GmBh
00005B	Eltec Elektronik GmBh
00005C	Telematics International Inc.
00005D	CS Telecom (formerely RCE)
00005E	USC Information Sciences—IANA—Internet Ass'ned Nos. Authority
00005F	Sumitomo Electric Ind., Ltd.
000060	Kontron Elektronik GmBh
000061	Gateway Communications Inc. (then Megahertz & now 3Com)
000062	BULL HN Information Systems—Honeywell
000063	Dr.Ing.Seufert GmBh—Hewlett-Packard LanProbe seen under this OUI
000064	Yokogawa Digital Computer Corp
000065	Network Associates, Inc. (formerly Network General)
000066	Talaris Systems, Inc.
000067	Soft * Rite, Inc.
000068	Rosemount Controls
000069	Concord Communications, Inc.
00006A	Computer Consoles Inc. Computer Products Division
00006B	Silicon Graphics Inc. (SGI)/MIPS
00006D	Cray Communications Inc.—Case seen under this OUI
00006E	Artisoft, Inc.
00006F	Madge Networks Ltd.- Token-ring adapters—Madge sends bit reversed as 0000F6 on Token Ring
000070	HCL Limited Research & Development Unit
000071	Adra Systems Inc.
000072	Miniware Technology
000073	Siecor Corporation—DuPont seen under this OUI
000074	Ricoh Company Ltd.
000075	Northern Telecom Bell Northern Research (BNR) (now Nortel)
000076	Abekas Video System
000077	Interphase Corporation-(Used in other systems, e.g. MIPS, Motorola)
000078	Labtam Limited Australia
000079	Networth Incorporated (now owned by Compaq, used in Netelligent series)
00007A	Dana Computer Inc.—Ardent Communications Corp. seen under this OUI
00007B	Research Machines

OUI	Vendor/Comment
00007C	Ampere Incorporated
00007D	Cray Research Superservers,Inc (Also Harris (3M) (old))
00007E	Clustrix Corporation—NetFRAME multiprocessor network servers seen under this OUI—No IEEE OUI assigned to NetFrame
00007F	Linotype-Hell AG—Linotronic typesetters
000080	Cray Communications A/S (formerly Dowty Network Services)— Harris (3M) and DATACO in Denmark seen under this OUI
000081	SynOptics Comunications Inc. (then Bay Networks, now Nortel)
000082	Lectra Systemes SA
000083	Tadpole Technology PLC—Olicom seen on Token Ring—bit reveresed of assigned OUI 0000C1
000084	Supernet—Aquila and ADI Systems Inc. seen under this OUI
000085	Canon Inc.
000086	Gateway Communications Inc. (then Megahertz & now 3Com)
000087	Hitachi, Ltd.
000088	Computer Network Tech. Corp.
000089	Cayman Systems Inc.-Gatorbox
00008A	Datahouse Information Systems
00008B	Infotron
00008C	Alloy Computer Products, Inc.
00008D	Verdix Corporation
00008E	Solbourne Computer Inc.
00008F	Raytheon Company
000090	Microcom
000091	Anritsu Corporation
000092	Cogent Data Technologies—Unisys seen under this OUI
000093	Proteon, Inc.—Proteon sends bit reversed as 0000C9 on Token Ring
000094	Asante Technologies—MAC
000095	Sony Tektronix Corp.
000096	Marconi Electronics Ltd.
000097	Epoch Systems
000098	CrossComm Corporation
000099	Memorex Telex Corporation (MTX, Inc.)
00009A	RC Computer A/S
00009B	Information International, Inc.
00009C	ROLM MIL-SPEC Computers
00009D	Locus Computing Corporation
00009E	Marli S.A.
00009F	Ameristar Technologies Inc.
0000A0	Tokyo Sanyo Electic Co. Ltd.
0000A1	Marquette Electric Co.
0000A2	Wellfleet (then Bay Networks now Nortel)—Wellfleet sends bit reversed as 000045 on Token Ring
0000A3	Network Application Technology (NAT)
0000A4	Acorn Computers Limited
0000A5	Compatible Systems Corporation
0000A6	Network General (internal assignment, not for products)—now NAI
0000A7	Network Computing Devices Inc. (NCD) X-terminals
0000A8	Stratus Computer, Inc.
0000A9	Network Systems Corp.
0000AA	Xerox Corp.— Xerox machines—Milan seen on Token Ring—bit reversed of assigned OUI 0040C8
0000AB	Logic Modeling Corporation
0000AC	Conware Computer Consulting
0000AD	Bruker Instruments Inc.
0000AE	Dassault Electronique
0000AF	Nuclear Data Instrumentation—Acquisition Interface Modules (AIM)
0000B0	RND (RND-RAD Network Devices)
0000B1	Alpha Microsystems Inc.
0000B2	Televideo Systems, Inc.
0000B3	CIMLinc Incorporated
0000B4	Edimax Computer Company
0000B5	Datability Software Systems Inc.—Terminal Servers
0000B6	Micro-matic Research
0000B7	Dove Computer Corp.—Fastnet
0000B8	Seikosha Co., Ltd.—Cabletron seen on Token Ring—bit reversed of assigned OUI 00001D
0000B9	McDonnell Douglas Computer Sys
0000BA	SIIG, Inc.
0000BB	TRI-DATA Systems Inc.— Netway products, 3274 emulators
0000BC	Allen-Bradley Co. Inc.
0000BD	Mitsubishi Cable Company
0000BE	The NTI Group
0000BF	Symmetric Computer Systems
0000C0	Western Digital now SMC (Standard Microsystems Corp.)
0000C1	Olicom A/S—Olicom sends bit reversed as 000083 on Token Ring
0000C2	Information Presentation Tech.
0000C3	Harris Corp Computer Sys Div
0000C4	Waters Div. Of Millipore
0000C5	Farallon Computing Inc/Netopia
0000C6	Hewlett Packard Intelligent Networks Operation (formerly Eon Systems)
0000C7	Arix Corporation
0000C8	Altos Computer Systems
0000C9	Emulex Corp.—Terminal Servers, Print Servers—Proteon seen on Token Ring—bit reversed of assigned OUI 000093
0000CA	Applitek then LANcity Cable Modems (now owned by BayNetworks owned by Nortel)
0000CB	Compu-Shack Electronic GmBh
0000CC	Densan Co., Ltd.
0000CD	Teltrend (NZ) Limited—Industrial Research Limited seen under this OUI
0000CE	Megadata Corp.
0000CF	Hayes Microcomputer Products (Canada) Ltd.
0000D0	Develcon Electronics, Ltd.
0000D1	Adaptec, Inc.—"Nodem" product
0000D2	SBE Inc.
0000D3	Wang Laboratories Inc.
0000D4	Pure Data Limited
0000D5	Micrognosis International
0000D6	Punch Line Holding
0000D7	Dartmouth College (NED Router)
0000D8	Novell—old NE1000's (before about 1987)—3Com seen under this OUI
0000D9	Nippon Telegraph & Telephone Corporation (NTT)
0000DA	Atex
0000DB	British Telecommunications PLC
0000DC	Hayes Microcomputer Products
0000DD	TCL Incorporated—Gould seen under this OUI
0000DE	CETIA—Unigraph seen under this OUI
0000DF	Bell & Howell Pub Sys Div
0000E0	Quadram Corp.
0000E1	GRID Systems—Hitachi (laptop built-in) seen under this OUI
0000E2	Acer Technologies Corp.
0000E3	Integrated Micro Products Ltd
0000E4	IN2 Groupe Intertechnique
0000E5	Sigmex Ltd.
0000E6	Aptor Produits De Comm Indust
0000E7	Star Gate Technologies
0000E8	Accton Technology Corporation
0000E9	ISICAD, Inc.
0000EA	Upnod AB
0000EB	Matsushita Comm. Ind. Co. Ltd.
0000EC	Microprocess
0000ED	April
0000EE	Network Designers Limited (also KNX Ltd, a former division)
0000EF	KTI—Alantec (now owned by ForeSystems) seen under this OUI
0000F0	Samsung Electronics Co. Ltd.
0000F1	Magna Computer Corporation
0000F2	Spider Communications— (Montreal, not Spider Systems)
0000F3	Gandalf Data Ltd. - Canada
0000F4	Allied Telesyn International
0000F5	Diamond Sales Limited
0000F6	Applied Microsystems Corp.—Madge Networks Ltd. sends on Token Ring—bit revereseed of assigned OUI 00006F
0000F7	Youth Keep Enterprise Co. Ltd.
0000F8	Digital Equipment Corp.
0000F9	Quotron Systems Inc.
0000FA	Microsage Computer Systems Inc.
0000FB	Rechner zur Kommunikation
0000FC	Meiko
0000FD	High Level Hardware (Orion, UK)
0000FE	Annapolis Micro Systems
0000FF	Camtec Electronics (UK) Ltd.
000104	Andrew seen on Token Ring—bit reversed of assigned OUI 008020—Not an IEEE assigned OUI

OUI	Vendor/Comment
000134	Agilis Corp seen on Token Ring—bit reversed of assigned OUI 00805C—Not an IEEE assigned OUI
000184	Newbridge seen on Token Ring—bit reversed of assigned OUI 008021—Not an IEEE assigned OUI
0001C8	Thomas Conrad Corp.(now owned by Compaq)
0001CB	IBM 8235's seen on Token Ring—Not an IEEE assigned OUI
0001E3	Xircom pocket adapters seen on Token Ring—bit reversed of assinged OUI 0080C7—Not an IEEE assigned OUI
0001FA	Compaq (PageMarq printers)—Compaq sends bit reversed of assigned OUI 00805F on Token Ring—Not an IEEE assigned OUI
0001FE	Digital Equipment Corp.
000216	Extended Systems, Inc (ESI) print server on Token Ring—bit reversed of assigned OUI 004068— Not an IEEE assigned OUI
000230	Silcom Manufacturing—bit reveresed of assigned OUI 40CC—Not an IEEE assigned OUI
00024C	Digital Communications Associates (DCA) (sold to Microdyne and Attachmate) seen on Token Ring—bit reversed of assigned OUI 004032—Not an IEEE assigned OUI
000267	Node Runner Inc.
000288	Global Village Communications (PCcard in Mac portable)
0002CC	Addtron Technology Corp.sends on Token Ring—bit reversed of assigned OUI 004033—Not an IEEE assigned OUI
0002F5	Digital Products, Inc.—Jet Print 100—bit reversed of assigned OUI 0040AF—Not an IEEE assigned OUI
00030F	Kingston Technology Corp.—bit reveresed of assigned OUI 00C0F0—Not an IEEE asigned OUI
0003E8	Forte Networks—makes the Fluke LAN Meters—bit reversed of assigned OUI 00C017—Not an IEEE assigned OUI
000400	Lexmark International (Print Server)
00045B	Xylan—Token Ring Switches—bit reversed of assigned OUI 0020DA—Not an IEEE assigned OUI
0004AC	IBM—PCMCIA Ethernet adapter
000500	Centillion Networks, bought by Bay Networks, bought by Nortel—bit reversed of assigned OUI 005000—Not an IEEE assigned OUI
000502	Apple Computer (PCI Mac)
0005A8	Wyle Electronics—PowerComputing—Mac clones seen under this OUI. IEEE assigned PowerComputing OUI 006012
00060D	Hewlett Packard, Inc. newer LaserJets—bit reversed of assigned OUI 0060B0—Not an IEEE assigned OUI
000629	IBM RISC6000 system
00067C	Cisco Systems, Inc.
0006C1	Cisco Systems, Inc.
000701	Racal-Datacom—Lan Internetworking Division
00070D	Cisco Systems, Inc.— 2511 Token Ring—bit reversal of assigned OUI 00E0B0—Not an IEEE assigned OUI
000800	Multitech Systems, Inc.
0008C7	Compaq Computer Corp.
001000	Cable Television Laboraties, Inc.
001001	MCK Communications
001002	ACTIA
001003	Imatron, Inc.
001004	The Brantley Coile Company, Inc.
001005	UEC Commercial
001006	Racal Recorders Ltd.
001007	Cisco Systems, Inc.—Catalyst 1900
001008	Vienna Systems Corporation
001009	Horo Quartz
00100A	Williams Communications Group Advanced Technologies
00100B	Cisco Systems, Inc.
00100C	ITO Co., Ltd.
00100D	Cisco Systems, Inc.—Catalyst 2924-XL
00100E	Micro Linear Coporation
00100F	Industrial CPU Systems
001010	Initio Corporation
001011	Cisco Systems, Inc.—Cisco 75xx
001012	Processor Systems (I) PVT Ltd.
001013	Industrial Computer Source
001014	Cisco Systems, Inc.
001015	Oomon Inc.
001016	T.Sqware
001017	MICOS GmBh
001018	Broadcom Corporation
001019	Sirona Dental Systems GmBh & Co., KG
00101A	Picturetel Corp.
00101B	Cornet Technology, Inc
00101C	Ohm Technologies Intl, LLC
00101D	Winbond Electronics Corp.
00101E	Matsushita Electronic
00101F	Cisco Systems, Inc.—Catalyst 2901
001020	Welch Allyn, Data Collection
001021	Encanto Networks, Inc.
001022	Satcom Media Corporation
001023	Flowwise Networks, Inc.
001024	Nagoya Electric Works Co., Ltd.
001025	Grayhill Inc.
001026	Accelerated Networks, Inc.
001027	L-3 Communications East
001028	Computer Technica, Inc.
001029	Cisco Systems, Inc.—Catalyst 5000
00102A	ZF Microsystems, Inc.
00102B	Umax Data Systems, Inc.
00102C	PnP Technology A/S
00102D	Hitachi Software Engineering
00102E	Network Systems & Technologies Pvt. Ltd.
00102F	Cisco Systems, Inc.—Cisco 5000
001030	Wi-Lan, Inc.
001031	Objective Communications, Inc.
001032	Alta Technology
001033	AccessLAN Communications, Inc.
001034	GNP Computers
001035	Elitegroup Computer Systems Co., Ltd.
001036	Inter-Tel Integrated Systems
001037	Cyq've Technology Co., Ltd.
001038	Micro Research Institute, Inc.
001039	Vectron Systems GmBh
00103A	Diamond Network Tech
00103B	HIPPI Networking Forum
00103C	IC Ensemble, Inc.
00103D	Phasecom, Ltd.
00103E	NetSchools Corporation
00103F	Tollgrade Communications, Inc.
001040	Intermec Corporation
001041	Bristol Babcock, Inc.
001042	Alacritech
001043	A2 Corporation
001044	Innolabs Corporation
001045	Cambrian Systems Corporation
001046	Alcorn McBride Inc.
001047	Echo Eletric Co. Ltd.
001048	HTRC Automation, Inc.
001049	Shoreline Teleworks, Inc.
00104A	The Parvuc Corporation
00104B	3Com Corporation—3C905-TX PCI
00104C	Computer Access Technology
00104D	Surtec Industries, Inc.
00104E	Ceologic
00104F	Storage Technology Corporation
001050	Rion Co., Ltd.
001051	Cmicro Corporation
001052	Mettler-Toledo (Albstadt) GmBh
001053	Computer Technology Corp.
001054	Cisco Systems, Inc.
001055	Fujitsu Microelectronics, Inc.
001056	Sodick Co., Ltd.
001057	Corel Computer Corporation
001058	Arrowpoint Communications, Inc.
001059	Diablo Research Co. LLC
00105A	3Com Corporation—Fast Etherlink XL in a Gateway 2000
00105B	Net Insight AB
00105C	Quantum Designs (H.K.) Ltd.
00105D	Drager, Business Unit Monitoring
00105E	Hekimian Laboratories, Inc.

OUI	Vendor/Comment
00105F	IN-SNEC
001060	Billionton—Novell NE200 Compatible
001061	Hostlink Corp.
001062	NX Server, ILNC.
001063	Starguide Digital Networks
001064	Digital Equipment Corp.
001065	Radyne Corporation
001066	Advanced Control Systems, Inc.
001067	Redback Networks, Inc.
001068	Comos Telecom
001069	Helioss Communications, Inc.
00106A	Digital Microwave Corporation
00106B	Sonus Networks, Inc.
00106C	Infratec Plus GmBh
00106D	Integrity Communications, Inc.
00106E	Tadiran Com. Ltd.
00106F	Trenton Technology Inc.
001070	Caradon Trend Ltd.
001071	Advanet Inc.
001072	GVN Technologies, Inc.
001073	Technobox, Inc.
001074	Aten International Co., Ltd.
001075	Creative Design Solutions, Inc.
001076	Eurem GmBh
001077	Saf Drive Systems, Ltd.
001078	Nuera Communications, Inc.
001079	Cisco Systems, Inc—5500 Router
00107A	Ambicom, Inc.
00107B	Cisco Systems, Inc.
00107C	P-COM, Inc.
00107D	Aurora Communications, Ltd.
00107E	Bachmann Electronic GmBh
00107F	Crestron Electronics, Inc.
001080	Metawave Communications
001081	DPS, Inc.
001082	JNA Telecommunications Limited
001083	Hewlett-Packard Company—HP-UX E 9000
001084	K-Bot Communications
001085	Polaris Communications, Inc.
001086	Atto Technology, Inc.
001087	EPL, Ltd.
001088	American Networks Inc.
001089	Websonic
00108A	Teralogic, Inc.
00108B	Laseranimation Sollinger GmBh
00108C	Fujitsu Telecommunications
00108D	Johnson Controls, Inc.
00108E	Hugh Symons Concept Technologies, Ltd.
00108F	Raptor Systems
001090	Cimetrics, Inc.
001091	No Wires Needed BV
001092	Netcore Inc.
001093	CMS Computers, Ltd.
001094	Adtech, Inc.
001095	Thomson Consumer Electronics
001096	Tracewell Systems, Inc.
001097	Winnet Metropolitan Communications Systems, Inc.
001098	Starnet Technologies, Inc.
001099	Innomedia, Inc.
00109A	Netline
00109B	Vixel Corporation
00109C	M-System Co., Ltd.
00109D	Clarinet Systems, Inc.
00109E	Aware, Inc.
00109F	Pavo, Inc.
0010A0	Innovex Technologies, Inc.
0010A1	Kendin Semiconductor, Inc.
0010A2	Hyundai Network Systems
0010A3	Omnitronix, Inc.
0010A4	Xircom—RealPort 10
0010A5	Oxford Instruments
0010A6	Cisco Systems, Inc.
0010A7	Unex Technology Corporation
0010A8	Reliance Computer Corp.
0010A9	Adhoc Technologies
0010AA	Media4, Inc.
0010AB	Koito Industries, Ltd.
0010AC	IMCI Technologies
0010AD	Softronics USB, Inc.
0010AE	Shinko Electric Industries Co.
0010AF	TAC Systems, Inc.
0010B0	Meridian Technology Corp.
0010B1	For-A Co., Ltd.
0010B2	Coactive Aesthetics
0010B3	Nokia Satellite Systems AB
0010B4	Atmosphere Networks
0010B5	Accton Technology Corporation
0010B6	Entrata Communications Corp.
0010B7	Coyote Technologies, LLC
0010B8	Ishigaki Computer System Co.
0010B9	Maxtor Corp.
0010BA	Martinho-Davis Systems, Inc.
0010BB	Data & Information Technology
0010BC	Aptis Communications, Inc.
0010BD	The Telecommunication
0010BE	Telexis Corp.
0010BF	Inter Air Wireless
0010C0	Arma, Inc.
0010C1	OI Electric Co., Ltd.
0010C2	Willnet, Inc.
0010C3	CSI-Control Systems
0010C4	Media Links Co., Ltd.
0010C5	Protocol Technologies, Inc.
0010C6	USI
0010C7	Data Transmission Network
0010C8	Communications Electronics
0010C9	Mitsubishi Electronics Logistic Support Co.
0010CA	Integral Access
0010CB	Facit K.K.
0010CC	CLP Computer Logistik
0010CD	Interface Concept
0010CE	Volamp, Ltd.
0010CF	Fiberlane Communications
0010D0	Witcom, Ltd.
0010D1	Blazenet, Inc.
0010D2	Nitto Tsushinki Co., Ltd.
0010D3	Grips Electronic GmBh
0010D4	Storage Computer Corporation
0010D5	Imasde Canarias, S.A.
0010D6	ITT A/CD
0010D7	Argosy Research Inc.—EN 220 Fast Ethernet PCMCIA
0010D8	Calista
0010D9	IBM Japan, Fujisawa MT+D
0010DA	Motion Engineering, Inc.
0010DB	Netscreen Technologies, Inc.
0010DC	Micro-Star International Co., Ltd.
0010DD	Enable Semiconductor, Inc.
0010DE	International Datacasting
0010DF	Rise Computer Inc.
0010E0	Cobalt Microserver, Inc.
0010E1	S.I.Tech, Inc.
0010E2	Arraycomm, Inc.
0010E3	Compaq Computer Corporation
0010E4	NSI Corporation
0010E5	Solectron Texas
0010E6	Applied Intelligent Systems, Inc.
0010E7	Breezecom
0010E8	Telocity, Incorporated (formerly Aspen Internet System, Inc.)
0010E9	Raidtec Ltd.
0010EA	Adept Technology
0010EB	Selsius Systems, ILNC.
0010EC	RPCG, LLC

OUI	Vendor/Comment
0010ED	Sundance Technology, Inc.
0010EE	CTI Products, Inc.
0010EF	DB Networks, Inc.
0010F0	Rittal-Werk Rudolf LOH GmBh & Co. KG
0010F1	I-O Corporation
0010F2	Antec
0010F3	Nexcom International Co., Ltd.
0010F4	Vertical Networks, Inc.
0010F5	Amherst Systems, Inc.
0010F6	Cisco Systems, Inc.
0010F7	Iriichi Technologies Incorporation
0010F8	Kenwood TMI Corporation
0010F9	Unique Systems, Inc.
0010FA	Zayante, Inc.
0010FB	Zida Technologies Limited
0010FC	Broadband Networks, Inc.
0010FD	Cocom A/S
0010FE	Digital Equipment Corporation
0010FF	Cisco Systems, Inc.
001700	Kabel Rheydt AG—Not an IEEE assigned OUI—IEEE assigned OUI 00C029
001C7C	Perq Systems Corporation
002000	Lexmark (Printers and Print Servers)
002001	DSP Solutions, Inc.
002002	Seritech Enterprise Co., Ltd.
002003	Pixel Power Ltd.
002004	Yamatake-Honeywell Co., Ltd.
002005	Simple Technology
002006	Garrett Communications, Inc.
002007	SFA, Inc.
002008	Cable & Computer Technology
002009	Packard Bell Elec., Inc.
00200A	Source-Comm Corp.
00200B	Octagon Systems Corp.
00200C	Adastra Systems Corp.
00200D	Carl Zeiss
00200E	Satellite Technology Mgmt, Inc.
00200F	Tanbac Co., Ltd.
002010	Jeol System Technology Co. Ltd.
002011	Canopus Co., Ltd.
002012	Camtronics Medical Systems
002013	Diversified Technology, Inc.
002014	Global View Co., Ltd.
002015	Actis Computer SA
002016	Showa Electric Wire & Cable Co.
002017	Orbotech
002018	CIS Technology Inc.
002019	Ohler GmBh
00201A	N-Base Switch Communications
00201B	Northern Telecom/Network Systems Corporation (Nortel)
00201C	Excel, Inc.
00201D	Katana Products
00201E	Netquest Corporation
00201F	Best Power Technology, Inc.
002020	Megatron Computer Industries Pty, Ltd.
002021	Algorithms Software PVT. Ltd.
002022	Teknique, Inc.
002023	T.C. Technologies PTY. Ltd
002024	Pacific Communication Sciences
002025	Control Technology Inc. (Industrial Controls and Network Interfaces)
002026	Amkly Systems, Inc.
002027	Ming Fortune Industry Co., Ltd.
002028	West Egg Systems, Inc.
002029	TeleProcessing Products Inc.—CSU
00202A	N.V. Dzine
00202B	Advanced Telecommunications Modules, Ltd. (ATML)
00202C	Welltronix Co., Ltd.
00202D	Taiyo Corporation
00202E	Daystar Digital
00202F	Zeta Communications, Ltd.
002030	Analog & Digital Systems
002031	Ertec GmBh
002032	Alcatel Taisel
002033	Synapse Technologies, Inc.
002034	Rotec Industrieautomation GmBh
002035	IBM Corporation (International Business Machines)
002036	BMC Software
002037	Seagate Technology
002038	VME Microsystems International
002039	SCINets
00203A	Digital Biometrics Inc.
00203B	Wisdm Ltd.
00203C	Eurotime AB
00203D	Novar Electronics Corporation
00203E	Logican Technologies, Inc.
00203F	Juki Corporation
002040	General Instrument Corporation of Delaware GI Communications Division
002041	Data Net
002042	Datametrics Corp.
002043	Neuron Company Limited
002044	Genitech PTY Ltd.
002045	SolCom Systems Limited
002046	Ciprico, Inc.
002047	Steinbrecher Corp.
002048	Fore Systems Inc.
002049	Comtron, Inc.
00204A	Pronet GmBh
00204B	Autocomputer Co., Ltd.
00204C	Mitron Computer PTE Ltd.
00204D	Inovis GmBh
00204E	Network Security Systems, Inc.
00204F	Deutsche Aerospace AG
002050	Korea Computer Inc.
002051	Phoenix Data Commununications Corp.
002052	Ragula Systems
002053	Huntsville Microsystems, Inc.
002054	Eastern Research, Inc.
002055	Altech Co., Ltd.
002056	Neoproducts
002057	Titze Datentechnik GmBh
002058	Allied Signal Inc.
002059	Miro Computer Products AG
00205A	Computer Identics
00205B	Skyline Technology
00205C	Internet Systems/ Florida Inc.
00205D	Nanomatic OY
00205E	Castle Rock, Inc.
00205F	Gammadata Computer GmBh
002060	Alcatel Italia S.P.A.
002061	Dynatech Communications Inc.
002062	Scorpion Logic, Ltd.
002063	Wipro Infotech Ltd.
002064	Protec Microsystems, Inc.
002065	Supernet Networking Inc.
002066	General Magic, Inc.
002067	IEEE OUI ADMINISTRATOR
002068	Isdyne
002069	ISDN Systems Corporation
00206A	Osaka Computer Corp.
00206B	Minolta Co., Ltd.—Network printers
00206C	Evergreen Technology Corp.
00206D	Data Race, Inc.
00206E	Xact, Inc.
00206F	Flowpoint Corporation
002070	Hynet, Ltd.
002071	IBR GmBh
002072	Worklink Innovations
002073	Fusion Systems Corporation
002074	Sungwoon Systems
002075	Motorola Communication Israel
002076	Reudo Corporation
002077	Kardios Systems Corp.
002078	Runtop, Inc.

OUI	Vendor/Comment
002079	Mikron GmBh
00207A	Wise Communications, Inc.
00207B	Level One Communications
00207C	Autec GmBh
00207D	Advanced Computer Applications
00207E	Finecom Co., Ltd.
00207F	Kyoei Sangyo Co., Ltd.
002080	Synergy (UK) Ltd.
002081	Titan Electronics
002082	Oneac Corporation
002083	Presticom Incorporated
002084	OCE Printing Systems, GmBh
002085	Exide Electronics—3Com Corporation SuperStack II UPS management module seen under this OUI
002086	Microtech Electronics Limited
002087	Memotec Communications Corp.
002088	Global Village Communication
002089	T3Plus Networking, Inc.
00208A	Sonix Communications, Ltd.
00208B	Lapis Technologies, Inc.—Focus Enhancements seen under this OUI
00208C	Galaxy Networks, Inc.
00208D	CMD Technology
00208E	Chevin Software Eng. Ltd.
00208F	ECI Telecom Ltd.
002090	Advanced Compression
002091	J125, National Security Agency
002092	Chess Engineering B.V.
002093	Landings Technology Corp.
002094	Cubix Corporation
002095	Riva Electronics
002096	Siebe Environmental Controls
002097	Applied Signal Technology
002098	Hectronic AB
002099	Bon Electric Co., Ltd.
00209A	The 3DO Company
00209B	Ersat Electronic GmBh
00209C	Primary Access Corp.
00209D	Lippert Automationstechnik
00209E	Brown's Operating System Services, Ltd.
00209F	Mercury Computer Systems, Inc.
0020A0	OA Laboratory Co., Ltd.
0020A1	Dovatron Products Division
0020A2	Galcom Networking Ltd. (now owned by N-Base)
0020A3	Divicom Inc.
0020A4	Multipoint Networks
0020A5	API Engineering—Newer Technology seen under this OUI—IEEE assigned OUI 00507E
0020A6	Proxim, Inc.
0020A7	Pairgain Technologies, Inc.
0020A8	Sast Technology Corp.
0020A9	White Horse Industrial
0020AA	Digimedia Vision Ltd.
0020AB	Micro Industries Corp.
0020AC	Interflex Datensysteme GmBh
0020AD	LINQ Systems
0020AE	Ornet Data Communication
0020AF	3Com Corporation (newer Token Ring Adapters w/ TI-C60 and 3Com MAC)
0020B0	Gateway Devices, Inc.
0020B1	Comtech Research Inc.
0020B2	GKD Gesellschaft fur Kommunikation und Datentechnik—CSP (Printline Multiconnectivity converter) seen under this OUI
0020B3	Scltec Communications Systems
0020B4	Terma Elektronik AS
0020B5	Yaskawa Electric Corporation
0020B6	Agile Networks, Inc.
0020B7	Namaqua Computerware
0020B8	Prime Option, Inc.
0020B9	Metricom, Inc.
0020BA	Center For High Performance Computing of WPI
0020BB	Zax Corporation
0020BC	Jtec PTY. Ltd.
0020BD	Niobrara R & D Corporation
0020BE	Lan Access Corp.
0020BF	Aehr Test Systems
0020C0	Pulse Electronics, Inc.
0020C1	Taiko Electric Works, Ltd.
0020C2	Texas Memory Systems, Inc.
0020C3	Counter Solutions Ltd.
0020C4	INET, Inc.
0020C5	Eagle Technology— NE2000
0020C6	Nectec
0020C7	Akai Electric Co., Ltd.
0020C8	Larscom Incorporated
0020C9	Victron BV
0020CA	Digital Ocean
0020CB	Pretec Electronics Corp.
0020CC	Digital Services, Ltd.
0020CD	Hybrid Networks, Inc.
0020CE	Logical Design Group, Inc.
0020CF	Test & Measurement Systems Inc.
0020D0	Versalynx Corporation— "The One Port" terminal server
0020D1	Microcomputer Systems (M) SDN.
0020D2	RAD Data Communications Ltd, then Armon now owned by Bay Networks, owned by Nortel
0020D3	OST (Ouet Standard Telematique)
0020D4	Cabletron - Zeittnet Inc.
0020D5	Vipa GmBh
0020D6	Breezecom
0020D7	Japan Minicomputer Systems Co.
0020D8	Netwave Technologies, Inc.
0020D9	Panasonic Technologies, Inc./Mieco-US
0020DA	Xylan Corporation—sends bit reversed as 00045B on Token Ring
0020DB	Xnet Technology, Inc.
0020DC	Densitron Taiwan Ltd.
0020DD	AWA Ltd.
0020DE	Japan Digital Laborat'y Co.Ltd. (JDL)
0020DF	Kyosan Electric Mfg. Co., Ltd.
0020E0	PreMax Electronics, Inc.—PE-200 (PCMCIA NE2000-clone card, sold by InfoExpress)
0020E1	Alamar Electronics
0020E2	Information Resource Engineering
0020E3	MCD Kencom Corporation
0020E4	Hsing Tech Enterprise Co., Ltd.
0020E5	Apex Data, Inc.
0020E6	Lidkoping Machine Tools AB
0020E7	B&W Nuclear Service Company
0020E8	Datatrek Corporation
0020E9	Dantel
0020EA	Efficient Networks, Inc.
0020EB	Cincinnati Microwave, Inc.
0020EC	Techware Systems Corp.
0020ED	Giga-Byte Technology Co., Ltd.
0020EE	Gtech Corporation
0020EF	USC Corporation
0020F0	Universal Microelectronics Co.
0020F1	Altos India Limited
0020F2	Sun Microsystems, Inc. (formerly Maximum Strategy Inc.)
0020F3	Raynet Corporation
0020F4	Spectrix Corporation
0020F5	Pandatel AG (formerly PAN DACOM TELECOM'CATIONS GmBh)
0020F6	Net Tek And Karlnet, Inc.
0020F7	Cyberdata
0020F8	Carrera Computers, Inc.
0020F9	Paralink Networks, Inc.
0020FA	GDE Systems, Inc.
0020FB	Octel Communications Corp.
0020FC	Matrox
0020FD	ITV Technologies, Inc.
0020FE	Topware Inc. / Grand Computer Corporation
0020FF	Symmetrical Technologies
003000	Allwell Technology Corp.
003001	SMP

OUI	Vendor/Comment
003006	Superpower Computer Electronics Co., Ltd.
003007	Opti, Inc.
003008	Avio Digital, Inc.
00300A	Aztech Systems Ltd.
00300C	Congruency, Ltd.
003010	Visionetics International
003011	HMS Fieldbus Systems AB
003012	Digital Engineering Ltd.
003014	Divio, Inc.
003015	CP Clare Corp.
003016	Ishida Co., Ltd.
003017	Terastack Ltd.
003019	Cisco Systems, Inc.
00301A	SmartBridges PTE. Ltd.
00301C	Altvater Airdata Systems GmBh & Co. KG
00301D	Skystream, Inc.
00301F	Optical Networks, Inc.
003021	Shing Tech. Enterprise Co., Ltd.
003023	Cogent Computer Systems, Inc.
003024	Cisco Systems, Inc.
003025	Checkout Computer Systems, Ltd.
003026	Heitel Kommunikationselektronik GmBh
003027	Kerbango, Inc.
003029	Opicom
00302A	Southern Information System, Inc.
00302B	Inalp Networks, Inc.
00302C	Sylantro Systems Corporation
00302D	Quantum Bridge Communications
003030	Harmonix Corporation
003031	Lightwave Communications, Inc.
003032	Magicram, Inc.
003033	Orient Telecom Co., Ltd.
003036	RMP Elektroniksysteme GmBh
003038	XCP, Inc.
003039	Softbook Press
00303A	Maatel
00303C	Onnto Corp.
00303D	IVA Corporation
003040	Cisco Systems, Inc.
003041	Saejin T & M Co., Ltd.
003043	Idream Technologies, PTE. Ltd.
003047	Nissei Electric Co., Ltd.
003049	Bryant Technology, Ltd.
00304A	Fraunhofer Institute IMS
00304B	Orbacom Systems, Inc.
00304C	Appian Communications, Inc.
00304E	Bustec Production Ltd.
003051	Orbit Avionic & Communication Systems, Ltd.
003052	Elastic Networks
003054	Castlenet Technology, Inc.
003058	API Motion
003059	Digital-Logic AG
00305A	Telgen Corporation
00305B	Module Department
00305D	Digitra Systems, Inc.
00305F	Imacon APS
003060	Starmatix, Inc.
003062	Path 1 Network Technol's Inc.
003063	Santera Systems, Inc.
003064	Adlink Technology, Inc.
003065	Apple Computer, Inc.
003066	Digital Wireless Corporation
003067	Biostar Microtech Int'l Corp.
003068	Cybernetics Tech. Co., Ltd.
003069	Impacct Technology Corp.
00306A	Penta Media Co., Ltd.
00306B	CMOS Systems, Inc.
00306D	Lucent Technologies
00306E	Hewlett-Packard
00306F	Seyeon Tech. Co., Ltd.
003072	Intellibyte Inc.
003074	Equiinet Ltd.
003076	N-Cubed.Net
003077	Onprem Networks
003078	Cqos, Inc.
00307C	Adid SA
00307D	GRE America, Inc.
00307F	Irlan Ltd.
003080	Cisco Systems, Inc.
003081	Altos C&C
003083	Vedia Technology, Ltd.
003084	Allied Telesyn Internaional
003085	Cisco Systems, Inc.
003087	Vega Grieshaber KG
00308A	Nicotra Sistemi S.P.A.
00308C	Advanced Digital Information
00308D	Pinnacle Systems, Inc.
00308E	Cross Match Technologies, Inc.
003090	Cyra Technologies, Inc.
003091	Taiwan First Line Elec. Corp.
003093	Sonnet Technologies, Inc.
003096	Cisco Systems, Inc.
003097	Exomatic AB
003099	Boenig und Kallenbach OHG
00309A	Astro Terra Corp.
00309F	Amber Networks
0030A0	Tyco Submarine Systems, Ltd.
0030A1	Opti Tech Co., Ltd.
0030A3	Cisco Systems, Inc.
0030A5	Active Power
0030A6	Vianet Technologies, Ltd.
0030A7	Schweitzer Engineering Laboratories, Inc.
0030A8	Ol'e Communications, Inc.
0030AA	Axus Microsystems, Inc.
0030AB	Delta Networks, Inc.
0030AD	Shanghai Communication
0030B1	GOC Gesellschaft fur Optische Communication MBH
0030B2	Wescam - Healdsburg
0030B4	Intersil Corp.
0030B6	Cisco Systems, Inc.
0030B9	Ectel
0030BB	AC&T System Co., Ltd.
0030BD	Belkin Components
0030BF	Multidata GmBh
0030C1	Hewlett-Packard
0030C3	Flueckiger Elektronik AG
0030C5	Cadence Design Systems
0030C6	Control Solutions, Inc.
0030C7	Macromate Corp.
0030C8	Gad Line, Ltd.
0030CB	Omni Flow Computers, Inc.
0030CD	Conexant Systems, Inc.
0030CF	Two Technologies, Inc.
0030D1	Inova Corporation
0030D4	Comtier
0030D6	MSC Vertriebs GmBh
0030D8	Sitek
0030D9	Datacore Software Corp.
0030DA	Comtrend Co.
0030DB	Sederta Inc.
0030DC	Rightech Corporation
0030DD	Indigita Corporation
0030DF	KB/Tel Telecomunicaciones
0030E0	Oxford Semiconductor Ltd.
0030E1	Acrotron Systems, Inc.
0030E3	Netpoint Corporation
0030E4	Chiyoda System Riken
0030E6	Siemens Medical Systems
0030E7	CNF Mobile Solutions, Inc.
0030E9	GMA Communication Manufact'g Marketing, (1991) Ltd.
0030EA	Intelect Communictions, Inc.
0030EB	Turbonet Communications, Inc.

OUI	Vendor/Comment
0030EC	Borgardt
0030EF	Neon Technology, Inc.
0030F2	Cisco Systems, Inc.
0030F4	Stardot Technologies
0030F6	Securelogix Corporation
0030F7	Ramix Inc.
0030FA	Telica, Inc.
0030FD	Integrated Systems Design
0030FF	Datafab Systems, Inc.
004000	PCI Componentes da Amzonia Ltd
004001	Zyxel Communications, Inc.—Zero One Technology Co Ltd. seen under this OUI—IEEE has not assigned any OUIs to Zero One
004002	Perle Systems Limited
004003	Westinghouse Process Control Division
004004	ICM CO. LTD.
004005	ANI Communications Inc.—TRENDware International Inc.; Linksys; Simple Net all seen under this OUI
004006	Sampo Technology Corporation
004007	Telmat Informatique
004008	A Plus Info Corporation
004009	Tachibana Tectron Co., Ltd. Systematic Equipment Division
00400A	Pivotal Technologies, Inc.
00400B	Crescendo (now owned by Cisco Systems, Inc.)
00400C	General Micro Systems, Inc.
00400D	LANNET Data Communications
00400E	Memotec Communications, Inc.
00400F	Datacom Technologies
004010	Sonic—Mac Ethernet interfaces
004011	Andover Controls Corporation
004012	Windata, Inc.
004013	NTT Data Communication Systems Corp
004014	Comsoft GmBh
004015	Ascom Infrasys AG
004016	Hadax Electronics, Inc.
004017	XCD Inc.— XJet - HP printer server card
004018	Adobe Systems, Inc.
004019	Aeon Systems, Inc.
00401A	Fuji Electric Co., Ltd.
00401B	Printer Systems Corp
00401C	AST Research, Inc.—Pentium
00401D	Invisible Software, Inc.
00401E	ICC
00401F	Colorgraph Ltd
004020	Pinacl Communication Systems Limited
004021	Raster Graphics
004022	Klever Computers, Inc.
004023	Logic Corporation
004024	COMPAC Inc.
004025	Molecular Dynamics
004026	Melco Inc
004027	Standard Microsystems Corp—SMC Massachusetts
004028	Netcomm Limited
004029	Compex
00402A	Canoga-Perkins
00402B	TriGem Computer, Inc.
00402C	ISIS Distributed Systems, Inc.
00402D	Harris Adacom Corporation
00402E	Precision Software, Inc.
00402F	Xlnt Designs Inc (XDI)
004030	GK Computer
004031	Kokusai Electric Co., Ltd
004032	Digital Communications Associates (DCA) (sold to Microdyne and Attachmate)—sends bit reversed as 00024C on Token Ring
004033	Addtron Technology Co., Ltd.—sends bit reversed as 0002CC on Token Ring
004034	Bustek Corporation
004035	OPCOM
004036	Tribe Computer Works, Inc.
004037	Sea-Ilan, Inc.
004038	Talent Electric Incorporated
004039	Optec Daiichi Denko Co Ltd
00403A	Impact Technologies
00403B	Synerjet International Corp
00403C	Forks, Inc.
00403D	Teradata
00403E	Raster Ops Corporation
00403F	Ssangyong Computer Systems Corporation
004040	Ring Access, Inc.
004041	Fujikura Ltd.
004042	N.A.T. GmBh
004043	Nokia Telecommunications
004044	Qnix Computer Co., Ltd.
004045	Twinhead Corporation
004046	UDC Research Limited
004047	Wind River Systems
004048	SMD Informatica S.A.
004049	Tegimenta AG
00404A	West Australian Department of Employment (Devet)
00404B	Maple Computer Systems
00404C	Hypertec Pty Ltd.
00404D	Telecommunications Techniques
00404E	Fluent, Inc
00404F	Space & Naval Warfare Systems
004050	Ironics, Incorporated
004051	Gracilis, Inc.
004052	Star Technologies Inc
004053	Ampro Computers—Datum (Bancomm Division)—TymServe 2000 seen under this OUI
004054	Connection Machines Services (formerly Thinking Machines Corporation)
004055	Metronix GmBh
004056	MCM Japan Ltd.
004057	Lockheed-Sanders
004058	Kronos, Inc.
004059	Yoshida Kogyo K.K.
00405A	Goldstar Information & Comm.
00405B	Funasset Limited
00405C	Future Systems, Inc.
00405D	Star-Tek Inc.
00405E	North Hills Israel
00405F	AFE Computers Ltd.
004060	Comendec Ltd.
004061	Datatech Enterprises Co., Ltd.
004062	E-Systems, Inc./Garland Div.
004063	Via Technologies, Inc.
004064	KLA Instruments Corporation
004065	GTE Spacenet
004066	Hitachi Cable, Ltd.
004067	Omnibyte Corporation
004068	Extended Systems, Inc—ESI sends bit reversed as 000216 on Token Ring
004069	Lemcom Systems Inc
00406A	Kentek Information Systems Inc
00406B	Sysgen
00406C	Copernique
00406D	Lanco, Inc
00406E	Corollary, Inc.
00406F	Sync Research Inc
004070	Interware Co., Ltd.
004071	ATM Computer GmBh
004072	Applied Innovation Inc.
004073	Bass Associates
004074	Cable and Wireless
004075	M-Trade (UK) Ltd
004076	AMP Incorporated
004077	Maxton Technology Corporation
004078	Wearnes Automation PTE Ltd
004079	Juko Manufacture Company, Ltd.
00407A	Societe d'Exploitation du Cnit
00407B	Scientific Atlanta
00407C	Qume Corporation
00407D	Extension Technology Corp.
00407E	Evergreen Systems, Inc.
00407F	Agema Infrared Systems AB
004080	Athenix Corporation

OUI	Vendor/Comment
004081	Mannesmann Scangraphic GmBh
004082	Laboratory Equipment Corp
004083	TDA Industria de Produtos
004084	Honeywell Inc.
004085	SAAB Instruments AB
004086	Michels & Kleberhoff Computer
004087	Ubitrex Corporation
004088	Mobuis—NuBus (Mac) combination video
004089	Meidensha Corporation
00408A	TPS Teleprocessing Sys. GmBh
00408B	Raylan Corporation
00408C	Axis Communications AB
00408D	The Goodyear Tire & Rubber Co.
00408E	CXR—IEEE assigned to Digilog, Inc.
00408F	WM-Data Minfo AB
004090	Ansel Communications—PC NE2000 compatible twisted-pair ethernet cards
004091	Procomp Industria Eletronica
004092	ASP Computer Products, Inc.
004093	Paxdata Networks Ltd.
004094	Shographics Inc
004095	R.P.T. Intergroups Int'l Ltd.—Eagle Technology and UMC seen under this OUI
004096	Telesystems SLW Inc.
004097	Datex Division of Instrumentarium Corp.
004098	Dressler GmBh & Co.
004099	Newgen Systems Corp.
00409A	Network Express Inc.
00409B	Hal Computer Systems Inc.
00409C	Transware
00409D	DigiBoard—Ethernet-ISDN bridges
00409E	Concurrent Technologies Ltd.
00409F	Lancast/Casat Technology, Inc.
0040A0	Goldstar Co., Ltd.
0040A1	Ergo Computing
0040A2	Kingstar Technology Inc.
0040A3	Microunity Systems Engineering
0040A4	Rose Electronics
0040A5	Clinicomp Intl.
0040A6	Cray Research Inc.
0040A7	Itautec Informatica S.A.
0040A8	IMF International Ltd.
0040A9	Datacom Inc.
0040AA	Valmet Automation Inc.
0040AB	Roland DG Corporation
0040AC	Super Workstation, Inc.
0040AD	SMA Regelsysteme GmBh
0040AE	Delta Controls, Inc.
0040AF	Digital Products, Inc. (DPI).—sends bit reversed as 0002F5 on Token Ring
0040B0	Bytex Corporation, Engineering
0040B1	Codonics Inc.
0040B2	Systemforschung
0040B3	PAR Microsystems Corporation
0040B4	3Com Corporation (formerly Nextcom K.K.)
0040B5	Video Technology Computers Ltd.
0040B6	Computerm Corporation
0040B7	Stealth Computer Systems
0040B8	IDEA Associates
0040B9	MACQ Electronique SA
0040BA	Alliant Computer Systems Corp.
0040BB	Goldstar Cable Co., Ltd.
0040BC	Algorithmics Ltd.
0040BD	Starlight Networks Inc
0040BE	Boeing Defense & Space
0040BF	Channel Systems Intern'l Inc.
0040C0	Vista Controls Corporation
0040C1	Bizerba-Werke Wilheim Kraut
0040C2	Applied Computing Devices
0040C3	Fischer and Porter Co.
0040C4	Kinkei System Corporation
0040C5	Micom Communications Corp.
0040C6	Fibernet Research, Inc.
0040C7	Ruby Tech Corporation—Danpex Corporation seen under this OUI

OUI	Vendor/Comment
0040C8	Milan Technology Corp.—sends bit reversed as 0000AA on Token Ring
0040C9	NCUBE
0040CA	First Internat'l Computer, Inc.
0040CB	LANWAN Technologies
0040CC	Silcom Manufacturing Technology Inc.—sends bit reversed as 000230 on Token Ring
0040CD	Tera Microsystems, Inc.
0040CE	Net-Source, Inc.
0040CF	Strawberry Tree Inc
0040D0	Digital Equipment Corp. (DEC)—IEEE assigned to Mitac International Corp.
0040D1	Fukuda Denshi Co., Ltd.
0040D2	Pagine Corporation
0040D3	Kimpsion International Corp.
0040D4	Gage Talker Corp.
0040D5	Sartorius AG
0040D6	Locamation B.V.
0040D7	Studio Gen Inc.
0040D8	Ocean Office Automation Ltd.
0040D9	American Megatrends Inc.
0040DA	Telspec Ltd.
0040DB	Advanced Technical Solutions
0040DC	Tritec Electronic GmBh
0040DD	Hong Technologies
0040DE	Elettronica San Giorgio Elsag S.P.A.
0040DF	Digalog Systems, Inc.
0040E0	Atomwide Ltd.
0040E1	Marner International Inc.
0040E2	Mesa Ridge Technologies Inc.
0040E3	Quin Systems Ltd.
0040E4	E-M Technology, Inc.
0040E5	Sybus Corporation
0040E6	C.A.E.N.
0040E7	Arnos Instruments & Computer
0040E8	Charles River Data Systems, Inc.
0040E9	Accord Systems, Inc.
0040EA	Plain Tree Systems Inc.
0040EB	Martin Marietta Corporation
0040EC	Mikasa System Engineering Co., Ltd.
0040ED	Network Controls International Inc.
0040EE	OPTIMEM
0040EF	Hypercom, Inc.
0040F0	Micro Systems Inc.
0040F1	Chuo Electronics Co., Ltd.
0040F2	Janich & Klass Computertechnik
0040F3	NETCOR
0040F4	Cameo Communications, Inc.
0040F5	OEM Engines
0040F6	Katron Computers Inc.
0040F7	Polaroid Medical Imaging Sys.
0040F8	Systemhaus Discom Distributed Computing GmBh
0040F9	Combinet
0040FA	Microboards Inc.
0040FB	Cascade Communications Corp.
0040FC	IBR Computer Technik GmBh
0040FD	LXE
0040FE	Symplex Communications
0040FF	Telebit Corporation—Personal NetBlazer
004854	Digital SemiConductor—21143/2 based 10/100 seen under this OUI—Not an IEEE assigned OUI—No IEEE OUI assigned to Digital SemiConductor
004F49	Realtek Semiconductor Corp. seen under this OUI—Not an IEEE assigned OUI—IEEE assigned OUI 00E04C to Realtek
004F4B	Pine Technology Ltd. seen under this OUI—Not and IEEE assigned OUI—No IEEE OUI assigned to Pine Technology
005000	Nexo Communications, Inc.
005001	Yamashita Systems Corp.
005002	Omnisec AG
005003	Gretag Macbeth AG
005004	3Com Corporation—3C90X seen under this OUI—Not an IEEE assigned OUI
005006	Tac AB
005007	Siemens Telecommunication Systems Limited
005008	Tiva Microcomputer Corp. (TMC)

OUI	Vendor/Comment
005009	Philips Broadband Networks
00500A	Iris Technologies, Inc.
00500C	Etek Labs, Inc.
00500D	Satori Electoric Co., Ltd.
00500E	Chromatis Networks, Inc.
00500F	Cisco Systems, Inc.
005010	Novanet Learning, Inc.
005013	Chaparral Technologies, Inc.
005014	Cisco Systems, Inc.
005015	Bright Star Engineering
005016	SST/Woodhead Industries
005017	RSR S.R.L.
005018	Advanced Multimedia Internet Technology Inc.
005019	Spring Tide Networks, Inc.
00501A	Uisiqn
00501B	ABL Canada, Inc.
00501C	Jatom Systems, Inc.
00501E	Miranda Technologies, Inc.
00501F	MRG Systems, Ltd.
005020	Mediastar Co., Ltd.
005021	EIS International, Inc.
005022	Zonet Technology, Inc.
005023	PG Design Electronics, Inc.
005024	Navic Systems, Inc.
005026	Cosystems, Inc.
005027	Genicom Corporation
005028	Aval Communications
005029	1394 Printer Working Group
00502A	Cisco Systems, Inc.
00502B	Genrad Ltd.
00502C	Soyo Computer, Inc.
00502D	Accel, Inc.
00502E	Cambex Corporation
00502F	Tollbridge Technologies, Inc.
005030	Future Plus Systems
005031	Aeroflex Laboratories, Inc.
005033	Mayan Networks
005036	Netcam, Ltd.
005037	Koga Electronics Co.
005038	Dain Telecom Co., Ltd.
005039	Mariner Networks
00503A	Datong Electronics Ltd.
00503B	Mediafire Corporation
00503C	Tsinghua Novel Electronics Co., Ltd.
00503E	Cisco Systems, Inc.
005040	Emware, Inc.
005041	CTX Opto Electronic Corp.
005042	SCI Manufacturing Singapore PTE., Ltd.
005043	Marvell Semiconductor, Inc.
005044	Asaca Corporation
005045	Rioworks Solutions, Inc.
005046	Menicx International Co., Ltd.
005049	Ellacoya Networks, Inc.
00504A	Elteco A.S.
00504B	Barco N.V. BCS
00504C	Galil Motion Control, Inc.
00504F	Olencom Electronics
005050	Cisco Systems, Inc.
005051	Iwatsu Electric Co., Ltd.
005052	Tiara Networks, Inc.
005053	Cisco Systems, Inc.
005054	Cisco Systems, Inc.
005055	Doms A/S
005056	Vmware, Inc.
005057	Broadband Access Systems
005058	Vegastream Limited
005059	Suite Technology Systems Network
00505A	Network Alchemy, Inc.
00505B	Kawasaki Lsi U.S.A., Inc.
00505C	Tundo Corporation
00505E	Digitek Micrologic S.A.
00505F	Brand Innovators
005060	Tandberg Telecom AS
005062	Kouwell Electronics Corp.
005063	OY Comsel System AB
005064	CAE Electronics
005065	Nippon Electric Industry Co., Ltd.
005066	Atecom GmBh Advanced Telecommunication Modules
005067	Aerocomm, Inc.
005068	Electronic Industries Association (EIA)
005069	PixStream Incorporated
00506A	Edeva, Inc.
00506B	SPX-ATEG
00506C	G & L Beijer Electronics AB
00506D	Videojet Systems International Inc.
00506E	Corder Engineering Corporation
00506F	G-Connect
005070	Chaintech Computer Co., Ltd.
005071	Aiwa Co., Ltd.
005073	Cisco Systems, Inc.
005074	Advanced Hi-Tech Corp
005075	Kestrel Solutions
005076	IBM
005077	Prolific Technology, Inc.
005078	Megaton House, Ltd.
00507A	Xpeed, Inc.
00507B	Merlot Communications
00507C	Videocon AG
00507D	IFP
00507E	Newer Technology
005080	Cisco Systems, Inc.
005081	Murata Machinery, Ltd.
005082	Foresson Corporation
005083	Gilbarco, Inc.
005084	ATL Products
005086	Telkom SA, Ltd.
005087	Terasaki Electric Co., Ltd.
005088	Amano Corporation
005089	Safety Management Systems
00508C	RSI Systems
00508D	Abit Computer Corporation
00508E	Optimation, Inc.
00508F	Asita Technologies Int'l Ltd.
005090	DCTRI
005091	Netaccess, Inc.
005092	Rigaku Industrial Corporation
005093	Boeing
005094	Pace Micro Technology Plc
005095	Peracom Networks
005096	Salix Technologies, Inc.
005097	MMC-Embedded Computertechnik GmBh
005098	Globaloop, Ltd.
005099	3Com Europe, Ltd.
00509A	Tag Electronic Systems
00509B	Switchcore AB
00509C	Beta Research
00509D	The Industree B.V.
00509E	Les Technologies
00509F	Horizon Computer
0050A0	Delta Computer Systems, Inc.
0050A1	Carlo Gavazzi, Inc.
0050A2	Cisco Systems, Inc.
0050A3	Transmedia Communications, Inc.
0050A4	IO Tech, Inc.
0050A5	Capitol Business Systems, Ltd.
0050A6	Optronics
0050A7	Cisco Systems, Inc.
0050A8	Opencon Systems, Inc.
0050A9	Moldat Wireless Technolgies
0050AA	Konica Corporation
0050AB	Naltec, Inc
0050AC	Maple Computer Corporation

OUI	Vendor/Comment
0050AD	Communique Wireless Corp.
0050AF	Intergon, Inc
0050B0	Technology Atlanta Corporation
0050B1	Giddings & Lewis
0050B2	Brodel Automation
0050B3	Voiceboard Corporation
0050B4	Satchwell Control Systems, Ltd
0050B5	Fichet-Bauche
0050B6	Good Way Ind. Co., Ltd.
0050B7	Boser Technology Co., Ltd.
0050B8	Inova Computers GmBh & Co. KG
0050B9	Xitron Technologies, Inc.
0050BA	D-Link
0050BB	CMS Technologies
0050BC	Hammer Storage Solutions
0050BD	Cisco Systems, Inc.
0050BE	Fast Multimedia AG
0050BF	Mototech Inc.
0050C0	Gatan, Inc.
0050C1	Gemflex Networks, Ltd.
0050C2	IEEE Registration Authority
0050C4	IMD
0050C5	ADS Technologies, Inc.
0050C6	Loop Telecommunication International, Inc.
0050C8	Addonics Communications, Inc.
0050C9	Maspro Denkoh Corp.
0050CA	Net To Net Technologies
0050CB	Jetter
0050CC	Xyratex
0050CD	Digianswer A/S
0050CE	LG International Corp.
0050CF	Vanlink Communication
0050D0	Minerva Systems
0050D3	Digital Audio Processing Pty. Ltd.
0050D4	Joohong Information & Communications, Ltd.
0050D5	AD Systems Corp.
0050D6	Atlas Copco Tools AB
0050D7	Telstrat
0050D8	Unicorn Computer Corp.
0050D9	Engetron-Engenharia Eletronica Ind. E Com. Ltda.
0050DB	Contemporary Control Systems, Inc.
0050DC	Tas Telefonbau A. Schwabe GmBh & Co. KG
0050DD	Serra Soldadura, S.A.
0050DE	Signum Systems Corp.
0050DF	Airfiber, Inc.
0050E1	Ns Tech Electronics Sdn Bhd
0050E2	Cisco Systems, Inc.
0050E3	Telegate
0050E4	Apple Computer, Inc.
0050E6	Hakusan Corporation
0050E7	Paradise Innovations (Asia) Pte. Ltd.
0050EB	Alpha-Top Corporation
0050EC	Olicom A/S
0050ED	Anda Networks
0050EF	SPE Systemhaus GmBh
0050F1	Libit Signal Processing, Ltd.
0050F2	Microsoft Corp.
0050F3	Global Net Information Co., Ltd
0050F4	Sigmatek GmBh & Co. Kg
0050F6	Pan-International Industrial Corp.
0050F7	Venture Manufacturing (Singapore) Ltd.
0050F8	Entrega Technologies, Inc.
0050FA	Oxtel, Ltd.
0050FB	VSK Electronics
0050FC	Edimax Technology Co., Ltd.
0050FD	Isioncomm Co., Ltd.
0050FE	Pctvnet Asa
005500	Xerox reported—Intel sends on Token Ring bit reversed of assigned OUI 00AA00—Not an IEEE assigned OUI
006000	Xycom Inc.
006001	Innosys, Inc.

OUI	Vendor/Comment
006002	Screen Subtitling Systems, Ltd.
006003	Teraoka Weigh System Pte, Ltd.
006004	Computadores Modulares SA
006005	Feedback Data Ltd.
006006	Sotec Co., Ltd.
006007	Acres Gaming, Inc.
006008	3Com Corporation
006009	Cisco Systems, Inc—Catalyst 5000 Ethernet switch
00600A	Sord Computer Corporation
00600B	Logware GmBh
00600C	Applied Data Systems, Inc.
00600D	Microdesign GmBh
00600E	Wavenet International, Inc.
00600F	Westell, Inc.
006010	Network Machines, Inc.
006011	Crystal Semiconductor Corp.
006012	Power Computing Corporation
006013	Netstal Maschinen AG
006014	Edec Co., Ltd.
006015	Net2Net Corporation
006016	Clariion
006017	Tokimec Inc.
006018	Stellar One Corporation
006019	Boehringer Mannheim Corp.
00601A	Keithley Instruments
00601B	Mesa Electronics
00601C	Telxon Corporation
00601D	Lucent Technologies
00601E	Softlab, Inc.
00601F	Stallion Technologies
006020	Pivotal Networking, Inc.
006021	DSC Corporation
006022	Vicom Systems, Inc.
006023	Pericom Semiconductor Corp.
006024	Gradient Technologies, Inc.
006025	Active Imaging Plc
006026	Viking Components, Inc.
006027	General Technology, Inc.
006028	Macrovision Corporation
006029	Cary Peripherals Inc.
00602A	Symicron Computer Communications, Ltd.
00602B	Peak Audio
00602C	Linx Data Terminals, Inc.
00602D	Alerton Technologies, Inc.
00602E	Cyclades Corporation
00602F	Cisco Systems, Inc.
006030	Village Tronic—used on Amiga
006031	HRK Systems
006032	I-Cube, Inc.
006033	Acuity Imaging, Inc.
006034	Robert Bosch GmBh
006035	Dallas Semiconductor, Inc.
006036	Austrian Research Center
006037	Philips Semiconductors
006038	Nortel (Northern Telecom)
006039	Sancom Technology, Inc.
00603A	Quick Controls Ltd.
00603B	Amtec SPA
00603C	Hagiwara Sys-Com Co., Ltd.
00603D	3CX (formerly INTEGRATED MICRO SOLUTIONS, INC)
00603E	Cisco Systems, Inc—100Mbps interface
00603F	Patapsco Designs
006040	Netro Corp.
006041	3A International, Inc.
006042	TKS (USA), Inc.
006043	Comsoft Systems, Inc.
006044	Litton/Poly-Scientific
006045	Pathlight Technologies
006046	Vmetro, Inc.
006047	Cisco Systems, Inc.
006048	EMC Corporation

OUI	Vendor/Comment
006049	Vina Technologies
00604A	SAIC Ideas Group
00604B	Biodata GmBh
00604C	SAT
00604D	MMC Networks, Inc.
00604E	Cycle Computer Corporation (Sun Motherboard Replacements)
00604F	Suzuki Mfg. Co., Ltd.
006050	Internix Inc.
006051	Quality Semiconductor
006052	Realtek—(RTL 8029 == PCI NE2000)—IEEE assigned to PERIPHERALS ENTERPRISE CO., Ltd.
006053	Toyoda Machine Works, Ltd.
006054	Controlware GmBh
006055	Cornell University
006056	Network Tools, Inc.
006057	Murata Manufacturing Co., Ltd.
006058	Copper Mountain Communications, Inc.
006059	Technical Communications Corp.
00605A	Celcore, Inc.
00605B	Intraserver Technology Inc.
00605C	Cisco Systems, Inc.
00605D	Scanivalve Corp.
00605E	Liberty Technology Networking
00605F	Nippon Unisoft Corporation
006060	Dawning Technologies, Inc.
006061	Whistle Communications Corp.
006062	Telesync, Inc.
006063	Psion Dacom Plc.
006064	Netcomm Limited
006065	Bernecker & Rainer Industrie-Elektronic GmBh
006066	Lacroix Technolgie
006067	Acer Netxus Inc.
006068	Eicon Technology Corporation
006069	Brocade Communications Systems
00606A	Mitsubishi Wireless Comm. Inc.
00606B	Aichi Electronics Co., Ltd.
00606C	Arescom
00606D	Digital Equipment Corp.
00606E	Davicom Semiconductor, Inc.
00606F	Clarion Corporation Of America
006070	Cisco Systems, Inc—routers (2524 and 4500)
006071	Midas Lab, Inc.
006072	VXL Instruments, Limited
006073	Redcreek Communications, Inc.
006074	QSC Audio Products
006075	Pentek, Inc.
006076	Schlumberger Technologies
006077	Prisa Networks
006078	Power Measurement Ltd.
006079	Wavephore Networks, Inc.
00607A	DVS GmBh
00607B	Fore Systems, Inc.
00607C	Waveaccess, Ltd.
00607D	Sentient Networks Inc.
00607E	Gigalabs, Inc.
00607F	Aurora Technologies, Inc.
006080	Microtronix Datacom Ltd.
006081	TV/Com International
006082	Novalink Technologies, Inc.
006083	Cisco Systems, Inc.—3620
006084	Digital Video
006085	Storage Concepts
006086	Logic Replacement Tech. Ltd.
006087	Kansai Electric Co., Ltd.
006088	White Mountain DSP, Inc.
006089	XATA
00608A	Citadel Computer
00608B	Confertech International
00608C	3Com Corporation (1990 onwards)
00608D	Unipulse Corp.
00608E	HE Electronics, Technologie & Systemtechnik GmBh

OUI	Vendor/Comment
00608F	Tekram Technology Co., Ltd.
006090	Able Communications, Inc.
006091	First Pacific Networks, Inc.
006092	Micro/Sys, Inc.
006093	Varian
006094	IBM Corp.—AMD PCNET PCI—IEEE assigned to IBM
006095	Accu-Time Systems, Inc.
006096	T.S. Microtech Inc.
006097	3Com Corporation
006098	HT Communications
006099	Lan Media Corporation
00609A	NJK Techno Co.
00609B	Astro-Med, Inc.
00609C	Perkin-Elmer Corporation
00609D	PMI Food Equipment Group
00609E	X3 - Information Technology
00609F	Phast Corporation
0060A0	Switched Network
0060A1	VPNet
0060A2	Nihon Unisys Limited Co.
0060A3	Continuum Technology Corp.
0060A4	Grinaker System Technologies
0060A5	Performance Telecom Corp.
0060A6	Particle Measuring Systems
0060A7	Microsens GmBh & Co. KG
0060A8	Tidomat AB
0060A9	Gesytec MBh
0060AA	Intelligent Devices Inc. (IDI)
0060AB	Larscom Incorporated
0060AC	Resilience Corporation
0060AD	Megachips Corporation
0060AE	Trio Information Systems AB (formerly OBJECTA)
0060AF	Pacific Micro Data, Inc.
0060B0	Hewlett-Packard Co.—newer LaserJets.—sends bit reversed as 00060D on Token Ring
0060B1	Input/Output, Inc.
0060B2	Process Control Corp.
0060B3	Z-Com, Inc.
0060B4	Glenayre R&D Inc.
0060B5	KEBA GmBh
0060B6	Land Computer Co., Ltd.
0060B7	Channelmatic, Inc.
0060B8	Corelis Inc.
0060B9	Nitsuko Corporation
0060BA	Sahara Networks, Inc.
0060BB	Cabletron - Netlink, Inc.
0060BC	Keunyoung Electronics & Communication
0060BD	Hubbell-Pulsecom
0060BE	Webtronics
0060BF	Macraigor Systems, Inc.
0060C0	Nera AS
0060C1	Wavespan Corporation
0060C2	MPL AG
0060C3	Netvision Corporation
0060C4	Soliton Systems K.K.
0060C5	Ancot Corp.
0060C6	DCS AG
0060C7	Amati Communications Corp.
0060C8	Kuka Welding Systems & Robots
0060C9	Controlnet, Inc.
0060CA	Harmonic Systems Incorporated
0060CB	Hitachi Zosen Corporation
0060CC	Emtrak, Incorporated
0060CD	Videoserver, Inc.
0060CE	Acclaim Communications
0060CF	Alteon Networks, Inc.
0060D0	SNMP Research Incorporated
0060D1	Cascade Communications
0060D2	Lucent Technologies Taiwan
0060D3	AT&T
0060D4	Eldat Communication Ltd.

OUI	Vendor/Comment
0060D5	Miyachi Technos Corp.
0060D6	Novatel Wireless Technologies Ltd.
0060D7	Ecole Polytechnique Federale De Lausanne (EPFL)
0060D8	Elmic Systems, Inc.
0060D9	Transys Networks Inc.
0060DA	JBM Electronics Co.
0060DB	NTP Elektronik A/S
0060DC	Toyo Communication Equipment
0060DD	Myricom, Inc.
0060DE	Kayser-Threde GmBh
0060DF	Inrange Technologies Corp.
0060E0	Axiom Technology Co., Ltd.
0060E1	Orckit Communications Ltd.
0060E2	Quest Engineering & Dev.
0060E3	Arbin Instruments
0060E4	Compuserve, Inc.
0060E5	Fuji Automation Co., Ltd.
0060E6	Shomiti Systems Incorporated
0060E7	RanData
0060E8	Hitachi Computer Products
0060E9	Atop Technologies, Inc.
0060EA	Streamlogic
0060EB	Fourthtrack Systems
0060EC	Hermary Opto Electronics Inc
0060ED	Ricardo Test Automation Ltd.
0060EE	Apollo
0060EF	Flytech Technology Co., Ltd.
0060F0	Johnson & Johnson Medical, Inc.
0060F1	Exp Computer, Inc.
0060F2	Lasergraphics, Inc.
0060F3	Netcom Systems, Inc.
0060F4	Advanced Computer Solutions
0060F5	Phobos FastEthernet for Unix WS—IEEE assigned to Icon West, Inc.
0060F6	Nextest Communication Products, Inc.
0060F7	Datafusion Systems
0060F8	Loran International Techn. Inc.
0060F9	Diamond Lane Communications
0060FA	Educational Technology Resources, Inc.
0060FB	Packeteer, Inc.
0060FC	Conservation Through Innovation Ltd.
0060FD	Netics, Inc.
0060FE	Lynx System Developers, Inc.
0060FF	Quvis, Inc.
0070B0	M/A-Com Inc. Companies
0070B3	Data Recall Ltd.
008000	Multitech Systems Inc
008001	Periphonics Corporation
008002	Satelcom (Uk) Ltd.
008003	Hytec Electronics Ltd.
008004	Antlow Computers, Ltd.
008005	Cactus Computer Inc.
008006	Compuadd Corporation
008007	Dlog NC-Systeme
008008	Dynatech Computer Systems
008009	Jupiter Systems (older MX-600 series machines)
00800A	Japan Computer Corp.
00800B	CSK Corporation
00800C	Videcom Limited
00800D	Vosswinkel FU
00800E	Atlantix Corporation
00800F	Standard Microsystems Corp. (SMC)
008010	Commodore International
008011	Digital Systems Int'l. Inc.
008012	Integrated Measurement Systems Corp. (IMS) — IMS failure analysis tester
008013	Thomas Conrad Corp. (now owned by Compaq)
008014	Esprit Systems
008015	Seiko Systems Inc
008016	Wandel & Goltermann
008017	PFU Limited
008018	Kobe Steel, Ltd.
008019	Dayna Communications—"Etherprint" product
00801A	Bell Atlantic
00801B	Kodiak Technology
00801C	Cisco Systems, Inc—IEEE assigned to NEWPORT SYSTEMS SOLUTIONS
00801D	Integrated Inference Machines
00801E	Xinetron, Inc.
00801F	Krupp Atlas Electronik GmBh
008020	Andrew Coproration Network Products Div—Andrew sends bit reversed as 000104 on Token Ring
008021	Newbridge Research Corporation—Newbridge sends bit reversed as 000184 on Token Ring
008022	Scan-Optics
008023	Integrated Business Networks
008024	Kalpana Inc.
008025	Stollmann GmBh
008026	Network Products Corporation
008027	Adaptive Systems, Inc.
008028	Tradpost (HK) Ltd.
008029	Microdyne Corporation—IEEE assigned to EAGLE TECHNOLOGY, INC.
00802A	Test Systems & Simulations Inc
00802B	Integrated Marketing Co.
00802C	The Sage Group Plc
00802D	Xylogics, Inc.—Annex terminal servers
00802E	Plexcom, Inc.—IEEE assigned to CASTLE ROCK COMPUTING
00802F	National Instruments Corp.
008030	Nexus Electronics
008031	Basys, Corp.
008032	Access Co., Ltd.
008033	Formation Inc.
008034	SMT-Goupil
008035	Technology Works
008036	Reflex Manufacturing Systems
008037	Telefon Ab Lm Ericsson Corp.
008038	Data Research & Applications
008039	Alcatel STC Australia
00803A	Varityper, Inc.
00803B	APT Communications, Inc.
00803C	TVS Electronics Ltd.
00803D	Surigiken Co., Ltd.
00803E	Synernetics
00803F	Hyundai Electronics—IEEE assigned to TATUNG COMPANY
008040	John Fluke Manufacturing Co
008041	VEB Kombinat Robotron
008042	Force Computers
008043	Networld, Inc.
008044	Systech Computer Corp.
008045	Matsushita Electric Ind Co
008046	University of Toronto Dept. of Electrical Engin'ing
008047	In-Net Corp.
008048	Compex Incorporated, used by Commodore and DEC at least
008049	Nissin Electric Co., Ltd.
00804A	PRO-LOG
00804B	Eagle Technologies Pty. Ltd.
00804C	Contec Co., Ltd.
00804D	Cyclone Microsystems, Inc.
00804E	Apex Computer Company
00804F	Daikin Industries, Ltd.
008050	Ziatech Corporation
008051	ADC Fibermux
008052	Network Professor—IEEE assigned to TECHNICALLY ELITE CONCEPTS
008053	Intellicom, Inc.
008054	Frontier Technologies Corp.
008055	Fermilab
008056	Sphinx Elektronik GmBh
008057	Adsoft, Ltd.
008058	Printer Systems Corporation
008059	Stanley Electric Co., Ltd.
00805A	Tulip Computers International BV
00805B	Condor Systems, Inc.
00805C	Agilis Corporation—Agilis sends bit reversed as 00013A on Token Ring
00805D	Canstar
00805E	LSI Logic Corporation

OUI	Vendor/Comment
00805F	Compaq Computer Corporation—Compaq sends bit reversed as 0001FA on Token Ring
008060	Network Interface Corporation
008061	Litton Systems, Inc.
008062	Interface Co.
008063	Richard Hirschmann GmBh & Co.
008064	Wyse Technology
008065	Cybergraphic Systems Pty Ltd.
008066	Arcom Control Systems, Ltd.
008067	Square D Company
008068	Yamatech Scientific Ltd.
008069	Computone Systems
00806A	Empac Research Inc. (ERI)
00806B	Schmid Telecommunication
00806C	Cegelec Projects Ltd.
00806D	Century Systems Corp.
00806E	Nippon Steel Corporation
00806F	Onelan Ltd.
008070	Computadoras Micron
008071	SAI Technology
008072	Microplex Systems Ltd.
008073	DWB Associates
008074	Fisher Controls
008075	Parsytec GmBh
008076	MCNC
008077	Brother Industries, Ltd.
008078	Practical Peripherals, Inc.
008079	Microbus Designs Ltd.
00807A	Aitech Systems Ltd.
00807B	Artel Communications Corp.
00807C	FiberCom, Inc.
00807D	Equinox Systems Inc.
00807E	Southern Pacific Ltd.
00807F	DY-4 Incorporated
008080	Datamedia Corporation
008081	Kendall Square Research Corp.
008082	PEP Modular Computers GmBh
008083	Amdahl
008084	The Cloud Inc.
008085	H-Three Systems Corporation
008086	Computer Generation Inc.
008087	Oki Electric Industry Co., Ltd.—Okidata
008088	Victor Company Of Japan, Ltd
008089	Tecnetics (Pty) Ltd.
00808A	Summit Microsystems Corp.
00808B	Dacoll Limited
00808C	Netscout Systems (formerly Frontier Software Development)
00808D	Westcove Technology B.V.
00808E	Radstone Technology
00808F	C. Itoh Electronics, Inc.
008090	Microtek International, Inc.
008091	Tokyo Electric Co., Ltd.
008092	Japan Computer Industry, Inc.
008093	Xyron Corporation
008094	Sattcontrol AB—IEEE assigned to ALFA LAVAL AUTOMATION AB
008095	Basic Merton Handelsges.M.B.H.
008096	Human Designed Systems, Inc. (HDS) —X terminals
008097	Centralp Automatismes
008098	TDK Corporation
008099	Klockner Moeller IPC
00809A	Novus Networks Ltd.
00809B	Justsystem Corporation
00809C	Luxcom, Inc.
00809D	Datacraft Manufactur'g Pty Ltd
00809E	DATUS GmBh
00809F	Alcatel Business Systems
0080A0	Edisa Hewlett Packard S/A
0080A1	Microtest Inc. (disc port product)
0080A2	Creative Electronic Systems
0080A3	Lantronix
0080A4	Liberty Electronics

OUI	Vendor/Comment
0080A5	Speed International
0080A6	Republic Technology Inc.
0080A7	Measurex Corp.
0080A8	Vitacom Corporation
0080A9	Clearpoint Research
0080AA	Maxpeed
0080AB	Dukane Network Integration
0080AC	Imlogix, Division Of Genesys
0080AD	CNet Technology, Inc.—Used by Telebit (among others)
0080AE	Hughes Network Systems
0080AF	Allumer Co., Ltd.
0080B0	Advanced Information Technology, Inc.
0080B1	Softcom A/S
0080B2	NET (Network Equipment Technologies)
0080B3	Aval Data Corporation
0080B4	Sophia Systems
0080B5	United Networks Inc.
0080B6	Themis Computer
0080B7	Stellar Computer
0080B8	Bug, Incorporated
0080B9	Arche Technoligies Inc.
0080BA	Specialix (Asia) Pte, Ltd.
0080BB	Hughes Lan Systems
0080BC	Hitachi Engineering Co., Ltd.
0080BD	The Furukawa Electric Co., Ltd.
0080BE	Aries Research
0080BF	Takaoka Electric Mfg. Co. Ltd.
0080C0	Penril Datability Networks
0080C1	Lanex Corporation
0080C2	IEEE—802.1 Committee
0080C3	BICC Information Systems & SVC
0080C4	Document Technologies, Inc.
0080C5	Novellco De Mexico
0080C6	Soho—IEEE assigned to NATIONAL DATACOMM CORPORATION
0080C7	Xircom, Inc.—Xircom pocket adapters send bit reversed as 0001E3 on Token Ring
0080C8	D-Link-(also Solectek Pocket Adapters, and LinkSys PCMCIA)—IEEE assigned to D-LINK SYSTEMS, INC.
0080C9	Alberta Microelectronic Centre
0080CA	Netcom Research Incorporated
0080CB	Falco Data Products
0080CC	Microwave Bypass Systems
0080CD	Micronics Computer, Inc.
0080CE	Broadcast Television Systems
0080CF	Embedded Performance Inc.
0080D0	Computer Peripherals Inc.
0080D1	Kimtron Corporation
0080D2	Shinnihondenko Co., Ltd.
0080D3	Shiva Corp.—Appletalk-Ethernet interface
0080D4	Chase Research LTD.
0080D5	Cadre Technologies
0080D6	Apple Computer— Mac Portable(?)—IEEE assigned to NUVOTECH, INC.
0080D7	Fantum Engineering, Inc.
0080D8	Network Peripherals Inc.
0080D9	EMK Elektronik
0080DA	Bruel & Kjaer
0080DB	Graphon Corporation
0080DC	Picker International
0080DD	GMX Inc/GIMIX
0080DE	GIPSI S.A.
0080DF	ADC Codenoll Technology Corp.
0080E0	XTP Systems Inc
0080E1	STMicroelectronics
0080E2	T.D.I. Co., Ltd. Development Div.
0080E3	Coral Network Corporation (bought by Bay Networks, now Nortel)
0080E4	Northwest Digital Systems, Inc.
0080E5	Mylex Corporation
0080E6	Peer Networks, Inc.
0080E7	Lynwood Scientific Dev. Ltd.
0080E8	Cumulus Corporatiion
0080E9	Madge Networks

OUI	Vendor/Comment
0080EA	The Fiber Company
0080EB	Compcontrol B.V
0080EC	Supercomputing Solutions, Inc.
0080ED	IQ Technologies, Inc.
0080EE	Thomson CSF
0080EF	Rational
0080F0	Kyushu Matsushita Electric Co.
0080F1	Opus Systems
0080F2	Raycom Systems Inc.
0080F3	Sun Electronics Corp.
0080F4	Telemechanique Electrique
0080F5	Quantel Ltd.
0080F6	Synergy Microsystems
0080F7	Zenith Communications Products
0080F8	Mizar, Inc.
0080F9	Heurikon Corporation
0080FA	RWT GmBh
0080FB	BVM Limited
0080FC	Avatar Corporation
0080FD	Exsceed Corpration
0080FE	Azure Technologies, Inc.
0080FF	Soc. De Teleinformatique RTC
009000	Diamond Multimedia
009001	Nishimu Elctronics Industries
009002	Allgon Westcom AB
009003	Aplio
009004	3Com Europe, formerly BICC
009005	Protech Systems Co., Ltd.
009006	Hamamatsu Photonics K.K.
009007	Domex Technology Corp.
009008	Han A Systems, Inc.
009009	Hyundai Information Technology Co., Ltd.
00900A	Proton Electronic Industrial Co., Ltd.
00900B	Lanner Electronics, Inc.
00900C	Cisco Systems, Inc.
00900D	Overland Data Inc.
00900E	Handlink Technologies, Inc.
00900F	Kawasaki Heavy Industries, Ltd.
009010	Simulation Laboratories, Inc.
009011	Wavtrace, Inc.
009012	Globespan Semiconductor, Inc.
009013	Samsan Corp.
009014	Rotork Instruments, Ltd.
009015	Centigram Communications Corp.
009016	ZAC
009017	Zypcom, Inc.
009018	Ito Electric Industry Co, Ltd.
009019	Hermes Electronics Co., Ltd.
00901A	Redstone Communications
00901B	Digital Controls
00901C	MPS Software GmBh
00901D	PEC (NZ) Ltd.
00901E	Selesta Ingegne Ria S.P.A.
00901F	Adtec Productions, Inc.
009020	Philips Analytical X-Ray B.V.
009021	Cisco Systems, Inc.
009022	Ivex (formerly BRIVIDA, INC.)
009023	Zilog Inc.
009024	Pipelinks, Inc.
009025	Vision Systems Ltd. Pty
009026	Advanced Switching Communications, Inc.
009027	Intel Corporation
009028	Nippon Signal Co., Ltd.
009029	Crypto AG
00902A	Communication Devices, Inc.
00902B	Cisco Systems, Inc.—Ethernet Switches and Light Streams
00902C	Data & Control Equipment Ltd.
00902D	Data Electronics (Aust.) Pty, Ltd.
00902E	Namco Limited
00902F	Netcore Systems, Inc.
009030	Honeywell-Dating

OUI	Vendor/Comment
009031	Mysticom, Ltd.
009032	Pelcombe Group Ltd.
009033	Innovaphone GmBh
009034	Imagic, Inc.
009035	Alpha Telecom, Inc.
009036	ENS, Inc.
009037	Acucomm, Inc.
009038	Fountain Technologies, Inc.
009039	Shasta Networks (now Nortel Networks)
00903A	Nihon Media Tool Inc.
00903B	Triems Research Lab, Inc.
00903C	Atlantic Network Systems
00903D	Biopac Systems, Inc.
00903E	N.V. Philips Industrial Activities
00903F	Aztec Radiomedia
009040	Castle Networks, Inc.
009041	Applied Digital Access
009042	ECCS
009043	Nichibei Denshi Co., Ltd.
009044	Assured Digital, Inc.
009045	Mariposa Technology
009046	Dexdyne, Ltd.
009047	Giga Fast E. Ltd.
009048	Zeal Corporation
009049	Entridia Corporation
00904A	Concur System Technologies
00904B	Gemtek Technology Co., Ltd.
00904C	Epigram, Inc.
00904D	SPEC S.A.
00904E	DELEM BV
00904F	ABB Power T&D Company, Inc.
009050	Teleste OY
009051	Ultimate Technology Corp.
009052	Selcom Elettronica S.R.L.
009053	Daewoo Electronics Co., Ltd.
009054	Innovative Semiconductors, Inc.
009055	Parker Hannifin Corporation
009056	Telestream, Inc.
009057	Aanetcom, Inc.
009058	Ultra Electronics Ltd.
009059	Telecom Device K.K.
00905A	Dearborn Group, Inc.
00905B	Raymond And Lae Engineering
00905C	EDMI
00905D	Netcom Sicherheitstechnik Gmb
00905E	Rauland-Borg Corporation
00905F	Cisco Systems, Inc.
009060	System Create Corp.
009061	Pacific Research & Engineering Corporation
009062	ICP Vortex Computersysteme GbMh
009063	Coherent Communications
009064	Thomson Broadcast Systems
009065	Finisar Corporation
009066	Troika Design, Inc.
009067	Walkabout Computers, Inc.
009068	DVT Corp.
009069	Juniper Networks, Inc.
00906A	Turnstone Systems, Inc.
00906B	Applied Resources, Inc.
00906C	GWT Global Weighing Technologies GmBh
00906D	Cisco Systems, Inc.
00906E	Praxon, Inc.
00906F	Cisco Systems, Inc.
009070	Neo Networks, Inc
009071	Badger Technology, Inc.
009072	Simrad AS
009073	Gaio Technology
009074	Argon Networks, Inc.
009075	NEC Do Brasil S.A
009076	FMT Aircraft Gate Support Systems AB
009077	Advanced Fibre Communications

OUI	Vendor/Comment
009078	MER Telemanagement
009079	Clearone Inc.
00907A	Spectralink Corp.
00907B	E-Tech, Inc.
00907C	Digitalcast, Inc.
00907D	Home Wireless Networks
00907E	Vetronix Corp.
00907F	Watchguard Technologies, Inc.
009080	Not Limited, Inc.
009081	Aloha Networks, Inc.
009082	Force Institute
009083	Turbo Communication, Inc
009084	Atech System
009085	Golden Enterprises, Inc.
009086	Cisco Systems, Inc.
009087	ITIS
009088	Baxall Security Ltd.
009089	Softcom Microsystems, Inc.
00908A	Bayly Communications, Inc.
00908B	Cell Computing, Inc.
00908C	Etrend Electronics, Inc.
00908D	Vickers Electronics Systems
00908E	Promatory Communications
00908F	Audiocodes Ltd.
009090	I-Bus
009091	Digitalscape, Inc
009092	Cisco Systems, Inc.
009093	Nanao Corporation
009094	Osprey Technologies, Inc.
009095	Universal Avionics
009096	Askey Computer Corp.
009097	Sycamore Networks
009098	SBC Designs, Inc.
009099	Allied Telesis, K.K.
00909A	One World Systems, Inc.
00909B	Markpoint AB
00909C	Combox, Ltd.
00909D	GSE Systems, Inc.
00909E	Delphi Engineering Group
00909F	Digi-Data Corporation
0090A0	8x8 Inc.
0090A1	Flying Pig Systems, Ltd.
0090A2	Cybertan Technology, Inc.
0090A3	Medialincs Co., Ltd.
0090A4	Altiga Networks
0090A5	Spectra Logic
0090A6	Cisco Systems, Inc.
0090A7	Clientec Corporation
0090A8	Ninetiles Networks Ltd.
0090A9	Western Digital
0090AA	Indigo Active Vision Systems Limited
0090AB	Cisco Systems, Inc.
0090AC	Optivision, Inc.
0090AD	Aspect Electronics, Inc.
0090AE	Italtel SPA
0090AF	J. Morita Mfg. Corp.
0090B0	VADEM
0090B1	Cisco Systems, Inc.
0090B2	Avici Systems Inc.
0090B3	Agranat Systems
0090B4	Willowbrook Technologies
0090B5	Nikon Corporation Electr. Imaging Design Dept.
0090B6	Fibex Systems
0090B7	Digital Lightwave, Inc.
0090B8	Rohde & Schwarz GmBh & Co. KG
0090B9	Beran Instruments Ltd.
0090BA	Valid Networks, Inc.
0090BB	Tainet Communication System Corporation
0090BC	Telemann Co., Ltd.
0090BD	Omnia Communications, Inc
0090BE	IBC/Integrated Business Computers
0090BF	Cisco Systems, Inc.
0090C0	K.J. Law Engineers, Inc.
0090C1	Eda Industries
0090C2	JK Microsystems, Inc.
0090C3	Topic Semiconductor Corp.
0090C4	Javelin Systems, Inc.
0090C5	Internet Magic, Inc.
0090C6	Optim Systems, Inc.
0090C7	Icom Inc.
0090C8	Waverider Communications (Canada) Inc.
0090C9	Productivity Enhancement Products., Inc.
0090CA	Accord Video Telecommunications, Ltd.
0090CB	Wireless Online, Inc.
0090CC	Planex Communications, Inc.
0090CD	Ent-Empresa Nacional
0090CE	Tetra GmBh
0090CF	Nortel
0090D0	Alcatel Bell
0090D1	Leichu Enterprise Co., Ltd.
0090D2	Artel Video Systems
0090D3	Giesecke & Devrient GmBh
0090D4	Bindview Development Corp.
0090D5	Euphonix, Inc.
0090D6	Crystal Group
0090D7	Netboost Corp.
0090D8	Whitecross Systems
0090D9	Cisco Systems, Inc.
0090DA	Dynarc, Inc.
0090DB	Next Level Communications
0090DC	Teco Information Systems Co., Ltd.
0090DD	The Miharu Communications Co., Ltd.
0090DE	Cardkey Systems, Inc.
0090DF	Mitsubishi Chemical America, Inc.
0090E0	Systran Corp.
0090E1	Telena S.P.A.
0090E2	Distributed Processing Technology
0090E3	Avex Electronics Inc.
0090E4	NEC America, Inc.
0090E5	Teknema, Inc.
0090E6	Acer Laboratories, Inc.
0090E7	Horsch Elektronik Ag
0090E8	Moxa Technologies Corp., Ltd.
0090E9	Janz Computer Ag
0090EA	Alpha Technologies, Inc.
0090EB	Sentry Telecom Systems
0090EC	Pyrescom
0090ED	Central System Research Co., Ltd.
0090EE	Personal Communications Technologies
0090EF	Integrix, Inc.
0090F0	Harmonic Lightwaves, Ltd.
0090F1	Boxhill Systems Corporation
0090F2	Cisco Systems Inc.—Ethernet Switches and Light Streams
0090F3	Aspect Telecommunications Corp
0090F4	Lightning Instrumentation
0090F5	Clevo Co.
0090F6	Escalate Networks, Inc.
0090F7	Nbase Communications Ltd.
0090F8	Mediatrix Telecom
0090F9	Leitch
0090FA	Giganet, Inc.
0090FB	Portwell, Inc.
0090FC	Network Computing Devices
0090FD	Coppercom, Inc.
0090FE	Elecom Co., Ltd. (Laneed Div. Sumitomo Fudosan)
0090FF	Tellus Technology Inc.
009D8E	Cardiac Recorders, Inc.
00A000	Centillion Networks, bought by Bay Networks now Nortel—Ethernet switch—sends bit reversed as 000500 on Token Ring
00A001	Watkins-Johnson Company
00A002	Leeds & Northrup Australia Pty Ltd.
00A003	Staefa Control System

OUI	Vendor/Comment
00A004	Netpower, Inc.
00A005	Daniel Instruments, Ltd.
00A006	Image Data Processing
00A007	Apexx Technology, Inc.
00A008	Netcorp
00A009	Whitetree Network Technologies, Inc.
00A00A	R.D.C. Communication
00A00B	Computex Co., Ltd.
00A00C	Kingmax Technology, Inc.
00A00D	The Panda Project
00A00E	Visual Networks, Inc.
00A00F	Broadband Technologies
00A010	Syslogic Datentechnik Ag
00A011	Mutoh Industries Ltd.
00A012	B.A.T.M. Advanced Technologies
00A013	Teltrend Ltd.
00A014	Csir
00A015	Wyle
00A016	Micropolis Corp.
00A017	J B M Corporation
00A018	Creative Controllers, Inc.
00A019	Nebula Consultants, Inc.
00A01A	Binar Elektronik Ab
00A01B	Premisys Communications, Inc.
00A01C	Nascent Networks Corporation
00A01D	Sixnet
00A01E	Est Corporation
00A01F	Tricord Systems, Inc.
00A020	Citicorp/TTI
00A021	General Dynamics-Communication Systems
00A022	Centre For Development Of Advanced Computing
00A023	Applied Creative Technology, Inc.
00A024	3com Corporation
00A025	Redcom Labs Inc.
00A026	Teldat, S.A.
00A027	Firepower Systems, Inc.
00A028	Conner Peripherals
00A029	Coulter Corporation
00A02A	Trancell Systems
00A02B	Transitions Research Corp.
00A02C	Interwave Communications
00A02D	Skipstone, Inc.
00A02E	Brand Communications, Ltd.
00A02F	Pirelli Cavi
00A030	Captor Nv/Sa
00A031	Hazeltine Corporation
00A032	Ges Singapore Pte. Ltd.
00A033	Imc Mess-Systeme GmBh
00A034	Axel
00A035	Cylink Corporation
00A036	Applied Network Technology
00A037	Datascope Corporation
00A038	Email Electronics
00A039	Ross Technology, Inc.
00A03A	Kubotek Corporation
00A03B	Toshin Electric Co., Ltd.
00A03C	EG&G Nuclear Instruments
00A03D	Opto - 22
00A03E	ATM Forum
00A03F	Computer Society Microproces'r & Micropro'r Stds Committee
00A040	Apple Computer
00A041	Leybold-Inficon
00A042	Spur Products Corp.
00A043	American Technology Labs, Inc.
00A044	Ntt Intelligent Technology
00A045	Phoenix Contact GmBh & Co.
00A046	Scitex Corp. Ltd.
00A047	Integrated Fitness Corp.
00A048	Questech, Ltd.
00A049	Digitech Industries, Inc.
00A04A	Nisshin Electric Co., Ltd.
00A04B	Sonic Systems Inc.—EtherFE 10—IEEE assigned to TFL LAN INC.
00A04C	Innovative Systems & Tech. Inc
00A04D	Eda Instruments, Inc.
00A04E	Voelker Technologies, Inc.
00A04F	Ameritec Corp.
00A050	Cypress Semiconductor
00A051	Angia Communications. Inc.
00A052	Stanilite Electronics Pty. Ltd
00A053	Compact Devices, Inc.
00A054	Gassan Salman
00A055	Linktech, Inc.
00A056	Micropross
00A057	Elsa AG
00A058	Glory, Ltd.
00A059	Hamilton Hallmark
00A05A	Kofax Image Products
00A05B	Marquip, Inc.
00A05C	Inventory Conversion, Inc./Nekotech Division
00A05D	CS Computer Systeme GmBh
00A05E	Myriad Logic Inc.
00A05F	BTG Engineering Bv
00A060	Acer Peripherals, Inc.
00A061	Puritan Bennett
00A062	Aes Prodata
00A063	JRL Systems, Inc.
00A064	KVB/Analect
00A065	Nexland, Inc.
00A066	Isa Co., Ltd.
00A067	Network Services Group
00A069	Truetime
00A06A	Verilink Corp.
00A06B	DMS Dorsch Mikrosystem GmBh
00A06D	Mannesmann Tally Corporation
00A06E	Austron, Inc.
00A06F	The Appcon Group, Inc.
00A070	Coastcom
00A071	Video Lottery Technologies, Inc
00A072	Ovation Systems Ltd.
00A073	Com21, Inc.
00A074	Perception Technology
00A075	Micron Technology, Inc.
00A076	Cardware Lab, Inc.
00A077	Fujitsu Nexion, Inc.
00A078	Cellaccess Technology, Inc.
00A079	Alps Electric (Usa), Inc.
00A07A	Advanced Peripherals Technologies, Inc.
00A07B	Dawn Computer Incorporation
00A07C	Tonyang Nylon Co., Ltd.
00A07D	Seeq Technology, Inc.
00A07E	Avid Technology, Inc.
00A07F	Gsm-Syntel, Ltd.
00A080	Antares Microsystems
00A081	Alcatel Data Networks
00A082	NKT Elektronik A/S
00A083	Intel Corporation—IEEE assigned to ASIMMPHONY TURKEY ELECTRONIC LTD.
00A084	Dataplex Pty. Ltd.
00A085	IEEE STANDARDS DEPARTMENT
00A086	Amber Wave Systems, Inc.
00A087	Mitel Semiconductor, Ltd.
00A088	Essential Communications
00A089	Xpoint Technologies, Inc.
00A08A	Brooktrout Technology, Inc.
00A08B	Aston Electronic Designs Ltd.
00A08C	Multimedia Lans, Inc.
00A08D	Jacomo Corporation
00A08E	Ipsilon Networks, Inc.
00A08F	Desknet Systems, Inc.
00A090	Timestep Corporation
00A091	Applicom International
00A092	Intermate International—(LAN printer interfaces)—IEEE assigned to H. BOLLMANN MANUFACTURERS, LTD

OUI	Vendor/Comment
00A093	B/E Aerospace
00A094	Comsat Corporation
00A095	Acacia Networks, Inc.
00A096	Mitsumi Electric Co., Ltd.
00A097	JC Information Systems
00A098	Network Appliance Corp.
00A099	K-Net Ltd.
00A09A	Nihon Kohden America
00A09B	Qpsx Communications, Ltd.
00A09C	Xyplex, Inc.
00A09D	Johnathon Freeman Technologies
00A09E	IVTV
00A09F	Commvision Corp.
00A0A0	Compact Data, Ltd.
00A0A1	Epic Data Inc.
00A0A2	Digicom S.P.A.
00A0A3	Reliable Power Meters
00A0A4	Micros Systems, Inc.
00A0A5	Teknor Microsysteme, Inc.
00A0A6	M.I. Systems, K.K.
00A0A7	Vorax Corporation
00A0A8	Renex Corporation
00A0A9	GN Nettest (Canada) Inc.
00A0AA	Spacelabs Medical
00A0AB	Netcs Informationstechnik GmBh
00A0AC	Gilat Satellite Networks, Ltd.
00A0AD	Marconi Spa
00A0AE	Network Peripherals, Inc.—IEEE assigned to NUCOM SYSTEMS, INC.
00A0AF	WMS Industries
00A0B0	I-O Data Device, Inc.
00A0B1	First Virtual Corporation
00A0B2	Shima Seiki
00A0B3	Zykronix
00A0B4	Texas Microsystems, Inc.
00A0B5	3H Technology
00A0B6	Sanritz Automation Co., Ltd.
00A0B7	Cordant, Inc.
00A0B8	Symbios Logic Inc.
00A0B9	Eagle Technology, Inc.
00A0BA	Patton Electronics Co.
00A0BB	Hilan GmBh
00A0BC	Viasat, Incorporated
00A0BD	I-Tech Corp.
00A0BE	Integrated Circuit Systems, Inc. Communications Group
00A0BF	Wireless Data Group Motorola
00A0C0	Digital Link Corp.
00A0C1	Ortivus Medical AB
00A0C2	R.A. Systems Co., Ltd.
00A0C3	Unicomputer GmBh
00A0C4	Cristie Electronics Ltd.
00A0C5	Zyxel Communication
00A0C6	Qualcomm Incorporated
00A0C7	Tadiran Telecommunications
00A0C8	Adtran Inc.
00A0C9	Intel Corporation (PRO100B and PRO100+)—(used on Cisco PIX firewall among others)
00A0CA	Fujitsu Denso Ltd.
00A0CB	Ark Telecommunications, Inc.
00A0CC	Lite-On Communications, Inc.
00A0CD	Dr. Johannes Heidenhain GmBh
00A0CE	Astrocom Corporation
00A0CF	Sotas, Inc.
00A0D0	Ten X Technology, Inc.
00A0D1	Inventec Corporation
00A0D2	Allied Telesis International Corporation
00A0D3	Instem Computer Systems, Ltd.
00A0D4	Radiolan, Inc.
00A0D5	Sierra Wireless Inc.
00A0D6	SBE, Inc.
00A0D7	Kasten Chase Applied Research
00A0D8	Spectra - Tek
00A0D9	Convex Computer Corporation
00A0DA	Integrated Systems
00A0DB	Fisher & Paykel Production
00A0DC	O.N. Electronic Co., Ltd.
00A0DD	Azonix Corporation
00A0DE	Yamaha Corporation
00A0DF	STS Technologies, Inc.
00A0E0	Tennyson Technologies Pty Ltd
00A0E1	Westport Research Associates, Inc.
00A0E2	Keisoku Giken Corp.
00A0E3	XKL Systems Corp.
00A0E4	Optiquest
00A0E5	NHC Communications
00A0E6	Dialogic Corporation
00A0E7	Central Data Corporation
00A0E8	Reuters Holdings Plc
00A0E9	Electronic Retailing Systems International
00A0EA	Ethercom Corp.
00A0EB	Fastcomm Communications Corp.
00A0EC	Transmitton Ltd.
00A0ED	Pri Automation
00A0EE	Nashoba Networks
00A0EF	Lucidata Ltd.
00A0F0	Toronto Microelectronics Inc.
00A0F1	MTI
00A0F2	Infotek Communications, Inc.
00A0F3	Staubli
00A0F4	GE
00A0F5	Radguard Ltd.
00A0F6	Autogas Systems, Inc.
00A0F7	V.I Computer Corp.
00A0F8	Symbol Technologies, Inc.
00A0F9	Bintec Communications GmBh
00A0FA	Ant Nachrichtentechnik GmBh
00A0FB	Toray Engineering Co., Ltd.
00A0FC	Image Sciences, Inc.
00A0FD	Scitex Digital Printing, Inc.
00A0FE	Boston Technology, Inc.
00A0FF	Tellabs Operations, Inc.
00AA00	Intel Corporation—sends bit reversed as 005500 on Token Ring
00AA01	Intel Corporation
00AA02	Intel Corporation
00AA3C	Olivetti Telecom SPA (Olteco)
00B0D0	Computer Products International—Not an IEEE assigned OUI
00BB01	Octothorpe Corp.
00BBF0	Ungermann-Bass Inc.
00C000	Lanoptics, Ltd.
00C001	Diatek Patient Managment
00C002	Sercomm Corporation
00C003	Globalnet Communications
00C004	Japan Business Computer Co.Ltd
00C005	Livingston Enterprises Inc—Portmaster (OEMed by Cayman)
00C006	Nippon Avionics Co., Ltd.
00C007	Pinnacle Data Systems, Inc.
00C008	Seco Srl
00C009	Kt Technology (S) Pte Ltd
00C00A	Micro Craft
00C00B	Norcontrol A.S.
00C00C	ARK PC Technology, Inc.—IEEE assigned to RELIA TECHNOLGIES
00C00D	Advanced Logic Research Inc
00C00E	Psitech Inc
00C00F	QNX Software Systems Ltd.—(also Quantum Software Systems Ltd)—IEEE assigned to QUANTUM SOFTWARE SYSTEMS
00C010	Hirakawa Hewtech Corp.
00C011	Interactive Computing Devices
00C012	Netspan Corporation
00C013	Netrix
00C014	Telematics Calabasas Int'l, Inc
00C015	New Media Corporation
00C016	Electronic Theatre Controls

OUI	Vendor/Comment
00C017	John Fluke Mfg.—IEEE assigned to Forte Networks which makes the Fluke LAN Meters—sends bit reversed as 0003E8 on Token Ring
00C018	Lanart Corp
00C019	Leap Technology, Inc.
00C01A	Corometrics Medical Systems
00C01B	Socket Communications, Inc.
00C01C	Interlink Communications Ltd.
00C01D	Grand Junction Networks, Inc.—(Cisco Catalyst also reported)
00C01E	La Francaise Des Jeux
00C01F	S.E.R.C.E.L.
00C020	Arco Electronic, Control Ltd.
00C021	Netexpress
00C022	Lasermaster Technologies, Inc.
00C023	Tutankhamon Electronics
00C024	Eden Sistemas De Computacao SA
00C025	Dataproducts Corporation
00C026	LANS Technology Co., Ltd.
00C027	Cipher Systems, Inc.
00C028	Jasco Corporation
00C029	Kabel Rheydt AG
00C02A	Ohkura Electric Co
00C02B	Gerloff Gesellschaft Fur
00C02C	Centrum Communications, Inc.
00C02D	Fuji Photo Film Co., Ltd.
00C02E	Netwiz
00C02F	Okuma Corp
00C030	Integrated Engineering B.V.
00C031	Design Research Systems, Inc.
00C032	I-Cubed Limited
00C033	Telebit Communications APS
00C034	Dale Computer Corporation
00C035	Quintar Company
00C036	Raytech Electronic Corp
00C037	Dynatem
00C038	Raster Image Processing System
00C039	Silicon Systems—IEEE assigned to TDK SEMICONDUCTOR CORPORATION
00C03A	Men-Mikro Elektronik GmBh
00C03B	Multiaccess Computing Corp
00C03C	Tower Tech S.R.L.
00C03D	Wiesemann & Theis GmBh
00C03E	FA. Gebr. Heller GmBh
00C03F	Stores Automated Systems Inc
00C040	ECCI
00C041	Digital Transmission Systems
00C042	Datalux Corp.
00C043	Stratacom
00C044	Emcom Corporation
00C045	Isolation Systems Inc
00C046	Kemitron Ltd
00C047	Unimicro Systems Inc
00C048	Bay Technical Associates
00C049	US Robotics Total Control (tm) NETServer Card (now owned by 3Com)
00C04A	Group 2000 Ag
00C04B	Creative Microsystems
00C04C	Department Of Foreign Affairs & Trade
00C04D	Mitec, Inc.
00C04E	Comtrol Corporation
00C04F	Dell Computer Corporation
00C050	Toyo Denki Seizo K.K.
00C051	Advanced Integration Research
00C052	Burr-Brown
00C053	Davox Corporation
00C053	Network Peripherals, Ltd.
00C055	Modular Computing Technologies
00C056	Somelec
00C057	Myco Electronics
00C058	Dataexpert Corp.
00C059	Nippon Denso Co., Ltd.
00C05A	Semaphore Communications Corp.
00C05B	Networks Northwest, Inc.
00C05C	Elonex Plc
00C05D	L&N Technologies
00C05E	Vari-Lite Inc
00C05F	Fine-Pal Company Limited
00C060	ID Scandinavia AS
00C061	Solectek Corporation
00C062	Impulse Technology
00C063	Morning Star Technologies Inc
00C064	General Datacomm Ind Inc
00C065	Scope Communications Inc
00C066	Docupoint, Inc.
00C067	United Barcode Industries
00C068	Philp Drake Electronics Ltd
00C069	California Microwave Inc
00C06A	Zahner-Elektrik GmBh & Co KG
00C06B	OSI Plus Corporation
00C06C	SVEC Computer Corp
00C06D	Boca Research, Inc.
00C06E	Haft Technology, Inc.
00C06F	Komatsu Ltd.
00C070	Sectra Secure-Transmission AB
00C071	Areanex Communications, Inc.
00C072	KNX Ltd
00C073	Xedia Corporation
00C074	Toyoda Automatic Loom Works Ltd
00C075	Xante Corporation
00C076	I-Data International A-S
00C077	Daewoo Telecom Ltd
00C078	Computer Systems Engineering
00C079	Fonsys Co Ltd
00C07A	Priva BV
00C07B	Ascend Communications—ISDN bridges
00C07C	Hightech Information
00C07D	RISC Developments Ltd.
00C07E	Kubota Corporation Electronic Device Dept.
00C07F	Nupon Computing Corp.
00C080	Netstar Inc.
00C081	Metrodata Ltd.
00C082	Moore Products Co.
00C083	Trace Mountain Products, Inc.
00C084	Data Link Corp Ltd.
00C085	Canon—IEEE assigned to ELECTRONICS FOR IMAGING, INC.
00C086	The Lynk Corporation
00C087	UUNET Technologies Inc
00C088	EKF Elektronik GmBh
00C089	Telindus Distribution
00C08A	Lauterbach Datentechnik GmBh
00C08B	RISQ Modular Systems Inc.
00C08C	Performance Technologies Inc.
00C08D	Tronix Product Development
00C08E	Network Information Technology
00C08F	Matsushita Electric Works, Ltd.
00C090	Praim S.R.L.
00C091	Jabil Circuit, Inc.
00C092	Mennen Medical Inc
00C093	Alta Research Corp.
00C094	VMX INC.
00C095	Znyx (Network Appliance)—Jupiter Systems (MX-700)—Apple (G3)—all seen—IEEE assigned to ZNYX
00C096	Tamura Corporation
00C097	Archipel SA
00C098	Chuntex Electronic Co., Ltd.
00C099	Yoshiki Industrial Co.,Ltd.
00C09A	Photonics Corporation
00C09B	Reliance Comm/Tec, R-Tec
00C09C	TOA Electronic Ltd.
00C09D	Distributed Systems Int'l, Inc.
00C09E	Cache Computers, Inc.
00C09F	Quanta Computer Inc
00C0A0	Advance Micro Research, Inc.
00C0A1	Tokyo Denshi Sekei Co
00C0A2	Intermedium A/S

OUI	Vendor/Comment
00C0A3	Dual Enterprises Corporation
00C0A4	Unigraf OY
00C0A5	Dickens Data Systems
00C0A6	Exicom Australia Pty. Ltd
00C0A7	SEEL Ltd.
00C0A8	GVC Corporation
00C0A9	Barron McCann Ltd
00C0AA	Silicon Valley Computer
00C0AB	Jupiter Technology Inc
00C0AC	Gambit Computer Communications
00C0AD	Computer Communication Systems—IEEE asssigned to MARBEN COMMUNICATION SYSTEMS
00C0AE	Towercom Co Inc DBA PC House
00C0AF	Teklogix Inc.
00C0B0	GCC Technologies,Inc.
00C0B1	Genius Net Co.
00C0B2	Norand Corporation
00C0B3	Comstat Datacomm Corporation
00C0B4	Myson Technology Inc.
00C0B5	Corporate Network Systems Inc
00C0B6	Meridian Data Inc.
00C0B7	American Power Conversion Corp.
00C0B8	Fraser's Hill Ltd.
00C0B9	Funk Software Inc.
00C0BA	Netvantage
00C0BB	Forval Creative Inc.
00C0BC	Telecom Australia/CSSC
00C0BD	Inex Technologies, Inc.
00C0BE	Alcatel - Sel
00C0BF	Technology Concepts Ltd.
00C0C0	Shore Microsystems Inc.
00C0C1	Quad/Graphics Inc.
00C0C2	Infinite Networks Ltd.
00C0C3	Acuson Computed Sonography
00C0C4	Computer Operational
00C0C5	SID Informatica
00C0C6	Personal Media Corp.
00C0C7	Sparktrum Microsystems, Inc.
00C0C8	Micro Byte Pty Ltd
00C0C9	ELSAG Bailey Controls Co.
00C0CA	Alfa, Inc.
00C0CB	Control Technology Inc.
00C0CC	Telesciences Co Systems, Inc.
00C0CD	Comelta, S.A.
00C0CE	CEI Systems & Engineering Pte
00C0CF	Imatran Voima Oy
00C0D0	Ratoc System Inc.
00C0D1	Comtree Technology Corporation (EFA also reported)—IEEE assigned to COMTREE TECHNOLOGY CORPORATION
00C0D2	Syntellect Inc
00C0D3	Olympus Image Systems, Inc.
00C0D4	Axon Networks Inc.
00C0D5	Quancom Electronic GmBh
00C0D6	J1 Systems, Inc.
00C0D7	Taiwan Trading Center DBA TTC Computer Products
00C0D8	Universal Data Systems
00C0D9	Quinte Network Confidentiality Equipment Inc
00C0DA	Nice Systems Ltd.
00C0DB	IPC Corporation (Pte) Ltd
00C0DC	EOS Technologies, Inc.
00C0DD	Ancor Communications
00C0DE	ZComm Inc.
00C0DF	Kye Systems Corp.
00C0E0	DSC Communication Corp.
00C0E1	Sonic Solutions
00C0E2	Calcomp, Inc.
00C0E3	Ositech Communications Inc.
00C0E4	Landis & Gyr Powers Inc.
00C0E5	GESPAC S.A.
00C0E6	TXPORT
00C0E7	Fiberdata AB

OUI	Vendor/Comment
00C0E8	Plexcom, Inc.
00C0E9	Oak Solutions Ltd.
00C0EA	Array Technology Ltd.
00C0EB	SEH Computertechnik GmBh
00C0EC	Dauphin Technology
00C0ED	US Army Electronic Proving Ground
00C0EE	Kyocera Corporation
00C0EF	Abit Corporation
00C0F0	Kingston Technology Corporation—sends bit reversed as 00030F on Token Ring
00C0F1	Shinko Electric Co Ltd
00C0F2	Transition Engineering Inc
00C0F3	Network Communications Corp
00C0F4	Interlink System Co., Ltd.
00C0F5	Metacomp Inc
00C0F6	Celan Technology Inc.
00C0F7	Engage Communication, Inc.
00C0F8	About Computing Inc.
00C0F9	Harris And Jeffries, Inc.
00C0FA	Canary Communications Inc
00C0FB	Advanced Technology Labs
00C0FC	ASDG Incorporated—IEEE assigned to ELASTIC REALITY, INC.
00C0FD	Prosum
00C0FE	Aptec Computer Systems, Inc.
00C0FF	Box Hill Systems Corporation
00CF1C	Communication Machinery Corp.
00DD00	Ungermann-Bass Inc.—IBM RS6000
00DD01	Ungermann-Bass Inc.
00DD02	Ungermann-Bass Inc.
00DD03	Ungermann-Bass Inc.
00DD04	Ungermann-Bass Inc.
00DD05	Ungermann-Bass Inc.
00DD06	Ungermann-Bass Inc.
00DD07	Ungermann-Bass Inc.
00DD08	Ungermann-Bass Inc.
00DD09	Ungermann-Bass Inc.
00DD0A	Ungermann-Bass Inc.
00DD0B	Ungermann-Bass Inc.
00DD0C	Ungermann-Bass Inc.
00DD0D	Ungermann-Bass Inc.
00DD0E	Ungermann-Bass Inc.
00DD0F	Ungermann-Bass Inc.
00D010	Convergent Networks, Inc.
00D011	Prism Video, Inc.
00D012	Gateworks Corp.
00D013	Primex Aerospace Company
00D014	Root, Inc.
00D015	Univex Microtechnology Corp.
00D016	Scm Microsystems, Inc.
00D017	Syntech Information Co., Ltd.
00D018	Qwes. Com, Inc.
00D019	Dainippon Screen Corporate
00D01A	Urmet Sud S.P.A.
00D01B	Mimaki Engineering Co., Ltd.
00D01C	SBS Technologies, Connectivity Products
00D01D	Furuno Electric Co., Ltd.
00D01E	Pingtel Corp.
00D01F	Ctam Pty. Ltd.
00D020	Aim System, Inc.
00D021	Regent Electronics Corp.
00D022	Incredible Technologies, Inc.
00D023	Infortrend Technology, Inc.
00D024	Cognex
00D025	Xrosstech, Inc.
00D026	Hirschmann Austria GmBh
00D027	Applied Automation, Inc.
00D028	Omneon Video Networks
00D029	Wakefern Food Corporation
00D02A	Flexion Systems
00D02B	Jetcell, Inc.
00D02C	Campbell Scientific, Inc.
00D02D	Ademco

OUI	Vendor/Comment
00D02E	Communication Automation Corp.
00D02F	Vlsi Technology Inc.
00D030	Safetran Systems Corp.
00D031	Industrial Logic Corporation
00D032	Yano Electric Co., Ltd.
00D033	Dalian Daxian Network
00D034	Ormec Systems Corp.
00D035	Behavior Tech. Computer Corp.
00D036	Technology Atlanta Corp.
00D037	Philips-Dvs-Lo Bdr
00D038	Fivemere, Ltd.
00D039	Utilicom, Inc.
00D03A	Zoneworx, Inc.
00D03B	Vision Products Pty. Ltd.
00D03C	Power Micro Research
00D03D	Rocketchips, Inc.
00D03E	American Communication Technoligies International Inc.
00D03F	Sysmate Co., Ltd.
00D041	Amigo Technology Co., Ltd.
00D042	Mahlo GmBh & Co. Ug
00D043	Zonal Retail Data Systems
00D044	Alidian Networks, Inc.
00D045	Kvaser Ab
00D046	Dolby Laboratories, Inc.
00D047	Xn Technologies
00D048	Ecton, Inc.
00D049	Impresstek Co., Ltd.
00D04A	Presence Technology GmBh + Co. KG
00D04B	La Cie Group S.A.
00D04C	Eurotel Telecom Ltd.
00D04D	Div Of Research & Statistics
00D04E	Logibag
00D04F	Bitronics, Inc.
00D050	Iskratel
00D051	02 Micro, Inc.
00D052	Ascend Communications, Inc.
00D053	Connected Systems
00D054	Sas Institute Inc.
00D055	Kathrein-Werke Kg
00D056	Somat Corporation
00D057	Ultrak, Inc.
00D058	Cisco Systems, Inc.
00D059	Ambit Microsystems Corp.
00D05A	Symbionics, Ltd.
00D05B	Acroloop Motion Control Systems
00D05C	Technotrend Systemtechnik GmBh
00D05D	Intelliworxx, Inc.
00D05E	Stratabeam Technology, Inc.
00D05F	Valcom, Inc.
00D060	Panasonic European
00D061	Tremon Enterprises Co., Ltd.
00D062	Digigram
00D063	Cisco Systems, Inc.
00D064	Multitel
00D065	Toko Electric
00D066	Wintriss Engineering Corp.
00D067	Campio Communications
00D068	Iwill Corporation
00D069	Technologic Systems
00D06A	Linkup Systems Corporation
00D06B	Sr Telecom Inc.
00D06C	Sharewave, Inc.
00D06D	Acrison, Inc.
00D06E	Trendview Recorders Ltd.
00D06F	Kmc Controls
00D070	Long Well Electronics Corp.
00D071	Echelon Corp.
00D072	Broadlogic
00D073	ACN Advanced Communications
00D074	Taqua Systems, Inc.
00D075	Alaris Medical Systems, Inc.
00D076	Merrill Lynch & Co., Inc.
00D077	Lucent Technologies
00D078	Eltex Of Sweden AB
00D079	Cisco Systems, Inc.
00D07A	Amaquest Computer Corp.
00D07B	Comcam International Ltd.
00D07C	Koyo Electronics Inc. Co., Ltd.
00D07D	Cosine Communications
00D07E	Keycorp Ltd.
00D07F	Strategy & Technology, Limited
00D080	Exabyte Corporation
00D081	Real Time Devices Usa, Inc.
00D082	Iowave Inc.
00D083	Invertex, Inc.
00D084	Nexcomm Systems, Inc.
00D085	Otis Elevator Company
00D086	Foveon, Inc.
00D087	Microfirst Inc.
00D088	Mainsail Networks, Inc.
00D089	Dynacolor, Inc.
00D08A	Photron USA
00D08B	Storage Area Networks, Ltd.
00D08C	Genoa Technology, Inc.
00D08D	Phoenix Group, Inc.
00D08E	Nvision Inc.
00D08F	Ardent Technologies, Inc.
00D090	Cisco Systems, Inc.
00D091	Smartsan Systems, Inc.
00D092	Glenayre Western Multiplex
00D093	Tq - Components GmBh
00D095	Xylan Corporation
00D096	3com Europe Ltd.
00D097	Cisco Systems, Inc.
00D098	Ips Automation
00D099	Elcard OY
00D09A	Filanet Corporation
00D09B	Spectel Ltd.
00D09C	Kapadia Communications
00D09D	Veris Industries
00D09E	2wire, Inc.
00D09F	Novtek Test Systems
00D0A0	Mips Denmark
00D0A1	Oskar Vierling GmBh + Co. KG
00D0A2	Integrated Device
00D0A3	Vocal Data, Inc.
00D0A4	Alantro Communications
00D0A5	American Arium
00D0A6	Lanbird Technology Co., Ltd.
00D0A7	Tokyo Sokki Kenkyujo Co., Ltd.
00D0A8	Network Engines, Inc.
00D0A9	Shinano Kenshi Co., Ltd.
00D0AA	Chase Communications
00D0AB	Deltakabel Telecom CV
00D0AC	Grayson Wireless
00D0AD	TI Industries
00D0AE	Oresis Communications, Inc.
00D0AF	Cutler-Hammer, Inc.
00D0B0	Bitswitch Ltd.
00D0B1	Omega Electronics SA
00D0B2	Xiotech Corporation
00D0B3	DRS Flight Safety And Communications
00D0B4	Katsujima Co., Ltd.
00D0B5	Dotcom
00D0B6	Crescent Networks, Inc.
00D0B7	Intel Corp.
00D0B8	Iomega Corp.
00D0B9	Microtek International, Inc.
00D0BA	Cisco Systems, Inc.
00D0BB	Cisco Systems, Inc.
00D0BC	Cisco Systems, Inc.
00D0BD	Sican GmBh

OUI	Vendor/Comment
00D0BE	Emutec Inc.
00D0BF	Pivotal Technologies
00D0C0	Cisco Systems, Inc.
00D0C1	Harmonic Data Systems, Ltd.
00D0C2	Balthazar Technology Ab
00D0C3	Vivid Technology Pte, Ltd.
00D0C4	Teratech Corporation
00D0C5	Computational Systems, Inc.
00D0C6	Thomas & Betts Corp.
00D0C7	Pathway, Inc.
00D0C8	I/O Consulting A/S
00D0C9	Advantech Co., Ltd.
00D0CA	Intrinsyc Software Inc.
00D0CB	Dasan Co., Ltd.
00D0CC	Technologies Lyre Inc.
00D0CD	Atan Technology Inc.
00D0CE	Asyst Electronic
00D0CF	Moreton Bay
00D0D0	Zhongxing Telecom Ltd.
00D0D1	Sirocco Systems, Inc.
00D0D2	Epilog Corporation
00D0D3	Cisco Systems, Inc.
00D0D4	V-Bits, Inc.
00D0D5	Grundig AG
00D0D6	Aethra Telecomunicazioni
00D0D7	B2c2, Inc.
00D0D8	Nomadic Technologies
00D0D9	Dedicated Microcomputers Group, Ltd.
00D0DA	Taicom Data Systems Co., Ltd.
00D0DB	Mcquay International
00D0DC	Modular Mining Systems, Inc.
00D0DD	Sunrise Telecom, Inc.
00D0DE	Philips Multimedia Network
00D0DF	Kuzumi Electronics, Inc.
00D0E0	Dooin Electronics Co.
00D0E1	Avionitek Israel Inc.
00D0E2	Mrt Micro, Inc.
00D0E3	Ele-Chem Engineering Co., Ltd.
00D0E4	Cisco Systems, Inc.
00D0E5	Solidum Systems Corp.
00D0E6	Ibond Inc.
00D0E7	Vcon Telecommunication Ltd.
00D0E8	Mac System Co., Ltd.
00D0E9	Advantage Century
00D0EA	Nextone Communications, Inc.
00D0EB	Lightera Networks, Inc.
00D0EC	Nakayo Telecommunications, Inc
00D0ED	Xiox
00D0EE	Dictaphone Corporation
00D0EF	IGT
00D0F0	Convision Technology GmBh
00D0F1	Sega Enterprises, Ltd.
00D0F2	Monterey Networks
00D0F3	Solari Di Udine SPA
00D0F4	Carinthian Tech Institute
00D0F5	Orange Micro, Inc.
00D0F6	Northchurch Communications Inc.
00D0F7	Next Nets Corporation
00D0F8	Fujian Star Terminal Device Co., Ltd.
00D0F9	Acute Communications Corp.
00D0FA	Racal Guardata
00D0FB	Tek Microsystems, Incorporated
00D0FC	Granite Microsystems
00D0FD	Optima Tele.Com, Inc.
00D0FE	Astral Point Communications, Inc.
00D0FF	Cisco Systems, Inc.
00DD00	Ungermann-Bass Inc.
00DD01	Ungermann-Bass Inc.
00DD02	Ungermann-Bass Inc.
00DD03	Ungermann-Bass Inc.
00DD04	Ungermann-Bass Inc.

OUI	Vendor/Comment
00DD05	Ungermann-Bass Inc.
00DD06	Ungermann-Bass Inc.
00DD07	Ungermann-Bass Inc.
00DD08	Ungermann-Bass Inc.
00DD09	Ungermann-Bass Inc.
00DD0A	Ungermann-Bass Inc.
00DD0B	Ungermann-Bass Inc.
00DD0C	Ungermann-Bass Inc.
00DD0D	Ungermann-Bass Inc.
00DD0E	Ungermann-Bass Inc.
00DD0F	Ungermann-Bass Inc.
00E000	Fujitsu, Ltd
00E001	Strand Lighting Limited
00E002	Crossroads Systems, Inc.
00E003	Nokia Wireless Business Commun
00E004	Pmc-Sierra, Inc.
00E005	Technical Corp.
00E006	Silicon Integrated Sys. Corp.
00E007	Network Alchemy Ltd.
00E008	Amazing Controls! Inc.
00E009	Marathon Technologies Corp.
00E00A	Diba, Inc.
00E00B	Rooftop Communications Corp.
00E00C	Motorola
00E00D	Radiant Systems
00E00E	Avalon Imaging Systems, Inc.
00E00F	Shanghai Baud Data Communication Development Corp.
00E010	Hess Sb-Automatenbau GmBh
00E011	Uniden San Diego
00E012	Pluto Technologies
00E013	Eastern Electronic Co., Ltd.
00E014	Cisco Systems, Inc.
00E015	Heiwa Corporation
00E016	Rapid-City Communications (bought by Bay Networks bought by Nortel)
00E017	EXXACT GmBh
00E018	Asustek Computer Inc.—Intel 82558-based Integrated Fast Ethernet for WIM
00E019	Ing. Giordano Elettronica
00E01A	Comtec Systems. Co., Ltd.
00E01B	Sphere Communications, Inc.
00E01C	Mobility Electronicsy
00E01D	Webtv Networks, Inc.
00E01E	Cisco Systems, Inc.
00E01F	Avidia Systems, Inc.
00E020	Tecnomen Oy
00E021	Freegate Corp.
00E022	Medialight Inc.
00E023	Telrad
00E024	Gadzoox Networks
00E025	Dit Co., Ltd.
00E026	Eastman Kodak Co.
00E027	Dux, Inc.
00E028	Aptix Corporation
00E029	Standard Microsystems Corp.
00E02A	Tandberg Television As
00E02B	Extreme Networks
00E02C	AST - built into 5166M PC motherboard (win95 id's as Intel)
00E02D	Innomedialogic, Inc.
00E02E	SPC Electronics Corporation
00E02F	Mcns Holdings, L.P.
00E030	Melita International Corp.
00E031	Hagiwara Electric Co., Ltd.
00E032	Misys Financial Systems, Ltd.
00E033	E.E.P.D. GmBh
00E034	Cisco Systems, Inc.
00E035	Loughborough Sound Images, Plc
00E036	Pioneer Digital Technologies, Inc.
00E037	Century Corporation
00E038	Proxima Corporation
00E039	Paradyne Corp.
00E03A	Cabletron Systems, Inc.
00E03B	Prominet Corporation

OUI	Vendor/Comment
00E03C	Advansys
00E03D	Focon Electronic Systems A/S
00E03E	Alfatech, Inc.
00E03F	Jaton Corporation
00E040	Deskstation Technology, Inc.
00E041	CSPI
00E042	Pacom Data Ltd.
00E043	Vitalcom
00E044	Lsics Corporation
00E045	Touchwave, Inc.
00E046	Bently Nevada Corp.
00E047	Infocus Systems
00E048	SDL Communications, Inc.
00E049	Microwi Electronic GmBh
00E04A	Enhanced Messaging Systems, Inc
00E04B	Jump Industrielle
00E04C	Realtek Semiconductor Corp.
00E04D	Internet Initiative Japan, Inc
00E04E	Sanyo Denki Co., Ltd.
00E04F	Cisco Systems, Inc.
00E050	Executone Information Systems, Inc.
00E051	Talx Corporation
00E052	Foundry Networks, Inc.
00E053	Cellport Labs, Inc.
00E054	Kodai Hitec Co., Ltd.
00E055	Ingenieria Electronica
00E056	Holontech Corporation
00E057	Han Microtelecom. Co., Ltd.
00E058	Phase One Denmark A/S
00E059	Controlled Environments, Ltd.
00E05A	Galea Network Security
00E05B	West End Systems Corp.
00E05C	Matsushita Kotobuki Electronics Industries, Ltd.
00E05D	Unitec Co., Ltd.
00E05E	Japan Aviation Electronics Industry, Ltd.
00E05F	E-Net, Inc.
00E060	Sherwood
00E061	Edgepoint Networks, Inc.
00E062	Host Engineering
00E063	Cabletron - Yago Systems, Inc.
00E064	Samsung Electronics
00E065	Optical Access International
00E066	Promax Systems, Inc.
00E067	Eac Automation-Consulting GmBh
00E068	Merrimac Systems Inc.
00E069	Jaycor Networks, Inc.
00E06A	Kapsch AG
00E06B	W&G Special Products
00E06C	Baltimore Technologies, Ltd.
00E06D	Compuware Corporation
00E06E	Far Systems SPA
00E06F	Terayon Corp.
00E070	DH Technology
00E071	Epis Microcomputer
00E072	Lynk
00E073	National Amusement Network, Inc.
00E074	Tiernan Communications, Inc.
00E075	Atlas Computer Equipment, Inc.
00E076	Development Concepts, Inc.
00E077	Webgear, Inc.
00E078	Berkeley Networks
00E079	A.T.N.R.
00E07A	Mikrodidakt AB
00E07B	Bay Networks
00E07C	Mettler-Toledo, Inc.
00E07D	Netronix, Inc.
00E07E	Walt Disney Imagineering
00E07F	Logististem Srl
00E080	Control Resources Corporation
00E081	Tyan Computer Corp.—Onboard Intel 82558 10
00E082	Anerma

OUI	Vendor/Comment
00E083	Jato Technologies, Inc.
00E084	Compulite R&D
00E085	Global Maintech, Inc.
00E086	Cybex Computer Products
00E087	Lecroy Networking Products Division
00E088	Ltx Corporation
00E089	Microframe Inc.
00E08A	Gec Avery, Ltd.
00E08B	Qlogic Corp.
00E08C	Neoparadigm Labs, Inc.
00E08D	Pressure Systems, Inc.
00E08E	Utstarcom
00E08F	Cisco Systems, Inc.
00E090	Beckman Lab. Automation Div.
00E091	Lg Electronics, Inc.
00E092	Admtek Incorporated
00E093	Ackfin Networks
00E094	OSAI SRL
00E095	Advanced-Vision Technolgies Corp.
00E096	Shimadzu Corporation
00E097	Carrier Access Corporation
00E098	Linksys—PCMCIA card—IEEE assigned to ABOCOM SYSTEMS, INC.
00E099	Samson AG
00E09A	Positron Industries, Inc.
00E09B	Engage Networks, Inc.
00E09C	MII
00E09D	Sarnoff Corporation
00E09E	Quantum Corporation
00E09F	Pixel Vision
00E0A0	Wiltron Co.
00E0A1	Hima Paul Hildebrandt
00E0A2	Microslate Inc.
00E0A3	Cisco Systems, Inc.—Catalyst 1924
00E0A4	Esaote S.P.A.
00E0A5	Comcore Semiconductor, Inc.
00E0A6	Telogy Networks, Inc.
00E0A7	Ipc Information Systems, Inc.
00E0A8	Sat GmBh&Co
00E0A9	Funai Electric Co., Ltd.
00E0AA	Electrosonic Ltd.
00E0AB	Dimat S.A.
00E0AC	Midsco, Inc.
00E0AD	Ees Technology, Ltd.
00E0AE	Xaqti Corporation
00E0AF	General Dynamics Information Systems
00E0B0	Cisco Systems, Inc.—Various systems reported
00E0B1	Packet Engines, Inc.
00E0B2	Telmax Communications Corp.
00E0B3	Etherwan Systems, Inc.
00E0B4	Techno Scope Co., Ltd.
00E0B5	Ardent Communications Corp.
00E0B6	Osicom Technologies
00E0B7	PI Group, Ltd.
00E0B8	Gateway 2000—AMD PCNet in a Gateway 2000
00E0B9	Byas Systems
00E0BA	Berghof Automationstechnik GmBh
00E0BB	NBX Corporation
00E0BC	Symon Communications, Inc.
00E0BD	Interface Systems, Inc.
00E0BE	Genroco International, Inc.
00E0BF	Torrent Networking
00E0C0	Seiwa Erectric Mfg. Co., Ltd.
00E0C1	Memorex Telex Japan, Ltd.
00E0C2	Necsy SPA
00E0C3	Sakai System Development Corp.
00E0C4	Horner Electric, Inc.
00E0C5	Bcom Electronics Inc.
00E0C6	Link2it, L.L.C.
00E0C7	Eurotech SRL
00E0C8	Virtual Access, Ltd.
00E0C9	Automatedlogic Corporation

OUI	Vendor/Comment
00E0CA	Best Data Products
00E0CB	Reson, Inc.
00E0CC	Hero Systems, Ltd.
00E0CD	Sensis Corporation
00E0CE	ARN
00E0CF	Integrated Device Technology, Inc.
00E0D0	Netspeed, Inc.
00E0D1	Telsis Limited
00E0D2	Versanet Communications, Inc.
00E0D3	Datentechnik GmBh
00E0D4	Excellent Computer
00E0D5	Arcxel Technologies, Inc.
00E0D6	Computer & Communication
00E0D7	Sunshine Electronics, Inc.
00E0D8	Lanbit Computer, Inc.
00E0D9	Tazmo Co., Ltd.
00E0DA	Assured Access
00E0DB	Viavideo Communications
00E0DC	Nexware Corp.
00E0DD	ZENITH ELECTRONICS CORPORATION
00E0DE	Datax NV
00E0DF	Ke Kommunikations-Electronik GmBh
00E0E0	Si Electronics, Ltd.
00E0E1	G2 Networks, IInc.
00E0E2	Innova Corp.
00E0E3	SK-Elektronik GmBh
00E0E4	Fanuc Robotics North America, Inc.
00E0E5	Cinco Networks, Inc.
00E0E6	Incaa Datacom B.V.
00E0E7	Raytheon E-Systems, Inc.
00E0E8	Gretacoder Data Systems AG
00E0E9	Data Labs, Inc.
00E0EA	Innovat Communications, Inc.
00E0EB	Digicom Systems, Incorporated
00E0EC	Celestica Inc.
00E0ED	Silicom, Ltd.
00E0EE	Marel Hf
00E0EF	Dionex
00E0F0	Abler Technology, Inc.
00E0F1	That Corporation
00E0F2	Arlotto Comnet, Inc.
00E0F3	Websprint Communications, Inc.
00E0F4	Inside Technology A/S
00E0F5	Teles AG
00E0F6	Decision Europe
00E0F7	Cisco Systems, Inc.
00E0F8	Diana Control AB
00E0F9	Cisco Systems, Inc.
00E0FA	Trl Technology, Ltd.
00E0FB	Leightronix, Inc.
00E0FC	Huawei Technologies Co., Ltd.
00E0FD	A-Trend Technology Co., Ltd.
00E0FE	Cisco Systems, Inc.
00E0FF	Security Dynamics Technologies
00E6D3	Nixdorf Computer Corp.
020701	Racal-Datacom
021C7C	Perq Systems Corporation
026086	Satelcom MegaPac (UK)—IEEE assigned to LOGIC REPLACEMENT TECH. LTD.
02608C	3Com Corporation—IBM PC; Imagen; Valid; Cisco; Macintosh—IEEE assigned to 3COM CORPORATION—on Token Ring old 4 Mb/s adapters
027001	Racal-Datacom
0270B0	M/A-Com Inc. Companies
0270B3	Data Recall Ltd.
029D8E	Cardiac Recorders, Inc.
02AA3C	Olivetti Telecomm SPA (Olteco)
02BB01	Octothorpe Corp.
02C08C	3com Corporation
02CF1C	Communication Machinery Corp.
02E6D3	Bus-Tech, Inc. (BTI)—IBM Mainframes—IEEE has no OUI for Bus-Tech, Inc. Has a BusTek Corp w/ OUI 004034—IEEE assigned to NIXDORF COMPUTER CORPORATION
040AE0	Xmit AG Computer Networks
048845	Bay Networks—token ring line card—Bay says "not ours!"—IEEE assigned to TRIUMPH-ADLER AG
080001	Computervision Corporation
080002	Bridge Communications (now owned by 3Com)
080003	Advanced Computer Communications (ACC)
080004	Cromemco Incorporated
080005	Symbolics—Symbolics LISP machines
080006	Siemens AG—PC clone
080007	Apple Computer—sends bit reversed as 1000E0 on Token Ring
080008	Bolt Beranek and Newman, Inc. (BBN)
080009	Hewlett-Packard—sends bit reversed as 100090 on Token Ring
08000A	Nestar Systems
08000B	Unisys also Ascom-Timeplex (former Unisys subsidiary)
08000C	Miklyn Development Co.
08000D	ICL (International Computers, Ltd.)
08000E	National Cash Register (NCR) now owned by AT&T—sends bit reversed as 100070 on Token Ring
08000F	Standard Microsystems Corp. (SMC)—probably typo of assigned OUI 00800F—IEEE assigned to MITEL CORPORATION
080010	AT&T—probably typo of 800010—Not an IEEE assigned OUI
080011	Tektronix, Inc. (network printers)—sends bit reversed as 100088 on Token Ring
080012	Bell Atlantic Integrated Syst.
080013	Exxon
080014	Excelan—BBN Butterfly, Masscomp, Silicon Graphics—IEEE assigned to EXCELAN
080015	STC Business Systems
080016	Barrister Info Sys Corp
080017	National Semiconductor Corp. (used to have Network System Corp., wrong NSC)—IEEE assigned to NATIONAL SEMICONDUCTOR
080018	Pirelli Focom Networks
080019	General Electric Corporation
08001A	Tiara/ 10net
08001B	Data General
08001C	KDD-Kokusai Debnsin Denwa Co. Fuji Xerox Co., Ltd. Telegraph
08001D	Able Communications, Inc.
08001E	Apollo (now owned by H-P)
08001F	Sharp
080020	Sun Microsystems Inc.
080021	3M Company
080022	NBI (Nothing But Initials)
080023	Matsushita Graphics Comm Sys Inc.
080024	10NET Communications/DCA
080025	Control Data Corporation—CDC
080026	Norsk Data (Nord)
080027	PCS Computer Systems GmBh—IEEE assigned to CADMUS COMPUTER SYSTEMS
080028	Texas Instruments, Inc.—Explorer
080029	Megatek Corporation
08002A	Mosaic Technologies Inc.
08002B	Digital Equipment Corporation
08002C	Britton Lee Inc.
08002D	Lan-Tec Inc.
08002E	Metaphor Computer Systems
08002F	Prime Computer—Prime 50-Series LHC300
080030	CERN
080030	Network Research Corporation
080030	Royal Melbourne Inst Of Tech
080031	Little Machines Inc.
080032	Tigan Incorporated
080033	Bausch & Lomb
080034	Filenet Corporation
080035	Microfive Corporation
080036	Intergraph—CAE stations
080037	Fuji Xerox
080038	CII Honeywell Bull
080039	Spider Systems Ltd.
08003A	Orcatech Inc.
08003B	Torus Systems Limited
08003C	Schlumberger Well Services/Austin Engineering Services

OUI	Vendor/Comment
08003D	Cadnetix Corporations
08003E	Codex-(now owned by Motorola)—VME bus processor modules
08003F	Fred Koschara Enterprises
080040	Ferranti Computer Sys. Limited
080041	Digital Communications Associates (DCA)—IEEE assigned to RACAL-MILGO INFORMATION SYS..
080042	Japan Macnics Corp.
080043	Pixel Computer Inc.
080044	DAVID Systems, Inc. (DSI)
080045	(maybe Xylogics, but they claim not to know this number)—IEEE assigned to CONCURRENT COMPUTER CORP.
080046	Sony
080047	Sequent
080048	Eurotherm Gauging Systems
080049	Univation
08004A	Banyan Systems Inc.
08004B	Planning Research Corp.
08004C	Encore—IEEE assigned to HYDRA COMPUTER SYSTEMS INC.
08004D	Corvus Systems Inc.
08004E	BICC—(now owned by 3Com, so may appear on 3com equipment as well)—IEEE assigned to 3COM EUROPE
08004F	Cygnet Systems
080050	Daisy Systems Corp.
080051	Experdata
080052	Insystec
080053	Middle East Tech. University
080055	Stanford Telecomm. Inc.
080056	Stanford Linear Accel. Center
080057	Evans & Sutherland
080058	Systems Concepts
080059	A/S Mycron
08005A	IBM Corporation
08005B	Vta Technologies Inc.
08005C	Four Phase Systems
08005D	Gould Inc.
08005E	Counterpoint Computer Inc.
08005F	Saber Technology Corp.
080060	Industrial Networking Inc.
080061	Jarogate Ltd.
080062	General Dynamics
080063	Plessey
080064	Autophon AG
080065	Genrad Inc.
080066	AGFA—printers, phototypesetters etc.
080067	Comdesign
080068	Ridge Computers
080069	Silicon Graphics Inc.
08006A	AT&T Bell Laboratories
08006B	Accel Technologies Inc.
08006C	Suntek Technology Int'l
08006D	Whitechapel Computer Works
08006E	Masscomp
08006F	Philips Apeldoorn B.V.
080070	Mitsubishi Electric Corp.
080071	Matra (DSIE)
080072	Xerox Corp Univ Grant Program
080073	Tecmar Inc.
080074	Casio Computer Co. Ltd.
080075	Dansk Data Electronik
080076	PC Lan Technologies
080077	TSL Communications Ltd.
080078	Accell Corporation
080079	Silicon Graphics—probably a typo of SGI assigned OUI 080069—IEEE assigned to THE DROID WORKS
08007A	Indata
08007B	Sanyo Electric Co. Ltd.
08007C	Vitalink—TransLAN III
08007E	Amalgamated Wireless(Aus) Ltd
08007F	Carnegie-Mellon University (CMU)
080080	XIOS—IEEE assigned to AES DATA INC.
080081	Crosfield Electronics—IEEE assigned to ASTECH INC.
080082	Veritas Software
080083	Seiko Instrum. And Electronics
080084	Tomen Electronics Corp.
080085	Elxsi
080086	Imagen Corporation
080087	Xyplex—terminal servers
080088	McDATA Corporation
080089	Kinetics—AppleTalk-Ethernet interface
08008A	Performance Technology
08008B	Pyramid Technology Corp.
08008C	Network Research Corporation
08008D	XyVision—XyVision machines
08008E	Tandem Computers
08008F	Chipcom Corp. (now owned by 3Com)
080090	Retix, Inc.—Bridges
08BBCC	AK-Nord Edv Vertriebsges. MbH
09006A	AT&T—used in smart hub—Not an IEEE assigned OUI
100000	PUBLIC use assigned by IEEE
100014	Texas Instruments (Token Lite Manufacturing Kit)—bit reversed of assigned OUI 080028—Not an IEEE assigned OUI
100024	Digital Communications Associates (DCA) (sold to Microdyne and Attachmate)—seen on Token Ring—Not an IEEE assigned OUI
100058	Data General—seen on Token Ring—bit reversed of assigned OUI 08001A—Not an IEEE assigned OUI
10005A	IBM Corporation
100070	National Cash Register (now owned by AT&T)—seen on Token Ring—bit reversed of assigned OUI 08000E—Not an IEEE assigned OUI
100088	Tektronix, Inc. (network printers)—seen on Token Ring—bit reversed of assigned OUI 08000E—Not an IEEE assigned OUI
100090	Hewlett Packard—seen on Token Ring—bit reversed of assigned OUI 080009—Not an IEEE assigned OUI
1000D8	Data General—seen on Token Ring—bit reversed of assigned OUI 08001B—Not an IEEE assigned OUI
1000E0	Apple Computer—seen on Token Ring—bit reversed of assigned OUI 08007—Not an IEEE assigned OUI
1000E8	National Semiconductor
800010	AT&T Bell Laboratories
1100AA	IEEE STANDARDS OFFICE
40000C	Cisco Systems, Inc—Token Ring Locally Administered Address of OUI 00000C
400022	Visual Technology Inc.—Token Ring Locally Administered Address of OUI 000022
40002A	TRW—Token Ring Locally Administered Address of OUI 00002A
400065	Network General Inc. (now Network Associates Inc—NAI)—Token Ring Locally Administered Address of OU! 000065
40009F	Ameristar Technologies Inc.—Token Ring Locally Administered Address of OUI 00009F
4000A9	Network Systems Corp.(NSC)—Token Ring Locally Administered Address of OUI 0000A9
4000AA	Xerox Corp Office Systems Div—Token Ring Locally Administered Address of OUI 0000AA
4000B3	CIMLinc Incorporated —Token Ring Locally Administered Address of OUI 0000B3
4000C0	Western Digital (now owned by SMC)—Token Ring Locally Administered Address of OUI 0000C0
4000C9	Proteon—Token Ring Locally Administered Address of OUI 000093—bit reversed
4000F6	Madge Networks Ltd.—Token Ring Locally Administered Address of OUI 00006F—bit reveresed
420701	Racal Datacom—old Interlan interface—Token Ring Locally Administered Address of OUI 020701
42608C	3Com Corp.—Token Ring Locally Administered Address of OUI 02608C
480002	Bridge Communications, Inc—old 4Mb/s (now owned by 3Com)—Token Ring Locally Administered Address of OUI 080002
480003	ADVANCED COMPUTER COMM. (ACC)—Token Ring Locally Administered Address of OUI 080003
480005	SYMBOLICS INC.—Token Ring Locally Administered Address of OUI 080005
480008	BOLT BERANEK AND NEWMAN INC (BBN).—Token Ring Locally Administered Address of OUI 080008

OUI	Vendor/Comment
480009	Hewlett Packard—Token Ring Locally Administered Address of OUI 080009
48000A	NESTAR SYSTEMS INCORPORATED—Token Ring Locally Administered Address of OUI 08000A
480014	Excelan—Token Ring Locally Administered Address of OUI 080014
48001E	Apollo Computer Inc. (now owned by H-P)—Token Ring Locally Administered Address of OUI 08001E
480020	SUN MICROSYSTEMS INC.—Token Ring Locally Administered Address of OUI 080020
480025	Control Data Corp. (CDC)—Token Ring Locally Administered Address of OUI 080025
480028	Texas Instruments (TI) Token Ring Locally Administered Address of OUI 080028
48002B	Digital Equipment Corp. (DEC)—Token Ring Locally Administered Address of OUI 08002B
480036	Intergraph Corp.—Token Ring Locally Administered Address of OUI 080036
480039	Spider Systems Ltd.—Token Ring Locally Administered Address of OUI 080039
480047	SEQUENT COMPUTER SYSTEMS INC.—Token Ring Locally Administered Address of OUI 080047
480049	Univation—Token Ring Locally Administered Address of OUI 080049
48004E	BICC (now owned by 3Com)—Token Ring Locally Administered Address of OUI 08004E
480067	Comdesign—Token Ring Locally Administered Address of OUI 080067
480068	Ridge Computers—Token Ring Locally Administered Address of OUI 080068
480069	Silicon Graphics Inc. (SGI)—Token Ring Locally Administered Address of OUI 080069
48006A	AT&T BELL LABORATORIES—Token Ring Locally Administered Address of OUI 08006A
48007C	Vitalink Communications Corp.—Token Ring Locally Administered Address of OUI 08007C
480089	Kinetics—Token Ring Locally Administered Address of OUI 080089
48008B	Pyramid Technology Corp.—Token Ring Locally Administered Address of OUI 08008B
48008D	Xyvision Inc.—Token Ring Locally Administered Address of OUI 08008D
50005A	IBM—Token Ring Group Address of OUI 10005A
800010	AT&T Bell Labs
AA0000	Digital Equipment Corp.—obsolete
AA0001	Digital Equipment Corp.—obsolete
AA0002	Digital Equipment Corp.—obsolete
AA0003	Digital Equipment Corp. (DEC)—Global physical Address for some DEC machines
AA0004	Digital Equipment Corp. (DEC)—Local logical Address for DECNET systems
EA0003	Digital Equipment Corp. (DEC)—Token Ring Group Address of OUI AA0003
EA0004	Digital Equipment Corp. (DEC)—Token Ring Group Address of OUI AA0004

OUI by Vendor

OUI	Vendor/Comment
080045	(maybe Xylogics, but they claim not to know this number)—IEEE assigned to CONCURRENT COMPUTER CORP.
080024	10NET Communications/DCA
005029	1394 Printer Working Group
00D09E	2wire, Inc.
006041	3A International, Inc.
42608C	3Com Corp.—Token Ring Locally Administered Address of OUI 02608C
006008	3Com Corporation
006097	3Com Corporation
00A024	3com Corporation
02C08C	3com Corporation
0040B4	3Com Corporation (formerly Nextcom K.K.)
00608C	3Com Corporation (1990 onwards)
0020AF	3Com Corporation (newer Token Ring Adapters w/ TI-C60 and 3Com MAC)
00104B	3Com Corporation—3C905-TX PCI
005004	3Com Corporation—3C90X seen under this OUI—Not an IEEE assigned OUI
00105A	3Com Corporation—Fast Etherlink XL in a Gateway 2000
02608C	3Com Corporation—IBM PC; Imagen; Valid; Cisco; Macintosh—IEEE assigned to 3COM CORPORATION—on Token Ring old 4 Mb/s adapters
00D096	3com Europe Ltd.
009004	3Com Europe, formerly BICC
005099	3Com Europe, Ltd.
00603D	3CX (formerly INTEGRATED MICRO SOLUTIONS, INC)
00A0B5	3H Technology
080021	3M Company
0090A0	8x8 Inc.
004008	A Plus Info Corporation
00E079	A.T.N.R.
080059	A/S Mycron
001043	A2 Corporation
009057	Aanetcom, Inc.
000023	ABB Industrial Systems AB
00904F	ABB Power T&D Company, Inc.
000076	Abekas Video System
00508D	Abit Computer Corporation
00C0EF	Abit Corporation
00501B	ABL Canada, Inc.
006090	Able Communications, Inc.
08001D	Able Communications, Inc.
00E0F0	Abler Technology, Inc.
00C0F8	About Computing Inc.
0030BB	AC&T System Co., Ltd.
00A095	Acacia Networks, Inc.
08006B	Accel Technologies Inc.
00502D	Accel, Inc.
001026	Accelerated Networks, Inc.
080078	Accell Corporation
008032	Access Co., Ltd.
001033	AccessLAN Communications, Inc.
0060CE	Acclaim Communications
0040E9	Accord Systems, Inc.
0090CA	Accord Video Telecommunications, Ltd.
0000E8	Accton Technology Corporation
0010B5	Accton Technology Corporation
006095	Accu-Time Systems, Inc.
0090E6	Acer Laboratories, Inc.
006067	Acer Netxus Inc.
00A060	Acer Peripherals, Inc.

OUI	Vendor/Comment
0000E2	Acer Technologies Corp.
00E093	Ackfin Networks
00D073	ACN Advanced Communications
0000A4	Acorn Computers Limited
006007	Acres Gaming, Inc.
00D06D	Acrison, Inc.
00D05B	Acroloop Motion Control Systems
0030E1	Acrotron Systems, Inc.
001002	ACTIA
002015	Actis Computer SA
006025	Active Imaging Plc
0030A5	Active Power
009037	Acucomm, Inc.
006033	Acuity Imaging, Inc.
00C0C3	Acuson Computed Sonography
00D0F9	Acute Communications Corp.
0050D5	AD Systems Corp.
0000D1	Adaptec, Inc.—"Nodem" product
008027	Adaptive Systems, Inc.
00200C	Adastra Systems Corp.
0080DF	ADC Codenoll Technology Corp.
00004A	ADC Codenoll Technology Corp.
008051	ADC Fibermux
0050C8	Addonics Communications, Inc.
004033	Addtron Technology Co., Ltd.—sends bit reversed as 0002CC on Token Ring
0002CC	Addtron Technology Corp.sends on Token Ring—bit reversed of assigned OUI 004033—Not an IEEE assigned OUI
00D02D	Ademco
0010EA	Adept Technology
0010A9	Adhoc Technologies
00307C	Adid SA
003064	Adlink Technology, Inc.
00E092	Admtek Incorporated
004018	Adobe Systems, Inc.
000071	Adra Systems Inc.
0050C5	ADS Technologies, Inc.
008057	Adsoft, Ltd.
00901F	Adtec Productions, Inc.
001094	Adtech, Inc.
00A0C8	Adtran Inc.
00C0A0	Advance Micro Research, Inc.
002090	Advanced Compression
00207D	Advanced Computer Applications
480003	ADVANCED COMPUTER COMM. (ACC)—Token Ring Locally Administered Address of OUI 080003
080003	Advanced Computer Communications (ACC)
0060F4	Advanced Computer Solutions
001066	Advanced Control Systems, Inc.
00308C	Advanced Digital Information
009077	Advanced Fibre Communications
005074	Advanced Hi-Tech Corp
0080B0	Advanced Information Technology, Inc.
00C051	Advanced Integration Research
00C00D	Advanced Logic Research Inc
00001A	Advanced Micro Devices (AMD)—Racore Computer Products Inc. seen on Token Ring—bit reversed of assigned OUI 000058
005018	Advanced Multimedia Internet Technology Inc.
00A07A	Advanced Peripherals Technologies, Inc.
009026	Advanced Switching Communications, Inc.
0040DB	Advanced Technical Solutions
00C0FB	Advanced Technology Labs
00202B	Advanced Telecommunications Modules, Ltd. (ATML)
00E095	Advanced-Vision Technolgies Corp.
001071	Advanet Inc.
00E03C	Advansys
00D0E9	Advantage Century
00D0C9	Advantech Co., Ltd.

OUI	Vendor/Comment
0020BF	Aehr Test Systems
004019	Aeon Systems, Inc.
005067	Aerocomm, Inc.
005031	Aeroflex Laboratories, Inc.
00A062	Aes Prodata
00D0D6	Aethra Telecomunicazioni
00405F	AFE Computers Ltd.
00407F	Agema Infrared Systems AB
080066	AGFA—printers, phototypesetters etc.
0020B6	Agile Networks, Inc.
000134	Agilis Corp seen on Token Ring—bit reversed of assigned OUI 00805C—Not an IEEE assigned OUI
00805C	Agilis Corporation—Agilis sends bit reversed as 00013A on Token Ring
0090B3	Agranat Systems
00606B	Aichi Electronics Co., Ltd.
00D020	Aim System, Inc.
0050DF	Airfiber, Inc.
00807A	Aitech Systems Ltd.
005071	Aiwa Co., Ltd.
0020C7	Akai Electric Co., Ltd.
08BBCC	AK-Nord Edv Vertriebsges. MbH
001042	Alacritech
0020E1	Alamar Electronics
00D0A4	Alantro Communications
00D075	Alaris Medical Systems, Inc.
0080C9	Alberta Microelectronic Centre
00C0BE	Alcatel - Sel
0090D0	Alcatel Bell
00809F	Alcatel Business Systems
00A081	Alcatel Data Networks
002060	Alcatel Italia S.P.A.
008039	Alcatel STC Australia
002032	Alcatel Taisel
001046	Alcorn McBride Inc.
00602D	Alerton Technologies, Inc.
00C0CA	Alfa, Inc.
00E03E	Alfatech, Inc.
0040BC	Algorithmics Ltd.
002021	Algorithms Software PVT. Ltd.
00D044	Alidian Networks, Inc.
0000BC	Allen-Bradley Co. Inc.
009002	Allgon Westcom AB
0040BA	Alliant Computer Systems Corp.
002058	Allied Signal Inc.
00A0D2	Allied Telesis International Corporation
009099	Allied Telesis, K.K.
003084	Allied Telesyn Internaional
0000F4	Allied Telesyn International
00008C	Alloy Computer Products, Inc.
0080AF	Allumer Co., Ltd.
003000	Allwell Technology Corp.
009081	Aloha Networks, Inc.
0000B1	Alpha Microsystems Inc.
0090EA	Alpha Technologies, Inc.
009035	Alpha Telecom, Inc.
0050EB	Alpha-Top Corporation
00A079	Alps Electric (Usa), Inc.
00C093	Alta Research Corp.
001032	Alta Technology
002055	Altech Co., Ltd.
0060CF	Alteon Networks, Inc.
0090A4	Altiga Networks
003081	Altos C&C
0000C8	Altos Computer Systems
0020F1	Altos India Limited
00301C	Altvater Airdata Systems GmBh & Co. KG
08007E	Amalgamated Wireless (Aus) Ltd

OUI	Vendor/Comment
005088	Amano Corporation
00D07A	Amaquest Computer Corp.
0060C7	Amati Communications Corp.
00E008	Amazing Controls! Inc.
00309F	Amber Networks
00A086	Amber Wave Systems, Inc.
00107A	Ambicom, Inc.
00D059	Ambit Microsystems Corp.
008083	Amdahl
00D0A5	American Arium
00D03E	American Communication Technoligies International Inc.
0040D9	American Megatrends Inc.
001088	American Networks Inc.
00C0B7	American Power Conversion Corp.
00A043	American Technology Labs, Inc.
00009F	Ameristar Technologies Inc.
40009F	Ameristar Technologies Inc.—Token Ring Locally Administered Address of OUI 00009F
00A04F	Ameritec Corp.
0010F5	Amherst Systems, Inc.
00D041	Amigo Technology Co., Ltd.
002026	Amkly Systems, Inc.
004076	AMP Incorporated
00007C	Ampere Incorporated
00004E	Ampex Corporation
004053	Ampro Computers—Datum (Bancomm Division)—TymServe 2000 seen under this OUI
00603B	Amtec SPA
002030	Analog & Digital Systems
00C0DD	Ancor Communications
0060C5	Ancot Corp.
0050ED	Anda Networks
004011	Andover Controls Corporation
008020	Andrew Coproration Network Products Div—Andrew sends bit reversed as 000104 on Token Ring
000104	Andrew seen on Token Ring—bit reversed of assigned OUI 008020—Not an IEEE assigned OUI
00E082	Anerma
00A051	Angia Communications. Inc.
004005	ANI Communications Inc.—TRENDware International Inc.; Linksys; Simple Net all seen under this OUI
0000FE	Annapolis Micro Systems
000091	Anritsu Corporation
004090	Ansel Communications—PC NE2000 compatible twisted-pair ethernet cards
00A0FA	Ant Nachrichtentechnik GmbH
00A080	Antares Microsystems
0010F2	Antec
008004	Antlow Computers, Ltd.
00804E	Apex Computer Company
0020E5	Apex Data, Inc.
00A007	Apexx Technology, Inc.
0020A5	API Engineering—Newer Technology seen under this OUI—IEEE assigned OUI 00507E
003058	API Motion
009003	Aplio
0060EE	Apollo
08001E	Apollo (now owned by H-P)
48001E	Apollo Computer Inc. (now owned by H-P)—Token Ring Locally Administered Address of OUI 08001E
00304C	Appian Communications, Inc.
00A040	Apple Computer
000502	Apple Computer (PCI Mac)
080007	Apple Computer Inc.—sends bit reversed as 1000E0 on Token Ring
0080D6	Apple Computer— Mac Portable(?)—IEEE assigned to NUVOTECH, INC.
003065	Apple Computer, Inc.
0050E4	Apple Computer, Inc.
1000E0	Apple Computer—seen on Token Ring—bit reversed of assigned OUI 08007—Not an IEEE assigned OUI

OUI	Vendor/Comment
00A091	Applicom International
000040	Applicon, Inc.
00D027	Applied Automation, Inc.
0040C2	Applied Computing Devices
00A023	Applied Creative Technology, Inc.
00600C	Applied Data Systems, Inc.
009041	Applied Digital Access
000019	Applied Dynamics International
004072	Applied Innovation Inc.
0010E6	Applied Intelligent Systems, Inc.
0000F6	Applied Microsystems Corp.—Madge Networks Ltd. sends on Token Ring—bit revereset of assigned OUI 00006F
00A036	Applied Network Technology
00906B	Applied Resources, Inc.
002097	Applied Signal Technology
0000CA	Applitek then LANcity Cable Modems (now owned by BayNetworks owned by Nortel)
000049	Apricot Computers Ltd.
0000ED	April
00803B	APT Communications, Inc.
00C0FE	Aptec Computer Systems, Inc.
0010BC	Aptis Communications, Inc.
00E028	Aptix Corporation
0000E6	Aptor Produits De Comm Indust
0060E3	Arbin Instruments
0080B9	Arche Technoligies Inc.
00C097	Archipel SA
00C020	Arco Electronic, Control Ltd.
008066	Arcom Control Systems, Ltd.
00E0D5	Arcxel Technologies, Inc.
00E0B5	Ardent Communications Corp.
00D08F	Ardent Technologies, Inc.
00C071	Areanex Communications, Inc.
00606C	Arescom
009074	Argon Networks, Inc.
0010D7	Argosy Research Inc.—EN 220 Fast Ethernet PCMCIA
0080BE	Aries Research
0000C7	Arix Corporation
00C00C	ARK PC Technology, Inc.—IEEE assigned to RELIA TECHNOLGIES
00A0CB	Ark Telecommunications, Inc.
00E0F2	Arlotto Comnet, Inc.
0010C0	Arma, Inc.
00E0CE	ARN
0040E7	Arnos Instruments & Computer
00C0EA	Array Technology Ltd.
0010E2	Arraycomm, Inc.
001058	Arrowpoint Communications, Inc.
00807B	Artel Communications Corp.
0090D2	Artel Video Systems
00006E	Artisoft, Inc.
005044	Asaca Corporation
000094	Asante Technologies—MAC
00D052	Ascend Communications, Inc.
00C07B	Ascend Communications, Inc.—ISDN bridges
004015	Ascom Infrasys AG
00C0FC	ASDG Incorporated—IEEE assigned to ELASTIC REALITY, INC.
00508F	Asita Technologies Int'l Ltd.
009096	Askey Computer Corp.
004092	ASP Computer Products, Inc.
0090AD	Aspect Electronics, Inc.
0090F3	Aspect Telecommunications Corp
00E0DA	Assured Access
009044	Assured Digital, Inc.
00E02C	AST - built into 5166M PC motherboard (win95 id's as Intel)
00401C	AST Research, Inc.—Pentium
00A08B	Aston Electronic Designs Ltd.
00D0FE	Astral Point Communications, Inc.

OUI	Vendor/Comment
00309A	Astro Terra Corp.
00A0CE	Astrocom Corporation
00609B	Astro-Med, Inc.
00E018	Asustek Computer Inc.—Intel 82558-based Integrated Fast Ethernet for WIM
00D0CE	Asyst Electronic
0060D3	AT&T
08006A	AT&T Bell Laboratories
800010	AT&T Bell Laboratories
48006A	AT&T BELL LABORATORIES—Token Ring Locally Administered Address of OUI 08006A
800010	AT&T Bell Labs
080010	AT&T—probably typo of 800010—Not an IEEE assigned OUI
09006A	AT&T—used in smart hub—Not an IEEE assigned OUI
00D0CD	Atan Technology Inc.
000036	Atari Corporation
009084	Atech System
005066	Atecom GmBh Advanced Telecommunication Modules
001074	Aten International Co., Ltd.
0000DA	Atex
004080	Athenix Corporation
005084	ATL Products
00903C	Atlantic Network Systems
00800E	Atlantix Corporation
00E075	Atlas Computer Equipment, Inc.
0050D6	Atlas Copco Tools AB
004071	ATM Computer GmBh
00A03E	ATM Forum
0010B4	Atmosphere Networks
0040E0	Atomwide Ltd.
0060E9	Atop Technologies, Inc.
00E0FD	A-Trend Technology Co., Ltd.
001086	Atto Technology, Inc.
00908F	Audiocodes Ltd.
00107D	Aurora Communications, Ltd.
00607F	Aurora Technologies, Inc.
00003C	Auspex Systems Inc.
006036	Austrian Research Center
00A06E	Austron, Inc.
00207C	Autec GmBh
00204B	Autocomputer Co., Ltd.
00A0F6	Autogas Systems, Inc.
00E0C9	Automatedlogic Corporation
080064	Autophon AG
00002C	Autotote Limited—Network Resources Corporation (NRC) - MultiGate Hub1+, Hub2, etc.sends bit reversed of assigned OUI 000034)
005028	Aval Communications
0080B3	Aval Data Corporation
00E00E	Avalon Imaging Systems, Inc.
0080FC	Avatar Corporation
0090E3	Avex Electronics Inc.
0090B2	Avici Systems Inc.
00A07E	Avid Technology, Inc.
00E01F	Avidia Systems, Inc.
003008	Avio Digital, Inc.
00D0E1	Avionitek Israel Inc.
0020DD	AWA Ltd.
00109E	Aware, Inc.
00A034	Axel
0060E0	Axiom Technology Co., Ltd.
00408C	Axis Communications AB
00C0D4	Axon Networks Inc.
0030AA	Axus Microsystems, Inc.
00A0DD	Azonix Corporation
00903F	Aztec Radiomedia
00300A	Aztech Systems Ltd.
0080FE	Azure Technologies, Inc.
0020E7	B&W Nuclear Service Company
00A012	B.A.T.M. Advanced Technologies
00A093	B/E Aerospace
00D0D7	B2c2, Inc.
00107E	Bachmann Electronic GmBh
009071	Badger Technology, Inc.
00D0C2	Balthazar Technology Ab
00E06C	Baltimore Technologies, Ltd.
08004A	Banyan Systems Inc.
00504B	Barco N.V. BCS
080016	Barrister Info Sys Corp
00C0A9	Barron McCann Ltd
008095	Basic Merton Handelsges.M.B.H.
004073	Bass Associates
008031	Basys, Corp.
080033	Bausch & Lomb
009088	Baxall Security Ltd.
00E07B	Bay Networks
048845	Bay Networks—token ring line card—Bay says "not ours!"—IEEE assigned to TRIUMPH-ADLER AG
00C048	Bay Technical Associates
00908A	Bayly Communications, Inc.
00E0C5	Bcom Electronics Inc.
00E090	Beckman Lab. Automation Div.
00D035	Behavior Tech. Computer Corp.
0030BD	Belkin Components
0000DF	Bell & Howell Pub Sys Div
00801A	Bell Atlantic
080012	Bell Atlantic Integrated Syst.
00001C	Bell Technologies (formerly JDR Microdevices—generic, NE2000 drivers)
00E046	Bently Nevada Corp.
0090B9	Beran Instruments Ltd.
00E0BA	Berghof Automationstechnik GmBh
00E078	Berkeley Networks
006065	Bernecker & Rainer Industrie-Elektronic GmBh
00E0CA	Best Data Products
00201F	Best Power Technology, Inc.
00509C	Beta Research
48004E	BICC (now owned by 3Com)—Token Ring Locally Administered Address of OUI 08004E
0080C3	BICC Information Systems & SVC
08004E	BICC—(now owned by 3Com, so may appear on 3com equipment as well)—IEEE assigned to 3COM EUROPE
001060	Billionton—Novell NE200 Compatible
00A01A	Binar Elektronik Ab
0090D4	Bindview Development Corp.
00A0F9	Bintec Communications GmBh
00604B	Biodata GmBh
00903D	Biopac Systems, Inc.
003067	Biostar Microtech Int'l Corp.
00D04F	Bitronics, Inc.
00D0B0	Bitswitch Ltd.
0040C1	Bizerba-Werke Wilheim Kraut
0010D1	Blazenet, Inc.
002036	BMC Software
00C06D	Boca Research, Inc.
006019	Boehringer Mannheim Corp.
005093	Boeing
0040BE	Boeing Defense & Space
003099	Boenig und Kallenbach OHG
480008	BOLT BERANEK AND NEWMAN INC (BBN).—Token Ring Locally Administered Address of OUI 080008
080008	Bolt Beranek and Newman, Inc. (BBN)
002099	Bon Electric Co., Ltd.
0030EC	Borgardt
0050B7	Boser Technology Co., Ltd.
00A0FE	Boston Technology, Inc.
00C0FF	Box Hill Systems Corporation

OUI	Vendor/Comment
0090F1	Boxhill Systems Corporation
00A02E	Brand Communications, Ltd.
00505F	Brand Innovators
0010E7	Breezecom
0020D6	Breezecom
080002	Bridge Communications (now owned by 3Com)
480002	Bridge Communications, Inc—old 4Mb/s (now owned by 3Com)—Token Ring Locally Administered Address of OUI 080002
005015	Bright Star Engineering
001041	Bristol Babcock, Inc.
0000DB	British Telecommunications PLC
08002C	Britton Lee Inc.
005057	Broadband Access Systems
0010FC	Broadband Networks, Inc.
00A00F	Broadband Technologies
0080CE	Broadcast Television Systems
001018	Broadcom Corporation
00D072	Broadlogic
006069	Brocade Communications Systems
0050B2	Brodel Automation
00A08A	Brooktrout Technology, Inc
008077	Brother Industries, Ltd.
00209E	Brown's Operating System Services, Ltd.
0080DA	Bruel & Kjaer
0000AD	Bruker Instruments Inc.
003049	Bryant Technology, Ltd.
00A05F	BTG Engineering Bv
0080B8	Bug, Incorporated
000062	BULL HN Information Systems—Honeywell
00C052	Burr-Brown
00304E	Bustec Production Ltd.
02E6D3	Bus-Tech, Inc. (BTI)—IBM Mainframes—IEEE has no OUI for Bus-Tech, Inc. Has a BusTek Corp w/ OUI 004034—IEEE assigned to NIXDORF COMPUTER CORPORATION
004034	Bustek Corporation
0080FB	BVM Limited
00E0B9	Byas Systems
0040B0	Bytex Corporation, Engineering
00808F	C. Itoh Electronics, Inc.
0040E6	C.A.E.N.
002008	Cable & Computer Technology
004074	Cable and Wireless
001000	Cable Television Laboraties, Inc.
0060BB	Cabletron - Netlink, Inc.
00E063	Cabletron - Yago Systems, Inc.
0020D4	Cabletron - Zeittnet Inc.
00E03A	Cabletron Systems, Inc.
00001D	Cabletron Systems, Inc.—Cabletron sends bit reversed as 0000B8 on Token Ring
00C09E	Cache Computers, Inc.
008005	Cactus Computer Inc.
0030C5	Cadence Design Systems
08003D	Cadnetix Corporations
0080D5	Cadre Technologies
005064	CAE Electronics
00C0E2	Calcomp, Inc.
00C069	California Microwave Inc
0010D8	Calista
00502E	Cambex Corporation
001045	Cambrian Systems Corporation
0040F4	Cameo Communications, Inc.
000013	Camex
00D02C	Campbell Scientific, Inc.
00D067	Campio Communications
0000FF	Camtec Electronics (UK) Ltd.
002012	Camtronics Medical Systems
00C0FA	Canary Communications Inc

OUI	Vendor/Comment
00402A	Canoga-Perkins
000085	Canon Inc.
00C085	Canon—IEEE assigned to ELECTRONICS FOR IMAGING, INC.
002011	Canopus Co., Ltd.
00805D	Canstar
0050A5	Capitol Business Systems, Ltd.
00A030	Captor Nv/Sa
001070	Caradon Trend Ltd.
029D8E	Cardiac Recorders, Inc.
009D8E	Cardiac Recorders, Inc.
0090DE	Cardkey Systems, Inc.
00A076	Cardware Lab, Inc.
00D0F4	Carinthian Tech Institute
00200D	Carl Zeiss
0050A1	Carlo Gavazzi, Inc.
08007F	Carnegie-Mellon University (CMU)
0020F8	Carrera Computers, Inc.
00E097	Carrier Access Corporation
006029	Cary Peripherals Inc.
0060D1	Cascade Communications
0040FB	Cascade Communications Corp.
080074	Casio Computer Co. Ltd.
000044	Castelle Corporation—Castelle sends bit reversed as 000022 on Token Ring
009040	Castle Networks, Inc.
00205E	Castle Rock, Inc.
003054	Castlenet Technology, Inc.
000089	Cayman Systems Inc.-Gatorbox
00806C	Cegelec Projects Ltd.
00C0CE	CEI Systems & Engineering Pte
00C0F6	Celan Technology Inc.
00605A	Celcore, Inc.
00E0EC	Celestica Inc.
00908B	Cell Computing, Inc.
00A078	Cellaccess Technology, Inc.
00E053	Cellport Labs, Inc.
0020BA	Center For High Performance Computing of WPI
009015	Centigram Communications Corp.
00A000	Centillion Networks, bought by Bay Networks now Nortel—Ethernet switch—sends bit reversed as 000500 on Token Ring
000500	Centillion Networks, bought by Bay Networks, bought by Nortel—bit reversed of assigned OUI 005000—Not an IEEE assigned OUI
00A0E7	Central Data Corporation
0090ED	Central System Research Co., Ltd.
008097	Centralp Automatismes
00A022	Centre For Development Of Advanced Computing
00C02C	Centrum Communications, Inc.
00E037	Century Corporation
00806D	Century Systems Corp.
00104E	Ceologic
080030	CERN
0000DE	CETIA—Unigraph seen under this OUI
005070	Chaintech Computer Co., Ltd.
0040BF	Channel Systems Intern'l Inc.
0060B7	Channelmatic, Inc.
005013	Chaparral Technologies, Inc.
0040E8	Charles River Data Systems, Inc.
00D0AA	Chase Communications
0080D4	Chase Research LTD.
003025	Checkout Computer Systems, Ltd.
002092	Chess Engineering B.V.
00208E	Chevin Software Eng. Ltd.
08008F	Chipcom Corp. (now owned by 3Com)
0030E4	Chiyoda System Riken
00002D	Chromatics Inc.
00500E	Chromatis Networks, Inc.
00C098	Chuntex Electronic Co., Ltd.
0040F1	Chuo Electronics Co., Ltd.

OUI	Vendor/Comment
00003A	Chyron Corporation
080038	CII Honeywell Bull
001090	Cimetrics, Inc.
0000B3	CIMLinc Incorporated
4000B3	CIMLinc Incorporated —Token Ring Locally Administered Address of OUI 0000B3
0020EB	Cincinnati Microwave, Inc.
00E0E5	Cinco Networks, Inc.
00C027	Cipher Systems, Inc.
002046	Ciprico, Inc.
002018	CIS Technology Inc.
0090F2	Cisco Systems Inc.—Ethernet Switches and Light Streams
00067C	Cisco Systems, Inc.
0006C1	Cisco Systems, Inc.
00100B	Cisco Systems, Inc.
001014	Cisco Systems, Inc.
001054	Cisco Systems, Inc.
00107B	Cisco Systems, Inc.
0010A6	Cisco Systems, Inc.
0010F6	Cisco Systems, Inc.
0010FF	Cisco Systems, Inc.
003019	Cisco Systems, Inc.
003024	Cisco Systems, Inc.
003040	Cisco Systems, Inc.
0030A3	Cisco Systems, Inc.
00500F	Cisco Systems, Inc.
005014	Cisco Systems, Inc.
00502A	Cisco Systems, Inc.
00503E	Cisco Systems, Inc.
005050	Cisco Systems, Inc.
005053	Cisco Systems, Inc.
005054	Cisco Systems, Inc.
005073	Cisco Systems, Inc.
005080	Cisco Systems, Inc.
0050A2	Cisco Systems, Inc.
0050BD	Cisco Systems, Inc.
0050E2	Cisco Systems, Inc.
00602F	Cisco Systems, Inc.
006047	Cisco Systems, Inc.
00605C	Cisco Systems, Inc.
00900C	Cisco Systems, Inc.
009021	Cisco Systems, Inc.
00905F	Cisco Systems, Inc.
00906D	Cisco Systems, Inc.
00906F	Cisco Systems, Inc.
009086	Cisco Systems, Inc.
009092	Cisco Systems, Inc.
0090A6	Cisco Systems, Inc.
0090AB	Cisco Systems, Inc.
0090B1	Cisco Systems, Inc.
0090BF	Cisco Systems, Inc.
0090D9	Cisco Systems, Inc.
00D058	Cisco Systems, Inc.
00D063	Cisco Systems, Inc.
00D079	Cisco Systems, Inc.
00D090	Cisco Systems, Inc.
00D097	Cisco Systems, Inc.
00D0BA	Cisco Systems, Inc.
00D0BB	Cisco Systems, Inc.
00D0BC	Cisco Systems, Inc.
00D0C0	Cisco Systems, Inc.
00D0D3	Cisco Systems, Inc.
00D0E4	Cisco Systems, Inc.
00D0FF	Cisco Systems, Inc.
00E014	Cisco Systems, Inc.
00E01E	Cisco Systems, Inc.
00E034	Cisco Systems, Inc.
00E04F	Cisco Systems, Inc.
00E08F	Cisco Systems, Inc.
00E0F7	Cisco Systems, Inc.
00E0F9	Cisco Systems, Inc.
00E0FE	Cisco Systems, Inc.
003080	Cisco Systems, Inc.
003085	Cisco Systems, Inc.
003096	Cisco Systems, Inc.
0030B6	Cisco Systems, Inc.
0030F2	Cisco Systems, Inc.
0050A7	Cisco Systems, Inc.
00070D	Cisco Systems, Inc.— 2511 Token Ring—bit reversal of assigned OUI 00E0B0—Not an IEEE assigned OUI
006083	Cisco Systems, Inc.—3620
001007	Cisco Systems, Inc.—Catalyst 1900
00E0A3	Cisco Systems, Inc.—Catalyst 1924
00101F	Cisco Systems, Inc.—Catalyst 2901
00100D	Cisco Systems, Inc.—Catalyst 2924-XL
001029	Cisco Systems, Inc.—Catalyst 5000
00102F	Cisco Systems, Inc.—Cisco 5000
001011	Cisco Systems, Inc.—Cisco 75xx
00000C	Cisco Systems, Inc.—Cisco sends bit reversed as 000030 on Token Ring
00902B	Cisco Systems, Inc.—Ethernet Switches and Light Streams
00E0B0	Cisco Systems, Inc.—Various systems reported
00603E	Cisco Systems, Inc—100Mbps interface
001079	Cisco Systems, Inc—5500 Router
006009	Cisco Systems, Inc—Catalyst 5000 Ethernet switch
00801C	Cisco Systems, Inc—IEEE assigned to NEWPORT SYSTEMS SOLUTIONS
006070	Cisco Systems, Inc—routers (2524 and 4500)
40000C	Cisco Systems, Inc—Token Ring Locally Administered Address of OUI 00000C
00608A	Citadel Computer
00A020	Citicorp/TTI
006016	Clariion
00109D	Clarinet Systems, Inc.
00606F	Clarion Corporation Of America
009079	Clearone Inc.
0080A9	Clearpoint Research
0090F5	Clevo Co.
0090A7	Clientec Corporation
0040A5	Clinicomp Intl.
0010CC	CLP Computer Logistik
00007E	Clustrix Corporation—NetFRAME multiprocessor network servers seen under this OUI—No IEEE OUI assigned to NetFrame
00208D	CMD Technology
001051	Cmicro Corporation
00306B	CMOS Systems, Inc.
001093	CMS Computers, Ltd.
0050BB	CMS Technologies
0080AD	CNet Technology, Inc.—Used by Telebit (among others)
0030E7	CNF Mobile Solutions, Inc.
0010B2	Coactive Aesthetics
00A070	Coastcom
0010E0	Cobalt Microserver, Inc.
0010FD	Cocom A/S
08003E	Codex-(now owned by Motorola)—VME bus processor modules
0040B1	Codonics Inc.
003023	Cogent Computer Systems, Inc.
000092	Cogent Data Technologies—Unisys seen under this OUI
00D024	Cognex
009063	Coherent Communications
00401F	Colorgraph Ltd
00A073	Com21, Inc.
0040F9	Combinet
00909C	Combox, Ltd.
00D07B	Comcam International Ltd.
00E0A5	Comcore Semiconductor, Inc.
080067	Comdesign

OUI	Vendor/Comment
480067	Comdesign—Token Ring Locally Administered Address of OUI 080067
00C0CD	Comelta, S.A.
004060	Comendec Ltd.
000055	Commissariat a l'Energie Atom.—AT&T seen under this OUI
008010	Commodore International
00D02E	Communication Automation Corp.
00902A	Communication Devices, Inc.
00CF1C	Communication Machinery Corp.
02CF1C	Communication Machinery Corp.
0010C8	Communications Electronics
0050AD	Communique Wireless Corp.
00A09F	Commvision Corp.
001068	Comos Telecom
004024	COMPAC Inc.
00A0A0	Compact Data, Ltd.
00A053	Compact Devices, Inc.
0001FA	Compaq (PageMarq printers)—Compaq sends bit reversed of assigned OUI 00805F on Token Ring—Not an IEEE assigned OUI
0008C7	Compaq Computer Corp.
0010E3	Compaq Computer Corporation
00805F	Compaq Computer Corporation—Compaq sends bit reversed as 0001FA on Token Ring
0000A5	Compatible Systems Corporation
0080EB	Compcontrol B.V
004029	Compex
008048	Compex Incorporated, used by Commodore and DEC at least
008006	Compuadd Corporation
000053	Compucorp
00E084	Compulite R&D
0060E4	Compuserve, Inc.
0000CB	Compu-Shack Electronic GmBh
008070	Computadoras Micron
006004	Computadores Modulares SA
00D0C5	Computational Systems, Inc.
00E0D6	Computer & Communication
00104C	Computer Access Technology
00C0AD	Computer Communication Systems—IEEE asssigned to MARBEN COMMUNICATION SYSTEMS
00006A	Computer Consoles Inc. Computer Products Division
008086	Computer Generation Inc.
00205A	Computer Identics
000088	Computer Network Tech. Corp.
00C0C4	Computer Operational
0080D0	Computer Peripherals Inc.
00B0D0	Computer Products International—Not an IEEE assigned OUI
00A03F	Computer Society Microproces'r & Micropro'r Stds Committee
00C078	Computer Systems Engineering
001028	Computer Technica, Inc.
001053	Computer Technology Corp.
0040B6	Computerm Corporation
080001	Computervision Corporation
00A00B	Computex Co., Ltd.
008069	Computone Systems
00E06D	Compuware Corporation
00A094	Comsat Corporation
004014	Comsoft GmBh
006043	Comsoft Systems, Inc.
00C0B3	Comstat Datacomm Corporation
00E01A	Comtec Systems. Co., Ltd.
0020B1	Comtech Research Inc.
0030D4	Comtier
00C0D1	Comtree Technology Corporation (EFA also reported)—IEEE assigned to COMTREE TECHNOLOGY CORPORATION
0030DA	Comtrend Co.
00C04E	Comtrol Corporation
002049	Comtron, Inc.
000069	Concord Communications, Inc.

OUI	Vendor/Comment
00904A	Concur System Technologies
00409E	Concurrent Technologies Ltd.
00805B	Condor Systems, Inc.
0030CD	Conexant Systems, Inc.
00608B	Confertech International
00300C	Congruency, Ltd.
000024	Connect AS (now owned by Olicom)
00D053	Connected Systems
004054	Connection Machines Services (formerly Thinking Machines Corporation)
00A028	Conner Peripherals
0060FC	Conservation Through Innovation Ltd.
00804C	Contec Co., Ltd.
0050DB	Contemporary Control Systems, Inc.
0060A3	Continuum Technology Corp.
480025	Control Data Corp. (CDC)—Token Ring Locally Administered Address of OUI 080025
080025	Control Data Corporation—CDC
00E080	Control Resources Corporation
0030C6	Control Solutions, Inc.
00C0CB	Control Technology Inc.
002025	Control Technology Inc. (Industrial Controls and Network Interfaces)
00E059	Controlled Environments, Ltd.
0060C9	Controlnet, Inc.
006054	Controlware GmBh
00D010	Convergent Networks, Inc.
00A0D9	Convex Computer Corporation
00D0F0	Convision Technology GmBh
0000AC	Conware Computer Consulting
00406C	Copernique
006058	Copper Mountain Communications, Inc.
0090FD	Coppercom, Inc.
0080E3	Coral Network Corporation (bought by Bay Networks, now Nortel)
00A0B7	Cordant, Inc.
00506E	Corder Engineering Corporation
001057	Corel Computer Corporation
0060B8	Corelis Inc.
006055	Cornell University
00101B	Cornet Technology, Inc
00406E	Corollary, Inc.
00C01A	Corometrics Medical Systems
00C0B5	Corporate Network Systems Inc
08004D	Corvus Systems Inc.
00D07D	Cosine Communications
005026	Cosystems, Inc.
00A029	Coulter Corporation
0020C3	Counter Solutions Ltd.
08005E	Counterpoint Computer Inc.
0010B7	Coyote Technologies, LLC
003015	CP Clare Corp.
003078	Cqos, Inc.
000080	Cray Communications A/S (formerly Dowty Network Services)—Harris (3M) and DATACO in Denmark seen under this OUI
00006D	Cray Communications Inc.—Case seen under this OUI
0040A6	Cray Research Inc.
00007D	Cray Research Superservers, Inc (Also Harris (3M) (old))
00A018	Creative Controllers, Inc.
001075	Creative Design Solutions, Inc.
0080A2	Creative Electronic Systems
00C04B	Creative Microsystems
00400B	Crescendo (now owned by Cisco Systems, Inc.)
00D0B6	Crescent Networks, Inc.
00107F	Crestron Electronics, Inc.
00002B	Crisp Automation, Inc.
00A0C4	Cristie Electronics Ltd.
080004	Cromemco Incorporated
080081	Crosfield Electronics—IEEE assigned to ASTECH INC.
00308E	Cross Match Technologies, Inc.

OUI	Vendor/Comment
000098	CrossComm Corporation
00E002	Crossroads Systems, Inc.
00001F	Cryptall LP
009029	Crypto AG
0090D6	Crystal Group
006011	Crystal Semiconductor Corp.
00A05D	CS Computer Systeme GmBh
00005D	CS Telecom (formerely RCE)
0010C3	CSI-Control Systems
00A014	Csir
00800B	CSK Corporation
00E041	CSPI
000038	CSS LABS
00D01F	Ctam Pty. Ltd.
0010EE	CTI Products, Inc.
005041	CTX Opto Electronic Corp.
002094	Cubix Corporation
0080E8	Cumulus Corporatiion
00D0AF	Cutler-Hammer, Inc.
00408E	CXR—IEEE assigned to Digilog, Inc.
0020F7	Cyberdata
008065	Cybergraphic Systems Pty Ltd.
003068	Cybernetics Tech. Co., Ltd.
0090A2	Cybertan Technology, Inc.
00E086	Cybex Computer Products
00602E	Cyclades Corporation
00604E	Cycle Computer Corporation (Sun Motherboard Replacements)
00804D	Cyclone Microsystems, Inc.
08004F	Cygnet Systems
00A035	Cylink Corporation
00A050	Cypress Semiconductor
001037	Cyq've Technology Co., Ltd.
003090	Cyra Technologies, Inc.
00808B	Dacoll Limited
009053	Daewoo Electronics Co., Ltd.
00C077	Daewoo Telecom Ltd
00804F	Daikin Industries, Ltd.
005038	Dain Telecom Co., Ltd.
00D019	Dainippon Screen Corporate
080050	Daisy Systems Corp.
00C034	Dale Computer Corporation
00D033	Dalian Daxian Network
006035	Dallas Semiconductor, Inc.
00007A	Dana Computer Inc.—Ardent Communications Corp. seen under this OUI
00A005	Daniel Instruments, Ltd.
080075	Dansk Data Electronik
0020E9	Dantel
0000D7	Dartmouth College (NED Router)
00D0CB	Dasan Co., Ltd.
0000AE	Dassault Electronique
00902C	Data & Control Equipment Ltd.
0010BB	Data & Information Technology
00902D	Data Electronics (Aust.) Pty, Ltd.
08001B	Data General
100058	Data General—seen on Token Ring—bit reversed of assigned OUI 08001A—Not an IEEE assigned OUI
1000D8	Data General—seen on Token Ring—bit reversed of assigned OUI 08001B—Not an IEEE assigned OUI
00E0E9	Data Labs, Inc.
00C084	Data Link Corp Ltd.
002041	Data Net
00206D	Data Race, Inc.
0070B3	Data Recall Ltd.
0270B3	Data Recall Ltd.
008038	Data Research & Applications
0010C7	Data Transmission Network
0000B5	Datability Software Systems Inc.—Terminal Servers

OUI	Vendor/Comment
0040A9	Datacom Inc.
00400F	Datacom Technologies
0030D9	Datacore Software Corp.
00809D	Datacraft Manufactur'g Pty Ltd
00C058	Dataexpert Corp.
0030FF	Datafab Systems, Inc.
0060F7	Datafusion Systems
00008A	Datahouse Information Systems
00C042	Datalux Corp.
008080	Datamedia Corporation
002042	Datametrics Corp.
00A084	Dataplex Pty. Ltd.
000015	Datapoint Corporation—Stratus Corp seen on Token Ring—bit reversed of assigned OUI 0000A8
00C025	Dataproducts Corporation
00A037	Datascope Corporation
004061	Datatech Enterprises Co., Ltd.
0020E8	Datatrek Corporation
00E0DE	Datax NV
00E0D3	Datentechnik GmBh
004097	Datex Division of Instrumentarium Corp.
00503A	Datong Electronics Ltd.
00809E	DATUS GmBh
00C0EC	Dauphin Technology
00606E	Davicom Semiconductor, Inc.
080044	DAVID Systems, Inc. (DSI)
00C053	Davox Corporation
00A07B	Dawn Computer Incorporation
006060	Dawning Technologies, Inc.
008019	Dayna Communications—"Etherprint" product
00202E	Daystar Digital
0010EF	DB Networks, Inc.
00004D	DCI Corporation
0060C6	DCS AG
005090	DCTRI
00905A	Dearborn Group, Inc.
00E0F6	Decision Europe
00D0D9	Dedicated Microcomputers Group, Ltd.
00904E	DELEM BV
00C04F	Dell Computer Corporation
00909E	Delphi Engineering Group
0050A0	Delta Computer Systems, Inc.
0040AE	Delta Controls, Inc.
0030AB	Delta Networks, Inc.
00D0AB	Deltakabel Telecom CV
0000CC	Densan Co., Ltd.
0020DC	Densitron Taiwan Ltd.
00C04C	Department Of Foreign Affairs & Trade
00C031	Design Research Systems, Inc.
00A08F	Desknet Systems, Inc.
00E040	Deskstation Technology, Inc.
00204F	Deutsche Aerospace AG
0000D0	Develcon Electronics, Ltd.
00E076	Development Concepts, Inc.
009046	Dexdyne, Ltd.
00E070	DH Technology
000020	DIAB (Data Intdustrier AB)
001059	Diablo Research Co. LLC
00A0E6	Dialogic Corporation
0060F9	Diamond Lane Communications
009000	Diamond Multimedia
00103A	Diamond Network Tech
0000F5	Diamond Sales Limited
00E0F8	Diana Control AB
00C001	Diatek Patient Managment
00E00A	Diba, Inc.
00C0A5	Dickens Data Systems

OUI	Vendor/Comment
00D0EE	Dictaphone Corporation
0040DF	Digalog Systems, Inc.
0050CD	Digianswer A/S
00409D	DigiBoard—Ethernet-ISDN bridges
00A0A2	Digicom S.P.A.
00E0EB	Digicom Systems, Incorporated
00909F	Digi-Data Corporation
00D062	Digigram
0020AA	Digimedia Vision Ltd.
0050D3	Digital Audio Processing Pty. Ltd.
00203A	Digital Biometrics Inc.
00024C	Digital Communications Associates (DCA) (sold to Microdyne and Attachmate) seen on Token Ring—bit reversed of assigned OUI 004032—Not an IEEE assigned OUI
100024	Digital Communications Associates (DCA) (sold to Microdyne and Attachmate)—seen on Token Ring—Not an IEEE assigned OUI
004032	Digital Communications Associates (DCA) (sold to Microdyne and Attachmate)—sends bit reversed as 00024C on Token Ring
080041	Digital Communications Associates (DCA)—IEEE assigned to RACAL-MILGO INFORMATION SYS..
00901B	Digital Controls
003012	Digital Engineering Ltd.
0000F8	Digital Equipment Corp.
0001FE	Digital Equipment Corp.
00606D	Digital Equipment Corp.
001064	Digital Equipment Corp.
AA0003	Digital Equipment Corp. (DEC)—Global physical Address for some DEC machines
0040D0	Digital Equipment Corp. (DEC)—IEEE assigned to Mitac International Corp.
AA0004	Digital Equipment Corp. (DEC)—Local logical Address for DECNET systems
EA0003	Digital Equipment Corp. (DEC)—Token Ring Group Address of OUI AA0003
EA0004	Digital Equipment Corp. (DEC)—Token Ring Group Address of OUI AA0004
48002B	Digital Equipment Corp. (DEC)—Token Ring Locally Administered Address of OUI 08002B
AA0000	Digital Equipment Corp.—obsolete
AA0001	Digital Equipment Corp.—obsolete
AA0002	Digital Equipment Corp.—obsolete
08002B	Digital Equipment Corporation
0010FE	Digital Equipment Corporation
0090B7	Digital Lightwave, Inc.
00A0C0	Digital Link Corp.
00106A	Digital Microwave Corporation
0020CA	Digital Ocean
0040AF	Digital Products, Inc. (DPI).—sends bit reversed as 0002F5 on Token Ring
0002F5	Digital Products, Inc.—Jet Print 100—bit reversed of assigned OUI 0040AF—Not an IEEE assigned OUI
004854	Digital SemiConductor—21143/2 based 10/100 seen under this OUI—Not an IEEE assigned OUI—No IEEE OUI assigned to Digital SemiConductor
0020CC	Digital Services, Ltd.
008011	Digital Systems Int'l. Inc.
00C041	Digital Transmission Systems
006084	Digital Video
003066	Digital Wireless Corporation
00907C	Digitalcast, Inc.
003059	Digital-Logic AG
009091	Digitalscape, Inc
00A049	Digitech Industries, Inc.
00505E	Digitek Micrologic S.A.
00305D	Digitra Systems, Inc.
00E0AB	Dimat S.A.
00E0EF	Dionex
0090E2	Distributed Processing Technology
00C09D	Distributed Systems Int'l, Inc.
00E025	Dit Co., Ltd.
00D04D	Div Of Research & Statistics
002013	Diversified Technology, Inc.
0020A3	Divicom Inc.

OUI	Vendor/Comment
003014	Divio, Inc.
0050BA	D-Link
0080C8	D-Link-(also Solectek Pocket Adapters, and LinkSys PCMCIA)—IEEE assigned to D-LINK SYSTEMS, INC.
008007	Dlog NC-Systeme
00A06B	DMS Dorsch Mikrosystem GmbH
0080C4	Document Technologies, Inc.
00C066	Docupoint, Inc.
00D046	Dolby Laboratories, Inc.
009007	Domex Technology Corp.
005055	Doms A/S
00D0E0	Dooin Electronics Co.
00D0B5	Dotcom
0020A1	Dovatron Products Division
0000B7	Dove Computer Corp.—Fastnet
001081	DPS, Inc.
000056	Dr. B. Struck
00A0CD	Dr. Johannes Heidenhain GmbH
000063	Dr. Ing. Seufert GmBh—Hewlett-Packard LanProbe seen under this OUI
00105D	Drager, Business Unit Monitoring
004098	Dressler GmBh & Co.
00D0B3	DRS Flight Safety And Communications
00C0E0	DSC Communication Corp.
006021	DSC Corporation
002001	DSP Solutions, Inc.
000016	Du Pont Pixel Systems
00C0A3	Dual Enterprises Corporation
0080AB	Dukane Network Integration
00E027	Dux, Inc.
00607A	DVS GmBh
009068	DVT Corp.
008073	DWB Associates
00807F	DY-4 Incorporated
00D089	Dynacolor, Inc.
0090DA	Dynarc, Inc.
002061	Dynatech Communications Inc.
008008	Dynatech Computer Systems
00C037	Dynatem
00E033	E.E.P.D. GmBh
00E067	Eac Automation-Consulting GmbH
00804B	Eagle Technologies Pty. Ltd.
0020C5	Eagle Technology— NE2000
00A0B9	Eagle Technology, Inc.
00E013	Eastern Electronic Co., Ltd.
002054	Eastern Research, Inc.
00E026	Eastman Kodak Co.
00C040	ECCI
009042	ECCS
00D071	Echelon Corp.
001047	Echo Eletric Co. Ltd.
00208F	ECI Telecom Ltd.
0060D7	Ecole Polytechnique Federale De Lausanne (EPFL)
0030B9	Ectel
00D048	Ecton, Inc.
0090C1	Eda Industries
00A04D	Eda Instruments, Inc.
006014	Edec Co., Ltd.
00C024	Eden Sistemas De Computacao SA
00506A	Edeva, Inc.
00E061	Edgepoint Networks, Inc.
0000B4	Edimax Computer Company
0050FC	Edimax Technology Co., Ltd.
0080A0	Edisa Hewlett Packard S/A
00905C	EDMI
0060FA	Educational Technology Resources, Inc.
00E0AD	Ees Technology, Ltd.
0020EA	Efficient Networks, Inc.

OUI	Vendor/Comment
00A03C	EG&G Nuclear Instruments
000033	Egan Machinery Company
006068	Eicon Technology Corporation
005021	EIS International, Inc.
00C088	EKF Elektronik GmBh
003052	Elastic Networks
00D099	Elcard OY
0060D4	Eldat Communication Ltd.
00D0E3	Ele-Chem Engineering Co., Ltd.
0090FE	Elecom Co., Ltd. (Laneed Div. Sumitomo Fudosan)
005068	Electronic Industries Association (EIA)
00A0E9	Electronic Retailing Systems International
00C016	Electronic Theatre Controls
00E0AA	Electrosonic Ltd.
0040DE	Elettronica San Giorgio Elsag S.P.A.
001035	Elitegroup Computer Systems Co., Ltd.
005049	Ellacoya Networks, Inc.
0060D8	Elmic Systems, Inc.
00C05C	Elonex Plc
00A057	Elsa AG
00C0C9	ELSAG Bailey Controls Co.
00005B	Eltec Elektronik GmBh
00504A	Elteco A.S.
00D078	Eltex Of Sweden AB
080085	Elxsi
0040E4	E-M Technology, Inc.
00A038	Email Electronics
0080CF	Embedded Performance Inc.
006048	EMC Corporation
00C044	Emcom Corporation
0080D9	EMK Elektronik
00806A	Empac Research Inc. (ERI)
0060CC	Emtrak, Incorporated
0000C9	Emulex Corp.—Terminal Servers, Print Servers—Proteon seen on Token Ring—bit reversed of assigned OUI 000093
00D0BE	Emutec Inc.
005040	Emware, Inc.
0010DD	Enable Semiconductor, Inc.
001021	Encanto Networks, Inc.
08004C	Encore—IEEE assigned to HYDRA COMPUTER SYSTEMS INC.
00E05F	E-Net, Inc.
00C0F7	Engage Communication, Inc.
00E09B	Engage Networks, Inc.
0050D9	Engetron-Engenharia Eletronica Ind. E Com. Ltda.
00E04A	Enhanced Messaging Systems, Inc
009036	ENS, Inc.
0090CD	Ent-Empresa Nacional
0010B6	Entrata Communications Corp.
0050F8	Entrega Technologies, Inc.
009049	Entridia Corporation
00C0DC	EOS Technologies, Inc.
00A0A1	Epic Data Inc.
00904C	Epigram, Inc.
00D0D2	Epilog Corporation
00E071	Epis Microcomputer
001087	EPL, Ltd.
000097	Epoch Systems
003074	Equiinet Ltd.
00807D	Equinox Systems Inc.
0040A1	Ergo Computing
00209B	Ersat Electronic GmBh
002031	Ertec GmBh
00E0A4	Esaote S.P.A.
0090F6	Escalate Networks, Inc.
008014	Esprit Systems
00A088	Essential Communications
00A01E	Est Corporation
004062	E-Systems, Inc./Garland Div.
00907B	E-Tech, Inc.
00500C	Etek Labs, Inc.
00A0EA	Ethercom Corp.
00E0B3	Etherwan Systems, Inc.
00908C	Etrend Electronics, Inc.
0090D5	Euphonix, Inc.
001076	Eurem GmBh
00E0C7	Eurotech SRL
00D04C	Eurotel Telecom Ltd.
080048	Eurotherm Gauging Systems
00203C	Eurotime AB
080057	Evans & Sutherland
00407E	Evergreen Systems, Inc.
00206C	Evergreen Technology Corp.
00D080	Exabyte Corporation
00201C	Excel, Inc.
080014	Excelan—BBN Butterfly, Masscomp, Silicon Graphics—IEEE assigned to EXCELAN
480014	Excelan—Token Ring Locally Administered Address of OUI 080014
00E0D4	Excellent Computer
00E050	Executone Information Systems, Inc.
00C0A6	Exicom Australia Pty. Ltd
002085	Exide Electronics—3Com Corporation SuperStack II UPS management module seen under this OUI
003097	Exomatic AB
0060F1	Exp Computer, Inc.
080051	Experdata
0080FD	Exsceed Corpration
000216	Extended Systems, Inc (ESI) print server on Token Ring—bit reversed of assigned OUI 004068— Not an IEEE assigned OUI
004068	Extended Systems, Inc—ESI sends bit reversed as 000216 on Token Ring
00407D	Extension Technology Corp.
00E02B	Extreme Networks
00E017	EXXACT GmBh
080013	Exxon
00C03E	FA. Gebr. Heller GmBh
0010CB	Facit K.K.
0080CB	Falco Data Products
0080D7	Fantum Engineering, Inc.
00E0E4	Fanuc Robotics North America, Inc.
00E06E	Far Systems SPA
0000C5	Farallon Computing Inc/Netopia
0050BE	Fast Multimedia AG
00A0EB	Fastcomm Communications Corp.
006005	Feedback Data Ltd.
008055	Fermilab
080040	Ferranti Computer Sys. Limited
00807C	FiberCom, Inc.
00C0E7	Fiberdata AB
0010CF	Fiberlane Communications
0040C6	Fibernet Research, Inc.
0090B6	Fibex Systems
00000D	Fibronics Ltd.
0050B5	Fichet-Bauche
00D09A	Filanet Corporation
080034	Filenet Corporation
00207E	Finecom Co., Ltd.
00C05F	Fine-Pal Company Limited
009065	Finisar Corporation
00A027	Firepower Systems, Inc.
0040CA	First Internat'l Computer, Inc.
006091	First Pacific Networks, Inc.
00A0B1	First Virtual Corporation
0040C3	Fischer and Porter Co.
00A0DB	Fisher & Paykel Production
008074	Fisher Controls

OUI	Vendor/Comment
00D038	Fivemere, Ltd.
00D02A	Flexion Systems
00206F	Flowpoint Corporation
001023	Flowwise Networks, Inc.
0030C3	Flueckiger Elektronik AG
00404E	Fluent, Inc
0090A1	Flying Pig Systems, Ltd.
0060EF	Flytech Technology Co., Ltd.
009076	FMT Aircraft Gate Support Systems AB
00E03D	Focon Electronic Systems A/S
00C079	Fonsys Co Ltd
0010B1	For-A Co., Ltd.
008042	Force Computers
009082	Force Institute
000045	Ford Aerospace & Comm. Corp.—Wellfleet seen on Token Ring—bit reversed of assigned OUI 0000A2
002048	Fore Systems Inc.
00607B	Fore Systems, Inc.
005082	Foresson Corporation
00403C	Forks, Inc.
008033	Formation Inc.
0003E8	Forte Networks—makes the Fluke LAN Meters—bit reversed of assigned OUI 00C017—Not an IEEE assigned OUI
00C0BB	Forval Creative Inc.
00E052	Foundry Networks, Inc.
009038	Fountain Technologies, Inc.
08005C	Four Phase Systems
0060EB	Fourthtrack Systems
00D086	Foveon, Inc.
00C0B8	Fraser's Hill Ltd.
00304A	Fraunhofer Institute IMS
08003F	Fred Koschara Enterprises
00E021	Freegate Corp.
008054	Frontier Technologies Corp.
0060E5	Fuji Automation Co., Ltd.
00401A	Fuji Electric Co., Ltd.
00C02D	Fuji Photo Film Co., Ltd.
080037	Fuji Xerox
00D0F8	Fujian Star Terminal Device Co., Ltd.
004041	Fujikura Ltd.
00A0CA	Fujitsu Denso Ltd.
00000E	Fujitsu Limited
001055	Fujitsu Microelectronics, Inc.
00A077	Fujitsu Nexion, Inc.
00108C	Fujitsu Telecommunications
00E000	Fujitsu, Ltd
0040D1	Fukuda Denshi Co., Ltd.
00E0A9	Funai Electric Co., Ltd.
00405B	Funasset Limited
00C0B9	Funk Software Inc.
00D01D	Furuno Electric Co., Ltd.
002073	Fusion Systems Corporation
005030	Future Plus Systems
00405C	Future Systems, Inc.
00506C	G & L Beijer Electronics AB
00E0E1	G2 Networks, IInc.
0030C8	Gad Line, Ltd.
00E024	Gadzoox Networks
0040D4	Gage Talker Corp.
009073	Gaio Technology
00208C	Galaxy Networks, Inc.
0020A2	Galcom Networking Ltd. (now owned by N-Base)
00E05A	Galea Network Security
00504C	Galil Motion Control, Inc.
00C0AC	Gambit Computer Communications
00205F	Gammadata Computer GmBh
0000F3	Gandalf Data Ltd. - Canada
002006	Garrett Communications, Inc.
00A054	Gassan Salman
0050C0	Gatan, Inc.
00E0B8	Gateway 2000—AMD PCNet in a Gateway 2000
000061	Gateway Communications Inc. (then Megahertz & now 3Com)
000086	Gateway Communications Inc. (then Megahertz & now 3Com)
0020B0	Gateway Devices, Inc.
00D012	Gateworks Corp.
00C0B0	GCC Technologies, Inc.
00506F	G-Connect
0020FA	GDE Systems, Inc.
00A0F4	GE
00E08A	Gec Avery, Ltd.
0050C1	Gemflex Networks, Ltd.
00904B	Gemtek Technology Co., Ltd.
00C064	General Datacomm Ind Inc
080062	General Dynamics
00E0AF	General Dynamics Information Systems
00A021	General Dynamics-Communication Systems
080019	General Electric Corporation
002040	General Instrument Corporation of Delaware GI Communications Division
002066	General Magic, Inc.
00400C	General Micro Systems, Inc.
006027	General Technology, Inc.
005027	Genicom Corporation
002044	Genitech PTY Ltd.
00C0B1	Genius Net Co.
00D08C	Genoa Technology, Inc.
080065	Genrad Inc.
00502B	Genrad Ltd.
00E0BE	Genroco International, Inc.
00C02B	Gerloff Gesellschaft Fur
00A032	Ges Singapore Pte. Ltd.
00C0E5	GESPAC S.A.
0060A9	Gesytec MBh
0050B1	Giddings & Lewis
0090D3	Giesecke & Devrient GmBh
009047	Giga Fast E. Ltd.
0020ED	Giga-Byte Technology Co., Ltd.
00607E	Gigalabs, Inc.
0090FA	Giganet, Inc.
00A0AC	Gilat Satellite Networks, Ltd.
005083	Gilbarco, Inc.
0080DE	GIPSI S.A.
004030	GK Computer
0020B2	GKD Gesellschaft fur Kommunikation und Datentechnik—CSP (Printline Multiconnectivity converter) seen under this OUI
0060B4	Glenayre R&D Inc.
00D092	Glenayre Western Multiplex
00E085	Global Maintech, Inc.
0050F3	Global Net Information Co., Ltd
002014	Global View Co., Ltd.
002088	Global Village Communication
000288	Global Village Communications (PCcard in Mac portable)
00C003	Globalnet Communications
005098	Globaloop, Ltd.
009012	Globespan Semiconductor, Inc.
00A058	Glory, Ltd.
0030E9	GMA Communication Manufact'g Marketing, (1991) Ltd.
0080DD	GMX Inc/GIMIX
00A0A9	GN Nettest (Canada) Inc.
001034	GNP Computers
0030B1	GOC Gesellschaft fur Optische Communication MBH
009085	Golden Enterprises, Inc.
0040BB	Goldstar Cable Co., Ltd.
0040A0	Goldstar Co., Ltd.
00405A	Goldstar Information & Comm.

OUI	Vendor/Comment
0050B6	Good Way Ind. Co., Ltd.
08005D	Gould Inc.
000032	GPT Limited (reassigned from GEC Computers Ltd)
004051	Gracilis, Inc.
006024	Gradient Technologies, Inc.
00C01D	Grand Junction Networks, Inc.—(Cisco Catalyst also reported)
00D0FC	Granite Microsystems
0080DB	Graphon Corporation
001025	Grayhill Inc.
00D0AC	Grayson Wireless
00307D	GRE America, Inc.
00E0E8	Gretacoder Data Systems AG
005003	Gretag Macbeth AG
0000E1	GRID Systems—Hitachi (laptop built-in) seen under this OUI
0060A4	Grinaker System Technologies
0010D3	Grips Electronic GmBh
00C04A	Group 2000 Ag
00D0D5	Grundig AG
00909D	GSE Systems, Inc.
00A07F	Gsm-Syntel, Ltd.
004065	GTE Spacenet
0020EE	Gtech Corporation
00C0A8	GVC Corporation
001072	GVN Technologies, Inc.
00906C	GWT Global Weighing Technologies GmBh
004016	Hadax Electronics, Inc.
00C06E	Haft Technology, Inc.
00E031	Hagiwara Electric Co., Ltd.
00603C	Hagiwara Sys-Com Co., Ltd.
0050E6	Hakusan Corporation
00409B	Hal Computer Systems Inc.
009006	Hamamatsu Photonics K.K.
00A059	Hamilton Hallmark
0050BC	Hammer Storage Solutions
009008	Han A Systems, Inc.
00E057	Han Microtelecom. Co., Ltd.
00900E	Handlink Technologies, Inc.
00D0C1	Harmonic Data Systems, Ltd.
0090F0	Harmonic Lightwaves, Ltd.
0060CA	Harmonic Systems Incorporated
003030	Harmonix Corporation
00402D	Harris Adacom Corporation
00C0F9	Harris And Jeffries, Inc.
0000C3	Harris Corp Computer Sys Div
0000DC	Hayes Microcomputer Products
0000CF	Hayes Microcomputer Products (Canada) Ltd.
00A031	Hazeltine Corporation
000070	HCL Limited Research & Development Unit
00608E	HE Electronics, Technologie & Systemtechnik GmBh
002098	Hectronic AB
003026	Heitel Kommunikationselektronik GmBh
00E015	Heiwa Corporation
00105E	Hekimian Laboratories, Inc.
001069	Helioss Communications, Inc.
000059	Hellige GmBh E-SW
0060EC	Hermary Opto Electronics Inc
009019	Hermes Electronics Co., Ltd.
00E0CC	Hero Systems, Ltd.
00E010	Hess Sb-Automatenbau GmBh
0080F9	Heurikon Corporation
00306E	Hewlett-Packard
0000C6	Hewlett-Packard Intelligent Networks Operation (formerly Eon Systems)
00060D	Hewlett Packard, Inc. newer LaserJets—bit reversed of assigned OUI 0060B0—Not an IEEE assigned OUI
100090	Hewlett Packard—seen on Token Ring—bit reversed of assigned OUI 080009—Not an IEEE assigned OUI
480009	Hewlett Packard—Token Ring Locally Administered Address of OUI 080009

OUI	Vendor/Comment
0030C1	Hewlett-Packard
0060B0	Hewlett-Packard Co.—newer LaserJets.—sends bit reversed as 00060D on Token Ring
001083	Hewlett-Packard Company—HP-UX E 9000
080009	Hewlett-Packard—sends bit reversed as 100090 on Token Ring
0000FD	High Level Hardware (Orion, UK)
00C07C	Hightech Information
00A0BB	Hilan GmBh
00E0A1	Hima Paul Hildebrandt
00103B	HIPPI Networking Forum
00C010	Hirakawa Hewtech Corp.
00D026	Hirschmann Austria GmBh
004066	Hitachi Cable, Ltd.
0060E8	Hitachi Computer Products
0080BC	Hitachi Engineering Co., Ltd.
00102D	Hitachi Software Engineering
0060CB	Hitachi Zosen Corporation
000087	Hitachi, Ltd.
003011	HMS Fieldbus Systems AB
000051	Hob Electronic GmBh & Co. KG
00E056	Holontech Corporation
00907D	Home Wireless Networks
004084	Honeywell Inc.
009030	Honeywell-Dating
0040DD	Hong Technologies
00509F	Horizon Computer
00E0C4	Horner Electric, Inc.
001009	Horo Quartz
0090E7	Horsch Elektronik Ag
00E062	Host Engineering
001061	Hostlink Corp.
006031	HRK Systems
0020E4	Hsing Tech Enterprise Co., Ltd.
006098	HT Communications
008085	H-Three Systems Corporation
001048	HTRC Automation, Inc.
00E0FC	Huawei Technologies Co., Ltd.
0060BD	Hubbell-Pulsecom
00108E	Hugh Symons Concept Technologies, Ltd.
0080BB	Hughes Lan Systems
000010	Hughes LAN Systems (formerly Sytek, Inc.)
0080AE	Hughes Network Systems
008096	Human Designed Systems, Inc. (HDS) —X terminals
002053	Huntsville Microsystems, Inc.
0020CD	Hybrid Networks, Inc.
002070	Hynet, Ltd.
0040EF	Hypercom, Inc.
00404C	Hypertec Pty Ltd.
008003	Hytec Electronics Ltd.
00003B	Hyundai Electronics
00803F	Hyundai Electronics—IEEE assigned to TATUNG COMPANY
009009	Hyundai Information Technology Co., Ltd.
0010A2	Hyundai Network Systems
00D0C8	I/O Consulting A/S
0090BE	IBC/Integrated Business Computers
005076	IBM
0001CB	IBM 8235's seen on Token Ring—Not an IEEE assigned OUI
006094	IBM Corp.—AMD PCNET PCI—IEEE assigned to IBM
08005A	IBM Corporation
10005A	IBM Corporation
002035	IBM Corporation (International Business Machines)
0010D9	IBM Japan, Fujisawa MT+D
000629	IBM RISC6000 system
0004AC	IBM—PCMCIA Ethernet adapter
50005A	IBM—Token Ring Group Address of OUI 10005A
00D0E6	Ibond Inc.
0040FC	IBR Computer Technik GmBh

OUI	Vendor/Comment
002071	IBR GmBh
009090	I-Bus
00103C	IC Ensemble, Inc.
00401E	ICC
000041	ICE Corporation
08000D	ICL (International Computers, Ltd.)
00004B	ICL Data OY— APT and Nokia Data Systems also seen or reported
004004	ICM CO. LTD.
0090C7	Icom Inc.
009062	ICP Vortex Computersysteme GbMh
006032	I-Cube, Inc.
00C032	I-Cubed Limited
00C060	ID Scandinavia AS
00C076	I-Data International A-S
0040B8	IDEA Associates
003043	Idream Technologies, PTE. Ltd.
002067	IEEE OUI ADMINISTRATOR
0050C2	IEEE Registration Authority
00A085	IEEE STANDARDS DEPARTMENT
1100AA	IEEE STANDARDS OFFICE
0080C2	IEEE—802.1 Committee
00507D	IFP
00D0EF	IGT
00305F	Imacon APS
00A006	Image Data Processing
00A0FC	Image Sciences, Inc.
080086	Imagen Corporation
009034	Imagic, Inc.
0010D5	Imasde Canarias, S.A.
00C0CF	Imatran Voima Oy
001003	Imatron, Inc.
00A033	Imc Mess-Systeme GmBh
0010AC	IMCI Technologies
000029	IMC—IMC Networks Corp.
0050C4	IMD
0040A8	IMF International Ltd.
0080AC	Imlogix, Division Of Genesys
003069	Impacct Technology Corp.
00403A	Impact Technologies
00D049	Impresstek Co., Ltd.
00C062	Impulse Technology
0000E4	IN2 Groupe Intertechnique
00302B	Inalp Networks, Inc.
00E0E6	Incaa Datacom B.V.
00D022	Incredible Technologies, Inc.
08007A	Indata
0030DD	Indigita Corporation
0090AA	Indigo Active Vision Systems Limited
001013	Industrial Computer Source
00100F	Industrial CPU Systems
00D031	Industrial Logic Corporation
080060	Industrial Networking Inc.
0020C4	INET, Inc.
00C0BD	Inex Technologies, Inc.
00C0C2	Infinite Networks Ltd.
00E047	Infocus Systems
00009B	Information International, Inc.
0000C2	Information Presentation Tech.
0020E2	Information Resource Engineering
000012	Information Technology Limited
00D023	Infortrend Technology, Inc.
00A0F2	Infotek Communications, Inc.
00008B	Infotron
00106C	Infratec Plus GmBh
00E019	Ing. Giordano Elettronica
00E055	Ingenieria Electronica
001010	Initio Corporation
008047	In-Net Corp.
001044	Innolabs Corporation
001099	Innomedia, Inc.
00E02D	Innomedialogic, Inc.
006001	Innosys, Inc.
00E0E2	Innova Corp.
009033	Innovaphone GmBh
00E0EA	Innovat Communications, Inc.
009054	Innovative Semiconductors, Inc.
00A04C	Innovative Systems & Tech. Inc
0010A0	Innovex Technologies, Inc.
0050B8	Inova Computers GmBh & Co. KG
0030D1	Inova Corporation
00204D	Inovis GmBh
0060B1	Input/Output, Inc.
0060DF	Inrange Technologies Corp.
00E0F4	Inside Technology A/S
00105F	IN-SNEC
00A0D3	Instem Computer Systems, Ltd.
080052	Insystec
0010CA	Integral Access
008023	Integrated Business Networks
00A0BE	Integrated Circuit Systems, Inc. Communications Group
00D0A2	Integrated Device
00E0CF	Integrated Device Technology, Inc.
00C030	Integrated Engineering B.V.
00A047	Integrated Fitness Corp.
00801D	Integrated Inference Machines
00802B	Integrated Marketing Co.
008012	Integrated Measurement Systems Corp. (IMS) — IMS failure analysis tester
0000E3	Integrated Micro Products Ltd
00A0DA	Integrated Systems
0030FD	Integrated Systems Design
00106D	Integrity Communications, Inc.
0090EF	Integrix, Inc.
00D0B7	Intel Corp.
009027	Intel Corporation
00AA01	Intel Corporation
00AA02	Intel Corporation
00A0C9	Intel Corporation (PRO100B and PRO100+)—(used on Cisco PIX firewall among others)
00A083	Intel Corporation—IEEE assigned to ASIMMPHONY TURKEY ELECTRONIC LTD.
00AA00	Intel Corporation—sends bit reversed as 005500 on Token Ring
0030EA	Intelect Communictions, Inc.
003072	Intellibyte Inc.
008053	Intellicom, Inc.
0060AA	Intelligent Devices Inc. (IDI)
00D05D	Intelliworxx, Inc.
0010BF	Inter Air Wireless
00C011	Interactive Computing Devices
008062	Interface Co.
0010CD	Interface Concept
00E0BD	Interface Systems, Inc.
0020AC	Interflex Datensysteme GmBh
0050AF	Intergon, Inc
480036	Intergraph Corp.—Token Ring Locally Administered Address of OUI 080036
080036	Intergraph—CAE stations
00C01C	Interlink Communications Ltd.
00C0F4	Interlink System Co., Ltd.
00A092	Intermate International—(LAN printer interfaces)—IEEE assigned to H. BOLLMANN MANUFACTURERS, LTD
001040	Intermec Corporation
00C0A2	Intermedium A/S
0010DE	International Datacasting
00E04D	Internet Initiative Japan, Inc
0090C5	Internet Magic, Inc.
00205C	Internet Systems/ Florida Inc.

OUI	Vendor/Comment
006050	Internix Inc.
000077	Interphase Corporation-(Used in other systems, e.g. MIPS, Motorola)
0030B4	Intersil Corp.
001036	Inter-Tel Integrated Systems
004070	Interware Co., Ltd.
00A02C	Interwave Communications
00605B	Intraserver Technology Inc.
00D0CA	Intrinsyc Software Inc.
00A0D1	Inventec Corporation
00A05C	Inventory Conversion, Inc./Nekotech Division
00D083	Invertex, Inc.
00401D	Invisible Software, Inc.
0010F1	I-O Corporation
00A0B0	I-O Data Device, Inc.
0050A4	IO Tech, Inc.
00D0B8	Iomega Corp.
00D082	Iowave Inc.
00C0DB	IPC Corporation (Pte) Ltd
00E0A7	Ipc Information Systems, Inc.
00D098	Ips Automation
00A08E	Ipsilon Networks, Inc.
0080ED	IQ Technologies, Inc.
0010F7	Iriichi Technologies Incorporation
00500A	Iris Technologies, Inc.
00307F	Irlan Ltd.
004050	Ironics, Incorporated
00A066	Isa Co., Ltd.
000046	ISC-Bunker Ramo, An Olivetti America Company
002069	ISDN Systems Corporation
002068	Isdyne
003016	Ishida Co., Ltd.
0010B8	Ishigaki Computer System Co.
0000E9	ISICAD, Inc.
0050FD	Isioncomm Co., Ltd.
00402C	ISIS Distributed Systems, Inc.
00D050	Iskratel
00C045	Isolation Systems Inc
0090AE	Italtel SPA
0040A7	Itautec Informatica S.A.
00A0BD	I-Tech Corp.
009087	ITIS
00100C	ITO Co., Ltd.
009018	Ito Electric Industry Co, Ltd.
0010D6	ITT A/CD
0020FD	ITV Technologies, Inc.
00303D	IVA Corporation
009022	Ivex (formerly BRIVIDA, INC.)
00A09E	IVTV
005051	Iwatsu Electric Co., Ltd.
00D068	iwill Corporation
00A017	J B M Corporation
0090AF	J. Morita Mfg. Corp.
00C0D6	J1 Systems, Inc.
002091	J125, National Security Agency
00C091	Jabil Circuit, Inc.
00A08D	Jacomo Corporation
0040F2	Janich & Klass Computertechnik
0090E9	Janz Computer Ag
00E05E	Japan Aviation Electronics Industry, Ltd.
00C004	Japan Business Computer Co.Ltd
00800A	Japan Computer Corp.
008092	Japan Computer Industry, Inc.
0020DE	Japan Digital Laborat'y Co.Ltd. (JDL)
080042	Japan Macnics Corp.
0020D7	Japan Minicomputer Systems Co.
000027	Japan Radio Company
080061	Jarogate Ltd.
00C028	Jasco Corporation
00E083	Jato Technologies, Inc.
00501C	Jatom Systems, Inc.
00E03F	Jaton Corporation
0090C4	Javelin Systems, Inc.
00E069	Jaycor Networks, Inc.
0060DA	JBM Electronics Co.
00A097	JC Information Systems
002010	Jeol System Technology Co. Ltd.
00D02B	Jetcell, Inc.
0050CB	Jetter
0090C2	JK Microsystems, Inc.
001082	JNA Telecommunications Limited
008040	John Fluke Manufacturing Co
00C017	John Fluke Mfg.—IEEE assigned to Forte Networks which makes the Fluke LAN Meters—sends bit reversed as 0003E8 on Token Ring
00A09D	Johnathon Freeman Technologies
0060F0	Johnson & Johnson Medical, Inc.
00108D	Johnson Controls, Inc.
0050D4	Joohong Information & Communications, Ltd.
00A063	JRL Systems, Inc.
0020BC	Jtec PTY. Ltd.
00203F	Juki Corporation
004079	Juko Manufacture Company, Ltd.
00E04B	Jump Industrielle
009069	Juniper Networks, Inc.
008009	Jupiter Systems (older MX-600 series machines)
00C0AB	Jupiter Technology Inc
00809B	Justsystem Corporation
0090C0	K.J. Law Engineers, Inc.
00C029	Kabel Rheydt AG
001700	Kabel Rheydt AG—Not an IEEE assigned OUI—IEEE assigned OUI 00C029
008024	Kalpana Inc.
006087	Kansai Electric Co., Ltd.
00D09C	Kapadia Communications
00E06A	Kapsch AG
002077	Kardios Systems Corp.
00A0D7	Kasten Chase Applied Research
00201D	Katana Products
00D055	Kathrein-Werke KG
0040F6	Katron Computers Inc.
00D0B4	Katsujima Co., Ltd.
00900F	Kawasaki Heavy Industries, Ltd.
00505B	Kawasaki Lsi U.S.A., Inc.
0060DE	Kayser-Threde GmBh
0030DF	KB/Tel Telecomunicaciones
001084	K-Bot Communications
08001C	KDD-Kokusai Debnsin Denwa Co. Fuji Xerox Co., Ltd. Telegraph
00E0DF	Ke Kommunikations-Electronik GmBh
0060B5	KEBA GmBh
00A0E2	Keisoku Giken Corp.
00601A	Keithley Instruments
00C046	Kemitron Ltd
008081	Kendall Square Research Corp.
0010A1	Kendin Semiconductor, Inc.
00406A	Kentek Information Systems Inc
0010F8	Kenwood TMI Corporation
003027	Kerbango, Inc.
005075	Kestrel Solutions
0060BC	Keunyoung Electronics & Communication
00D07E	Keycorp Ltd.
0040D3	Kimpsion International Corp.
0080D1	Kimtron Corporation
080089	Kinetics—AppleTalk-Ethernet interface
480089	Kinetics—Token Ring Locally Administered Address of OUI 080089
00A00C	Kingmax Technology, Inc.
0040A2	Kingstar Technology Inc.

OUI	Vendor/Comment
00030F	Kingston Technology Corp.—bit reveresed of assigned OUI 00C0F0—Not an IEEE asigned OUI
00C0F0	Kingston Technology Corporation—sends bit reversed as 00030F on Token Ring
0040C4	Kinkei System Corporation
004064	KLA Instruments Corporation
004022	Klever Computers, Inc.
008099	Klockner Moeller IPC
00D06F	Kmc Controls
00A099	K-Net Ltd.
00C072	KNX Ltd
008018	Kobe Steel, Ltd.
00E054	Kodai Hitec Co., Ltd.
00801B	Kodiak Technology
00A05A	Kofax Image Products
005037	Koga Electronics Co.
0010AB	Koito Industries, Ltd.
004031	Kokusai Electric Co., Ltd
00C06F	Komatsu Ltd.
0050AA	Konica Corporation
000060	Kontron Elektronik GmBh
002050	Korea Computer Inc.
005062	Kouwell Electronics Corp.
00D07C	Koyo Electronics Inc. Co., Ltd.
004058	Kronos, Inc.
00801F	Krupp Atlas Electronik GmBh
00C009	Kt Technology (S) Pte Ltd
0000EF	KTI—Alantec (now owned by ForeSystems) seen under this OUI
00C07E	Kubota Corporation Electronic Device Dept.
00A03A	Kubotek Corporation
0060C8	Kuka Welding Systems & Robots
00D0DF	Kuzumi Electronics, Inc.
00D045	Kvaser Ab
00A064	KVB/Analect
00C0DF	Kye Systems Corp.
00C0EE	Kyocera Corporation
00207F	Kyoei Sangyo Co., Ltd.
0020DF	Kyosan Electric Mfg. Co., Ltd.
0080F0	Kyushu Matsushita Electric Co.
00C05D	L&N Technologies
001027	L-3 Communications East
00D04B	La Cie Group S.A.
00C01E	La Francaise Des Jeux
004082	Laboratory Equipment Corp
000078	Labtam Limited Australia
006066	Lacroix Technolgie
0020BE	Lan Access Corp.
006099	Lan Media Corporation
00C018	Lanart Corp
00D0A6	Lanbird Technology Co., Ltd.
00E0D8	Lanbit Computer, Inc.
00409F	Lancast/Casat Technology, Inc.
00406D	Lanco, Inc
0060B6	Land Computer Co., Ltd.
002093	Landings Technology Corp.
00C0E4	Landis & Gyr Powers Inc.
0080C1	Lanex Corporation
00900B	Lanner Electronics, Inc.
00400D	LANNET Data Communications
00C000	Lanoptics, Ltd.
00C026	LANS Technology Co., Ltd.
08002D	Lan-Tec Inc.
0080A3	Lantronix
0040CB	LANWAN Technologies
00208B	Lapis Technologies, Inc.—Focus Enhancements seen under this OUI
0020C8	Larscom Incorporated
0060AB	Larscom Incorporated

OUI	Vendor/Comment
00108B	Laseranimation Sollinger GmBh
0060F2	Lasergraphics, Inc.
00C022	Lasermaster Technologies, Inc.
00C08A	Lauterbach Datentechnik GmBh
00C019	Leap Technology, Inc.
00E087	Lecroy Networking Products Division
000082	Lectra Systemes SA
00A002	Leeds & Northrup Australia Pty Ltd.
0090D1	Leichu Enterprise Co., Ltd.
00E0FB	Leightronix, Inc.
0090F9	Leitch
004069	Lemcom Systems Inc
00509E	Les Technologies
00207B	Level One Communications
002000	Lexmark (Printers and Print Servers)
000400	Lexmark International (Print Server)
00A041	Leybold-Inficon
00E091	Lg Electronics, Inc.
0050CE	LG International Corp.
0080A4	Liberty Electronics
00605E	Liberty Technology Networking
0050F1	Libit Signal Processing, Ltd.
0020E6	Lidkoping Machine Tools AB
00D0EB	Lightera Networks, Inc.
0090F4	Lightning Instrumentation
003031	Lightwave Communications, Inc.
00E0C6	Link2it, L.L.C.
00E098	Linksys—PCMCIA card—IEEE assigned to ABOCOM SYSTEMS, INC.
00A055	Linktech, Inc.
00D06A	Linkup Systems Corporation
00007F	Linotype-Hell AG—Linotronic typesetters
0020AD	LINQ Systems
00602C	Linx Data Terminals, Inc.
00209D	Lippert Automationstechnik
00A0CC	Lite-On Communications, Inc.
080031	Little Machines Inc.
008061	Litton Systems, Inc.
006044	Litton/Poly-Scientific
00C005	Livingston Enterprises Inc—Portmaster (OEMed by Cayman)
0040D6	Locamation B.V.
004057	Lockheed-Sanders
00009D	Locus Computing Corporation
00D04E	Logibag
004023	Logic Corporation
0000AB	Logic Modeling Corporation
006086	Logic Replacement Tech. Ltd.
0020CE	Logical Design Group, Inc.
00203E	Logican Technologies, Inc.
00004F	Logicraft 386-Ware P.C. Emulator
00E07F	Logististem Srl
00600B	Logware GmBh
00D070	Long Well Electronics Corp.
0050C6	Loop Telecommunication International, Inc.
0060F8	Loran International Techn. Inc.
00E035	Loughborough Sound Images, Plc
00805E	LSI Logic Corporation
00E044	Lsics Corporation
00E088	Ltx Corporation
00306D	Lucent Technologies
00601D	Lucent Technologies
00D077	Lucent Technologies
0060D2	Lucent Technologies Taiwan
00A0EF	Lucidata Ltd.
00809C	Luxcom, Inc.
0040FD	LXE
00E072	Lynk
0080E7	Lynwood Scientific Dev. Ltd.

OUI	Vendor/Comment
0060FE	Lynx System Developers, Inc.
00A0A6	M.I. Systems, K.K.
0070B0	M/A-Com Inc. Companies
0270B0	M/A-Com Inc. Companies
00303A	Maatel
00D0E8	Mac System Co., Ltd.
0040B9	MACQ Electronique SA
0060BF	Macraigor Systems, Inc.
0030C7	Macromate Corp.
006028	Macrovision Corporation
0080E9	Madge Networks
00006F	Madge Networks Ltd.- Token-ring adapters—Madge sends bit reversed as 0000F6 on Token Ring
4000F6	Madge Networks Ltd.—Token Ring Locally Administered Address of OUI 00006F—bit revereseed
003032	Magicram, Inc.
0000F1	Magna Computer Corporation
00D042	Mahlo GmBh & Co. Ug
00D088	Mainsail Networks, Inc.
004081	Mannesmann Scangraphic GmBh
00A06D	Mannesmann Tally Corporation
0050AC	Maple Computer Corporation
00404B	Maple Computer Systems
00E009	Marathon Technologies Corp.
000096	Marconi Electronics Ltd.
00A0AD	Marconi Spa
00E0EE	Marel Hf
005039	Mariner Networks
009045	Mariposa Technology
00909B	Markpoint AB
00009E	Marli S.A.
0040E1	Marner International Inc.
0000A1	Marquette Electric Co.
00A05B	Marquip, Inc.
0040EB	Martin Marietta Corporation
0010BA	Martinho-Davis Systems, Inc.
005043	Marvell Semiconductor, Inc.
0050C9	Maspro Denkoh Corp.
08006E	Masscomp
080071	Matra (DSIE)
00000B	Matrix Corporation
0020FC	Matrox
0000EB	Matsushita Comm. Ind. Co. Ltd.
008045	Matsushita Electric Ind Co
00C08F	Matsushita Electric Works, Ltd.
00101E	Matsushita Electronic
080023	Matsushita Graphics Comm Sys Inc.
00E05C	Matsushita Kotobuki Electronics Industries, Ltd.
0080AA	Maxpeed
004077	Maxton Technology Corporation
0010B9	Maxtor Corp.
005033	Mayan Networks
0020E3	MCD Kencom Corporation
080088	McDATA Corporation
0000B9	McDonnell Douglas Computer Sys
001001	MCK Communications
004056	MCM Japan Ltd.
008076	MCNC
00E02F	Mcns Holdings, L.P.
00D0DB	Mcquay International
0080A7	Measurex Corp.
0010C4	Media Links Co., Ltd.
0010AA	Media4, Inc.
00503B	Mediafire Corporation
00E022	Medialight Inc.
0090A3	Medialincs Co., Ltd.
005020	Mediastar Co., Ltd.
0090F8	Mediatrix Telecom
0060AD	Megachips Corporation
0000CE	Megadata Corp.
080029	Megatek Corporation
005078	Megaton House, Ltd.
002020	Megatron Computer Industries Pty, Ltd.
004089	Meidensha Corporation
0000FC	Meiko
004026	Melco Inc
00E030	Melita International Corp.
000099	Memorex Telex Corporation (MTX, Inc.)
00E0C1	Memorex Telex Japan, Ltd.
002087	Memotec Communications Corp.
00400E	Memotec Communications, Inc.
005046	Menicx International Co., Ltd.
00C03A	Men-Mikro Elektronik GmBh
00C092	Mennen Medical Inc
009078	MER Telemanagement
00209F	Mercury Computer Systems, Inc.
00C0B6	Meridian Data Inc.
0010B0	Meridian Technology Corp.
00507B	Merlot Communications
00D076	Merrill Lynch & Co., Inc.
00E068	Merrimac Systems Inc.
00601B	Mesa Electronics
0040E2	Mesa Ridge Technologies Inc.
00C0F5	Metacomp Inc
08002E	Metaphor Computer Systems
001080	Metawave Communications
000042	Metier Management Systems Ltd.
0020B9	Metricom, Inc.
00C081	Metrodata Ltd.
004055	Metronix GmBh
001052	Mettler-Toledo (Albstadt) GmBh
00E07C	Mettler-Toledo, Inc.
004086	Michels & Kleberhoff Computer
0040C5	Micom Communications Corp.
001017	MICOS GmBh
00C0C8	Micro Byte Pty Ltd
00C00A	Micro Craft
0020AB	Micro Industries Corp.
00100E	Micro Linear Coporation
001038	Micro Research Institute, Inc.
0040F0	Micro Systems Inc.
000043	Micro Technology
006092	Micro/Sys, Inc.
0040FA	Microboards Inc.
008079	Microbus Designs Ltd.
000090	Microcom
0020D1	Microcomputer Systems (M) SDN.
00600D	Microdesign GmBh
008029	Microdyne Corporation—IEEE assigned to EAGLE TECHNOLOGY, INC.
00D087	Microfirst Inc.
080035	Microfive Corporation
00E089	Microframe Inc.
0000D5	Micrognosis International
0000B6	Micro-matic Research
00A075	Micron Technology, Inc.
0080CD	Micronics Computer, Inc.
008072	Microplex Systems Ltd.
00A016	Micropolis Corp.
0000EC	Microprocess
00A056	Micropross
00A0A4	Micros Systems, Inc.
0000FA	Microsage Computer Systems Inc.
0060A7	Microsens GmBh & Co. KG
00E0A2	Microslate Inc.

OUI	Vendor/Comment
0050F2	Microsoft Corp.
0010DC	Micro-Star International Co., Ltd.
002086	Microtech Electronics Limited
008090	Microtek International, Inc.
00D0B9	Microtek International, Inc.
0080A1	Microtest Inc. (disc port product)
006080	Microtronix Datacom Ltd.
0040A3	Microunity Systems Engineering
0080CC	Microwave Bypass Systems
00E049	Microwi Electronic GmBh
006071	Midas Lab, Inc.
080053	Middle East Tech. University
00E0AC	Midsco, Inc.
00E09C	MII
0040EC	Mikasa System Engineering Co., Ltd.
08000C	Miklyn Development Co.
00E07A	Mikrodidakt AB
002079	Mikron GmBh
0040C8	Milan Technology Corp.—sends bit reversed as 0000AA on Token Ring
00D01B	Mimaki Engineering Co., Ltd.
0050D0	Minerva Systems
002027	Ming Fortune Industry Co., Ltd.
000072	Miniware Technology
00206B	Minolta Co., Ltd.—Network printers
00D0A0	Mips Denmark
00501E	Miranda Technologies, Inc.
002059	Miro Computer Products AG
00E032	Misys Financial Systems, Ltd.
00C04D	Mitec, Inc.
00A087	Mitel Semiconductor, Ltd.
00204C	Mitron Computer PTE Ltd.
0000BD	Mitsubishi Cable Company
0090DF	Mitsubishi Chemical America, Inc.
080070	Mitsubishi Electric Corp.
0010C9	Mitsubishi Electronics Logistic Support Co.
00606A	Mitsubishi Wireless Comm. Inc.
00A096	Mitsumi Electric Co., Ltd.
0060D5	Miyachi Technos Corp.
0080F8	Mizar, Inc.
00604D	MMC Networks, Inc.
005097	MMC-Embedded Computertechnik GmBh
00E01C	Mobility Electronicsy
004088	Mobuis—NuBus (Mac) combination video
000054	Modicon, Inc.
00C055	Modular Computing Technologies
00D0DC	Modular Mining Systems, Inc.
00305B	Module Department
0050A9	Moldat Wireless Technolgies
004025	Molecular Dynamics
00D0F2	Monterey Networks
00C082	Moore Products Co.
00D0CF	Moreton Bay
00C063	Morning Star Technologies Inc
08002A	Mosaic Technologies Inc.
0010DA	Motion Engineering, Inc.
00E00C	Motorola
002075	Motorola Communication Israel
0050BF	Mototech Inc.
0090E8	Moxa Technologies Corp., Ltd.
0060C2	MPL AG
00901C	MPS Software GmBh
00501F	MRG Systems, Ltd.
00D0E2	Mrt Micro, Inc.
0030D6	MSC Vertriebs GmBh
00109C	M-System Co., Ltd.
00A0F1	MTI
004075	M-Trade (UK) Ltd

OUI	Vendor/Comment
00C03B	Multiaccess Computing Corp
0030BF	Multidata GmBh
00A08C	Multimedia Lans, Inc.
0020A4	Multipoint Networks
008000	Multitech Systems Inc
000800	Multitech Systems, Inc.
00D064	Multitel
005081	Murata Machinery, Ltd.
006057	Murata Manufacturing Co., Ltd.
00A011	Mutoh Industries Ltd.
00C057	Myco Electronics
0080E5	Mylex Corporation
00A05E	Myriad Logic Inc.
0060DD	Myricom, Inc.
00C0B4	Myson Technology Inc.
009031	Mysticom, Ltd.
004042	N.A.T. GmBh
00202A	N.V. Dzine
00903E	N.V. Philips Industrial Activities
001024	Nagoya Electric Works Co., Ltd.
00D0EC	Nakayo Telecommunications, Inc
0050AB	Naltec, Inc
0020B7	Namaqua Computerware
00902E	Namco Limited
009093	Nanao Corporation
00205D	Nanomatic OY
00A01C	Nascent Networks Corporation
00A0EE	Nashoba Networks
00E073	National Amusement Network, Inc.
08000E	National Cash Register (NCR) now owned by AT&T—sends bit reversed as 100070 on Token Ring
100070	National Cash Register (now owned by AT&T)—seen on Token Ring—bit reversed of assigned OUI 08000E—Not an IEEE assigned OUI
00802F	National Instruments Corp.
1000E8	National Semiconductor
080017	National Semiconductor Corp. (used to have Network System Corp., wrong NSC)—IEEE assigned to NATIONAL SEMICONDUCTOR
005024	Navic Systems, Inc.
0090F7	Nbase Communications Ltd.
00201A	N-Base Switch Communications
080022	NBI (Nothing But Initials)
00E0BB	NBX Corporation
0040C9	NCUBE
003076	N-Cubed.Net
00A019	Nebula Consultants, Inc.
0090E4	NEC America, Inc.
00004C	NEC Corporation
009075	NEC Do Brasil S.A
00E0C2	Necsy SPA
0020C6	Nectec
009070	Neo Networks, Inc
0030EF	Neon Technology, Inc.
00E08C	Neoparadigm Labs, Inc.
002056	Neoproducts
0060C0	Nera AS
08000A	Nestar Systems
48000A	NESTAR SYSTEMS INCORPORATED—Token Ring Locally Administered Address of OUI 08000A
0080B2	NET (Network Equipment Technologies)
00105B	Net Insight AB
0020F6	Net Tek And Karlnet, Inc.
0050CA	Net To Net Technologies
006015	Net2Net Corporation
005091	Netaccess, Inc.
0090D7	Netboost Corp.
005036	Netcam, Ltd.
0080CA	Netcom Research Incorporated

OUI	Vendor/Comment
00905D	Netcom Sicherheitstechnik Gmb
0060F3	Netcom Systems, Inc.
006064	Netcomm Limited
004028	Netcomm Limited
0040F3	NETCOR
001092	Netcore Inc.
00902F	Netcore Systems, Inc.
00A008	Netcorp
00A0AB	Netcs Informationstechnik GmBh
00C021	Netexpress
0060FD	Netics, Inc.
00109A	Netline
0030E3	Netpoint Corporation
00A004	Netpower, Inc.
00201E	Netquest Corporation
00C013	Netrix
006040	Netro Corp.
000014	Netronix
00E07D	Netronix, Inc.
00103E	NetSchools Corporation
00808C	Netscout Systems (formerly Frontier Software Development)
0010DB	Netscreen Technologies, Inc.
0040CE	Net-Source, Inc.
00C012	Netspan Corporation
00E0D0	Netspeed, Inc.
006013	Netstal Maschinen AG
00C080	Netstar Inc.
00C0BA	Netvantage
0060C3	Netvision Corporation
0020D8	Netwave Technologies, Inc.
00C02E	Netwiz
00E007	Network Alchemy Ltd.
00505A	Network Alchemy, Inc.
00A098	Network Appliance Corp.
0000A3	Network Application Technology (NAT)
000065	Network Associates, Inc. (formerly Network General)
00C0F3	Network Communications Corp
0090FC	Network Computing Devices
0000A7	Network Computing Devices Inc. (NCD) X-terminals
0040ED	Network Controls International Inc.
0000EE	Network Designers Limited (also KNX Ltd, a former division)
00D0A8	Network Engines, Inc.
00409A	Network Express Inc.
0000A6	Network General (internal assignment, not for products)—now NAI
400065	Network General Inc. (now Network Associates Inc—NAI)—Token Ring Locally Administered Address of OUI 000065
00C08E	Network Information Technology
008060	Network Interface Corporation
006010	Network Machines, Inc.
0080D8	Network Peripherals Inc.
00A0AE	Network Peripherals, Inc.—IEEE assigned to NUCOM SYSTEMS, INC.
00C053	Network Peripherals, Ltd.
008026	Network Products Corporation
008052	Network Professor—IEEE assigned to TECHNICALLY ELITE CONCEPTS
080030	Network Research Corporation
08008C	Network Research Corporation
000034	Network Resources Corporation
00204E	Network Security Systems, Inc.
00A067	Network Services Group
00102E	Network Systems & Technologies Pvt. Ltd.
0000A9	Network Systems Corp.
4000A9	Network Systems Corp. (NSC)—Token Ring Locally Administered Address of OUI 0000A9
006056	Network Tools, Inc.
00C05B	Networks Northwest, Inc.
008043	Networld, Inc.
000079	Networth Incorporated (now owned by Compaq, used in Netelligent series)

OUI	Vendor/Comment
002043	Neuron Company Limited
00C015	New Media Corporation
008021	Newbridge Research Corporation—Newbridge sends bit reversed as 000184 on Token Ring
000184	Newbridge seen on Token Ring—bit reversed of assigned OUI 008021—Not an IEEE assigned OUI
00507E	Newer Technology
004099	Newgen Systems Corp.
0010F3	Nexcom International Co., Ltd.
00D084	Nexcomm Systems, Inc.
00A065	Nexland, Inc.
005000	Nexo Communications, Inc.
0090DB	Next Level Communications
00D0F7	Next Nets Corporation
00000F	NeXT, Inc.
0060F6	Nextest Communication Products, Inc.
00D0EA	Nextone Communications, Inc.
008030	Nexus Electronics
00E0DC	Nexware Corp.
00A0E5	NHC Communications
00C0DA	Nice Systems Ltd.
009043	Nichibei Denshi Co., Ltd.
000047	Nicolet Instruments Corp.
00308A	Nicotra Sistemi S.P.A.
00A09A	Nihon Kohden America
00903A	Nihon Media Tool Inc.
0060A2	Nihon Unisys Limited Co.
0090B5	Nikon Corporation Electr. Imaging Design Dept.
0090A8	Ninetiles Networks Ltd.
0020BD	Niobrara R & D Corporation
00C006	Nippon Avionics Co., Ltd.
00C059	Nippon Denso Co., Ltd.
005065	Nippon Electric Industry Co., Ltd.
009028	Nippon Signal Co., Ltd.
00806E	Nippon Steel Corporation
0000D9	Nippon Telegraph & Telephone Corporation (NTT)
00605F	Nippon Unisoft Corporation
009001	Nishimu Elctronics Industries
003047	Nissei Electric Co., Ltd.
00A04A	Nisshin Electric Co., Ltd.
008049	Nissin Electric Co., Ltd.
0060B9	Nitsuko Corporation
0010D2	Nitto Tsushinki Co., Ltd.
00E6D3	Nixdorf Computer Corp.
00609A	NJK Techno Co.
00A082	NKT Elektronik A/S
001091	No Wires Needed BV
000267	Node Runner Inc.
0010B3	Nokia Satellite Systems AB
004043	Nokia Telecommunications
00E003	Nokia Wireless Business Commun
00D0D8	Nomadic Technologies
00C0B2	Norand Corporation
00C00B	Norcontrol A.S.
000011	Normerel Systemes
080026	Norsk Data (Nord)
0090CF	Nortel
006038	Nortel (Northern Telecom)
00405E	North Hills Israel
00D0F6	Northchurch Communications Inc.
000075	Northern Telecom Bell Northern Research (BNR) (now Nortel)
00201B	Northern Telecom/Network Systems Corporation (Nortel)
0080E4	Northwest Digital Systems, Inc.
009080	Not Limited, Inc.
006082	Novalink Technologies, Inc.
005010	Novanet Learning, Inc.
00203D	Novar Electronics Corporation

OUI	Vendor/Comment
0060D6	Novatel Wireless Technologies Ltd.
00001B	Novell, Inc. (sold this business line to Eagle Technology)
0080C5	Novellco De Mexico
0000D8	Novell—old NE1000's (before about 1987)—3Com seen under this OUI
00D09F	Novtek Test Systems
00809A	Novus Networks Ltd.
0050E1	Ns Tech Electronics Sdn Bhd
0010E4	NSI Corporation
0060DB	NTP Elektronik A/S
004013	NTT Data Communication Systems Corp
00A044	Ntt Intelligent Technology
0000AF	Nuclear Data Instrumentation—Acquisition Interface Modules (AIM)
001078	Nuera Communications, Inc.
00C07F	Nupon Computing Corp.
00D08E	Nvision Inc.
001062	NX Server, ILNC.
00A0DC	O.N. Electronic Co., Ltd.
00D051	O2 Micro, Inc.
0020A0	OA Laboratory Co., Ltd.
00C0E9	Oak Solutions Ltd.
001031	Objective Communications, Inc.
002084	OCE Printing Systems, GmBh
0040D8	Ocean Office Automation Ltd.
00200B	Octagon Systems Corp.
0020FB	Octel Communications Corp.
00BB01	Octothorpe Corp.
02BB01	Octothorpe Corp.
0040F5	OEM Engines
00C02A	Ohkura Electric Co
002019	Ohler GmBh
00101C	Ohm Technologies Intl, LLC
0010C1	OI Electric Co., Ltd.
008087	Oki Electric Industry Co., Ltd.—Okidata
00C02F	Okuma Corp
0030A8	Ol'e Communications, Inc.
00504F	Olencom Electronics
0050EC	Olicom A/S
0000C1	Olicom A/S—Olicom sends bit reversed as 000083 on Token Ring
00AA3C	Olivetti Telecom SPA (Olteco)
02AA3C	Olivetti Telecomm SPA (Olteco)
00C0D3	Olympus Image Systems, Inc.
00D0B1	Omega Electronics SA
00D028	Omneon Video Networks
0030CB	Omni Flow Computers, Inc.
0090BD	Omnia Communications, Inc
004067	Omnibyte Corporation
005002	Omnisec AG
0010A3	Omnitronix, Inc.
00000A	Omron Tateisi Electronics Co. Section NFF, System R&D Labs. Research & Tech. Assessmnt Div.
00909A	One World Systems, Inc.
002082	Oneac Corporation
00806F	Onelan Ltd.
00303C	Onnto Corp.
003077	Onprem Networks
001015	Oomon Inc.
004035	OPCOM
0050A8	Opencon Systems, Inc.
003029	Opicom
004039	Optec Daiichi Denko Co Ltd
0030A1	Opti Tech Co., Ltd.
003007	Opti, Inc.
00E065	Optical Access International
000052	Optical Data Systems
00301F	Optical Networks, Inc.
0090C6	Optim Systems, Inc.
00D0FD	Optima Tele.Com, Inc.
00508E	Optimation, Inc.
0040EE	OPTIMEM
00A0E4	Optiquest
0090AC	Optivision, Inc.
00A03D	Opto - 22
0050A6	Optronics
0080F1	Opus Systems
00D0F5	Orange Micro, Inc.
00304B	Orbacom Systems, Inc.
003051	Orbit Avionic & Communication Systems, Ltd.
002017	Orbotech
08003A	Orcatech Inc.
0060E1	Orckit Communications Ltd.
00D0AE	Oresis Communications, Inc.
003033	Orient Telecom Co., Ltd.
00D034	Ormec Systems Corp.
0020AE	Ornet Data Communication
00A0C1	Ortivus Medical AB
00E094	OSAI SRL
00206A	Osaka Computer Corp.
00C06B	OSI Plus Corporation
00E0B6	Osicom Technologies
00C0E3	Ositech Communications Inc.
00D0A1	Oskar Vierling GmBh + Co. KG
009094	Osprey Technologies, Inc.
0020D3	OST (Ouet Standard Telematique)
00D085	Otis Elevator Company
00A072	Ovation Systems Ltd.
00900D	Overland Data Inc.
0010A5	Oxford Instruments
000037	Oxford Metrics Ltd
0030E0	Oxford Semiconductor Ltd.
0050FA	Oxtel, Ltd.
005063	OY Comsel System AB
005094	Pace Micro Technology Plc
002024	Pacific Communication Sciences
0060AF	Pacific Micro Data, Inc.
009061	Pacific Research & Engineering Corporation
002009	Packard Bell Elec., Inc.
00E0B1	Packet Engines, Inc.
0060FB	Packeteer, Inc.
00E042	Pacom Data Ltd.
0040D2	Pagine Corporation
0020A7	Pairgain Technologies, Inc.
00D060	Panasonic European
0020D9	Panasonic Technologies, Inc./Mieco-US
0020F5	Pandatel AG (formerly PAN DACOM TELECOM'CATIONS GmBh)
0050F6	Pan-International Industrial Corp.
0040B3	PAR Microsystems Corporation
0050E7	Paradise Innovations (Asia) Pte. Ltd.
00E039	Paradyne Corp.
0020F9	Paralink Networks, Inc.
009055	Parker Hannifin Corporation
008075	Parsytec GmBh
0060A6	Particle Measuring Systems
00603F	Patapsco Designs
003062	Path 1 Network Technol's Inc.
006045	Pathlight Technologies
00D0C7	Pathway, Inc.
00A0BA	Patton Electronics Co.
00109F	Pavo, Inc.
004093	Paxdata Networks Ltd.
080076	PC Lan Technologies
004000	PCI Componentes da Amzonia Ltd
00107C	P-COM, Inc.
080027	PCS Computer Systems GmBh—IEEE assigned to CADMUS COMPUTER SYSTEMS

OUI	Vendor/Comment
0050FE	Pctvnet Asa
00602B	Peak Audio
00901D	PEC (NZ) Ltd.
0080E6	Peer Networks, Inc.
009032	Pelcombe Group Ltd.
0080C0	Penril Datability Networks
00306A	Penta Media Co., Ltd.
006075	Pentek, Inc.
008082	PEP Modular Computers GmBh
005095	Peracom Networks
00A074	Perception Technology
00C08C	Performance Technologies Inc.
08008A	Performance Technology
0060A5	Performance Telecom Corp.
006023	Pericom Semiconductor Corp.
008001	Periphonics Corporation
00609C	Perkin-Elmer Corporation
004002	Perle Systems Limited
001C7C	Perq Systems Corporation
021C7C	Perq Systems Corporation
0090EE	Personal Communications Technologies
00C0C6	Personal Media Corp.
008017	PFU Limited
005023	PG Design Electronics, Inc.
00E058	Phase One Denmark A/S
00103D	Phasecom, Ltd.
00609F	Phast Corporation
009020	Philips Analytical X-Ray B.V.
08006F	Philips Apeldoorn B.V.
005009	Philips Broadband Networks
00D0DE	Philips Multimedia Network
006037	Philips Semiconductors
00D037	Philips-Dvs-Lo Bdr
00C068	Philp Drake Electronics Ltd
0060F5	Phobos FastEthernet for Unix WS—IEEE assigned to Icon West, Inc.
00A045	Phoenix Contact GmBh & Co.
002051	Phoenix Data Commununications Corp.
00D08D	Phoenix Group, Inc.
00C09A	Photonics Corporation
00D08A	Photron USA
00E0B7	PI Group, Ltd.
0080DC	Picker International
00101A	Picturetel Corp.
004020	Pinacl Communication Systems Limited
004F4B	Pine Technology Ltd. seen under this OUI—Not and IEEE assigned OUI—No IEEE OUI assigned to Pine Technology
00D01E	Pingtel Corp.
00C007	Pinnacle Data Systems, Inc.
00308D	Pinnacle Systems, Inc.
00E036	Pioneer Digital Technologies, Inc.
009024	Pipelinks, Inc.
00A02F	Pirelli Cavi
080018	Pirelli Focom Networks
006020	Pivotal Networking, Inc.
00D0BF	Pivotal Technologies
00400A	Pivotal Technologies, Inc.
080043	Pixel Computer Inc.
002003	Pixel Power Ltd.
00E09F	Pixel Vision
005069	PixStream Incorporated
0040EA	Plain Tree Systems Inc.
0090CC	Planex Communications, Inc.
08004B	Planning Research Corp.
080063	Plessey
00C0E8	Plexcom, Inc.
00802E	Plexcom, Inc.—IEEE assigned to CASTLE ROCK COMPUTING
00E012	Pluto Technologies

OUI	Vendor/Comment
00E004	Pmc-Sierra, Inc.
00609D	PMI Food Equipment Group
00102C	PnP Technology A/S
001085	Polaris Communications, Inc.
0040F7	Polaroid Medical Imaging Sys.
0090FB	Portwell, Inc.
00E09A	Positron Industries, Inc.
006012	Power Computing Corporation
006078	Power Measurement Ltd.
00D03C	Power Micro Research
008078	Practical Peripherals, Inc.
00C090	Praim S.R.L.
00906E	Praxon, Inc.
00402E	Precision Software, Inc.
0020E0	PreMax Electronics, Inc.—PE-200 (PCMCIA NE2000-clone card, sold by InfoExpress)
00A01B	Premisys Communications, Inc.
00D04A	Presence Technology GmBh + Co. KG
00E08D	Pressure Systems, Inc.
002083	Presticom Incorporated
0020CB	Pretec Electronics Corp.
00A0ED	Pri Automation
00209C	Primary Access Corp.
08002F	Prime Computer—Prime 50-Series LHC300
0020B8	Prime Option, Inc.
00D013	Primex Aerospace Company
00401B	Printer Systems Corp
008058	Printer Systems Corporation
006077	Prisa Networks
00D011	Prism Video, Inc.
00C07A	Priva BV
0060B2	Process Control Corp.
001012	Processor Systems (I) PVT Ltd.
004091	Procomp Industria Eletronica
000028	Prodigy Systems Corporation
0090C9	Productivity Enhancement Products., Inc.
005077	Prolific Technology, Inc.
00804A	PRO-LOG
00908E	Promatory Communications
00E066	Promax Systems, Inc.
00E03B	Prominet Corporation
00204A	Pronet GmBh
00C0FD	Prosum
002064	Protec Microsystems, Inc.
009005	Protech Systems Co., Ltd.
000093	Proteon, Inc.—Proteon sends bit reversed as 0000C9 on Token Ring
4000C9	Proteon—Token Ring Locally Administered Address of OUI 000093—bit reversed
0010C5	Protocol Technologies, Inc.
00900A	Proton Electronic Industrial Co., Ltd.
0020A6	Proxim, Inc.
00E038	Proxima Corporation
006063	Psion Dacom Plc.
00C00E	Psitech Inc
100000	PUBLIC use assigned by IEEE
0020C0	Pulse Electronics, Inc.
0000D6	Punch Line Holding
0000D4	Pure Data Limited
00A061	Puritan Bennett
08008B	Pyramid Technology Corp.
48008B	Pyramid Technology Corp.—Token Ring Locally Administered Address of OUI 08008B
0090EC	Pyrescom
00E08B	Qlogic Corp.
004044	Qnix Computer Co., Ltd.
00C00F	QNX Software Systems Ltd.—(also Quantum Software Systems Ltd)—IEEE assigned to QUANTUM SOFTWARE SYSTEMS

OUI	Vendor/Comment
000031	Qpsx Communications Pty Ltd.
00A09B	Qpsx Communications, Ltd.
006074	QSC Audio Products
00C0C1	Quad/Graphics Inc.
0000E0	Quadram Corp.
00A0C6	Qualcomm Incorporated
006051	Quality Semiconductor
00C0D5	Quancom Electronic GmBh
00C09F	Quanta Computer Inc
0080F5	Quantel Ltd.
00302D	Quantum Bridge Communications
00E09E	Quantum Corporation
00105C	Quantum Designs (H.K.) Ltd.
0060E2	Quest Engineering & Dev.
00A048	Questech, Ltd.
00603A	Quick Controls Ltd.
0040E3	Quin Systems Ltd.
00C035	Quintar Company
00C0D9	Quinte Network Confidentiality Equipment Inc
00407C	Qume Corporation
0000F9	Quotron Systems Inc.
0060FF	Quvis, Inc.
00D018	Qwes. Com, Inc.
00A0C2	R.A. Systems Co., Ltd.
00A00A	R.D.C. Communication
004095	R.P.T. Intergroups Int'l Ltd.—Eagle Technology and UMC seen under this OUI
420701	Racal Datacom—old Interlan interface—Token Ring Locally Administered Address of OUI 020701
00D0FA	Racal Guardata
001006	Racal Recorders Ltd.
020701	Racal-Datacom
027001	Racal-Datacom
000701	Racal-Datacom—Lan Internetworking Division
000058	Racore Computer Products Inc.
0020D2	RAD Data Communications Ltd, then Armon now owned by Bay Networks, owned by Nortel
00A0F5	Radguard Ltd.
00E00D	Radiant Systems
00A0D4	Radiolan, Inc.
000050	Radisys Corporation
00808E	Radstone Technology
001065	Radyne Corporation
002052	Ragula Systems
0010E9	Raidtec Ltd.
0030F7	Ramix Inc.
000025	Ramtek Corp.
0060E7	RanData
00E016	Rapid-City Communications (bought by Bay Networks bought by Nortel)
00108F	Raptor Systems
004021	Raster Graphics
00C038	Raster Image Processing System
00403E	Raster Ops Corporation
0080EF	Rational
00C0D0	Ratoc System Inc.
00905E	Rauland-Borg Corporation
0080F2	Raycom Systems Inc.
00408B	Raylan Corporation
00905B	Raymond And Lae Engineering
0020F3	Raynet Corporation
00C036	Raytech Electronic Corp
00008F	Raytheon Company
00E0E7	Raytheon E-Systems, Inc.
00009A	RC Computer A/S
00D081	Real Time Devices Usa, Inc.
00E04C	Realtek Semiconductor Corp.
004F49	Realtek Semiconductor Corp. seen under this OUI—Not an IEEE assigned OUI—IEEE assigned OUI 00E04C to Realtek
006052	Realtek—(RTL 8029 == PCI NE2000)—IEEE assigned to PERIPHERALS ENTERPRISE CO., Ltd.
0000FB	Rechner zur Kommunikation
001067	Redback Networks, Inc.
00A025	Redcom Labs Inc.
006073	Redcreek Communications, Inc.
00901A	Redstone Communications
008036	Reflex Manufacturing Systems
00D021	Regent Electronics Corp.
00A0A3	Reliable Power Meters
00C09B	Reliance Comm/Tec, R-Tec
0010A8	Reliance Computer Corp.
00A0A8	Renex Corporation
0080A6	Republic Technology Inc.
00007B	Research Machines
0060AC	Resilience Corporation
00E0CB	Reson, Inc.
080090	Retix, Inc.—Bridges
002076	Reudo Corporation
00A0E8	Reuters Holdings Plc
0060ED	Ricardo Test Automation Ltd.
008063	Richard Hirschmann GmBh & Co.
000074	Ricoh Company Ltd.
080068	Ridge Computers
480068	Ridge Computers—Token Ring Locally Administered Address of OUI 080068
005092	Rigaku Industrial Corporation
0030DC	Rightech Corporation
004040	Ring Access, Inc.
001050	Rion Co., Ltd.
005045	Rioworks Solutions, Inc.
00C07D	RISC Developments Ltd.
0010DF	Rise Computer Inc.
00C08B	RISQ Modular Systems Inc.
0010F0	Rittal-Werk Rudolf LOH GmBh & Co. KG
002095	Riva Electronics
003036	RMP Elektroniksysteme GmBh
0000B0	RND (RND-RAD Network Devices)
006034	Robert Bosch GmBH
00D03D	Rocketchips, Inc.
0090B8	Rohde & Schwarz GmBh & Co. KG
0040AB	Roland DG Corporation
00009C	ROLM MIL-SPEC Computers
00E00B	Rooftop Communications Corp.
00D014	Root, Inc.
0040A4	Rose Electronics
000068	Rosemount Controls
00A039	Ross Technology, Inc.
002034	Rotec Industrieautomation GmBh
009014	Rotork Instruments, Ltd.
080030	Royal Melbourne Inst Of Tech
0010EC	RPCG, LLC
00508C	RSI Systems
005017	RSR S.R.L.
0040C7	Ruby Tech Corporation—Danpex Corporation seen under this OUI
002078	Runtop, Inc.
0080FA	RWT GmBh
00C01F	S.E.R.C.E.L.
0010E1	S.I. Tech, Inc.
004085	SAAB Instruments AB
08005F	Saber Technology Corp.
003041	Saejin T & M Co., Ltd.
001077	Saf Drive Systems, Ltd.
00D030	Safetran Systems Corp.
005089	Safety Management Systems
0060BA	Sahara Networks, Inc.
008071	SAI Technology
00604A	SAIC Ideas Group

OUI	Vendor/Comment
00E0C3	Sakai System Development Corp.
005096	Salix Technologies, Inc.
004006	Sampo Technology Corporation
009013	Samsan Corp.
00E099	Samson AG
00E064	Samsung Electronics
0000F0	Samsung Electronics Co. Ltd.
006039	Sancom Technology, Inc.
00A0B6	Sanritz Automation Co., Ltd.
003063	Santera Systems, Inc.
00E04E	Sanyo Denki Co., Ltd.
08007B	Sanyo Electric Co. Ltd.
00E09D	Sarnoff Corporation
0040D5	Sartorius AG
00D054	Sas Institute Inc.
0020A8	Sast Technology Corp.
00604C	SAT
00E0A8	Sat GmBh&Co
0050B4	Satchwell Control Systems, Ltd
001022	Satcom Media Corporation
008002	Satelcom (Uk) Ltd.
026086	Satelcom MegaPac (UK)—IEEE assigned to LOGIC REPLACEMENT TECH. LTD.
00200E	Satellite Technology Mgmt, Inc.
00500D	Satori Electoric Co., Ltd.
008094	Sattcontrol AB—IEEE assigned to ALFA LAVAL AUTOMATION AB
009098	SBC Designs, Inc.
0000D2	SBE Inc.
00A0D6	SBE, Inc.
00D01C	SBS Technologies, Connectivity Products
000021	SC&C—Sureman Comp. & Commun. Corp.
00605D	Scanivalve Corp.
008022	Scan-Optics
006076	Schlumberger Technologies
08003C	Schlumberger Well Services/Austin Engineering Services
00806B	Schmid Telecommunication
0030A7	Schweitzer Engineering Laboratories, Inc.
005042	SCI Manufacturing Singapore PTE., Ltd.
00407B	Scientific Atlanta
002039	SCINets
00A046	Scitex Corp. Ltd.
000057	Scitex Corporation Ltd.
00A0FD	Scitex Digital Printing, Inc.
0020B3	Scltec Communications Systems
00D016	Scm Microsystems, Inc.
00C065	Scope Communications Inc
002062	Scorpion Logic, Ltd.
006002	Screen Subtitling Systems, Ltd.
00E048	SDL Communications, Inc.
002037	Seagate Technology
004037	Sea-Ilan, Inc.
00C008	Seco Srl
00C070	Sectra Secure-Transmission AB
0030F6	Securelogix Corporation
00E0FF	Security Dynamics Technologies
0030DB	Sederta Inc.
00C0A7	SEEL Ltd.
00A07D	Seeq Technology, Inc.
00D0F1	Sega Enterprises, Ltd.
00C0EB	SEH Computertechnik GmBh
000048	Seiko Epson Corporation
080083	Seiko Instrum. And Electronics
008015	Seiko Systems Inc
0000B8	Seikosha Co., Ltd.—Cabletron seen on Token Ring—bit reversed of assigned OUI 00001D
00E0C0	Seiwa Erectric Mfg. Co., Ltd.
009052	Selcom Elettronica S.R.L.
00901E	Selesta Ingegne Ria S.P.A.

OUI	Vendor/Comment
0010EB	Selsius Systems, ILNC.
00C05A	Semaphore Communications Corp.
00E0CD	Sensis Corporation
00607D	Sentient Networks Inc.
0090EB	Sentry Telecom Systems
080047	Sequent
480047	SEQUENT COMPUTER SYSTEMS INC.—Token Ring Locally Administered Address of OUI 080047
00C002	Sercomm Corporation
002002	Seritech Enterprise Co., Ltd.
0050DD	Serra Soldadura, S.A.
00306F	Seyeon Tech. Co., Ltd.
002007	SFA, Inc.
000026	Sha-Ken Co., Ltd.
00E00F	Shanghai Baud Data Communication Development Corp.
0030AD	Shanghai Communication
00D06C	Sharewave, Inc.
08001F	Sharp
009039	Shasta Networks (now Nortel Networks)
00E060	Sherwood
00A0B2	Shima Seiki
00E096	Shimadzu Corporation
00D0A9	Shinano Kenshi Co., Ltd.
003021	Shing Tech. Enterprise Co., Ltd.
00C0F1	Shinko Electric Co Ltd
0010AE	Shinko Electric Industries Co.
0080D2	Shinnihondenko Co., Ltd.
0080D3	Shiva Corp.—Appletalk-Ethernet interface
004094	Shographics Inc
0060E6	Shomiti Systems Incorporated
00C0C0	Shore Microsystems Inc.
001049	Shoreline Teleworks, Inc.
002016	Showa Electric Wire & Cable Co.
00E0E0	Si Electronics, Ltd.
00D0BD	Sican GmBh
00C0C5	SID Informatica
002096	Siebe Environmental Controls
000073	Siecor Corporation—DuPont seen under this OUI
080006	Siemens AG—PC clone
0030E6	Siemens Medical Systems
005007	Siemens Telecommunication Systems Limited
00A0D5	Sierra Wireless Inc.
0050F4	Sigmatek GmBh & Co. Kg
0000E5	Sigmex Ltd.
0050DE	Signum Systems Corp.
0000BA	SIIG, Inc.
0040CC	Silcom Manufacturing Technology Inc.—sends bit reversed as 000230 on Token Ring
000230	Silcom Manufacturing—bit revereseed of assigned OUI 40CC—Not an IEEE assigned OUI
00E0ED	Silicom, Ltd.
080069	Silicon Graphics Inc.
00006B	Silicon Graphics Inc. (SGI)/MIPS
480069	Silicon Graphics Inc. (SGI)—Token Ring Locally Administered Address of OUI 080069
080079	Silicon Graphics—probably a typo of SGI assigned OUI 080069—IEEE assigned to THE DROID WORKS
00E006	Silicon Integrated Sys. Corp.
00C039	Silicon Systems—IEEE assigned to TDK SEMICONDUCTOR CORPORATION
00C0AA	Silicon Valley Computer
00003E	Simpact
002005	Simple Technology
009072	Simrad AS
009010	Simulation Laboratories, Inc.
00D0D1	Sirocco Systems, Inc.
001019	Sirona Dental Systems GmBh & Co., KG
0030D8	Sitek

OUI	Vendor/Comment
00A01D	Sixnet
00E0E3	SK-Elektronik GmBh
00A02D	Skipstone, Inc.
00205B	Skyline Technology
00301D	Skystream, Inc.
0040AD	SMA Regelsysteme GmBh
00301A	SmartBridges PTE. Ltd.
00D091	Smartsan Systems, Inc.
004048	SMD Informatica S.A.
003001	SMP
008034	SMT-Goupil
0060D0	SNMP Research Incorporated
0080FF	Soc. De Teleinformatique RTC
00407A	Societe d'Exploitation du Cnit
00002E	Societe Evira
00C01B	Socket Communications, Inc.
001056	Sodick Co., Ltd.
000067	Soft * Rite, Inc.
003039	Softbook Press
0080B1	Softcom A/S
009089	Softcom Microsystems, Inc.
00601E	Softlab, Inc.
0010AD	Softronics USB, Inc.
0080C6	Soho—IEEE assigned to NATIONAL DATACOMM CORPORATION
00D0F3	Solari Di Udine SPA
00008E	Solbourne Computer Inc.
002045	SolCom Systems Limited
00C061	Solectek Corporation
0010E5	Solectron Texas
00D0E5	Solidum Systems Corp.
0060C4	Soliton Systems K.K.
00D056	Somat Corporation
00C056	Somelec
00C0E1	Sonic Solutions
00A04B	Sonic Systems Inc.—EtherFE 10—IEEE assigned to TFL LAN INC.
004010	Sonic—Mac Ethernet interfaces
00208A	Sonix Communications, Ltd.
003093	Sonnet Technologies, Inc.
00106B	Sonus Networks, Inc.
080046	Sony
000095	Sony Tektronix Corp.
0080B4	Sophia Systems
00600A	Sord Computer Corporation
00A0CF	Sotas, Inc.
006006	Sotec Co., Ltd.
00200A	Source-Comm Corp.
00302A	Southern Information System, Inc.
00807E	Southern Pacific Ltd.
00502C	Soyo Computer, Inc.
00404F	Space & Naval Warfare Systems
00A0AA	Spacelabs Medical
00C0C7	Sparktrum Microsystems, Inc.
00E02E	SPC Electronics Corporation
0050EF	SPE Systemhaus GmbH
00904D	SPEC S.A.
0080BA	Specialix (Asia) Pte, Ltd.
00D09B	Spectel Ltd.
00A0D8	Spectra - Tek
0090A5	Spectra Logic
000035	Spectragraphics Corporation (Lan Manufacturing Engineer)
00907A	Spectralink Corp.
0020F4	Spectrix Corporation
0080A5	Speed International
00E01B	Sphere Communications, Inc.
008056	Sphinx Elektronik GmBh
0000F2	Spider Communications— (Montreal, not Spider Systems)
080039	Spider Systems Ltd.

OUI	Vendor/Comment
480039	Spider Systems Ltd.—Token Ring Locally Administered Address of OUI 080039
005019	Spring Tide Networks, Inc.
00A042	Spur Products Corp.
00506B	SPX-ATEG
008067	Square D Company
00D06B	Sr Telecom Inc.
00403F	Ssangyong Computer Systems Corporation
005016	SST/Woodhead Industries
00A003	Staefa Control System
00601F	Stallion Technologies
00E029	Standard Microsystems Corp.
00800F	Standard Microsystems Corp. (SMC)
08000F	Standard Microsystems Corp. (SMC)—probably typo of assigned OUI 00800F—IEEE assigned to MITEL CORPORATION
004027	Standard Microsystems Corp—SMC Massachusetts
080056	Stanford Linear Accel. Center
080055	Stanford Telecomm. Inc.
00A052	Stanilite Electronics Pty. Ltd
008059	Stanley Electric Co., Ltd.
0000E7	Star Gate Technologies
004052	Star Technologies Inc
0030F4	Stardot Technologies
001063	Starguide Digital Networks
0040BD	Starlight Networks Inc
003060	Starmatix, Inc.
001098	Starnet Technologies, Inc.
00405D	Star-Tek Inc.
00A0F3	Staubli
080015	STC Business Systems
0040B7	Stealth Computer Systems
002047	Steinbrecher Corp.
0080B7	Stellar Computer
006018	Stellar One Corporation
0080E1	STMicroelectronics
008025	Stollmann GmBh
00D08B	Storage Area Networks, Ltd.
0010D4	Storage Computer Corporation
006085	Storage Concepts
00104F	Storage Technology Corporation
00C03F	Stores Automated Systems Inc
00E001	Strand Lighting Limited
00D05E	Stratabeam Technology, Inc.
00C043	Stratacom
00D07F	Strategy & Technology, Limited
0000A8	Stratus Computer, Inc.
0040CF	Strawberry Tree Inc
0060EA	Streamlogic
00A0DF	STS Technologies, Inc.
0040D7	Studio Gen Inc.
005059	Suite Technology Systems Network
00005F	Sumitomo Electric Ind., Ltd.
00808A	Summit Microsystems Corp.
0080F3	Sun Electronics Corp.
080020	Sun Microsystems Inc.
480020	Sun Microsystems, Inc.—Token Ring Locally Administered Address of OUI 080020
0020F2	Sun Microsystems, Inc. (formerly Maximum Strategy Inc.)
0010ED	Sundance Technology, Inc.
002074	Sungwoon Systems
00D0DD	Sunrise Telecom, Inc.
00E0D7	Sunshine Electronics, Inc.
08006C	Suntek Technology Int'l
0040AC	Super Workstation, Inc.
0080EC	Supercomputing Solutions, Inc.
002065	Supernet Networking Inc.
000084	Supernet—Aquila and ADI Systems Inc. seen under this OUI
003006	Superpower Computer Electronics Co., Ltd.
00803D	Surigiken Co., Ltd.

OUI	Vendor/Comment
00104D	Surtec Industries, Inc.
00604F	Suzuki Mfg. Co., Ltd.
00C06C	SVEC Computer Corp
00509B	Switchcore AB
0060A0	Switched Network
0040E5	Sybus Corporation
009097	Sycamore Networks
00302C	Sylantro Systems Corporation
00D05A	Symbionics, Ltd.
00A0B8	Symbios Logic Inc.
00A0F8	Symbol Technologies, Inc.
480005	SYMBOLICS INC.—Token Ring Locally Administered Address of OUI 080005
080005	Symbolics—Symbolics LISP machines
00602A	Symicron Computer Communications, Ltd.
0000BF	Symmetric Computer Systems
0020FF	Symmetrical Technologies
00E0BC	Symon Communications, Inc.
0040FE	Symplex Communications
002033	Synapse Technologies, Inc.
00406F	Sync Research Inc
002080	Synergy (UK) Ltd.
0080F6	Synergy Microsystems
00403B	Synerjet International Corp
00803E	Synernetics
000081	SynOptics Comunications Inc. (then Bay Networks, now Nortel)
00D017	Syntech Information Co., Ltd.
00C0D2	Syntellect Inc
00003F	Syntrex Inc.
00406B	Sysgen
00005A	Syskonnect—A Business Unit of Schneider & Koch Co. GmBh
00A010	Syslogic Datentechnik Ag
00D03F	Sysmate Co., Ltd.
008044	Systech Computer Corp.
009060	System Create Corp.
0040B2	Systemforschung
0040F8	Systemhaus Discom Distributed Computing GmBh
080058	Systems Concepts
0090E0	Systran Corp.
002023	T.C. Technologies PTY. Ltd
0080E2	T.D.I. Co., Ltd. Development Div.
006096	T.S. Microtech Inc.
001016	T.Sqware
002089	T3Plus Networking, Inc.
005006	Tac AB
0010AF	TAC Systems, Inc.
004009	Tachibana Tectron Co., Ltd. Systematic Equipment Division
00106E	Tadiran Com. Ltd.
00A0C7	Tadiran Telecommunications
000083	Tadpole Technology PLC—Olicom seen on Token Ring—bit reveresed of assigned OUI 0000C1
00509A	Tag Electronic Systems
00D0DA	Taicom Data Systems Co., Ltd.
0020C1	Taiko Electric Works, Ltd.
0090BB	Tainet Communication System Corporation
003091	Taiwan First Line Elec. Corp.
00C0D7	Taiwan Trading Center DBA TTC Computer Products
00202D	Taiyo Corporation
0080BF	Takaoka Electric Mfg. Co. Ltd.
000066	Talaris Systems, Inc.
004038	Talent Electric Incorporated
00E051	Talx Corporation
00C096	Tamura Corporation
00200F	Tanbac Co., Ltd.
005060	Tandberg Telecom AS
00E02A	Tandberg Television As
08008E	Tandem Computers
00D074	Taqua Systems, Inc.
0050DC	Tas Telefonbau A. Schwabe GmBh & Co. KG
00E0D9	Tazmo Co., Ltd.
0000DD	TCL Incorporated—Gould seen under this OUI
004083	TDA Industria de Produtos
008098	TDK Corporation
006059	Technical Communications Corp.
00E005	Technical Corp.
00E0B4	Techno Scope Co., Ltd.
001073	Technobox, Inc.
00D069	Technologic Systems
00D0CC	Technologies Lyre Inc.
00D036	Technology Atlanta Corp.
0050B0	Technology Atlanta Corporation
00C0BF	Technology Concepts Ltd.
008035	Technology Works
00D05C	Technotrend Systemtechnik GmBh
0020EC	Techware Systems Corp.
080073	Tecmar Inc.
008089	Tecnetics (Pty) Ltd.
00E020	Tecnomen Oy
0090DC	Teco Information Systems Co., Ltd.
004049	Tegimenta AG
00D0FB	Tek Microsystems, Incorporated
00C0AF	Teklogix Inc.
0090E5	Teknema, Inc.
002022	Teknique, Inc.
00A0A5	Teknor Microsysteme, Inc.
00608F	Tekram Technology Co., Ltd.
100088	Tektronix, Inc. (network printers)—seen on Token Ring—bit reversed of assigned OUI 08000E—Not an IEEE assigned OUI
080011	Tektronix, Inc. (network printers)—sends bit reversed as 100088 on Token Ring
00A026	Teldat, S.A.
00C033	Telebit Communications APS
0040FF	Telebit Corporation—Personal NetBlazer
00C0BC	Telecom Australia/CSSC
009059	Telecom Device K.K.
00404D	Telecommunications Techniques
008037	Telefon Ab Lm Ericsson Corp.
0050E3	Telegate
0090BC	Telemann Co., Ltd.
00C014	Telematics Calabasas Int'l, Inc
00005C	Telematics International Inc.
0080F4	Telemechanique Electrique
0090E1	Telena S.P.A.
002029	TeleProcessing Products Inc.—CSU
00E0F5	Teles AG
00C0CC	Telesciences Co Systems, Inc.
009050	Teleste OY
009056	Telestream, Inc.
006062	Telesync, Inc.
004096	Telesystems SLW Inc.
0000B2	Televideo Systems, Inc.
0010BE	Telexis Corp.
00305A	Telgen Corporation
0030FA	Telica, Inc.
00C089	Telindus Distribution
005086	Telkom SA, Ltd.
00A0FF	Tellabs Operations, Inc.
0090FF	Tellus Technology Inc.
004007	Telmat Informatique
00E0B2	Telmax Communications Corp.
0010E8	Telocity, Incorporated (formerly Aspen Internet System, Inc.)
00E0A6	Telogy Networks, Inc.
00E023	Telrad
00E0D1	Telsis Limited
00001E	Telsist Industria Electronica

OUI	Vendor/Comment
0040DA	Telspec Ltd.
0050D7	Telstrat
0000CD	Teltrend (NZ) Limited—Industrial Research Limited seen under this OUI
00A013	Teltrend Ltd.
00601C	Telxon Corporation
00A0D0	Ten X Technology, Inc.
00A0E0	Tennyson Technologies Pty Ltd
0040CD	Tera Microsystems, Inc.
00403D	Teradata
00108A	Teralogic, Inc.
006003	Teraoka Weigh System Pte, Ltd.
005087	Terasaki Electric Co., Ltd.
003017	Terastack Ltd.
00D0C4	Teratech Corporation
00E06F	Terayon Corp.
0020B4	Terma Elektronik AS
0020CF	Test & Measurement Systems Inc.
00802A	Test Systems & Simulations Inc
0090CE	Tetra GmBh
480028	Texas Instruments (TI) Token Ring Locally Administered Address of OUI 080028
100014	Texas Instruments (Token Lite Manufacturing Kit)—bit reversed of assigned OUI 080028—Not an IEEE assigned OUI
080028	Texas Instruments, Inc.—Explorer
0020C2	Texas Memory Systems, Inc.
00A0B4	Texas Microsystems, Inc.
00E0F1	That Corporation
00209A	The 3DO Company
00A06F	The Appcon Group, Inc.
001004	The Brantley Coile Company, Inc.
008084	The Cloud Inc.
0080EA	The Fiber Company
0080BD	The Furukawa Electric Co., Ltd.
00408D	The Goodyear Tire & Rubber Co.
00509D	The Industree B.V.
00C086	The Lynk Corporation
0090DD	The Miharu Communications Co., Ltd.
0000BE	The NTI Group
00A00D	The Panda Project
00104A	The Parvuc Corporation
00802C	The Sage Group Plc
0010BD	The Telecommunication
0080B6	Themis Computer
00D0C6	Thomas & Betts Corp.
0001C8	Thomas Conrad Corp. (now owned by Compaq)
008013	Thomas Conrad Corp. (now owned by Compaq)
009064	Thomson Broadcast Systems
001095	Thomson Consumer Electronics
0080EE	Thomson CSF
005052	Tiara Networks, Inc.
08001A	Tiara/ 10net
0060A8	Tidomat AB
00E074	Tiernan Communications, Inc.
080032	Tigan Incorporated
00002F	Timeplex Inc.
00A090	Timestep Corporation
002081	Titan Electronics
002057	Titze Datentechnik GmBh
005008	Tiva Microcomputer Corp. (TMC)
006042	TKS (USA), Inc.
00D0AD	TI Industries
00C09C	TOA Electronic Ltd.
006017	Tokimec Inc.
00D065	Toko Electric
00C0A1	Tokyo Denshi Sekei Co
008091	Tokyo Electric Co., Ltd.
0000A0	Tokyo Sanyo Electic Co. Ltd.
00D0A7	Tokyo Sokki Kenkyujo Co., Ltd.

OUI	Vendor/Comment
00502F	Tollbridge Technologies, Inc.
00103F	Tollgrade Communications, Inc.
080084	Tomen Electronics Corp.
00A07C	Tonyang Nylon Co., Ltd.
0090C3	Topic Semiconductor Corp.
0020FE	Topware Inc. / Grand Computer Corporation
00A0FB	Toray Engineering Co., Ltd.
00A0F0	Toronto Microelectronics Inc.
00E0BF	Torrent Networking
08003B	Torus Systems Limited
000039	Toshiba Corporation—Computer Division
00A03B	Toshin Electric Co., Ltd.
00E045	Touchwave, Inc.
00C03C	Tower Tech S.R.L.
00C0AE	Towercom Co Inc DBA PC House
0060DC	Toyo Communication Equipment
00C050	Toyo Denki Seizo K.K.
00C074	Toyoda Automatic Loom Works Ltd
006053	Toyoda Machine Works, Ltd.
00408A	TPS Teleprocessing Sys. GmBh
00D093	Tq - Components GmBh
00C083	Trace Mountain Products, Inc.
001096	Tracewell Systems, Inc.
008028	Tradpost (HK) Ltd.
00A02A	Trancell Systems
00C0F2	Transition Engineering Inc
00A02B	Transitions Research Corp.
0050A3	Transmedia Communications, Inc
00A0EC	Transmitton Ltd.
00409C	Transware
0060D9	Transys Networks Inc.
00D061	Tremon Enterprises Co., Ltd.
00D06E	Trendview Recorders Ltd.
00106F	Trenton Technology Inc.
004036	Tribe Computer Works, Inc.
00A01F	Tricord Systems, Inc.
0000BB	TRI-DATA Systems Inc.— Netway products, 3274 emulators
00903B	Triems Research Lab, Inc.
00402B	TriGem Computer, Inc.
0060AE	Trio Information Systems AB (formerly OBJECTA)
0040DC	Tritec Electronic GmBh
00E0FA	Trl Technology, Ltd.
009066	Troika Design, Inc.
00C08D	Tronix Product Development
00A069	Truetime
00002A	TRW—SEDD/INP
40002A	TRW—Token Ring Locally Administered Address of OUI 00002A
00503C	Tsinghua Novel Electronics Co., Ltd.
080077	TSL Communications Ltd.
00805A	Tulip Computers International BV
00505C	Tundo Corporation
009083	Turbo Communication, Inc
0030EB	Turbonet Communications, Inc.
00906A	Turnstone Systems, Inc.
00C023	Tutankhamon Electronics
006081	TV/Com International
00803C	TVS Electronics Ltd.
004045	Twinhead Corporation
0030CF	Two Technologies, Inc.
00C0E6	TXPORT
00E081	Tyan Computer Corp.—Onboard Intel 82558 10
0030A0	Tyco Submarine Systems, Ltd.
004087	Ubitrex Corporation
004046	UDC Research Limited
001005	UEC Commercial
00501A	Uisiqn
009051	Ultimate Technology Corp.

OUI	Vendor/Comment
009058	Ultra Electronics Ltd.
00D057	Ultrak, Inc.
00102B	Umax Data Systems, Inc.
0010A7	Unex Technology Corporation
00BBF0	Ungermann-Bass Inc.
00DD00	Ungermann-Bass Inc.
00DD01	Ungermann-Bass Inc.
00DD01	Ungermann-Bass Inc.
00DD02	Ungermann-Bass Inc.
00DD02	Ungermann-Bass Inc.
00DD03	Ungermann-Bass Inc.
00DD03	Ungermann-Bass Inc.
00DD04	Ungermann-Bass Inc.
00DD04	Ungermann-Bass Inc.
00DD05	Ungermann-Bass Inc.
00DD05	Ungermann-Bass Inc.
00DD06	Ungermann-Bass Inc.
00DD06	Ungermann-Bass Inc.
00DD07	Ungermann-Bass Inc.
00DD07	Ungermann-Bass Inc.
00DD08	Ungermann-Bass Inc.
00DD08	Ungermann-Bass Inc.
00DD09	Ungermann-Bass Inc.
00DD09	Ungermann-Bass Inc.
00DD0A	Ungermann-Bass Inc.
00DD0A	Ungermann-Bass Inc.
00DD0B	Ungermann-Bass Inc.
00DD0B	Ungermann-Bass Inc.
00DD0C	Ungermann-Bass Inc.
00DD0C	Ungermann-Bass Inc.
00DD0D	Ungermann-Bass Inc.
00DD0D	Ungermann-Bass Inc.
00DD0E	Ungermann-Bass Inc.
00DD0E	Ungermann-Bass Inc.
00DD0F	Ungermann-Bass Inc.
00DD0F	Ungermann-Bass Inc.
00DD00	Ungermann-Bass Inc.—IBM RS6000
00A0C3	Unicomputer GmBh
0050D8	Unicorn Computer Corp.
00E011	Uniden San Diego
00C0A4	Unigraf OY
00C047	Unimicro Systems Inc
00608D	Unipulse Corp.
0010F9	Unique Systems, Inc.
00003D	Unisys
08000B	Unisys also Ascom-Timeplex (former Unisys subsidiary)
00E05D	Unitec Co., Ltd.
00C067	United Barcode Industries
0080B5	United Networks Inc.
080049	Univation
480049	Univation—Token Ring Locally Administered Address of OUI 080049
009095	Universal Avionics
00C0D8	Universal Data Systems
0020F0	Universal Microelectronics Co.
008046	University of Toronto Dept. of Electrical Engin'ing
00D015	Univex Microtechnology Corp.
0000EA	Upnod AB
00D01A	Urmet Sud S.P.A.
00C0ED	US Army Electronic Proving Ground
00C049	US Robotics Total Control (tm) NETServer Card (now owned by 3Com)
0020EF	USC Corporation
00005E	USC Information Sciences—IANA—Internet Ass'ned Nos. Authority
0010C6	USI
00D039	Utilicom, Inc.
00E08E	Utstarcom
00C087	UUNET Technologies Inc
00A0F7	V.I Computer Corp.

OUI	Vendor/Comment
0090B0	VADEM
00D05F	Valcom, Inc.
0090BA	Valid Networks, Inc.
0040AA	Valmet Automation Inc.
0050CF	Vanlink Communication
006093	Varian
00C05E	Vari-Lite Inc
00803A	Varityper, Inc.
00D0D4	V-Bits, Inc.
00D0E7	Vcon Telecommunication Ltd.
008041	VEB Kombinat Robotron
001039	Vectron Systems GmBh
003083	Vedia Technology, Ltd.
003087	Vega Grieshaber KG
005058	Vegastream Limited
0050F7	Venture Manufacturing (Singapore) Ltd.
00008D	Verdix Corporation
00A06A	Verilink Corp.
00D09D	Veris Industries
080082	Veritas Software
0020D0	Versalynx Corporation—"The One Port" terminal server
00E0D2	Versanet Communications, Inc.
0010F4	Vertical Networks, Inc.
00907E	Vetronix Corp.
000030	VG Laboratory Systems Ltd.—Cisco Systems seen on Token Ring—bit reversed of assigned OUI 00000C
004063	Via Technologies, Inc.
0030A6	Vianet Technologies, Ltd.
00A0BC	Viasat, Incorporated
00E0DB	Viavideo Communications
00908D	Vickers Electronics Systems
006022	Vicom Systems, Inc.
008088	Victor Company Of Japan, Ltd
0020C9	Victron BV
00800C	Videcom Limited
00A071	Video Lottery Technologies, Inc
0040B5	Video Technology Computers Ltd.
00507C	Videocon AG
00506D	Videojet Systems International Inc.
0060CD	Videoserver, Inc.
001008	Vienna Systems Corporation
006026	Viking Components, Inc.
006030	Village Tronic—used on Amiga
006049	Vina Technologies
0020D5	Vipa GmBh
00E0C8	Virtual Access, Ltd.
00D03B	Vision Products Pty. Ltd.
009025	Vision Systems Ltd. Pty
003010	Visionetics International
0040C0	Vista Controls Corporation
00A00E	Visual Networks, Inc.
000022	Visual Technology Inc.—Castelle Corp seen on Token Ring—bit reversed of assigned OUI 000044
400022	Visual Technology Inc.—Token Ring Locally Administered Address of OUI 000022
0080A8	Vitacom Corporation
00E043	Vitalcom
48007C	Vitalink Communications Corp.—Token Ring Locally Administered Address of OUI 08007C
08007C	Vitalink—TransLAN III
00D0C3	Vivid Technology Pte, Ltd.
00109B	Vixel Corporation
00D02F	Vlsi Technology Inc.
002038	VME Microsystems International
006046	Vmetro, Inc.
005056	Vmware, Inc.
00C094	VMX INC.

OUI	Vendor/Comment
00D0A3	Vocal Data, Inc.
00A04E	Voelker Technologies, Inc.
0050B3	Voiceboard Corporation
0010CE	Volamp, Ltd.
00A0A7	Vorax Corporation
00800D	Vosswinkel FU
0060A1	VPNet
0050FB	VSK Electronics
08005B	Vta Technologies Inc.
006072	VXL Instruments, Limited
00E06B	W&G Special Products
00D029	Wakefern Food Corporation
009067	Walkabout Computers, Inc.
00E07E	Walt Disney Imagineering
008016	Wandel & Goltermann
0000D3	Wang Laboratories Inc.
00907F	Watchguard Technologies, Inc.
0000C4	Waters Div. Of Millipore
00A001	Watkins-Johnson Company
00607C	Waveaccess, Ltd.
00600E	Wavenet International, Inc.
006079	Wavephore Networks, Inc.
0090C8	Waverider Communications (Canada) Inc.
0060C1	Wavespan Corporation
009011	Wavtrace, Inc.
004078	Wearnes Automation PTE Ltd
00E077	Webgear, Inc.
001089	Websonic
00E0F3	Websprint Communications, Inc.
000018	Webster Computer Corporation—Appletalk
0060BE	Webtronics
00E01D	Webtv Networks, Inc.
001020	Welch Allyn, Data Collection
0000A2	Wellfleet (then Bay Networks now Nortel)—Wellfleet sends bit reversed as 000045 on Token Ring
00202C	Welltronix Co., Ltd.
0030B2	Wescam - Healdsburg
00404A	West Australian Department of Employment (Devet)
002028	West Egg Systems, Inc.
00E05B	West End Systems Corp.
00808D	Westcove Technology B.V.
00600F	Westell, Inc.
0090A9	Western Digital
4000C0	Western Digital (now owned by SMC)—Token Ring Locally Administered Address of OUI 0000C0
0000C0	Western Digital now SMC (Standard Microsystems Corp.)
004003	Westinghouse Process Control Division
00A0E1	Westport Research Associates, Inc.
006061	Whistle Communications Corp.
0020A9	White Horse Industrial
006088	White Mountain DSP, Inc.
08006D	Whitechapel Computer Works
0090D8	Whitecross Systems
00A009	Whitetree Network Technologies, Inc.
00C03D	Wiesemann & Theis GmBh
001030	Wi-Lan, Inc.
00100A	Williams Communications Group Advanced Technologies
0010C2	Willnet, Inc.
0090B4	Willowbrook Technologies
00E0A0	Wiltron Co.
00101D	Winbond Electronics Corp.
004047	Wind River Systems
004012	Windata, Inc.
001097	Winnet Metropolitan Communications Systems, Inc.
00D066	Wintriss Engineering Corp.
002063	Wipro Infotech Ltd.
00A0BF	Wireless Data Group Motorola

OUI	Vendor/Comment
0090CB	Wireless Online, Inc.
00203B	Wisdm Ltd.
00207A	Wise Communications, Inc.
0010D0	Witcom, Ltd.
00408F	WM-Data Minfo AB
00A0AF	WMS Industries
002072	Worklink Innovations
00A015	Wyle
0005A8	Wyle Electronics—PowerComputing—Mac clones seen under this OUI. IEEE assigned PowerComputing OUI 006012
008064	Wyse Technology
00609E	X3 - Information Technology
00206E	Xact, Inc.
00C075	Xante Corporation
00E0AE	Xaqti Corporation
006089	XATA
004017	XCD Inc.— XJet - HP printer server card
003038	XCP, Inc.
00C073	Xedia Corporation
4000AA	Xerox Corp Office Systems Div—Token Ring Locally Administered Address of OUI 0000AA
080072	Xerox Corp Univ Grant Program
0000AA	Xerox Corp.— Xerox machines—Milan seen on Token Ring—bit reversed of assigned OUI 0040C8
000000	Xerox Corporation
000009	Xerox Corporation
000004	Xerox Corporation—OFFICE SYSTEMS DIVISION
000005	Xerox Corporation—OFFICE SYSTEMS DIVISION
000006	Xerox Corporation—OFFICE SYSTEMS DIVISION
000007	Xerox Corporation—OFFICE SYSTEMS DIVISION
000008	Xerox Corporation—OFFICE SYSTEMS DIVISION
000001	Xerox Corporation—ZEROX SYSTEMS INSTITUTE
000002	Xerox Corporation—ZEROX SYSTEMS INSTITUTE
000003	Xerox Corporation—ZEROX SYSTEMS INSTITUTE—SMC seen on Token Ring—SMC assigned OUIs 00800F, 00E029, 004027
005500	Xerox reported—Intel sends on Token Ring bit reversed of assigned OUI 00AA00—Not an IEEE assigned OUI
00801E	Xinetron, Inc.
080080	XIOS—IEEE assigned to AES DATA INC.
00D0B2	Xiotech Corporation
00D0ED	Xiox
0001E3	Xircom pocket adapters seen on Token Ring—bit reversed of assinged OUI 0080C7—Not an IEEE assigned OUI
0080C7	Xircom, Inc.—Xircom pocket adapters send bit reversed as 0001E3 on Token Ring
0010A4	Xircom—RealPort 10
0050B9	Xitron Technologies, Inc.
00A0E3	XKL Systems Corp.
00402F	Xlnt Designs Inc (XDI)
040AE0	Xmit AG Computer Networks
00D047	Xn Technologies
0020DB	Xnet Technology, Inc.
00507A	Xpeed, Inc.
00A089	Xpoint Technologies, Inc.
00D025	Xrosstech, Inc.
0080E0	XTP Systems Inc
006000	Xycom Inc.
00D095	Xylan Corporation
0020DA	Xylan Corporation—sends bit reversed as 00045B on Token Ring
00045B	Xylan—Token Ring Switches—bit reversed of assigned OUI 0020DA—Not an IEEE assigned OUI
00802D	Xylogics, Inc.—Annex terminal servers
00A09C	Xyplex, Inc.
080087	Xyplex—terminal servers
0050CC	Xyratex
008093	Xyron Corporation
48008D	Xyvision Inc.—Token Ring Locally Administered Address of OUI 08008D

OUI	Vendor/Comment
08008D	XyVision—XyVision machines
00A0DE	Yamaha Corporation
005001	Yamashita Systems Corp.
002004	Yamatake-Honeywell Co., Ltd.
008068	Yamatech Scientific Ltd.
00D032	Yano Electric Co., Ltd.
0020B5	Yaskawa Electric Corporation
000064	Yokogawa Digital Computer Corp
004059	Yoshida Kogyo K.K.
00C099	Yoshiki Industrial Co.,Ltd.
0000F7	Youth Keep Enterprise Co.Ltd.
009016	ZAC
00C06A	Zahner-Elektrik GmBh & Co KG
0020BB	Zax Corporation
0010FA	Zayante, Inc.
0060B3	Z-Com, Inc.
00C0DE	ZComm Inc.
009048	Zeal Corporation
0080F7	Zenith Communications Products
00E0DD	ZENITH ELECTRONICS CORPORATION
00202F	Zeta Communications, Ltd.
00102A	ZF Microsystems, Inc.
00D0D0	Zhongxing Telecom Ltd.
008050	Ziatech Corporation
0010FB	Zida Technologies Limited
009023	Zilog Inc.
00C095	Znyx (Network Appliance)—Jupiter Systems (MX-700)—Apple (G3)—all seen—IEEE assigned to ZNYX
00D043	Zonal Retail Data Systems
005022	Zonet Technology, Inc.
00D03A	Zoneworx, Inc.
00A0B3	Zykronix
009017	Zypcom, Inc.
00A0C5	Zyxel Communication
004001	Zyxel Communications, Inc.—Zero One Technology Co Ltd. seen under this OUI—IEEE has not assigned any OUIs to Zero One

RMON MIB Assignments

1.3.6.1.2.1.16
iso.org.dod.internet.mgmt.mib-2.rmon

1. Statistics - The statistics group contains statistics measured by the probe for each monitored interface on this device. These statistics take the form of free running counters that start from zero when a valid entry is created. This group currently has statistics defined only for Ethernet interfaces. Each etherStatsEntry contains statistics for one Ethernet interface. The probe must create one etherStats entry for each monitored Ethernet interface on the device. In the future other groups will be defined for other media types including Token Ring and FDDI. These groups should follow the same model as the ethernet statistics group.

1. etherStatsTable
1. etherStatsIndex
2. etherStatsDataSource
3. etherStatsDropEvents
4. etherStatsOctets
5. etherStatsPkts
6. etherStatsBroadcastPkts
7. etherStatsMulticastPkts
8. etherStatsCRCAlignErrors
9. etherStatsUndersizePkts
10. etherStatsOversizePkts
11. etherStatsFragments
12. etherStatsJabbers
13. etherStatsCollisions
14. etherStatsPkts64Octets
15. etherStatsPkts65to127Octets
16. etherStatsPkts128to255Octets
17. etherStatsPkts256to511Octets
18. etherStatsPkts512to1023Octets
19. etherStatsPkts1024to1518Octets
20. etherStatsOwner
21. etherStatsStatus

2. tokenRingMLStatsTable
1. tokenRingMLStatsIndex
2. tokenRingMLStatsDataSource
3. tokenRingMLStatsDropEvents
4. tokenRingMLStatsMacOctets
5. tokenRingMLStatsMacPkts
6. tokenRingMLStatsRingPurgeEvents
7. tokenRingMLStatsRingPurgePkts
8. tokenRingMLStatsBeaconEvents
9. tokenRingMLStatsBeaconTime
10. tokenRingMLStatsBeaconPkts
11. tokenRingMLStatsClaimTokenEvents
12. tokenRingMLStatsClaimTokenPkts
13. tokenRingMLStatsNAUNChanges
14. tokenRingMLStatsLineErrors
15. tokenRingMLStatsInternalErrors
16. tokenRingMLStatsBurstErrors
17. tokenRingMLStatsACErrors
18. tokenRingMLStatsAbortErrors
19. tokenRingMLStatsLostFrameErrors
20. tokenRingMLStatsCongestionErrors
21. tokenRingMLStatsFrameCopiedErrors
22. tokenRingMLStatsFrequencyErrors
23. tokenRingMLStatsTokenErrors
24. tokenRingMLStatsSoftErrorReports
25. tokenRingMLStatsRingPollEvents
26. tokenRingMLStatsOwner
27. tokenRingMLStatsStatus

3. tokenRingPStatsTable
1. tokenRingPStatsIndex
2. tokenRingPStatsDataSource
3. tokenRingPStatsDropEvents
4. tokenRingPStatsDataOctets
5. tokenRingPStatsDataPkts
6. tokenRingPStatsDataBroadcastPkts
7. tokenRingPStatsDataMulticastPkts
8. tokenRingPStatsDataPkts18to63Octets
9. tokenRingPStatsDataPkts64to127Octets
10. tokenRingPStatsDataPkts128to255Octets
11. tokenRingPStatsDataPkts256to511Octets
12. tokenRingPStatsDataPkts512to1023Octets
13. tokenRingPStatsDataPkts1024to2047Octets
14. tokenRingPStatsDataPkts2048to4095Octets
15. tokenRingPStatsDataPkts4096to8191Octets
16. tokenRingPStatsDataPkts8192to18000Octets
17. tokenRingPStatsDataPktsGreaterThan18000Octets
18. tokenRingPStatsOwner
19. tokenRingPStatsStatus

2. History - The history group records periodic statistical samples from a network and stores them for later retrieval. The historyControlTable stores configuration entries that each define an interface, polling period, and other parameters. Once samples are taken, their data is stored in an entry in a media-specific table. Each such entry defines one sample, and is associated with the historyControlEntry that caused the sample to be taken. Currently the only media- specific table defined is the etherHistoryTable, for Ethernet networks. If the probe keeps track of the time of day, it should start the first sample of the history at a time such that when the next hour of the day begins, a sample is started at that instant. This tends to make more user-friendly reports, and enables comparison of reports from different probes that have relatively accurate time of day inputs. Upon initialization the monitor is encouraged to add two history control entries per monitored interface that describe a short term and a long term polling period. Suggested parameters are 30 seconds for the short term polling period and 30 minutes for the long term period.

1. historyControlTable
1. historyControlIndex
2. historyControlDataSource
3. historyControlBucketsRequested
4. historyControlBucketsGranted
5. historyControlInterval
6. historyControlOwner
7. historyControlStatus

2. etherHistoryTable
1. etherHistoryIndex
2. etherHistorySampleIndex
3. etherHistoryIntervalStart
4. etherHistoryDropEvents
5. etherHistoryOctets
6. etherHistoryPkts
7. etherHistoryBroadcastPkts
8. etherHistoryMulticastPkts
9. etherHistoryCRCAlignErrors
10. etherHistoryUndersizePkts
11. etherHistoryOversizePkts
12. etherHistoryFragments
13. etherHistoryJabbers
14. etherHistoryCollisions
15. etherHistoryUtilization

3. tokenRingMLHistoryTable
1. tokenRingMLHistoryIndex
2. tokenRingMLHistorySampleIndex
3. tokenRingMLHistoryIntervalStart
4. tokenRingMLHistoryDropEvents
5. tokenRingMLHistoryMacOctets
6. tokenRingMLHistoryMacPkts
7. tokenRingMLHistoryRingPurgeEvents
8. tokenRingMLHistoryRingPurgePkts
9. tokenRingMLHistoryBeaconEvents
10. tokenRingMLHistoryBeaconTime
11. tokenRingMLHistoryBeaconPkts
12. tokenRingMLHistoryClaimTokenEvents
13. tokenRingMLHistoryClaimTokenPkts
14. tokenRingMLHistoryNAUNChanges
15. tokenRingMLHistoryLineErrors
16. tokenRingMLHistoryInternalErrors
17. tokenRingMLHistoryBurstErrors
18. tokenRingMLHistoryACErrors
19. tokenRingMLHistoryAbortErrors
20. tokenRingMLHistoryLostFrameErrors
21. tokenRingMLHistoryCongestionErrors
22. tokenRingMLHistoryFrameCopiedErrors
23. tokenRingMLHistoryFrequencyErrors
24. tokenRingMLHistoryTokenErrors
25. tokenRingMLHistorySoftErrorReports
26. tokenRingMLHistoryRingPollEvents
27. tokenRingMLHistoryActiveStations

4. tokenRingPHistoryTable
1. tokenRingPHistoryIndex
2. tokenRingPHistorySampleIndex
3. tokenRingPHistoryIntervalStart
4. tokenRingPHistoryDropEvents
5. tokenRingPHistoryDataOctets
6. tokenRingPHistoryDataPkts
7. tokenRingPHistoryDataBroadcastPkts
8. tokenRingPHistoryDataMulticastPkts
9. tokenRingPHistoryDataPkts18to63Octets
10. tokenRingPHistoryDataPkts64to127Octets
11. tokenRingPHistoryDataPkts128to255Octets
12. tokenRingPHistoryDataPkts256to511Octets
13. tokenRingPHistoryDataPkts512to1023Octets
14. tokenRingPHistoryDataPkts1024to2047Octets
15. tokenRingPHistoryDataPkts2048to4095Octets
16. tokenRingPHistoryDataPkts4096to8191Octets
17. tokenRingPHistoryDataPkts8192to18000Octets
18. tokenRingPHistoryDataPktsGreaterThan18000Octets

3. Alarm - The Alarm group periodically takes statistical samples from variables in the probe and compares them to thresholds that have been configured. The alarm table stores configuration entries that each define a variable, polling period, and threshold parameters. If a sample is found to cross the threshold values, an event is generated. Only variables that resolve to an ASN.1 primitive type of INTEGER (INTEGER, Counter, Gauge, or TimeTicks) may be monitored in this way. This function has a hysteresis mechanism to limit the generation of events. This mechanism generates one event as a threshold is crossed in the appropriate direction. No more events are generated for that threshold until the opposite threshold is crossed. In the case of a sampling a delta Value, a probe may implement this mechanism with more precision if it takes a delta sample twice per period, each time comparing the sum of the latest two samples to the threshold. This allows the detection of threshold crossings that span the sampling boundary. Note that this does not require any special configuration of the threshold value. It is suggested that probes implement this more precise algorithm.

1. alarmTable
1. alarmIndex
2. alarmInterval
3. alarmVariable
4. alarmSampleType
5. alarmValue
6. alarmStartupAlarm
7. alarmRisingThreshold
8. alarmFallingThreshold
9. alarmRisingEventIndex
10. alarmFallingEventIndex
11. alarmOwner
12. alarmStatus

4. Host - The host group discovers new hosts on the network by keeping a list of source and destination MAC Addresses seen in good packets. For each of these addresses, the host group keeps a set of statistics. The hostControlTable controls which interfaces this function is performed on, and contains some information about the process. On behalf of each hostControlEntry, data is collected on an interface and placed in both the hostTable and the hostTimeTable. If the monitoring device finds itself short of resources, it may delete entries as needed. It is suggested that the device delete the least recently used entries first. The hostTable contains entries for each address discovered on a particular interface. Each entry contains statistical data about that host. This table is indexed by the MAC address of the host, through which a random access may be achieved. The hostTimeTable contains data in the same format as the hostTable, and must contain the same set of hosts, but is indexed using hostTimeCreationOrder rather than hostAddress. The hostTimeCreationOrder is an integer which reflects the relative order in which a particular entry was discovered and thus inserted into the table. As this order, and thus index, is among those entries currently in the table, the index for a particular entry may change if an (earlier) entry is deleted. Thus the association between hostTimeCreationOrder and hostTimeEntry may be broken at any time. The hostTimeTable has two important uses. The first is the fast download of this potentially large table. Because the index of this table runs from 1 to the size of the table, inclusive, its values are predictable. This allows very efficient packing of variables into SNMP PDU's and allows a table transfer to have multiple packets outstanding. These benefits increase transfer rates tremendously. The second use of the hostTimeTable is the efficient discovery by the management station of new entries added to the table. After the management station has downloaded the entire table, it knows that new entries will be added immediately after the end of the current table. It can thus detect new entries there and retrieve them easily. Because the association between hostTimeCreationOrder and hostTimeEntry may be broken at any time, the management station must monitor the related hostControlLastDeleteTime object. When the management station thus detects a deletion, it must assume that any such associations have been broken, and invalidate any it has stored locally. This includes restarting any download of the hostTimeTable that may have been in progress, as well as rediscovering the end of the hostTimeTable so that it may detect new entries. If the management station does not detect the broken association, it may continue to refer to a particular host by its creationOrder while unwittingly retrieving the data associated with another host entirely. If this happens while downloading the host table, the management station may fail to download all of the entries in the table.

1. hostControlTable
1. hostControlIndex
2. hostControlDataSource
3. hostControlTableSize
4. hostControlLastDeleteTime
5. hostControlOwner
6. hostControlStatus

2. hostTable
1. hostAddress
2. hostCreationOrder
3. hostIndex
4. hostInPkts
5. hostOutPkts
6. hostInOctets
7. hostOutOctets
8. hostOutErrors
9. hostOutBroadcastPkts
10. hostOutMulticastPkts

3. hostTimeTable
1. hostTimeAddress
2. hostTimeCreationOrder
3. hostTimeIndex
4. hostTimeInPkts
5. hostTimeOutPkts
6. hostTimeInOctets
7. hostTimeOutOctets
8. hostTimeOutErrors
9. hostTimeOutBroadcastPkts
10. hostTimeOutMulticastPkts

5. Host N Table - The Host Top N group is used to prepare reports that describe the hosts that top a list ordered by one of their statistics. The available statistics are samples of one of their base statistics, over an interval specified by the management station. Thus, these statistics are rate based. The management station also selects how many such hosts are reported. The hostTopNControlTable is used to initiate the generation of such a report. The management station may select the parameters of such a report, such as which interface, which statistic, how many hosts, and the start and stop times of the sampling. When the report is prepared, entries are created in the hostTopNTable associated with the relevant hostTopNControlEntry. These entries are static for each report after it has been prepared.

1. hostTopNControlTable
1. hostTopNControlIndex
2. hostTopNHostIndex
3. hostTopNRateBase
4. hostTopNTimeRemaining
5. hostTopNDuration
6. hostTopNRequestedSize
7. hostTopNGrantedSize
8. hostTopNStartTime
9. hostTopNOwner
10. hostTopNStatus

2. hostTopNTable
1. hostTopNReport
2. hostTopNIndex
3. hostTopNAddress
4. hostTopNRate

6. Matrix Group - The Matrix group consists of the matrixControlTable, matrixSDTable and the matrixDSTable. These tables store statistics for a particular conversation between two addresses. As the device detects a new conversation, including those to a non-unicast address, it creates a new entry in both of the matrix tables. It must only create new entries based on information received in good packets. If the monitoring device finds itself short of resources, it may delete entries as needed. It is suggested that the device delete the least recently used entries first.

1. matrixControlTable
1. matrixControlIndex
2. matrixControlDataSource
3. matrixControlTableSize
4. matrixControlLastDeleteTime
5. matrixControlOwner
6. matrixControlStatus

2. matrixSDTable
1. matrixSDSourceAddress
2. matrixSDDestAddress
3. matrixSDIndex
4. matrixSDPkts
5. matrixSDOctets
6. matrixSDErrors

3. matrixDSTable
1. matrixDSSourceAddress
2. matrixDSDestAddress
3. matrixDSIndex
4. matrixDSPkts
5. matrixDSOctets
6. matrixDSErrors

7. Filter - The Filter group allows packets to be captured with an arbitrary filter expression. A logical data and event stream or "channel" is formed by the packets that match the filter expression. This filter mechanism allows the creation of an arbitrary logical expression with which to filter packets. Each filter associated with a channel is OR'ed with the others. Within a filter, any bits checked in the data and status are AND'ed with respect to other bits in the same filter. The NotMask also allows for checking for inequality. Finally, the channelAcceptType object allows for inversion of the whole equation. The channel can be turned on or off, and can also generate events when packets pass through it.

1. filterTable
1. filterIndex
2. filterChannelIndex
3. filterPktDataOffset
4. filterPktData
5. filterPktDataMask
6. filterPktDataNotMask
7. filterPktStatus
8. filterPktStatusMask
9. filterPktStatusNotMask
10. filterOwner
11. filterStatus

2. channelTable
1. channelIndex
2. channelIfIndex
3. channelAcceptType
4. channelDataControl
5. channelTurnOnEventIndex
6. channelTurnOffEventIndex
7. channelEventIndex
8. channelEventStatus
9. channelMatches
10. channelDescription
11. channelOwner
12. channelStatus

8. Capture - The Packet Capture group allows packets to be captured upon a filter match. The bufferControlTable controls the captured packets output from a channel that is associated with it. The captured packets are placed in entries in the captureBufferTable. These entries are associated with the bufferControlEntry on whose behalf they were stored.

1. bufferControlTable
1. bufferControlIndex
2. bufferControlChannelIndex
3. bufferControlFullStatus
4. bufferControlFullAction
5. bufferControlCaptureSliceSize
6. bufferControlDownloadSliceSize
7. bufferControlDownloadOffset
8. bufferControlMaxOctetsRequested
9. bufferControlMaxOctetsGranted
10. bufferControlCapturedPackets
11. bufferControlTurnOnTime
12. bufferControlOwner
13. bufferControlStatus

2. captureBufferTable
1. captureBufferControlIndex
2. captureBufferIndex
3. captureBufferPacketID
4. captureBufferPacketData
5. captureBufferPacketLength
6. captureBufferPacketTime
7. captureBufferPacketStatus

9. Event - The Event group controls the generation and notification of events from this device. Each entry in the eventTable describes the parameters of the event that can be triggered. Each event entry is fired by an associated condition located elsewhere in the MIB. An event entry may also be associated with a function elsewhere in the MIB that will be executed when the event is generated. For example, a channel may be turned on or off by the firing of an event. Each eventEntry may optionally specify that a log entry be created on its behalf whenever the event occurs. Each entry may also specify that notification should occur by way of SNMP trap messages. In this case, the community for the trap message is given in the associated eventCommunity object. The enterprise and specific trap fields of the trap are determined by the condition that triggered the event. Three traps are defined in a companion document: risingAlarm, fallingAlarm, and packetMatch. If the eventTable is triggered by a condition specified elsewhere, the enterprise and specific trap fields must be specified for traps generated for that condition.

1. eventTable
1. eventIndex
2. eventDescription
3. eventType
4. eventCommunity
5. eventLastTimeSent
6. eventOwner
7. eventStatus

2. logTable
1. logEventIndex
2. logIndex
3. logTime
4. logDescription

10. Token Ring - defines some additional monitoring functions specifically for Token Ring. These are defined in the Ring Station Group, the Ring Station Order Group, the Ring Station Configuration Group, and the Source Routing Statistics Group.

1. ringStationControlTable
1. ringStationControlIfIndex
2. ringStationControlTableSize
3. ringStationControlActiveStations
4. ringStationControlRingState
5. ringStationControlBeaconSender
6. ringStationControlBeaconNAUN
7. ringStationControlActiveMonitor
8. ringStationControlOrderChanges
9. ringStationControlOwner
10. ringStationControlStatus

2. ringStationTable
1. ringStationIfIndex
2. ringStationMacAddress
3. ringStationLastNAUN
4. ringStationStationStatus
5. ringStationLastEnterTime
6. ringStationLastExitTime
7. ringStationDuplicateAddresses
8. ringStationInLineErrors
9. ringStationOutLineErrors
10. ringStationInternalErrors
11. ringStationInBurstErrors
12. ringStationOutBurstErrors
13. ringStationACErrors
14. ringStationAbortErrors
15. ringStationLostFrameErrors
16. ringStationCongestionErrors
17. ringStationFrameCopiedErrors
18. ringStationFrequencyErrors
19. ringStationTokenErrors
20. ringStationInBeaconErrors
21. ringStationOutBeaconErrors
22. ringStationInsertions

3. ringStationOrderTable
1. ringStationOrderIfIndex
2. ringStationOrderOrderIndex
3. ringStationOrderMacAddress

4. ringStationConfigControlTable
1. ringStationConfigControlIfIndex
2. ringStationConfigControlMacAddress
3. ringStationConfigControlRemove

5. ringStationConfigTable
1. ringStationConfigIfIndex
2. ringStationConfigMacAddress
3. ringStationConfigUpdateTime
4. ringStationConfigLocation
5. ringStationConfigMicrocode
6. ringStationConfigGroupAddress
7. ringStationConfigFunctionalAddress

6. sourceRoutingStatsTable
1. sourceRoutingStatsIfIndex
2. sourceRoutingStatsRingNumber
3. sourceRoutingStatsInFrames
4. sourceRoutingStatsOutFrames
5. sourceRoutingStatsThroughFrames
6. sourceRoutingStatsAllRoutesBroadcastFrames
7. sourceRoutingStatsSingleRouteBroadcastFrames
8. sourceRoutingStatsInOctets
9. sourceRoutingStatsOutOctets
10. sourceRoutingStatsThroughOctets
11. sourceRoutingStatsAllRoutesBroadcastOctets
12. sourceRoutingStatsSingleRoutesBroadcastOctets
13. sourceRoutingStatsLocalLLCFrames
14. sourceRoutingStats1HopFrames
15. sourceRoutingStats2HopsFrames
16. sourceRoutingStats3HopsFrames
17. sourceRoutingStats4HopsFrames
18. sourceRoutingStats5HopsFrames
19. sourceRoutingStats6HopsFrames
20. sourceRoutingStats7HopsFrames
21. sourceRoutingStats8HopsFrames
22. sourceRoutingStatsMoreThan8HopsFrames
23. sourceRoutingStatsOwner
24. sourceRoutingStatsStatus

Appendix

A – Ethertype Listing (Ethertypes as of 07/29/99)

The 13th and 14th octets of an Ethernet or IEEE802.3 packet (after the preamble) consist of the "Ethernet Type" or "IEEE802.3 Length" field. The "Ethernet Type" values are managed by XEROX. Some assignments are public (see + below), others private. Current information includes: Xerox Public Ethernet Packet Type documentation (Xerox Courier Vol. 3 Issue 4 October 1988); IEEE802.3 Std; NIC RFC1010; contributions from network managers and vendors.

Note	Ethertype (hex)	Protocol	Org.	Purpose
@	0000-05DC	802.3	IEEE	IEEE802.3 Length Field (0.:1500.)
+	0101-01FF			Experimental
	0200	PUP	Xerox	Conflicts with 802.3 Length Field range (see 0A00)
	0201	PUP	Xerox	Address Translation (conflicts, see 0A01)
	0400		Nixdorf	Nixdorf (conflicts with 802.3 Length Field)
+'*	0600	XNS	Xerox	IDP
	0601	XNS	Xerox	Address Translation (3Mb only)
+'*	0800	IP	DOD	Internet Protocol (IP)
+	0801	X.75	CCITT	Internet
+	0802		NBS	National Bureau of Standards Internet
+	0803		ECMA	European Computer Manufacturers Association Internet
+	0804	CHAOSnet	MIT	Massachusetts Institute of Technology CHAOSnet Internet
+	0805	X.25	CCITT	Level 3
+'*	0806	ARP	DOD	For IP and for CHAOSnet
	0807	XNS	Xerox	XNS Compatibility
	081C	Private	Symbolics	Private
+	0888-088A		Xyplex	Xyplex
	0900		U-B	Network Debugger
	0A00	PUP	Xerox	Xerox IEEE802.3 PUP
	0A01	PUP	Xerox	Xerox IEEE802.3 PUP Address Translation
	0BAD	VINES	Banyan	Banyan Systems
	0BAF	VINES	Banyan	VINES Echo
	1000		Berkley	Trailer Negotiation
	1001-100F	IP	Berkley	Trailer Encapsulation
	1234		DCA	Multicast
*	1600	Simnet	BBN	VALID system protocol
	1989		Arificial Horizons	"Aviator" dogfight simulator on Sun workstations
	1995	RCL	Datapoint	Lan Protocol
	3C00	NBP	3Com	Virtual circuit datagram (like XNS SPP) not registered
	3C01	NBP	3Com	System control datagram not registered
	3C02	NBP	3Com	Connect request (virtual cct) not registered
	3C03	NBP	3Com	Connect repsonse not registered
	3C04	NBP	3Com	Connect complete not registered
	3C05	NBP	3Com	Close request (virtual cct) not registered
	3C06	NBP	3Com	Close response not registered
	3C07	NBP	3Com	Datagram (like XNS IDP) not registered
	3C08	NBP	3Com	Datagram broadcast not registered
	3C09	NBP	3Com	Claim NetBIOS name not registered
	3C0A	NBP	3Com	Delete Netbios name not registered
	3C0B	NBP	3Com	Remote adaptor status request not registered
	3C0C	NBP	3Com	Remote adaptor response not registered
	3C0D	NBP	3Com	Reset not registered
	4242	BBP	PCS	PCS Basic Block Protocol
	424C			Information Modes Little Big LAN diagnostic
	4321		THD	THD - Diddle
	4C42			Information Modes Little Big LAN
%	5208	Simnet	BBN	Private
	6000		DEC	Unassigned, experimental
	6001	MOP	DEC	Maintenance Operation Protocol (MOP) Dump/Load Assistance
	6002	MOP	DEC	Maintenance Operation Protocol (MOP) Remote Console
	6003	DECNET	DEC	DECNET Phase IV, DNA Routing
	6004	LAT	DEC	Local Area Transport
	6005		DEC	Diagnostic protocol (at interface initialization)
	6006		DEC	Customer protocol
	6007	LVAC	DEC	Local Area VAX Cluster System Communication Architecture (SCA)
	6008	AMBER	DEC	AMBER
	6009	MUMPS	DEC	MUMPS
+	6010-6014		3Com	3Com Corporation
	7000		UB	Download

Note	Ethertype (hex)	Protocol	Org.	Purpose
	7001		UB	NIUs (network interface units)
	7002		UB	Diagnostic/loopback
	7003		UB	NMC (Network Management) to/from UB Bridge
	7005		UB	Bridge Spanning Tree
	7007		OS/9	Microware
	7009		OS/9	Net
+	7020-7029		LRT	LRT now Sintrom
	7030		Racal-Interlan	Racal-Interlan
	7031	NTS	Prime	Network Terminal Service
	7034		Cabletron	
	8003	VLN	Cronus	
	8004	DIRECT	Cronus	
	8005		HP	Probe protocol
+	8006		Nestar	Nestar
+	8008		AT&T-Stanford	AT&T/Stanford Univ.
	8010		Excelan	Excelan
+	8013		SGI	Diagnostics
+	8014		SGI	Network games
+	8015		SGI	Reserved
+	8016		SGI	XNS NameServer, bounce server
+	8019	DOMAIN	HP-Apollo	Apollo DOMAIN
+	802E		Tymshare	Tymshare
+	802F		Tigan	Tigan, Inc.
+	8035	RARP	DOD	Reverse Address Resolution Protocol
+	8036		Aeonic	Aeonic Systems
	8037	IPX	Novell	Novell Netware
	8038	BPDU	DEC	LanBridge Management
	8039		DEC	DSM/DDP
	803A		DEC	Argonaut Console
	803B	VAXLN	DEC	
	803C		DEC	DNS Naming Service
	803D		DEC	Ethernet CSMA/CD Encryption Protocol
	803E	DNA	DEC	Distributed Time Service
	803F		DEC	LAN Traffic Monitor Protocol
	8040		DEC	PATHWORKS DECnet NETBIOS Emulation
	8041	LAST	DEC	Local Area System Transport
	8042		DEC	Unassigned
+	8044		PRC	Planning Research Corp.
+	8046		AT&T	AT&T
+	8047		AT&T	AT&T
	8048	DECamds	DEC	Availability Manager for Distributed Systems
+	8049		ExperData	ExperData
+	805B	VMTP	IETF	Versatile Message Transaction Protocol, RFC-1045 (was Stanford V Kernel, experimental)
+	805C		Stanford	Stanford V Kernel, version 6.0
+	805D		E&S	Evans & Sutherland
+	8060		Little Machines	
+	8062		Counterpoint	Counterpoint Computers
+	8065		UMass	University of Mass. at Amherst
+	8066		Umass	University of Mass. at Amherst
+	8067		Veeco	Integrated Automation
+	8068		GenDynamics	General Dynamics
+	8069		AT&T	AT&T
+	806A		Autophon	Autophon
+	806C		ComDesign	ComDesign
+	806D		Compugraphic	Compugraphic Corporation
+	806E-8077		Landmark	Landmark Graphics Corporation
+	807A		Matra	Matra
+	807B		DanskData	Dansk Data Elektronik
+	807C		MERIT	Merit Internodal
+	807D-807F		Vitalink	Vitalink Communications
+	8080		Vitalink	Vitalink TransLAN III Management
+	8081-8083		Counterpoint	Counterpoint Computers
	8088-808A		Xyples	Xyplex
+	809B		Kinetics	EtherTalk (AppleTalk over Ethernet)
+	809C-809E		Datability	Datability
+	809F		Spider	Spider Systems Ltd.
+	80A3		Nixdorf	Nixdorf Computers
+	80A4-80B3		Siemens	Siemens Gammasonics Inc.
+	80C0-80C3	DEC	DCA	Digital Comm. Assoc. Data Exchange Cluster
+	80C6		Pacer	Pacer Software

Note	Ethertype (hex)	Protocol	Org.	Purpose
+	80C7		Applitek	Applitek Corporation
+	80C8-80CC		Intergraph	Intergraph Corporation
+	80CD-80CE		Harris	Harris Corporation
+	80CF-80D2		Taylor	Taylor Instrument
+	80D3-80D4		Rosemount	Rosemount Corporation
	80D5	SNA	IBM	SNA Services over Ethernet
+	80DD		Varian	Varian Associates
+	80DE-80DF	TRFS	IntegratedSolutions	Transparent Remote File System
+	80E0-80E3		Allen-Bradley	Allen-Bradley
+	80E4-80F0		Datability	Datability
+	80F2		Retix	Retix
+	80F3	AARP	Apple	AppleTalk Address Resolution Protocol (AARP)
+	80F4-80F5		Kinetics	Kinetics
+	80F7		HP-Apollo	Apollo Computer
+	80FF-8101		Wellfleet	Wellfleet Communications
+	8102	BOFL	Wellfleet	Breath OF Life pkts every 5-10 secs.
+	8103		Wellfleet	Wellfleet Communications
	8107-8109	Private	Symbolics	
	812B		Talaris	
+	8130		Waterloo	Waterloo Microsystems Inc.
+	8131		VGLab	VG Laboratory Systems
+	8137	IPX	Novell	Novell (old) NetWare IPX (ECONFIG E option)
+	8138		Novell	Novell, Inc.
+	8139-813D		KTI	KTI
	813F			M/MUMPS data sharing
	8145		Vrije	Vrije Universiteit (NL)
	8146		Vrije	Vrije Universiteit (NL)
	8147		Vrije	Vrije Universiteit (NL)
	814C	SNMP	IETF	SNMP over Ethernet (see RFC1089)
	814F		TEC	Technically Elite Concepts
	8191		PowerLan	PowerLAN
	817D		XTP	XTP
	81D6		Artisoft	Artisoft Lantastic
	81D7		Artisoft	Artisoft Lantastic
	8203-8205		QNX	QNX Software Systems Ltd.
	8390		Accton	Accton Technologies (unregistered)
	852B		Talaris	Multicast
	8582		Kalpana	
	86DD	IP	IETF	IP version 6
	8739		CTI	Control Technology Inc.
	873A		CTI	Control Technology Inc.
	873B		CTI	Control Technology Inc.
	873C		CTI	Control Technology Inc.
	8820		Hitachi	Hitachi Cable (Optoelectronic Systems Laboratory)
	8856		Axis	Axis Communications AB
	8888		HP	LanProbe test
+	9000	MOP	DEC	Loopback (Configuration Test Protocol)
	9001	XNS	3Com	(Formerly Bridge Communications), XNS Systems Management
	9002	IP	3Com	(Formerly Bridge Communications), TCP/IP Systems Managemen
	9003		3Com	(Formerly Bridge Communications), loopback detection
	AAAA	DECNET	DEC	
	FAF5		Sonix	Sonix Arpeggio
%	FF00		BBN	VITAL-LanBridge cache wakeups

Notes:

* These protocols use Ethernet broadcast, where multicast would be preferable.

% Possible Private Protocols, i.e. not registered

\+ These protocols are mentioned by Xerox in their October 1988 issue of COURIER (page 8-9) as the publicly assigned numbers. Only vendors are listed by Xerox, not what protocols.

@ According to the October 1988 issue of COURIER (page 8), "if it is less than 600H, the packet is assumed to be an 802.3 packet; if it is greater than 600H, the packet is flagged as an Ethernet packet."

Appendix B – Service Access Points (SAPs)

Address (Hex)	Protocol	Organization	Purpose
00	Null	IEEE	XID or TEST
02	LLC	ANSI	IEEE 802.1B: IEEE Standard for LAN/MAN Management
03	LLC	IEEE	Group Sublayer Mgmt Function
04	SNA	IBM	Path Control (individual)
05	SNA	IBM	Path Control (group)
06	IP	ANSI	Used in RFC 791: ARPANET/IP, Internet Protocol, DARPA Internet Program Protocol Specification, to identify the Internet Protocol
07	IP	DOD	
08	SNA	IBM	
09	SNA	IBM	
0C	SNA	IBM	
0D	SNA	IBM	
0E	IEC955	IEC	Used in IEC 955:1989, Process Data Highway, Type C (Proway C), for Distributed Process Control Systems to identify Network Management Maintenance and Initialization
10	IPX	Novell	
18		TI	Texas Instruments
20	CLNP	ISO	Network Layer
34	CLNP	ISO	
42	BPDU	ISO/IEC JTC1/SC6	Used in ISO/IEC 10038: Media Access Control (MAC) Bridges, to identify the Bridge Spanning Tree Protocol
43	BPDU		
4E	ISO 9506	ISO	Industrial Automation Systems - Manufacturing Message Specification to identify Manufacturing Message Service
7E	ISO 8208	ISO/IEC JTC1/SC6	X.25 over IEEE 802.2 Type 2 LLC
80	XNS	3Com	Xerox Network Systems
81	XNS	Xerox	
86	LLC	Nestar	
87	LLC		
8E	IEC955	IEC	Used in IEC 955:1989, Process Data Highway, Type C (Proway C), for Distributed Process Control Systems to identify Active Station list Maintenance
98	ARP	DOD	Address Resolution Protocol
AA	SNAP	IEEE	Used in IEEE Std 802-1990: LAN/MAN Overview and Architecture of Network Standards, to identify the SNAP/SAP
AB	SNAP	DOD	Sub-Network Access Protocol
BC	VIP	Banyan	
BD	VINES	Banyan	VINES
E0	IPX	Novell	Network Layer Routing
E1	XNS		
EC	CLNP	ISO	
F0	NetBios	IBM	
F1	NetBIOS		
F4	LNM	IBM	LAN Management (individual)
F5	LNM	IBM	LAN Management (group)
F8	RPL	IBM	Remote Program Load
F9	RPL		
FA		U-B	
FB		U-B	
FC	RPL		
FD	RPL		
FE	NLP	ISO/IEC JTC1/SC6	Used to identify ISO/IEC TR 9577
FF		IEEE	Global LSAP

Appendix C – Novell® Service Advertising Protocols

SAP-DEC	SAP-HEX	Service	Company
1	0001	USER	NOVELL - PROVO CORP HQ
2	0002	USER GROUP	NOVELL - PROVO CORP HQ
3	0003	NETWARE PRINT QUEUE OR PRINT GROUP	NOVELL - PROVO CORP HQ
4	0004	NETWARE FILE SERVER / NT SERVER WITH FPNW / WIN95 WITH FILE&PRNT SHARE FOR NW	NOVELL - PROVO CORP HQ
5	0005	JOB SERVER	NOVELL - PROVO CORP HQ
6	0006	GATEWAY	NOVELL - PROVO CORP HQ
7	0007	PRINT SERVER OR SILENT PRINT SERVER	NOVELL - PROVO CORP HQ
8	0008	ARCHIVE QUEUE	NOVELL - PROVO CORP HQ
9	0009	ARCHIVE SERVER	NOVELL - PROVO CORP HQ
10	000A	JOB QUEUE	NOVELL - PROVO CORP HQ
11	000B	ADMINSITRATION	NOVELL - PROVO CORP HQ
15	000F	INITIAL NOVELL ALLOCATIONS	NOVELL
23	0017	DIAGNOSTICS	FUJITSU NETWORKS INDUSTRY
32	0020	NETBIOS	NOVELL
33	0021	SNA GATEWAY	NAS
34	0022	COMM DRIVER	SPERRY CORP. COMPUTER SYSTEMS
35	0023	NACS2	NETPRO INC
36	0024	REMOTE BRIDGE SERVER	NOVELL - PROVO CORP HQ
38	0026	REMOTE BRIDGE SERVER 2 OR ASYNCHRONOUS BRIDGE SERVER	COMPUTER LOGICS
39	0027	TCP/IP GATEWAY	TMD CONSULTING
40	0028	POINT-POINT	EICON TECHNOLOGY
41	0029	MULTI-POINT	EICON TECHNOLOGY
42	002A		CHI CORP
44	002C		INTEL - AMERICAN FORK
45	002D	TIME SYNCHRONIZATION VAP	NOVELL - PROVO CORP HQ
46	002E	ARCHIVE SERVER DYNAMIC SAP /TARGET SERVICE AGENT	NOVELL - PROVO CORP HQ
71	0047	ADVERTISING PRINT SERVER	NOVELL - PROVO CORP HQ
72	0048	TCP/IP GATEWAY	MICOM INTERLAN
73	0049		BUSINESS RECORDS CORP
74	004A		PARADATA COMPUTER NETWORKS
75	004B	BTRIEVE SERVER VAP 5	NOVELL
76	004C	NETWARE SQL	NOVELL
77	004D	XTREE-NET(MVDFRM-23539-001 11/12/93)	CENTRAL POINT SOFTWARE
78	004E	ICL GATEWAY VAP	NOVELL - PROVO CORP HQ
79	004F	DATABASE SERVER	EMERALD BAY SYSTEMS
80	0050	BTRIEVE VAP 4.11	NOVELL - PROVO CORP HQ
81	0051	LAN SERVICES	NOVELL - PROVO CORP HQ
83	0053	PRINT QUEUE USER/MAC PROJECT	NOVELL
84	0054	VALUE ADDED FILE SYSTEM	NOVELL
85	0055	TERM EMULATOR	SYSTEMS ANALYSIS INC
86	0056	STOCKNET BROKER SAP TYPE	NOVELL - PROVO
87	0057	STOCKNET EXCHANGER SAP TYPE	NOVELL - PROVO
88	0058	MULTI-POINT X.25	EICON TECHNOLOGY
89	0059	LAN SERVICES	NOVELL - PROVO CORP HQ
90	005A		BUSINESS RECORDS CORP
91	005B		BUSINESS RECORDS CORP
92	005C		BUSINESS RECORDS CORP
93	005D		BUSINESS RECORDS CORP
94	005E		BUSINESS RECORDS CORP
95	005F		BUSINESS RECORDS CORP
96	0060	STOCKNET BROKER - STATIC	NOVELL - PROVO
97	0061	STOCKNET QUEUE TYPE	NOVELL - PROVO
98	0062	STOCKNET PLAYER TYPE	NOVELL - PROVO
99	0063		INTERACTIVE FINANCIAL SOL INC
100	0064		INTERACTIVE FINANCIAL SOL INC
101	0065		INTERACTIVE FINANCIAL SOL INC
102	0066	ARCSERVE 3.0	INTERACTIVE FINANCIAL SOL INC
103	0067		INTERACTIVE FINANCIAL SOL INC
104	0068		INTERACTIVE FINANCIAL SOL INC

SAP-DEC	SAP-HEX	Service	Company
105	0069		INTERACTIVE FINANCIAL SOL INC
106	006A		INTERACTIVE FINANCIAL SOL INC
107	006B		INTERACTIVE FINANCIAL SOL INC
108	006C		INTERACTIVE FINANCIAL SOL INC
109	006D	STOCKNET EXCHANGE - STATIC	NOVELL - PROVO CORP HQ
110	006E	NACS	NETPRO INC
111	006F	CONNECTION SERVICES (MVDFRM-31137 04/19/94)	RABBIT SOFTWARE CORP
112	0070	MIC SNA DFV SERVER	COMPUTERLAND - BONN
113	0071	TAPE DRIVE SERVER	DIGI DATA CORP
114	0072	WANCOPY UTILITY	NOVELL - PROVO CORP HQ
115	0073		NOVELL - PROVO CORP HQ
116	0074		NOVELL - PROVO CORP HQ
117	0075	NETWARE BTRIEVE	NOVELL - PROVO CORP HQ
118	0076	NETWARE SQL	NOVELL - PROVO CORP HQ
119	0077		NOVELL - PROVO CORP HQ
120	0078		NOVELL - PROVO CORP HQ
121	0079		NOVELL - PROVO CORP HQ
122	007A	TES - NETWARE VMS	NOVELL - PROVO CORP HQ
123	007B	CHGD FAN (MVDFRM 31270-901 12/29/93)	MERGENT INTERNATIONAL
124	007C		INTERACTIVE FINANCIAL SOL INC
125	007D		INTERACTIVE FINANCIAL SOL INC
126	007E		INTERACTIVE FINANCIAL SOL INC
127	007F		INTERACTIVE FINANCIAL SOL INC
128	0080		INTERACTIVE FINANCIAL SOL INC
129	0081		INTERACTIVE FINANCIAL SOL INC
130	0082		INTERACTIVE FINANCIAL SOL INC
131	0083		INTERACTIVE FINANCIAL SOL INC
132	0084		INTERACTIVE FINANCIAL SOL INC
133	0085		INTERACTIVE FINANCIAL SOL INC
134	0086		INTERACTIVE FINANCIAL SOL INC
135	0087		INTERACTIVE FINANCIAL SOL INC
136	0088		INTERACTIVE FINANCIAL SOL INC
137	0089		INTERACTIVE FINANCIAL SOL INC
138	008A		INTERACTIVE FINANCIAL SOL INC
139	008B		INTERACTIVE FINANCIAL SOL INC
140	008C		INTERACTIVE FINANCIAL SOL INC
141	008D	MAIL SERVER	MCGILL UNIVERSITY
142	008E		RATIONAL DATA SYSTEMS
143	008F	QUEUE TYPES	TATE ASSOCIATES INC
144	0090	TNET X.21 IDA BRIDGE	BRITISH TELECOM
145	0091	TNET X.21 BRIDGE	BRITISH TELECOM
146	0092	TAPE BACKUP SERVER	EMERALD
147	0093	WATCOM-SQL	WATCOM
148	0094	SILA COM SOFTWARE	NOVELL - PROVO
149	0095	VMS ROUTER CONTROL	INTERCONECTIONS
150	0096		MICRO DATA BASE SYSTEMS INC
151	0097	DART	COLLEGE HILL SYSTEMS
152	0098	NETWARE ACCESS SERVER	NOVELL - PROVO CORP HQ
153	0099	NETWORK COURIER	MICROSOFT WORKGROUP CANADA
154	009A	NAMED PIPES SERVER	NOVELL - PROVO
155	009B	JOB SERVER	DIS INC
156	009C		RAYLYNN KNIGHT
157	009D	CQ3270 LAN	CQ COMPUTER COMMUNICATIONS
158	009E	PORTABLE NETWARE/UNIX - PORTABLE GROUP	NOVELL - PROVO CORP HQ
159	009F	PROGRESS DATABASE	PROGRESS SOFTWARE CORP
160	00A0	GUPTA SQL BASE SERVER	GUPTA TECHNOLOGIES
161	00A1	POWERCHUTE—VAP/NLM SERVER POWER SUPPLY	AMERICAN POWER CONVERSION
162	00A2	AUDITOR PACKAGE	BLUE LANCE NETWORK INFO SYS
163	00A3	SECURITY	BLUE LANCE NETWORK INFO SYS
164	00A4	COREL DRIVER PRODUCT UNDER NOVELL 386	COREL SYSTEMS, OPTICAL DIV
165	00A5	ARCHIVE SERVER	GIGATREND INC
166	00A6	MENU PROGRAM	R&S DATA SYSTEMS
167	00A7	386 NLM	UNISYS - CAMARILLO

SAP-DEC	SAP-HEX	Service	Company
168	00A8	LAN 1 ROUTER	ATLANTA TECHNOLOGIES
169	00A9	OBJECT TYPE	COREL SYSTEMS, OPTICAL DIV
170	00AA	OBJECT TYPE	COREL SYSTEMS, OPTICAL DIV
171	00AB	REMOTE CONSOLE VAP	
172	00AC	IDA STATUS MONITOR (MVDFRM 31905 05/29/92)	COMPAQ COMPUTER CORP
173	00AD	LANPORT VIRTUAL EXTENSION OF PORTS	MICROTEST INC
256	0100	PIPES	PEER LOGIC
257	0101	R21PX	CROSSTALK
258	0102	INVENTORY MANAGER	INTEL - AMERICAN FORK
259	0103	SEQUELNET	ORACLE CORP
260	0104	VAP	PILLSBURY CO
261	0105	GATEWAY TO UNISYS	MARSHFIELD CLINIC
262	0106	GATEWAYS TO UNISYS	MARSHFIELD CLINIC
263	0107	RSPX SERVER /386 NETWARE	NOVELL - PROVO CORP HQ
264	0108	NETWARE FOR VM & MVS	
265	0109	NETWARE FOR VM & MVS	
266	010A	NETWARE FOR VM & MVS	
267	010B	NETWARE FOR VM & MVS	
268	010C	NET 3270	MCGILL UNIVERSITY COMPUTING CT
269	010D	IMAGE SERVER	FILE NET CORPORATION
270	010E		RTK
271	010F	SAP	NOVELL - PROVO CORP HQ
272	0110		ARTEFACT NETWORK SUPPORT
273	0111	TEST SERVER	NOVELL
274	0112	PRINT SERVER	HEWLETT PACKARD - BOISE
275	0113	COMMUNICATION SERVER	NOVELL - PROVO CORP HQ
276	0114	CSA MUX /COMMUNICATION SERVER	NOVELL - PROVO CORP HQ
277	0115	CSA LCA /COMMUNICATION SERVER	NOVELL - PROVO CORP HQ
278	0116	CSA CM /COMMUNICATION SERVER	NOVELL - PROVO CORP HQ
279	0117	CSA SMA /COMMUNICATION SERVER	NOVELL - PROVO CORP HQ
280	0118	CSA DBA /COMMUNICATION SERVER	NOVELL - PROVO CORP HQ
281	0119	CSA NMA /COMMUNICATION SERVER	NOVELL - PROVO CORP HQ
282	011A	CSA SSA /COMMUNICATION SERVER	NOVELL - PROVO CORP HQ
283	011B	CSA STATUS /COMMUNICATION SERVER	NOVELL - PROVO CORP HQ
284	011C	SAA DATA LINK AGENT	NOVELL - PROVO CORP HQ
285	011D	COMMUNICATION SERVER	NOVELL - PROVO CORP HQ
286	011E	CSA APPC /COMMUNICATION SERVER	NOVELL - PROVO CORP HQ
287	011F	COMMUNICATION SERVER	NOVELL - PROVO CORP HQ
288	0120	COMMUNICATION SERVER	NOVELL - PROVO CORP HQ
289	0121	COMMUNICATION SERVER	NOVELL - PROVO CORP HQ
290	0122	COMMUNICATION SERVER	NOVELL - PROVO CORP HQ
291	0123	COMMUNICATION SERVER	NOVELL - PROVO CORP HQ
292	0124	COMMUNICATION SERVER	NOVELL - PROVO CORP HQ
293	0125	COMMUNICATION SERVER	NOVELL - PROVO CORP HQ
294	0126	SNA TEST /COMMUNICATION SERVER	NOVELL - PROVO CORP HQ
295	0127	COMMUNICATION SERVER	NOVELL - PROVO CORP HQ
296	0128	COMMUNICATION SERVER	NOVELL - PROVO CORP HQ
297	0129	COMMUNICATION SERVER	NOVELL - PROVO CORP HQ
298	012A	CSA TRACE /COMMUNICATION SERVER	NOVELL - PROVO CORP HQ
299	012B	SUPER SNA AGENT	NOVELL - PROVO CORP HQ
300	012C	COMMUNICATION SERVER	NOVELL - PROVO CORP HQ
301	012D	COMMUNICATIONS SERVER	NOVELL - PROVO CORP HQ
302	012E	COMMUNICATIONS SERVER	NOVELL - PROVO CORP HQ
303	012F	COMMUNICATIONS SERVER	NOVELL - PROVO CORP HQ
304	0130	COMMUNICATIONS EXECUTIVE / COMMUNICATIONS SERVER	NOVELL - PROVO CORP HQ
305	0131	IMAGE SERVER	FILE NET CORPORATION
306	0132	BT X.25	BRITISH TELECOM
307	0133	NNS DOMAIN /ELMER PROJECT	NOVELL
308	0134	ELMER PROJECT	NOVELL
309	0135	NNS PROFILE /ELMER PROJECT	NOVELL
310	0136	ELMER PROJECT	NOVELL
311	0137	NNS QUEUE /ELMER PROJECT	NOVELL
312	0138	ELMER PROJECT	NOVELL
313	0139	ELMER PROJECT	NOVELL
314	013A	ELMER PROJECT	NOVELL
315	013B	ELMER PROJECT	NOVELL

SAP-DEC	SAP-HEX	Service	Company
316	013C	ELMER PROJECT	NOVELL
319	013F		NETWORK DESIGNERS LTD
320	0140	NETWORK MANAGEMENT SYSTEM	ACCUNETICS
321	0141	SOFTWARE GMBH	LUFTHANSA INFORMATION SERVICE
323	0143	CDROM	ONLINE COMPUTER SYSTEMS
324	0144		NETWISE INC
325	0145	COMMUNICATION PROCESSOR	EVERGREEN SYSTEMS
326	0146	XDB SERVER DATABASE	
327	0147	PIGGYBACK LOGIN NET_INC	MICRO ENHANCEMENT INC
329	0149	ADVERTISING REMOTE SERVER	ARTEFACT NETWORK SUPPORT
330	014A	ID 5001 WEATHER STATION	ZENITH DATA SYSTEMS
331	014B		NOVELL - PROVO CORP HQ
332	014C	NETFRAM NETWARE 386	NET FRAME SYSTEMS
333	014D	NETFRAM NETWARE 386	NET FRAME SYSTEMS
334	014E	NETFRAM NETWARE 386	NET FRAME SYSTEMS
335	014F	NETFRAM NETWARE 386	NET FRAME SYSTEMS
336	0150	MAXIBACK - VAP	SYSGEN
337	0151	DCS SYSTEM SERVER	COMPUTER CONCEPTS CORPORATION
338	0152	IPX COMMUNICATION PRODUCT	DCA(DIGITAL COMMUNICATIONS ASSOCIATES)
339	0153	IPX COMMUNICATION PRODUCT	DCA(DIGITAL COMMUNICATIONS ASSOCIATES)
340	0154	FORMS CAPABILITY	ROCHESTER TELEPHONE CORP
341	0155	FORMS CAPABILITY	ROCHESTER TELEPHONE CORP
342	0156	FORMS CAPABILITY	ROCHESTER TELEPHONE CORP
343	0157	FORMS CAPABILITY	ROCHESTER TELEPHONE CORP
344	0158	FORMS CAPABILITY	ROCHESTER TELEPHONE CORP
345	0159	FORMS CAPABILITY	ROCHESTER TELEPHONE CORP
346	015A	FORMS CAPABILITY	ROCHESTER TELEPHONE CORP
347	015B	FORMS CAPABILITY	ROCHESTER TELEPHONE CORP
348	015C		NETWORK COMPUTING INC - NCI
349	015D		NETWORK COMPUTING INC - NCI
350	015E		NETWORK COMPUTING INC - NCI
351	015F		NETWORK COMPUTING INC - NCI
352	0160		NETWORK COMPUTING INC - NCI
353	0161	ADVERTISING REMOTE SERVER	ARTEFACT NETWORK SUPPORT
354	0162	SYSTEM 9	HBF GROUP
355	0163	SYSTEM 9	HBF GROUP
356	0164	SYSTEM 9	HBF GROUP
357	0165	SYSTEM 9	HBF GROUP
358	0166	NW MANAGEMENT	NOVELL - PROVO
359	0167	INSTANTCOM COMMUNICATIONS SERVER	
360	0168	PICKIT (COMM SERVER)	INTEL CORP
361	0169	PIPES	PEER LOGIC
362	016A	OPEN IMAGE FOR NETWARE	
363	016B	OPEN IMAGE FOR NETWARE	
364	016C	OPEN IMAGE FOR NETWARE	
365	016D	OPEN IMAGE FOR NETWARE	
366	016E	OPEN IMAGE FOR NETWARE	
367	016F	LUMINAR OPTICAL SERVER	COREL SYSTEMS CORP
368	0170	TXD	THOMAS CONRAD CORP
369	0171	LANFAX REDIRECTOR	ALCOM INC
370	0172	FILE SHARE	COMPAQ COMPUTER CORP
371	0173	FILE SHARE	COMPAQ COMPUTER CORP
372	0174	COMPAQ SNMP AGENT /FILE SHARE	COMPAQ COMPUTER CORP
373	0175	FILE SHARE	COMPAQ COMPUTER CORP
374	0176	FILE SHARE	COMPAQ COMPUTER CORP
375	0177	LANWARE (MVDFRM 12638-7 01/20/92)	HORIZON TECHNOLOGY INC
376	0178	LANWARE (MVDFRM 12638-7 01/20/92)	HORIZON TECHNOLOGY INC
377	0179	LANWARE (MVDFRM 12638-7 01/20/92)	HORIZON TECHNOLOGY INC
378	017A	LANWARE (MVDFRM 12638-7 01/20/92)	HORIZON TECHNOLOGY INC
379	017B	LANWARE (MVDFRM 12638-7 01/20/92)	HORIZON TECHNOLOGY INC
380	017C	LANWARE (MVDFRM 12638-7 01/20/92)	HORIZON TECHNOLOGY INC
381	017D	LANWARE (MVDFRM 12638-7 01/20/92)	HORIZON TECHNOLOGY INC
382	017E	LANWARE (MVDFRM 12638-7 01/20/92)	HORIZON TECHNOLOGY INC
383	017F	LANWARE (MVDFRM 12638-7 01/20/92)	HORIZON TECHNOLOGY INC
384	0180	LANWARE	HORIZON TECHNOLOGY INC
385	0181	FILE SHARE	NETLINE INC
386	0182	FILE SHARE	NETLINE INC
387	0183	FILE SHARE	NETLINE INC

SAP-DEC	SAP-HEX	Service	Company
388	0184	FILE SHARE	NETLINE INC
389	0185	FILE SHARE	NETLINE INC
390	0186	FILE SHARE	NETLINE INC
391	0187	FILE SHARE	NETLINE INC
392	0188	SYSM/LAN2	H&W COMPUTER SYSTEMS
393	0189	XTREE SERVR (MVDFRM 23539-001 11/12/93)	CENTRAL POINT SOFTWARE
394	018A	ACCESS SERVER	DALLAS SEMICONDUCTOR
395	018B	ACCESS SERVER	DALLAS SEMICONDUCTOR
396	018C	ACCESS SERVER	DALLAS SEMICONDUCTOR
397	018D	ACCESS SERVER	DALLAS SEMICONDUCTOR
398	018E	PC METRO	CRYSTAL POINT
399	018F	PC METRO	CRYSTAL POINT
400	0190	SERVICE POINT	INTERPOINT SOFTWARE
401	0191	SERVICE POINT	INTERPOINT SOFTWARE
402	0192	NETWAY 2000	TRI DATA SYSTEMS
403	0193	NETWAY SNA	TRI DATA SYSTEMS
404	0194	MAXWAY 500	TRI DATA SYSTEMS
405	0195	TCP/IP GATEWAY	COMPUTERVISION SERVICES
406	0196		INTEGRATED TECHNOLOGIES INC
407	0197	SHARE MASTER	STORAGE DIMENSIONS
408	0198	LOCATE	ZENITH DATA SYSTEMS
409	0199	BOML	ZENITH DATA SYSTEMS
410	019A	RHOTHEOS	ZENITH DATA SYSTEMS
411	019B	APT NET REMOTE	AUTOMATED PROGRAMMING TECH
412	019C	APT NET REMOTE	AUTOMATED PROGRAMMING TECH
413	019D	APT NET REMOTE	AUTOMATED PROGRAMMING TECH
414	019E	APT NET REMOTE	AUTOMATED PROGRAMMING TECH
415	019F	MAILSLOTS (MVDFRM 08840-134 10/21/91)	IBM - FRANKLIN LAKES
416	01A0	TERMINAL GATEWAY FOR UNISYS MAINFRAME	
417	01A1	DB SERVER	LODGISTIX INC
418	01A2	SPARE	LODGISTIX INC
419	01A3	GATEWAY, COMPOSITE PAGE, & ETC SERVERS	TEKNOS SYSTEMS LTD
420	01A4	GATEWAY, COMPOSITE PAGE, & ETC SERVERS	TEKNOS SYSTEMS LTD
421	01A5	GATEWAY, COMPOSITE PAGE, & ETC SERVERS	TEKNOS SYSTEMS LTD
422	01A6	GATEWAY, COMPOSITE PAGE, & ETC SERVERS	TEKNOS SYSTEMS LTD
423	01A7	GATEWAY, COMPOSITE PAGE, & ETC SERVERS	TEKNOS SYSTEMS LTD
424	01A8	GATEWAY, COMPOSITE PAGE, & ETC SERVERS	TEKNOS SYSTEMS LTD
425	01A9	GATEWAY, COMPOSITE PAGE, & ETC SERVERS	TEKNOS SYSTEMS LTD
426	01AA	GATEWAY, COMPOSITE PAGE, & ETC SERVERS	TEKNOS SYSTEMS LTD
427	01AB	GATEWAY, COMPOSITE PAGE, & ETC SERVERS	TEKNOS SYSTEMS LTD
428	01AC	GATEWAY, COMPOSITE PAGE, & ETC SERVERS	TEKNOS SYSTEMS LTD
429	01AD	MENU PROGRAM	R&S DATA SYSTEMS
430	01AE	MENU PROGRAM	R&S DATA SYSTEMS
431	01AF	SET ITEM DESCRIPTION/OBJECT TYPE	
432	01B0	OBJECT TYPE FOR GARP SERVER	NET RESEARCH PTY LTD
433	01B1	LICENSING RESTRICTIONS/BINDERY TYPE	LAN SUPPORT GROUP, NETWORK RES
434	01B2	LICENSING RESTRICTIONS/BINDERY TYPE	LAN SUPPORT GROUP, NETWORK RES
435	01B3		MEDIA TOUCH SYSTEMS
436	01B4	NETWORK MANAGEMENT PRODUCT	NCR - COLUMBIA
437	01B5	NETWORK MANAGEMENT PRODUCT	NCR - COLUMBIA
438	01B6	NETWORK MANAGEMENT PRODUCT	NCR - COLUMBIA
439	01B7	NETWORK MANAGEMENT PRODUCT	NCR - COLUMBIA
440	01B8	NETWORK MANAGEMENT PRODUCT	NCR - COLUMBIA
441	01B9		BONSAI TECHNOLOGIES
443	01BB		BONSAI TECHNOLOGIES
444	01BC		BONSAI TECHNOLOGIES
445	01BD		BONSAI TECHNOLOGIES
446	01BE	SERVER TYPE	KM SYSTEMS
447	01BF	SERVER TYPE FOR VALUE ADDED PRODUCT	KM SYSTEMS
448	01C0	OBJECT TYPE	LA CITE COLLEGIALE
449	01C1	OBJECT TYPES	J&L INFORMATION SYSTEMS
450	01C2	OBJECT TYPES	J&L INFORMATION SYSTEMS
451	01C3	OBJECT TYPES	J&L INFORMATION SYSTEMS
452	01C4	OBJECT TYPES	J&L INFORMATION SYSTEMS
453	01C5	OBJECT TYPES	J&L INFORMATION SYSTEMS
454	01C6	DISTRIBUTED APPLICATION	FOLIO CORPORATION
455	01C7	BINDERY TYPE	MICROTEST INC
456	01C8	OBJECT BINDERY TYPE	MADGE NETWORKS LTD

SAP-DEC	SAP-HEX	Service	Company
457	01C9	OBJECT TYPES	FUNK SOFTWARE
458	01CA	OBJECT TYPES	FUNK SOFTWARE
459	01CB		SHIVA CORP
460	01CC		SHIVA CORP
461	01CD		SHIVA CORP
462	01CE		SHIVA CORP
463	01CF	E-MAIL QUEUE OBJECT-TYPE	C&D DATA SERVICES
464	01D0	E-MAIL SERVER OBJECT-TYPE	C&D DATA SERVICES
465	01D1	LANLORD PRODUCT	MICROCOM CLIENT SERVER TECHNOL
468	01D4	CREDIT CARD AUTH/FUND SETTLEMENT	MERCHANTEC INTERNATIONAL
469	01D5		CDC - CENTERS FOR DISEASE CONTROL
470	01D6	ON-QUEUE TASK QUEUE	NETPLUS SOFTWARE INC
471	01D7	ON-QUEUE TASK SERVER	NETPLUS SOFTWARE INC
472	01D8		CASTELLE INC
473	01D9		CASTELLE INC
474	01DA		CASTELLE INC
475	01DB		CASTELLE INC
476	01DC		CASTELLE INC
477	01DD		CASTELLE INC
478	01DE		CASTELLE INC
479	01DF		CASTELLE INC
480	01E0		CASTELLE INC
481	01E1		CASTELLE INC
482	01E2	AREA CODE & EXCHANGE LOOK-UP SERVER	EQUINOX INFORMATION SYSTEMS
483	01E3	SORTING SERVER	EQUINOX INFORMATION SYSTEMS
484	01E4		WALL DATA
486	01E6	X25 AUTOMATED BRIDGE MONITOR & CONTROL	AUTOMATED INTERACTIONS DIV OF KELLY GARRETT
487	01E7		RATIONAL DATA SYSTEMS
488	01E8		RATIONAL DATA SYSTEMS
489	01E9		RATIONAL DATA SYSTEMS
490	01EA		RATIONAL DATA SYSTEMS
491	01EB		MEDIA TOUCH SYSTEMS
492	01EC	POWERGRID NETWORK DAEMON	COGNOS INC
493	01ED		INTEGRALIS LTD
494	01EE		INTEGRALIS LTD
495	01EF	CHAT PROGRAM/IPX TALK	FELSINA SOFTWARE
496	01F0		LEGATO SYSTEMS
497	01F1		LEGATO SYSTEMS
498	01F2		LEGATO SYSTEMS
499	01F3		LEGATO SYSTEMS
500	01F4		LEGATO SYSTEMS
501	01F5		LEGATO SYSTEMS
502	01F6	FOUNDTN COOP PROC	ANDERSEN CONSULTING - CHICAGO
504	01F8	SBACKUP ENHANCEMENT PRODUCT	SYTRON CORP
505	01F9	FOR UNIBASE BV, BASED IN HOLLAND	INTEGRALIS LTD
507	01FB		NORTHEAST BROADCAST CONSULTANT
508	01FC		EXTENDED SYSTEMS
509	01FD		IBM - ENDICOTT
510	01FE	INTERFAX SERVER	
511	01FF	INTERFAX SERVER	
512	0200	NP/SQL SERVER	NOVELL - PROVO
513	0201	THE MAKE SERVER	NOVELL - PROVO
514	0202	GENERIC JOB SERVER	NOVELL - PROVO CORP HQ
515	0203	RMF2 UTILITY	NOVELL - PROVO CORP HQ
516	0204		NOVELL - PROVO CORP HQ
517	0205		NOVELL - PROVO CORP HQ
518	0206		NOVELL - PROVO CORP HQ
519	0207		NOVELL - PROVO CORP HQ
520	0208		NOVELL - PROVO CORP HQ
521	0209		NOVELL - PROVO CORP HQ
522	020A		NOVELL - PROVO CORP HQ
523	020B		NOVELL - PROVO CORP HQ
524	020C		NOVELL - PROVO CORP HQ
525	020D		NOVELL - PROVO CORP HQ
526	020E		NOVELL - PROVO CORP HQ
527	020F		NOVELL - PROVO CORP HQ
528	0210		NOVELL - PROVO CORP HQ

SAP-DEC	SAP-HEX	Service	Company
529	0211		NOVELL - PROVO CORP HQ
530	0212		NOVELL - PROVO CORP HQ
531	0213		NOVELL - PROVO CORP HQ
532	0214		NOVELL - PROVO CORP HQ
533	0215		NOVELL - PROVO CORP HQ
538	021A	OT_MESSAGING_SERVER	NOVELL - SAN JOSE
541	021D	NETWARE MOBILE IPX HOME ROUTER	NOVELL - FORTUNE
542	021E	NETWARE LONTALK GATEWAY	NOVELL - MONTERREY
545	0221	SERVER/SAP	
546	0222		NOVELL - PROVO
547	0223	COMMUNICATIONS SERVER	CSB SYSTEMS GMBH
548	0224	COMMUNICATIONS SERVER	CSB SYSTEMS GMBH
549	0225	COMMUNICATIONS SERVER	CSB SYSTEMS GMBH
550	0226	COMMUNICATIONS SERVER	CSB SYSTEMS GMBH
551	0227	COMMUNICATIONS SERVER	CSB SYSTEMS GMBH
552	0228	COMMUNICATIONS SERVER	CSB SYSTEMS GMBH
553	0229	COMMUNICATIONS SERVER	CSB SYSTEMS GMBH
554	022A	COMMUNICATIONS SERVER	CSB SYSTEMS GMBH
555	022B	COMMUNICATIONS SERVER	CSB SYSTEMS GMBH
556	022C	COMMUNICATIONS SERVER	CSB SYSTEMS GMBH
557	022D	COMMUNICATIONS SERVER	CSB SYSTEMS GMBH
558	022E	COMMUNICATIONS SERVER	CSB SYSTEMS GMBH
559	022F	COMMUNICATIONS SERVER	CSB SYSTEMS GMBH
560	0230	COMMUNICATIONS SERVER	CSB SYSTEMS GMBH
561	0231	COMMUNICATIONS SERVER	CSB SYSTEMS GMBH
562	0232	COMMUNICATIONS SERVER	CSB SYSTEMS GMBH
563	0233	NW MANAGEMENT - MNGMT AGENT 1.X	NOVELL - PROVO
564	0234	NETWORK MANAGEMENT INFO SERVER	NOVELL - PROVO
565	0235	BINDERY TYPE - TIRPC SERVICE	NOVELL
566	0236		NOVELL - SALT LAKE CITY
567	0237	NETWARE MANAGEMENT - NETEXPLORER / LANTERN	NOVELL - SAN JOSE
568	0238	NETWARE MANAGEMENT - LANTERN	NOVELL - SAN JOSE
569	0239	NETWARE MANAGEMENT - HMI HUB MNGMT	NOVELL - SAN JOSE
570	023A	NETWARE MANAGEMENT - LANALYZER AGENT	NOVELL - SAN JOSE
571	023B	BINDERY TYPE FOR BROADCAST	NOVELL - PROVO
572	023C	DOS TARGET SERVICE AGENT	NOVELL - PROVO
573	023D	SMS WORKSTATION NAME OBJECT	NOVELL - PROVO
574	023E	SMS TESTING & DEVELOPMENT	NOVELL - PROVO
575	023F	SMS TESTING & DEVELOPMENT	NOVELL - PROVO
576	0240		NOVELL - SAN JOSE
577	0241		NOVELL - SAN JOSE
578	0242		NOVELL - SAN JOSE
579	0243	NOVELL MHS DS GATEWAY FOR OCE	NOVELL - WALNUT CREEK
580	0244	NDS GATEWAY FOR OCE	NOVELL - WALNUT CREEK
581	0245	SUPERLAB FILE DISTRIBUTION SERVER	NOVELL - PROVO
582	0246	VERSION CONTROL QUEUE	NOVELL - PROVO
583	0247	NVT REMOTE LOGIN OVER SPX	NOVELL - SALT LAKE CITY
584	0248	QUEUE SERVER FOR IBM PSF/2	NOVELL - PROVO CORP HQ
585	0249	LANTERN RMON	NOVELL - SAN JOSE, NMPD
586	024A	LAT TRANSPORT SERVICE PROVIDER	NOVELL - SUNNYVALE
587	024B	LAT SESSION MANAGER	NOVELL - SUNNYVALE
588	024C	LAT NETWORK FROM NETWARE	NOVELL - SUNNYVALE
589	024D	ADDRESS SERVER	NOVELL - PROVO
590	024E	NETWARE CONNECT SERVER	
591	024F	APPLETALK REMOTE ACCESS SERVICE	
592	0250	NASI COMMUNICATION SERVER	
593	0251	REMOTE NODE SET	DST - DISTRIBUTED SYSTEMS TECH
594	0252	COMMUNICATION PORT	
595	0253	COMMUNICATION PORT GROUP	
596	0254	NETWARE ACCESS SERVICE	
597	0255	ASYNCHRONOUS I/O SERVICE	
598	0256	X.25 PAD SERVICE	RSJ SOFTWARE
599	0257	FOR FUTURE RELEASE	NOVELL
600	0258	FOR FUTURE RELEASE	NOVELL
601	0259	FOR FUTURE RELEASE	NOVELL
602	025A	FOR FUTURE RELEASE	NOVELL
603	025B	FOR FUTURE RELEASE	NOVELL

SAP-DEC	SAP-HEX	Service	Company
604	025C	FOR FUTURE RELEASE	NOVELL
605	025D	FOR FUTURE RELEASE	NOVELL
606	025E	XAPIA INTERFACE FOR NW 3.11	NOVELL - RICHMOND HILL
607	025F	X.400 PROTOCOL ACCESS MODULE	NOVELL - RICHMOND HILL
608	0260	SNADS PROTOCOL ACCESS MODULE	NOVELL - RICHMOND HILL
609	0261	SUPERLAB NETWORK SWITCH SERVER	NOVELL - PROVO
610	0262	HUB SERVICES	NOVELL - SUNNYVALE
611	0263	NETWARE MANAGEMENT AGENT	NOVELL - SUNNYVALE
612	0264	GLOBAL MHS	NOVELL - SUNNYVALE
613	0265	SNMP	NOVELL - SUNNYVALE
614	0266	VERSION CONTROL SERVER	NOVELL - PROVO
615	0267	APPLICATION RIGHTS PROGRAM	NOVELL - PROVO
616	0268	HP JETDIRECT USE THIS INSTEAD OF 0004	NOVELL - SAN JOSE
617	0269	SUPERLAB AUTOMATION SERVER	NOVELL - PROVO
618	026A	NETWORK MANAGEMENT SERVICES - NMS CONSOLE	NOVELL - SAN JOSE
619	026B	TIME SYNCHRONIZATION	NOVELL - PROVO
620	026C	UNIXWARE SNMP DAEMON	UNIVEL
621	026D	ADVERTISING JOB SERVER	NOVELL - PROVO
626	0272	DATALINK SWITCHING (DLSW)	NOVELL - SAN JOSE
627	0273	NEST DEVICE	NOVELL - MONTERREY
628	0274	GROUPWISE MESSAGE MULTIPLE SERVERS	NOVELL - OREM
629	0275	SAMPLE CODE FROM DEV. SUPPORT	NOVELL
631	0277	SAP SERVER TYPE	NOVELL - SAN JOSE
632	0278	NDS REPLICA /TREE /DIRECTORY SERVER	NOVELL - PROVO
635	027B	NW MANAGEMENT AGENT 2.1	NOVELL - SAN JOSE
640	0280		NOVELL - WALNUT CREEK
641	0281	DOMAIN APPLICATION SERVICES	NOVELL - SUNNYVALE
642	0282	NDPS SERVICE REGISTRY SERVICE / DOMAIN APPLICATION SERVICES	NOVELL - SUNNYVALE
643	0283	DOMAIN APPLICATION SERVICES	NOVELL - SUNNYVALE
644	0284	DOMAIN APPLICATION SERVICES	NOVELL - SUNNYVALE
645	0285	DOMAIN APPLICATION SERVICES	NOVELL - SUNNYVALE
646	0286	DOMAIN APPLICATION SERVICES	NOVELL - SUNNYVALE
647	0287	DOMAIN APPLICATION SERVICES	NOVELL - SUNNYVALE
648	0288	DOMAIN APPLICATION SERVICES	NOVELL - SUNNYVALE
650	028A	MPR - IPX ADDRESS MAPPING GATEWAY	NOVELL - SAN JOSE
654	028E	NETWARE NT ADMINISTRATOR / TABASCO	NOVELL - PROVO
759	02F7	IPX - IP GATEWAY	NOVELL
767	02FF		NOVELL - PROVO CORP HQ
768	0300	VAP - ADVERTISING SERVICES	FIREFOX COMMUNICATIONS LTD
769	0301	VAP - ADVERTISING SERVICES	FIREFOX COMMUNICATIONS LTD
770	0302	VAP - ADVERTISING SERVICES	FIREFOX COMMUNICATIONS LTD
771	0303	VAP - ADVERTISING SERVICES	FIREFOX COMMUNICATIONS LTD
772	0304	VAP - ADVERTISING SERVICES	FIREFOX COMMUNICATIONS LTD
773	0305	VAP - ADVERTISING SERVICES	FIREFOX COMMUNICATIONS LTD
774	0306	NPS (NETWARE PRINT SERVICES)	INTER CONNECTIONS INC
775	0307	NPS SPOOL CLIENT	INTER CONNECTIONS INC
776	0308	HP NS UTIL	HEWLETT PACKARD - BOISE
777	0309	DOCUMENT MANAGEMENT PACKAGE	PERFECT SOLUTIONS CORPORATION
778	030A	BBS SERVER	GALACTICOMM INC
780	030C	HP LASERJET /QUICK SILVER	HEWLETT PACKARD - BOISE
781	030D	BROADCAST OPERATION SUP	REMUERA CORP
782	030E	FAULT TOLERANT CONTROL	CLONE STAR SOFTWARE
783	030F	CD ROM SERVER	CROSS INTERNATIONAL CORP
784	0310	CD ROM SERVER	CROSS INTERNATIONAL CORP
785	0311	CD ROM SERVER	CROSS INTERNATIONAL CORP
786	0312	CD ROM SERVER	CROSS INTERNATIONAL CORP
787	0313	CD ROM SERVER	CROSS INTERNATIONAL CORP
788	0314	CD ROM SERVER	CROSS INTERNATIONAL CORP
789	0315	CD ROM SERVER	CROSS INTERNATIONAL CORP
790	0316	CD ROM SERVER	CROSS INTERNATIONAL CORP
791	0317	BATCH PROCESSOR	
800	0320		ATTACHMATE CORPORATION
801	0321		CHICAGO RESEARCH & TRADING
802	0322	NETWARE MANAGEMENT APPLICATION	FRYE COMPUTER SYSTEMS
803	0323	IMAGE SERVER	WANG LABORATORIES
804	0324	IMAGE SERVER	WANG LABORATORIES

SAP-DEC	SAP-HEX	Service	Company
805	0325	X.500 DSA SERVER	AAC SYSTEMS
806	0326	NOVELL REMOTE ISDN ROUTER	LANWORKS
807	0327	BOOTWARE/MSD	LANWORKS
808	0328	WATCOM-SQL	WATCOM
809	0329	AVL NLM DATABASE	AETNA LIFE & CASUALTY
810	032A	LU6.2 GATEWAY	AETNA LIFE & CASUALTY
811	032B	FAX SERVER	DIGITAL VISIONS CORP
812	032C	VOICE SERVER	DIGITAL VISIONS CORP
813	032D	INTERPROCESS EXCHANGE SERVER	DIGITAL VISIONS CORP
814	032E	APPLICATION SERVER	NATIONSBANK APPL SYSTEMS SUPP
815	032F	INTERFAX QUEUE	
816	0330	SAS SHARE SERVER	SAS INSTITUTE
817	0331	SAS CONNECT	SAS INSTITUTE
818	0332		ARCHETYPE
819	0333		ARCHETYPE
820	0334		AETNA LIFE & CASUALTY
821	0335	COMMUNICATIONS SERVER	MULTITECH
822	0336	COMMUNICATIONS SERVER	MULTITECH
823	0337	APPL IPX/SPX COMMUNICATIONS	MAGEE ENTERPRISES INC
824	0338	APPL IPX/SPX COMMUNICATIONS	MAGEE ENTERPRISES INC
825	0339	APPL IPX/SPX COMMUNICATIONS	MAGEE ENTERPRISES INC
826	033A	APPL IPX/SPX COMMUNICATIONS	MAGEE ENTERPRISES INC
827	033B	APPL IPX/SPX COMMUNICATIONS	MAGEE ENTERPRISES INC
828	033C	APPL IPX/SPX COMMUNICATIONS	MAGEE ENTERPRISES INC
829	033D	APPL IPX/SPX COMMUNICATIONS	MAGEE ENTERPRISES INC
830	033E	APPL IPX/SPX COMMUNICATIONS	MAGEE ENTERPRISES INC
831	033F	APPL IPX/SPX COMMUNICATIONS	MAGEE ENTERPRISES INC
832	0340	APPL IPX/SPX COMMUNICATIONS	MAGEE ENTERPRISES INC
833	0341	DATA SERVICE TO WORKSTATION/SCHOOL ADMIN	CHANCERY SOFTWARE LTD
834	0342	BINDERY TYPE	MICROTEST INC
835	0343	BINDERY TYPE	MICROTEST INC
836	0344	BINDERY TYPE	MICROTEST INC
837	0345	BINDERY TYPE	MICROTEST INC
838	0346	BINDERY TYPE	MICROTEST INC
839	0347	BINDERY TYPE	MICROTEST INC
840	0348	BINDERY TYPE	MICROTEST INC
841	0349	BINDERY TYPE	MICROTEST INC
842	034A		PREFERRED SYSTEMS INC.
843	034B		PREFERRED SYSTEMS INC.
844	034C		PREFERRED SYSTEMS INC.
845	034D		PREFERRED SYSTEMS INC.
846	034E		PREFERRED SYSTEMS INC.
847	034F		PREFERRED SYSTEMS INC.
848	0350		PREFERRED SYSTEMS INC.
849	0351		PREFERRED SYSTEMS INC.
850	0352	BINDERY TYPE	FUJITSU LTD
851	0353	BINDERY TYPE	FUJITSU LTD
852	0354	BINDERY TYPE	FUJITSU LTD
853	0355	MAYNESTREAM	ARCADA SOFTWARE
854	0356		LANOVATION
855	0357		LANOVATION
856	0358	BINDERY OBJECT TYPE	CBIS INC
858	035A	MULIT-PROCESSOR CONTROLLER	MBAC
859	035B	RIGHT-HAND-MAN, E-MAIL/SCHEDULING PKG	LAN ACES INC
860	035C	FAX SERVER	TRANSFAX CORPORATION
861	035D	FAX PRINT SERVER	TRANSFAX CORPORATION
862	035E	FAX MERGE SERVER	TRANSFAX CORPORATION
863	035F	NETWORK MANAGEMENT SERVER	TRANSFAX CORPORATION
864	0360	OBJECT TYPE	FUNK SOFTWARE
866	0362	OBJECT-TYPE, PSEUDO PEER-TO-PEER	
867	0363	PRINT SERVER - LASER JET	EXTENDED SYSTEMS
868	0364	SERVER-TYPE FOR LAN TIMES JAPAN ARTICLE	LAN TIMES JAPAN, SOFTBANK CORP
869	0365	QUEUE-TYPE FOR LAN TIMES JAPAN ARTICLE	LAN TIMES JAPAN, SOFTBANK CORP
870	0366	MGATE - COMMUNICATION GATEWAY/ LANS + VAX	COEFFICIENT SYSTEMS CORP
871	0367		EXCALIBUR TECHNOLOGIES CORP

SAP-DEC	SAP-HEX	Service	Company
872	0368		EXCALIBUR TECHNOLOGIES CORP
873	0369		EXCALIBUR TECHNOLOGIES CORP
874	036A		EXCALIBUR TECHNOLOGIES CORP
875	036B	BINDERY TYPE	AETNA LIFE & CASUALTY
876	036C	IPX LAYER PEER-TO-PEER COMMUNICATIONS	SEWELL DEVELOPMENT
877	036D		MICRO INTEGRATION
878	036E	NLM SERVER	PRAXIS
879	036F	BINDERY TYPE	AVAIL SYSTEMS CORP
880	0370	BINDARY TYPE	
881	0371	SCAN DIRECTORY ENTRY FOR SPACE USAGE	
882	0372		DIGITAL EQUIPMENT - MERRIMACK
883	0373	NLM ADVERTISING FOR UPS INFO	BRAINSTORM ENGINEERING
884	0374	SHAREWARE COMMUNICATIONS SERVER	CHERRY TREE SOFTWARE
885	0375	ENTERPRISE ECS	INTEL CORP
886	0376	ENTERPRISE INITIALIZATION MODE	INTEL CORP
887	0377	COMM SVR - NET BIOS IPX (MVDFRM 27341)	US ROBOTICS SOFTWARE
888	0378	NLM ADVERTISING FOR UPS INFO	BRAINSTORM ENGINEERING
889	0379	FAX SERVER	EXTENDED SYSTEMS
890	037A	FAX SERVER	EXTENDED SYSTEMS
891	037B	FAX SERVER	EXTENDED SYSTEMS
892	037C	FAX SERVER	EXTENDED SYSTEMS
893	037D	GATEWAY MANAGEMENT	WALL DATA
894	037E	POWERCHUTE ALERT - UPS MONITORING	AMERICAN POWER CONVERSION
895	037F	ALERT SERVER NLM (MVDFRM 23539-001 11/12/93)	CENTRAL POINT SOFTWARE
896	0380	FAX SERVER	OPTUS INFORMATION SYSTEMS
897	0381	TNS TRANSPORT NETWORK SUBSTRATE, VERS2	ORACLE
898	0382	ID FOR LASERMASTER PRINTER PRODUCTS	LASER MASTER CORP
899	0383	POWERCHUTE ADMINISTRATIVE	AMERICAN POWER CONVERSION
900	0384	SEQUEL LINK CLIENT-SERVER MIDDLEWARE	TECHGNOSIS INC
901	0385	MAIL SYSTEMS	SYNECTIC SYSTEMS LTD
902	0386	HEWLETT PACKARD BRIDGES	HEWLETT PACKARD - ROSEVILLE
903	0387	HEWLETT PACKARD HUBS	HEWLETT PACKARD - ROSEVILLE
904	0388	WORKSTATION PEER-TO-PEER COMMUNICATIONS	IBM - RESEARCH TRIANGLE PARK
905	0389		DATANEX CORPORATION
916	0394	NETWARE CONNECT NCS (MAC) SERVER	NOVELL
922	039A	HP OPEN MAIL & PORTABLE NETWARE	HEWLETT PACKARD - BERKSHIRE
923	039B		IRIS ASSOCIATES
924	039C	COMMUNICATIONS SERVER BINDERY	DATOR 3 SPOL SRO
925	039D	COMMUNICATIONS SERVER SAP	DATOR 3 SPOL SRO
926	039E	FAX SERVER	FERRARI ELECTRONIC GMBH
927	039F		COMPUTERVISION SERVICES
928	03A0	BINDERY TYPE FOR CD SERVER	JOSTENS LEARNING CORP
929	03A1		NEUMEIER & WALCH SYSTEMTECHNIK
930	03A2		HYPROTECH LTD
931	03A3		KYOCERA CORP, YOHAGA OFFICE
932	03A4		KYOCERA CORP, YOHAGA OFFICE
933	03A5		KYOCERA CORP, YOHAGA OFFICE
934	03A6		KYOCERA CORP, YOHAGA OFFICE
935	03A7	OBJECT TYPE FOR A GROUP	STADT PFORZHEIM POA
936	03A8	OBJECT TYPE FOR A QUEUE	STADT PFORZHEIM POA
937	03A9	OBJECT TYPE FOR A SERVER	STADT PFORZHEIM POA
938	03AA	RESET PRINT SERVERS	UDS MOTOROLA
939	03AB	RESET PRINT SERVERS	UDS MOTOROLA
940	03AC	RESET PRINT SERVERS	UDS MOTOROLA
941	03AD	RESET PRINT SERVERS	UDS MOTOROLA
942	03AE	DBMS LOCK MANAGER	RAIMA CORP
943	03AF	COPY PROTECTION SERVER	PACE SOFTWARE SYSTEMS INC
944	03B0	TNA COMMUNICATION WITH 2 NLM'S	PALINDROME
945	03B1	ETHERNET LAN CONTROLLER FOR NETWARE	BUS TECH INC
946	03B2	SERVER ID FOR NLM	NETWORK DESIGNERS
947	03B3	FILE MANAGEMENT SERVICES	SYSTEMS AXIS PLC
948	03B4	QUEUE MANAGEMENT SERVICES	SYSTEMS AXIS PLC
949	03B5	OBJECT TYPE	IWI
950	03B6		LANTECH SERVICES
952	03B8	MODEM PROTOCALL - SHARING SERIAL PORTS	LANSOURCE TECHNOLOGIES

SAP-DEC	SAP-HEX	Service	Company
953	03B9	GLOBAL INFO APPLICATION EXEC ENVIRO	FUNK SOFTWARE
954	03BA	MAGIX DATABASE SERVER	ADVANCED SOFTWARE TECHNOLOGIES
955	03BB	PERFORMANCE MONITOR	AMERIDATA
956	03BC	NETPORT ADVERTISING	INTEL PCED
957	03BD	WAN CONNECTION SERVER	IDEASSOCIATION
958	03BE	WICAT	JOSTENS LEARNING CORP
959	03BF	WICAT SERVER	JOSTENS LEARNING CORP
960	03C0	EMBEDDED IN OEM PLOTTER PRODUCT	CAL COMP
961	03C1	EMBEDDED IN OEM PLOTTER PRODUCT	CAL COMP
962	03C2	EMBEDDED IN OEM PLOTTER PRODUCT	CAL COMP
963	03C3	EMBEDDED IN OEM PLOTTER PRODUCT	CAL COMP
964	03C4	ARCSERVE 4.0 /ADVERTISING ARC SERVER	
965	03C5	FAX SERVER PRODUCT	ASCOM TELECOMMUNICATION LTD
966	03C6	ANTI-VIRUS PRODUCT	
967	03C7	LAN SPOOL 3.5	INTEL - AMERICAN FORK
968	03C8	NETWORK MANAGEMENT	MADGE NETWORKS LTD
970	03CA	POINT-OF-SALE SERVER	OPTICAL MARK SYSTEMS LTD
971	03CB	SOFTWARE ACCESS CONTROL SERVER	U OF PLYMOUTH
972	03CC	AXNETWAR - ENVELOPE/LABEL PRINT SERVER	PSI ASSOCIATES
973	03CD	ARC SERVER - MAC SERVE NLM	
974	03CE	ARC SERVER - UNIX SERVE NLM	
975	03CF	DATABASE ENGINE	BLUE LANCE NETWORK INFO SYS
976	03D0	REPORT ENGINE	BLUE LANCE NETWORK INFO SYS
977	03D1	JOB SERVER	BLUE LANCE NETWORK INFO SYS
978	03D2	OBJECT TYPE	BLUE LANCE NETWORK INFO SYS
979	03D3	OBJECT TYPE	BLUE LANCE NETWORK INFO SYS
980	03D4	VISINET NLM ID#	TECHNOLOGY DYNAMICS INC
981	03D5	PRINT SERVERS	LEXMARK INTERNATIONAL INC
982	03D6	PRINT SERVERS	LEXMARK INTERNATIONAL INC
983	03D7	PRINT SERVERS	LEXMARK INTERNATIONAL INC
984	03D8	PRINT SERVERS	LEXMARK INTERNATIONAL INC
985	03D9	SERVER MONITORING PROGRAM	TRELLIS
986	03DA	MULTIPLE SERVICES & APPLICATIONS	THINK SYSTEMS CORP
987	03DB	MULTIPLE SERVICES & APPLICATIONS	THINK SYSTEMS CORP
988	03DC	MULTIPLE SERVICES & APPLICATIONS	THINK SYSTEMS CORP
989	03DD	DEVELOPING SERVER PERFORMANCE ANALIZER	BANYAN SYSTEMS INC
990	03DE	NW SQL /GUPTA SEQUEL BASE SERVER	GUPTA TECHNOLOGIES
991	03DF	REMOTE DATABASE SERVICES	INTERACTIVE DATA
992	03E0	OBJECT STORE SERVER	OBJECT DESIGN
993	03E1	UNIXWARE APPLICATION SERVER /UNIVEL SERVER TYPE	UNIVEL
994	03E2	UNIXWARE APPLICATION INSTALL SERVER /UNIVEL SERVER TYPE	UNIVEL
995	03E3	UNIVEL SERVER TYPE	UNIVEL
996	03E4	UNIXWARE PEER-TO-PEER ENABLED /UNIVEL SERVER TYPE	UNIVEL
997	03E5	UNIVEL SERVER TYPE	UNIVEL
998	03E6	UNIXWARE REMOTE APPL. SERVER /UNIVEL SERVER TYPE	UNIVEL
999	03E7	UNIVEL SERVER TYPE	UNIVEL
1000	03E8	UNIVEL SERVER TYPE	UNIVEL
1001	03E9	UNIVEL SERVER TYPE	UNIVEL
1002	03EA	UNIVEL SERVER TYPE	UNIVEL
1003	03EB	UNIVEL SERVER TYPE	UNIVEL
1004	03EC	UNIVEL SERVER TYPE	UNIVEL
1005	03ED	UNIVEL SERVER TYPE	UNIVEL
1006	03EE	UNIVEL SERVER TYPE	UNIVEL
1007	03EF	UNIVEL SERVER TYPE	UNIVEL
1008	03F0	UNIVEL SERVER TYPE	UNIVEL
1009	03F1	FIRST CALL	THOMSON FINANCIAL
1010	03F2	ACCESS RIGHTS SERVER	
1011	03F3	VITAL SIGNS/LAN SERVER	BLUELINE SOFTWARE INC
1012	03F4	LAA SERVER BINDERY	SABER SOFTWARE
1013	03F5	MICROSOFT SQL SERVER IPX/SPX SUPPORT	MICROSOFT
1014	03F6	ASYNCHRONOUS SERIAL COMMUNICATIONS	BLACK CREEK INTEGRATED SYSTEMS
1015	03F7	ASYNCHRONOUS SERIAL COMMUNICATIONS	BLACK CREEK INTEGRATED SYSTEMS
1016	03F8	ASYNCHRONOUS SERIAL COMMUNICATIONS	BLACK CREEK INTEGRATED SYSTEMS

SAP-DEC	SAP-HEX	Service	Company
1017	03F9	ASYNCHRONOUS SERIAL COMMUNICATIONS	BLACK CREEK INTEGRATED SYSTEMS
1018	03FA	WATSON - COMMUNICATIONS SERVER	PRODIGY SERVICES
1019	03FB	NETPORT ADVERTISING	INTEL PCED
1020	03FC	NETPORT ADVERTISING	INTEL PCED
1021	03FD	NETPORT ADVERTISING	INTEL PCED
1022	03FE	NETPORT ADVERTISING	INTEL PCED
1023	03FF		MODULAR SOFTWARE CORPORATION
1024	0400	BINDERY OBJECT TYPES	ARTEFACT NETWORK SUPPORT
1025	0401	BINDERY OBJECT TYPES	ARTEFACT NETWORK SUPPORT
1026	0402	BINDERY OBJECT TYPES	ARTEFACT NETWORK SUPPORT
1027	0403	BINDERY OBJECT TYPES	ARTEFACT NETWORK SUPPORT
1028	0404	BINDERY OBJECT TYPES	ARTEFACT NETWORK SUPPORT
1029	0405	BINDERY OBJECT TYPES	ARTEFACT NETWORK SUPPORT
1030	0406	BINDERY OBJECT TYPES	ARTEFACT NETWORK SUPPORT
1031	0407	BINDERY OBJECT TYPES	ARTEFACT NETWORK SUPPORT
1032	0408	BINDERY OBJECT TYPES	ARTEFACT NETWORK SUPPORT
1033	0409	BINDERY OBJECT TYPES	ARTEFACT NETWORK SUPPORT
1034	040A	IMAGE APPLICATION	LASER DATA
1035	040B	IMAGE APPLICATION	LASER DATA
1036	040C	IMAGE APPLICATION	LASER DATA
1037	040D	IMAGE APPLICATION	LASER DATA
1038	040E	IMAGE APPLICATION	LASER DATA
1039	040F	IMAGE APPLICATION	LASER DATA
1040	0410	IMAGE APPLICATION	LASER DATA
1041	0411	IMAGE APPLICATION	LASER DATA
1042	0412	IMAGE APPLICATION	LASER DATA
1043	0413	IMAGE APPLICATION	LASER DATA
1044	0414	NETSPRINT	DIGITAL PRODUCTS INC
1045	0415	REMOTE DATABASE SERVICES - BINDERY	INTERACTIVE DATA
1046	0416	DEALING ROOM SERVERS	TEKNOS SYSTEMS LTD
1047	0417	DEALING ROOM SERVERS	TEKNOS SYSTEMS LTD
1048	0418	DEALING ROOM SERVERS	TEKNOS SYSTEMS LTD
1049	0419	DEALING ROOM SERVERS	TEKNOS SYSTEMS LTD
1050	041A	DEALING ROOM SERVERS	TEKNOS SYSTEMS LTD
1051	041B	DEALING ROOM SERVERS	TEKNOS SYSTEMS LTD
1052	041C	DEALING ROOM SERVERS	TEKNOS SYSTEMS LTD
1053	041D	DEALING ROOM SERVERS	TEKNOS SYSTEMS LTD
1054	041E	DEALING ROOM SERVERS	TEKNOS SYSTEMS LTD
1055	041F	DEALING ROOM SERVERS	TEKNOS SYSTEMS LTD
1056	0420	DEALING ROOM SERVERS	TEKNOS SYSTEMS LTD
1057	0421	DEALING ROOM SERVERS	TEKNOS SYSTEMS LTD
1058	0422	DEALING ROOM SERVERS	TEKNOS SYSTEMS LTD
1059	0423	DEALING ROOM SERVERS	TEKNOS SYSTEMS LTD
1060	0424	DEALING ROOM SERVERS	TEKNOS SYSTEMS LTD
1061	0425	DEALING ROOM SERVERS	TEKNOS SYSTEMS LTD
1062	0426	DEALING ROOM SERVERS	TEKNOS SYSTEMS LTD
1063	0427	DEALING ROOM SERVERS	TEKNOS SYSTEMS LTD
1064	0428	DEALING ROOM SERVERS	TEKNOS SYSTEMS LTD
1065	0429	DEALING ROOM SERVERS	TEKNOS SYSTEMS LTD
1066	042A	FULL TEXT RETRIEVAL CLIENT/SERVER DB ENG	IMPACT ITALIANA SRL
1067	042B	GATEWAY SOFTWARE	DATEV EG
1068	042C	GATEWAY SOFTWARE	DATEV EG
1069	042D	CLIENT-SERVER DRIVER FOR IPX/SPX	REFERENCE POINT SOFTWARE
1070	042E		INTRAK INC
1071	042F	LOADER BINDERY TYPE	NETWORK TECHNICAL SOLUTIONS
1072	0430	FINDER BINDERY TYPE	NETWORK TECHNICAL SOLUTIONS
1073	0431	DDE SERVER	
1074	0432	FILEMAKER PRO	CLARIS CORP
1075	0433	NETWORKING HUB	SYNOPTICS
1076	0434	NETWORK TERMINAL EMULATOR	IDE ASSOCIATION
1077	0435	ADMINISTRATION SERVER	MCGILL UNIVERSITY FAC OF ENGIN
1078	0436	NETWORK DYNAMIC DATA EXCHANGE	NETLOGIC INC
1079	0437	ASYNCH COMM SVR (MVDFRM 27341-1 06/22/94)	US ROBOTICS SOFTWARE
1080	0438	BACK UP PRODUCT	COREL SYSTEMS, OPTICAL DIV
1081	0439	BACK UP PRODUCT	COREL SYSTEMS, OPTICAL DIV
1082	043A	MISCELLANEOUS SOFTWARE COMMUNICATIONS	TENTERA COMPUTER SERVICES

SAP-DEC	SAP-HEX	Service	Company
1083	043B	ENTERPRISE IN MAINTENANCE MODE	INTEL CORP
1084	043C	CONNECTION STATION SERVICE TYPE	COROLLARY INC
1085	043D	CONNECTION STATION SERVICE TYPE	COROLLARY INC
1086	043E	CONNECTION STATION SERVICE TYPE	COROLLARY INC
1087	043F	CONNECTION STATION SERVICE TYPE	COROLLARY INC
1088	0440	CONNECTION STATION SERVICE TYPE	COROLLARY INC
1089	0441	CONNECTION STATION SERVICE TYPE	COROLLARY INC
1090	0442	CONNECTION STATION SERVICE TYPE	COROLLARY INC
1091	0443	CONNECTION STATION SERVICE TYPE	COROLLARY INC
1092	0444	SNA GATEWAY / MICROSOFT SNA SERVER	MICROSOFT
1093	0445	WORKSTATION TERMINAL ACCESS	HSD HARDWARE SOFTWARE DEVELOPM
1094	0446		DE INTERNATIONAL LTD
1095	0447	DISTRIBUTIVE CACHE PRODUCT	RAIMA CORP
1096	0448	ARC SERVE DATABASE AGENT	
1097	0449	ARCHIVE	
1098	044A	ARCHIVE	
1099	044B	ARCHIVE	
1100	044C	ARCHIVE	
1101	044D	IBM HOST GATEWAY	IDEA COURIER
1102	044E	PROJECT ID	URBAN SCIENCE APPLICATIONS
1103	044F	COMMON COMMUNICATION INTERFACE	COMPUTER ASSOCIATES
1104	0450	COMMUNICATIONS SERVER SDD SYNTHESIZER	SCANDINAVIAN AIRLINES DATA
1105	0451	TAPE BACK-UP FOR NLM APPLICATIONS	MOUNTAIN NETWORK SOLUTIONS INC
1106	0452	TAPE BACK-UP FOR NLM APPLICATIONS	MOUNTAIN NETWORK SOLUTIONS INC
1107	0453	TAPE BACK-UP FOR NLM APPLICATIONS	MOUNTAIN NETWORK SOLUTIONS INC
1108	0454	TAPE BACK-UP FOR NLM APPLICATIONS	MOUNTAIN NETWORK SOLUTIONS INC
1109	0455	TAPE BACK-UP FOR NLM APPLICATIONS	MOUNTAIN NETWORK SOLUTIONS INC
1110	0456	TAPE BACK-UP FOR NLM APPLICATIONS	MOUNTAIN NETWORK SOLUTIONS INC
1111	0457	CANON PERIPHERAL SERVER	CANON INFORMATION SYSTEMS
1112	0458	NETWARE SERVER PRODUCT	INTEL CORP
1113	0459	OBJECT ORIENTED DATABASE SYSTEM	ONTOS INC
1114	045A	QMS PRINTER - REMOTE CONFIGURATION	QMS
1115	045B	CLIENT SERVER MONITORING UTILITY	DELL COMPUTER
1116	045C	BINDERY TYPE - APPLICATION DEFINITION	LANOVATION
1117	045D	FAX SERVER	FERRARI ELECTRONIC GMBH
1118	045E	FAX SERVER	FERRARI ELECTRONIC GMBH
1119	045F	FAX SERVER	FERRARI ELECTRONIC GMBH
1120	0460	FAX SERVER	FERRARI ELECTRONIC GMBH
1121	0461	COMMUNICATIONS GATEWAY - OSI	EICON TECHNOLOGY
1122	0462	COMMUNICATIONS GATEWAY - SNA	EICON TECHNOLOGY
1123	0463	ARCHIVE QUEUE SERVER	
1124	0464	BATCHFILER APPLICATION	JOVANDI INTERNATIONAL INC
1125	0465	NLM ON FILE SERVER	JOVANDI INTERNATIONAL INC
1126	0466	TIME SYNCHRONIZATION SERVER	JOVANDI INTERNATIONAL INC
1127	0467	ELECTRIC MAIL SYSTEM	TOBIT SOFTWARE GMBH
1128	0468	TELEPHONE ANSWERING SYSTEM	A&M COMMUNICATIONS
1129	0469	NETWORK DISC BACK-UP SOFTWARE	FORTUNET INC
1130	046A	FAX SERVER	EXTENDED SYSTEMS
1131	046B	CADENCE TIME SERVICE	POLYGON INC
1132	046C	POLY-PORTAL FOR NETWARE/ETHERNET GATEWAY	
1133	046D	POLY-LINK PC TO VAX NETWORKING UTILITIES	
1134	046E	LAT SERVICE	NOVELL - SUNNYVALE
1135	046F	AUTOMATIC TRACKING SYSTEM	
1136	0470	MEASURESERVERS AND MEASURECLIENTS	ADVANTECH BENELUX BV
1137	0471	DISK MONITOR	STORAGE DEMINSIONS
1138	0472	ENTERPRISE NETWORK SERVICES	BANYAN SYSTEMS INC
1139	0473	ADMINISTRATIVE AGENT SERVER	MCGILL UNIVERSITY FAC OF ENGIN
1140	0474	SYBASE SQL SERVER	SYBASE INC
1141	0475	SYBASE SQL SERVER CONSOLE	SYBASE INC
1142	0476	SYBASE SQL SERVER MONITOR	SYBASE INC
1143	0477	SYBASE SQL SERVER BACK-UP	SYBASE INC
1144	0478	SYBASE OPEN SERVER	SYBASE INC
1145	0479	SYBASE OPEN SERVER CONSOLE	SYBASE INC
1148	047C	REMOTE PRINTER CONSOLE	PEERLESS GROUP
1149	047D	AUTO-ON/OFF CONTROL NLM	MITSUBISI DENKI COMPUTER
1150	047E	ASCOM FAX SERVER	ASCOM TELECOMMUNICATION LTD
1151	047F	ASCOM ADVERTISING FAX SERVER	ASCOM TELECOMMUNICATION LTD
1152	0480	ASCOM FAX QUEUE	ASCOM TELECOMMUNICATION LTD

SAP-DEC	SAP-HEX	Service	Company
1153	0481	JOB SERVER	HIGH ASPECT DEVELOPMENT
1154	0482	JOB QUEUE	HIGH ASPECT DEVELOPMENT
1155	0483	FINANCE	THINK SYSTEMS CORP
1156	0484	FORCASTING	THINK SYSTEMS CORP
1157	0485	SCHEMA	THINK SYSTEMS CORP
1158	0486	FAIL SAFE ANALYSIS	THINK SYSTEMS CORP
1159	0487		THINK SYSTEMS CORP
1160	0488	COMMUNICATION BETWEEN CLIENT AND SERVER	CDC - OFFICE OF PROGRAM SUPPRT
1161	0489	STORE NAME OF SPECIAL FILE ON FILESERVER	DELTA INFORMATION SYSTEMS
1162	048A	MAIN NLM SERVER FOR CALENDAR MANAGER	RUSSELL INFORMATION SCIENCES
1163	048B	ARRAY MONITOR SERVER	CORE INTERNATIONAL
1164	048C	DOCUMENT PROCESSING SERVER NLM	BOSS LOGIC INC
1165	048D	DOCUMENT PROCESSING SERVER NLM	BOSS LOGIC INC
1166	048E	DOCUMENT PROCESSING SERVER NLM	BOSS LOGIC INC
1167	048F	DOCUMENT PROCESSING SERVER NLM	BOSS LOGIC INC
1168	0490	POWER PRODUCT FOR FILE SERVER	BRAINSTORM ENGINEERING
1169	0491	NETBLAZER COMMUNICATION SERVER	TELEBIT CORPORATION
1170	0492	RTS TERMINAL EMULATION	DATA RESEARCH & APPLICATIONS
1171	0493	RSCF CLIENT-SERVER API	DATA RESEARCH & APPLICATIONS
1172	0494	CD NETWORKER BINDARY TYPE	UNKNOWN CUSTOMER
1173	0495	REMOTE BACK-UP DEVICE	ASTORA SOFTWARE INC
1174	0496	SQL SERVER IPX/SPX HIDDEN SERVER	MICROSOFT
1175	0497	DATABASE LOCK SERVER	HIGH ASPECT DEVELOPMENT
1176	0498	METER SERVER	POLYMETER RESPONSE LTD
1177	0499	LANCORP EOMS	LANCORP PTY LTD
1178	049A	BULL HN SDM	LANCORP PTY LTD
1179	049B	NETWORK MANAGEMENT AGENT	EICON TECHNOLOGY
1180	049C	ICOT SNA GATEWAY	ICOT CORP
1181	049D	SOFTWARE NLM (MVDFRM-008 06/07/94)	INTERCONECTIONS
1182	049E	INTERNET GATEWAY	METASCYBE SYSTEMS LTD
1183	049F	EMAIL & CALENDARING COMMUNICATION SERVER	ATTACHMATE CANADA
1184	04A0	INTELLIGENT HOST GATEWAY	AMERICAN AIRLINES DECISION TEC
1185	04A1	AUTOMATION INFORMATION ROUTER	KURT MANUFACTURING
1186	04A2	XBASE RECORD ENGINE	EXTENDED SYSTEMS
1187	04A3	REMOTE PROCEDURE CALL SYSTEM	FORTUNET INC
1188	04A4	DELVRY OF VALUABLE INFO W/USAGE TRACKING	WAVE SYSTEMS CORPORATION
1189	04A5	REMOTE ACCESS SERVER	TRAVELING SOFTWARE
1190	04A6	BINDERY FOR PROGRAM METERING DATABASE	PILOTT SYSTEMS
1191	04A7	LT AUDITOR 4.00 PLUS	BLUE LANCE INC
1192	04A8	BINDERY SERVICE	FUJITSU LTD
1193	04A9	BINDERY SERVICE	FUJITSU LTD
1194	04AA	SAP SERVICE	FUJITSU LTD
1195	04AB	SAP SERVICE	FUJITSU LTD
1196	04AC	CALENDAR SERVER	CAMPBELL SERVICES INC
1197	04AD	NETPRINT PRINT SERVER	INTERLINK COMMUNICATIONS LTD
1198	04AE	LED SERVER	INOVA CORPORATION
1199	04AF	ANTI-VIRUS NLM	
1200	04B0	CDNET SERVER	MERIDIAN DATA CORP
1201	04B1	POLICY ENGINE	EMERALD SYSTEMS
1202	04B2	POLICY ENGINE	EMERALD SYSTEMS
1203	04B3	POLICY ENGINE	EMERALD SYSTEMS
1204	04B4	POLICY ENGINE	EMERALD SYSTEMS
1205	04B5	POLICY ENGINE	EMERALD SYSTEMS
1206	04B6	POLICY ENGINE	EMERALD SYSTEMS
1207	04B7	LAN ASSIST PLUS REMOTE CONTROL	MICROTEST INC
1208	04B8	LAN ASSIST PLUS REMOTE CONTROL	MICROTEST INC
1209	04B9	LAN ASSIST PLUS REMOTE CONTROL	MICROTEST INC
1210	04BA	LAN ASSIST PLUS REMOTE CONTROL	MICROTEST INC
1211	04BB	MAP ASSIST PEER-TO-PEER	MICROTEST INC
1212	04BC	MAP ASSIST PEER-TO-PEER	MICROTEST INC
1213	04BD	MAP ASSIST PEER-TO-PEER	MICROTEST INC
1214	04BE	MAP ASSIST PEER-TO-PEER	MICROTEST INC
1215	04BF	JETNET DRIVER FOR NOVELL IPX PROTOCOLS	JETSTREAM TECHNOLOGY LTD
1216	04C0	DATABASE SERVER RUNNING AS NLM	LODESTAR DATA SYSTEMS

SAP-DEC	SAP-HEX	Service	Company
1217	04C1	ASYNCHRONOUS COMMUNICATIONS SERVERS	US ROBOTICS SOFTWARE
1218	04C2	DATABASE SERVER	FAIR COM
1219	04C3	TAURUS DATABASE SERVER	DCI
1220	04C4	TAURUS SERIAL SERVER	DCI
1221	04C5	CASPER QUEUE	NETWORK TECHNICAL SOLUTIONS
1222	04C6	CASPER GHOST	NETWORK TECHNICAL SOLUTIONS
1223	04C7	CONEFRENCING SERVICE	FUJITSU NETWORKS INDUSTRY
1224	04C8	MAIL SYSTEM QUEUE SERVICE	MITSUBISHI ELECTRIC ENGINEERIN
1225	04C9	VIDEO SERVER	NOVELL - MULTI MEDIA
1226	04CA	MESSAGE EXPRESS PRODUCT	HORIZON STRATEGIES INC
1227	04CB	CD ROM SERVER	CBIS INC
1228	04CC	COST RECOVERY SERVER	VINCENT LARSEN
1229	04CD	PC-BASED SNA GATEWAY	UNGERMANN BASS
1230	04CE	SECURITY PASSWORD PROPAGATION	NOVELL - OREM
1231	04CF	USER RESTRICTIONS	VAL LABORATORY CO LTD
1232	04D0	USER RESTRICTIONS	VAL LABORATORY CO LTD
1233	04D1	SAP ADVERTISING ON PRINT SERVER	NISSIN ELECTRIC CO LTD
1234	04D2	OT_ELLIPSE_SERVER	BACHMAN INFORMATION SYSTEMS
1235	04D3	ASYNCHRONOUS ACCESS SERVER	SKYLINE TECHNOLOGY
1236	04D4	ENTERPRISE DESCRIPTION OBJECT	HANS SPATZIER
1237	04D5	PRINT SERVER	FORESYTE TECHNOLOGIES
1238	04D6	SERVER TYPE FOR NLM	
1239	04D7	CUBIX QL SERVER	CUBIX CORP
1240	04D8	CUBIX QL CLIENT	CUBIX CORP
1241	04D9	JOB SERVER	STORAGE DIMENSIONS
1242	04DA	COMMUNICATION SERVER	FIREFOX COMMUNICATIONS LTD
1243	04DB	COMMUNICATION SERVER	FIREFOX COMMUNICATIONS LTD
1244	04DC	COMMUNICATION SERVER	FIREFOX COMMUNICATIONS LTD
1245	04DD	COMMUNICATION SERVER	FIREFOX COMMUNICATIONS LTD
1246	04DE	COMMUNICATION SERVER	FIREFOX COMMUNICATIONS LTD
1247	04DF	COMMUNICATION SERVER	FIREFOX COMMUNICATIONS LTD
1248	04E0	COMMUNICATION SERVER	FIREFOX COMMUNICATIONS LTD
1249	04E1	COMMUNICATION SERVER	FIREFOX COMMUNICATIONS LTD
1250	04E2	COMMUNICATION SERVER	FIREFOX COMMUNICATIONS LTD
1251	04E3	COMMUNICATION SERVER	FIREFOX COMMUNICATIONS LTD
1252	04E4	COMMUNICATION SERVER	FIREFOX COMMUNICATIONS LTD
1253	04E5	COMMUNICATION SERVER	FIREFOX COMMUNICATIONS LTD
1254	04E6	COMMUNICATION SERVER	FIREFOX COMMUNICATIONS LTD
1255	04E7	COMMUNICATION SERVER	FIREFOX COMMUNICATIONS LTD
1256	04E8	COMMUNICATION SERVER	FIREFOX COMMUNICATIONS LTD
1257	04E9	COMMUNICATION SERVER	FIREFOX COMMUNICATIONS LTD
1258	04EA	COMMUNICATION SERVER	FIREFOX COMMUNICATIONS LTD
1259	04EB	COMMUNICATION SERVER	FIREFOX COMMUNICATIONS LTD
1260	04EC	COMMUNICATION SERVER	FIREFOX COMMUNICATIONS LTD
1261	04ED	COMMUNICATION SERVER	FIREFOX COMMUNICATIONS LTD
1262	04EE	REMOTE ACCESS SERVER	DIGITAL COMMUNICATIONS ASSOC (DCA)
1263	04EF	STATISTIC MANAGEMENT	MULTITECH
1264	04F0	STATISTIC MANAGEMENT	MULTITECH
1265	04F1	REMOTE CONTROL SOFTWARE	MULTITECH
1266	04F2	REMOTE CONTROL SOFTWARE	MULTITECH
1267	04F3	MULTICOM NET - MODEM SHARING SOFTWARE	MULTITECH
1268	04F4	NETLYNX COMMUNICATION SERVER	ANDREW CORPORATION
1269	04F5	NETLYNX COMMUNICATION SERVER	ANDREW CORPORATION
1270	04F6	NETLYNX COMMUNICATION SERVER	ANDREW CORPORATION
1271	04F7	NETLYNX COMMUNICATION SERVER	ANDREW CORPORATION
1272	04F8	NETLYNX COMMUNICATION SERVER	ANDREW CORPORATION
1273	04F9	NETLYNX COMMUNICATION SERVER	ANDREW CORPORATION
1274	04FA	NETLYNX COMMUNICATION SERVER	ANDREW CORPORATION
1275	04FB	NETLYNX COMMUNICATION SERVER	ANDREW CORPORATION
1276	04FC	NETLYNX COMMUNICATION SERVER	ANDREW CORPORATION
1277	04FD	NETLYNX COMMUNICATION SERVER	ANDREW CORPORATION
1278	04FE	NETLYNX COMMUNICATION SERVER	ANDREW CORPORATION
1279	04FF	NETLYNX COMMUNICATION SERVER	ANDREW CORPORATION
1280	0500	NETLYNX COMMUNICATION SERVER	ANDREW CORPORATION
1281	0501	NETLYNX COMMUNICATION SERVER	ANDREW CORPORATION
1282	0502	NETLYNX COMMUNICATION SERVER	ANDREW CORPORATION
1283	0503	NETLYNX COMMUNICATION SERVER	ANDREW CORPORATION
1284	0504	DESKVIEW X BINDERY TYPE	QUARTERDECK OFFICE SYSTEMS

SAP-DEC	SAP-HEX	Service	Company
1285	0505	PRINT SERVER ADD-ON	INTEL - AMERICAN FORK
1286	0506	INDEX SEQUENTIAL ACCESS NLM	INFOPOINT SYSTEMS
1287	0507	ASSOCIATIVE INDEX SERVER	INFOPOINT SYSTEMS
1288	0508	NETSCRIBE SERVER	MERIDIAN DATA CORP
1289	0509	PRINT SERVER FOR REMOTE WORKSTATIONS	FUJI XEROX CO LTD
1290	050A	NETMAGIC BINDERY ID	NET MAGIC SYSTEMS INC
1291	050B	FINANCIAL MARKET INFORMATION SERVER	AT FINANCIAL
1292	050C	NETWORK MODEM	NANAGRAM
1293	050D	NETWORK MODEM	NANAGRAM
1294	050E	DOCUMENT MANAGEMENT SERVICE	IMAGERY SOFTWARE INC
1295	050F	IMAGE MANAGEMENT SERVICE	IMAGERY SOFTWARE INC
1296	0510	MASS STORAGE SERVICE	IMAGERY SOFTWARE INC
1297	0511	DATABASE SERVER	TOBIT SOFTWARE GMBH
1298	0512	TELI-LINK VOICE SERVER	COMPUTER & COMMUNICATIONS CO
1299	0513	PRINT SERVER	EMULEX CORPORATION
1300	0514	1012 HUB AGENT	ASANTE TECHNOLOGIES
1301	0515	1012 HUB AGENT	ASANTE TECHNOLOGIES
1302	0516	1012 HUB AGENT	ASANTE TECHNOLOGIES
1303	0517	1012 HUB AGENT	ASANTE TECHNOLOGIES
1304	0518	1012 HUB AGENT	ASANTE TECHNOLOGIES
1305	0519	1012 HUB AGENT	ASANTE TECHNOLOGIES
1306	051A	1012 HUB AGENT	ASANTE TECHNOLOGIES
1307	051B	1012 HUB AGENT	ASANTE TECHNOLOGIES
1308	051C	1012 HUB AGENT	ASANTE TECHNOLOGIES
1309	051D	1012 HUB AGENT	ASANTE TECHNOLOGIES
1310	051E	ADVERTISING NETWORK MODEMS	LANSOURCE TECHNOLOGIES
1311	051F	APPLICATION PROGRAMS	U OF PLYMOUTH
1312	0520	BLOOMBERG AUDIO SERVER	BLOOMBERG LP
1313	0521	BLOOMBERG PROCESS SERVER	BLOOMBERG LP
1314	0522	NET MODEM SERVER	PRACTICAL PERIPHERALS INC
1317	0525	AGENT FOR HUB MANAGEMENT	IDEA COURIER
1318	0526	NAMED PIPE COMMUNICATIONS SERVICE	SYMANTEC PETER NORTON GROUP
1319	0527	PRINT SERVER/REMOTE PRINTER	PACIFIC DATA PRODUCTS
1320	0528	PRINT SERVER	SEIKO EPSON CORP
1321	0529	INTERSERVER FILE COPYING	BANKERS TRUST CO
1322	052A	FAX & VOICE SERVER	DIGITAL COMMUNICATIONS ASSOC (DCA)
1323	052B	ANGORA ENTERPRISE GATEWAY	RABBIT SOFTWARE CORP
1324	052C	EZCON/EPSCON LANTRONIX /REMOTE LOGIN TERMINAL	GORDIAN
1325	052D	APPLICATION SERVER	CITRIX SYSTEMS
1326	052E	OT_BLOOMBERG	MEDIA TOUCH SYSTEMS
1327	052F	MICROCOM REMOTE ACCESS SERVER	MICROCOM INC
1328	0530	REMARK VOICE SERVER	SIMPACT & ASSOC INC
1329	0531	IPX COMMUNICATION SERVER	SYMANTEC PETER NORTON GROUP
1330	0532	ACCESS SERVER FOR MODEM SHARING	BAY TECHNICAL ASSOCIATES
1331	0533	GATEWAY BETWEEN NETWORK & VISA POS-PORT	HEMKO SYSTEMS CORP
1332	0534	DEVICE LOCATION VIA REMOTE CONFIG TOOL	MILAN TECHNOLOGY CORP
1333	0535	DEVICE LOCATION VIA REMOTE CONFIG TOOL	MILAN TECHNOLOGY CORP
1334	0536	DEVICE LOCATION VIA REMOTE CONFIG TOOL	MILAN TECHNOLOGY CORP
1335	0537	DEVICE LOCATION VIA REMOTE CONFIG TOOL	MILAN TECHNOLOGY CORP
1336	0538	DEVICE LOCATION VIA REMOTE CONFIG TOOL	MILAN TECHNOLOGY CORP
1337	0539	DEVICE LOCATION VIA REMOTE CONFIG TOOL	MILAN TECHNOLOGY CORP
1338	053A	DEVICE LOCATION VIA REMOTE CONFIG TOOL	MILAN TECHNOLOGY CORP
1339	053B	DEVICE LOCATION VIA REMOTE CONFIG TOOL	MILAN TECHNOLOGY CORP
1340	053C	DEVICE LOCATION VIA REMOTE CONFIG TOOL	MILAN TECHNOLOGY CORP
1341	053D	DEVICE LOCATION VIA REMOTE CONFIG TOOL	MILAN TECHNOLOGY CORP
1342	053E	REMOTED SMS SERVER	NEW ERA SYSTEMS SERVICES LTD
1343	053F	LISTENER/NETWARE	STERLING TEFEN LAB
1344	0540	LISTENER/DOS	STERLING TEFEN LAB
1345	0541	LISTENER/WINDOWS	STERLING TEFEN LAB
1346	0542	LISTENER/OS2	STERLING TEFEN LAB
1347	0543	LISTENER/MAC	STERLING TEFEN LAB
1348	0544	LISTENER/UNIX	STERLING TEFEN LAB
1349	0545	TECHRA CLIENT/SERVER RDBMS	KVATRO AS
1350	0546	DOCRA CLIENT/SERVER DOC MGMT SYSTEM	KVATRO AS
1351	0547	VOICE/FAX RESPONDING MACHINE	SYSTEM SOPHIA
1352	0548	I4/LS NAMING SERVICE	GRADIENT TECHNOLOGIES

SAP-DEC	SAP-HEX	Service	Company
1353	0549	TNSI NETWORK UTILITIES	TOADALLY NETWORK SYSTEMS INC
1354	054A	RM3 CONFIGURATION	CAYMAN SYSTEMS INC
1355	054B	SNMP CONFIGURATION	CAYMAN SYSTEMS INC
1357	054D	IMAGESOLVE_OFS	IMAGESOLVE INTERNATIONAL
1358	054E	DISTRICT COMMUNICATION GATEWAY	CHANCERY SOFTWARE LTD
1359	054F	DISTRICT LINK	CHANCERY SOFTWARE LTD
1360	0550	EDM CLIENT/PC	COMPUTER VISION
1362	0552	SECURITY OBJECT	BANK OF NEW ZEALAND
1365	0555	PRINTERS, PLOTTERS & ROUTERS	SEIKO INSTRUMENTS INC
1366	0556	FAX SERVICES	OAZ
1367	0557	FAX SERVICES	OAZ
1368	0558	FAX SERVICES	OAZ
1369	0559	AT&T JOINT VENTURE TELEPHONY SERVER	NOVELL - PROVO
1370	055A	AT&T JOINT VENTURE TELEPHONY SERVER	NOVELL - PROVO
1371	055B	AT&T JOINT VENTURE TELEPHONY SERVER	NOVELL - PROVO
1372	055C	AT&T JOINT VENTURE TELEPHONY SERVER	NOVELL - PROVO
1373	055D	HOSTVIEW UTILITY	ATTACHMATE CORPORATION
1374	055E	PRINT SERVER	LEXMARK INTERNATIONAL INC
1375	055F	PRINT SERVER	LEXMARK INTERNATIONAL INC
1376	0560	PRINT SERVER	LEXMARK INTERNATIONAL INC
1377	0561	PRINT SERVER	LEXMARK INTERNATIONAL INC
1378	0562	STATISTICS GATHERING AGENT	NETCRAFT SOFTWARE DEVELOPMENT
1379	0563	ERL DATABASE SERVER	SILVER PLATTER INFORMATION LTD
1380	0564	ERL DIRECTORY SERVER	SILVER PLATTER INFORMATION LTD
1381	0565	NEWSWIRE NOTIFICATION	GENERATION TECHNOLOGIES CORP
1382	0566	ZIFF PROPRIETARY SERVICES	ZIFF INFORMATION SERVICES
1383	0567	TIME SERVER NLM	MEINBERG FUNKUHREN
1384	0568	X-BASE DATABASE SERVER	EXTENDED SYSTEMS
1385	0569	ADVERTISING MEDIA-DB SERVER	RAVI TECHNOLOGIES INC
1386	056A	OEM REMOTE ACCESS FOR ETHERNET	SHIVA CORP
1387	056B	OEM REMOTE ACCESS FOR TOKENRING	SHIVA CORP
1388	056C	INTEGRATED REMOTE ACCESS FOR ETHERNET	SHIVA CORP
1389	056D	INTEGRATED REMOTE ACCESS FOR TOKENRING	SHIVA CORP
1391	056F	MED STATION SERVER	PYXIS CORPORATION
1392	0570	TELNET IPX ROUTER	TELETROL SYSTEMS INC
1393	0571	NLM SERVER	SR ASSOCIATES/CYBERMEDIA
1394	0572	ADVERTISING NETMODEM SERVER	PRACTICAL PERIPHERALS INC
1395	0573	SATELLITE CONNECTIONS	ULF ZIMMERMANN
1396	0574	LAN/CD SERVER	LOGICRAFT
1398	0576	PEER COMMUNICATIONS TOOL	INTELEC SYSTEMS CORPORATION
1399	0577	COMMUNICATION SERVER	NORMAN DATA DEFANSE SYSTEMS
1400	0578	MODEM SHARING SOFTWARE	SEG
1401	0579	HOPS DATABASE SERVER	HOPS INTERNATIONAL
1402	057A	DATAFILE ACCESS HANDLER	LMS
1403	057B	NOVUS GATEWAY PRODUCT	FIREFOX COMMUNICATIONS LTD
1404	057C	LGS	INTERCONECTIONS
1405	057D	LGS-PFT	INTERCONECTIONS
1406	057E	LGS-PPS	INTERCONECTIONS
1407	057F	TN 3270 GATEWAY	BUS TECH INC
1408	0580	MCAFEE VIRUS PATTERN SERVER	MCAFEE ASSOCIATES
1409	0581	OPTICAL SERVER	OPTISYS
1410	0582	PRORIDE X.500 LDAP SUPPORT	CONTROL DATA SYSTEMS
1411	0583	NET TRAX ALARM MONITOR	NET X CORP
1412	0584	NET TRAX FILE SERVER AGENT	NET X CORP
1413	0585	NET TRAX WORKSTATION AGENT	NET X CORP
1414	0586	NET TRAX BRIDGE AGENT	NET X CORP
1415	0587	TEAM OFFICE PRODUCT	ICL PERSONAL SYSTEMS OY
1417	0589	HYPERDESK DISTRIBUTE OBJECT MGMT SYSTEM	HYPERDESK
1418	058A	HYPERDESK DATABASE TYPE	HYPERDESK
1419	058B	TCP/SUPERLAT HOST PRINT SAP	MERIDIAN TECHNOLOGY CORP
1420	058C	OCS SAFESERVER	OMNITECH CORPORATE SOLUTIONS
1421	058D	JUKEBOX USER GROUP	TODD ENTERPRISES INC
1422	058E	FULL SCREEN LOGIN	CONFIRM
1423	058F	MEETING SPACE SERVER	WORLD BENDERS INC
1424	0590	LAN PRODUCT SERVER	BMC SOFTWARE INC
1425	0591	CBT SERVER	FIRST CLASS SYSTEMS
1426	0592	NET TUNE SERVICE	HAWKNET INC

SAP-DEC	SAP-HEX	Service	Company
1427	0593	PRINT SERVER	RICOH COMPANY LTD
1428	0594	REMOTE PRINTER	RICOH COMPANY LTD
1429	0595	SLATHP STATUS NLM	MERIDIAN TECHNOLOGY CORP
1431	0597	IPX NAMED PIPES COMMUNICATION SERVICE	SYMANTEC PETER NORTON GROUP
1432	0598	APPMETER FOR SUITES	FUNK SOFTWARE
1436	059C	SERVERLOG	DIAMOND SOFTWARE INC
1440	05A0	SERVICE LOCATION PROTOCOL	EICON TECHNOLOGY
1441	05A1	NMS ICONS	NETWORTH INC
1442	05A2	NMS ICONS	NETWORTH INC
1443	05A3	NMS ICONS	NETWORTH INC
1444	05A4	NMS ICONS	NETWORTH INC
1445	05A5	NMS ICONS	NETWORTH INC
1446	05A6	NMS ICONS	NETWORTH INC
1447	05A7	NMS ICONS	NETWORTH INC
1448	05A8	NMS ICONS	NETWORTH INC
1449	05A9	NMS ICONS	NETWORTH INC
1450	05AA	NMS ICONS	NETWORTH INC
1451	05AB	NMS ICONS	NETWORTH INC
1452	05AC	NMS ICONS	NETWORTH INC
1453	05AD	NMS ICONS	NETWORTH INC
1454	05AE	NMS ICONS	NETWORTH INC
1455	05AF	NMS ICONS	NETWORTH INC
1456	05B0	NMS ICONS	NETWORTH INC
1457	05B1	NMS ICONS	NETWORTH INC
1458	05B2	NMS ICONS	NETWORTH INC
1459	05B3	MULTI-SYSTEM MANAGER - NETVIEW INTERFACE	IBM - RESEARCH TRIANGLE PARK
1460	05B4	ONLINE TRANSACTION TRANSPORTATION SYSTEM	ABG TECHNOLOGY/UNITED CARD SVC
1462	05B6	DATASTAR NLM	APERTUS TECHNOLOGIES
1463	05B7	DS_LCC - LOGICAL LINK CONTROLLER	APERTUS TECHNOLOGIES
1464	05B8	DATABASE MANAGEMENT SERVICES	REVELATION TECHNOLOGIES
1465	05B9	DISTRIBUTION SERVICES DISCOVERY	IBM - ROME
1466	05BA	MANAGING HARDWARE ROUTERS	COMPATIBLE SYSTEMS CORP
1472	05C0	CALENDAR SERVER	CAMPBELL SERVICES INC
1473	05C1	NETWORK MANAGEMENT APPLICATION	XIRCOM
1474	05C2	NETWORK MANAGEMENT APPLICATION	XIRCOM
1475	05C3	NETWORK MANAGEMENT APPLICATION	XIRCOM
1476	05C4	NETWORK MANAGEMENT APPLICATION	XIRCOM
1477	05C5	NETWORK MANAGEMENT APPLICATION	XIRCOM
1478	05C6	NETWORK MANAGEMENT APPLICATION	XIRCOM
1497	05D9	ETHERNET-MANAGED STACKABLE HUB	IBM - RESEARCH TRIANGLE PARK
1508	05E4	FILENET NETWORK CLEARINGHOUSE	FILE NET CORPORATION
1509	05E5	HPCS DATA-BASED PRODUCTS	TOBIT SOFTWARE GMBH
1510	05E6	AST SNMP INSTRUMENTED SERVER	AST RESEARCH INC
1515	05EB	OFFICE EXTEND SERVER	FRANSEN KING
1516	05EC	WINDOWS NT NAMED PIPE SERVER	OPTUS INFORMATION SYSTEMS
1517	05ED	GATEWAY SERVER	FUJITSU LTD
1518	05EE	BINDERY FILE	FUJITSU LTD
1519	05EF	FAX WORKSTATION OBJECT	SOFNET INC
1522	05F2	ROUTER ADMINISTRATION SERVER	LIVINGSTON ENTERPRISES
1523	05F3	CLIENT SERVER ARRAY MONITOR PROGRAM	ALLODYNE INC
1524	05F4	EVERGREEN MANAGEMENT AGENT	GOODALL SOFTWARE
1525	05F5	WINDOWS BULLETIN BOARD SYSTEM	PACER SOFTWARE
1528	05F8	UPS SNMP MONITOR NLM	COMPUTER SITE TECHNOLOGIES INC
1530	05FA	GOODALL VIRTUAL PROTOCOL ADAPTOR	GOODALL SOFTWARE
1539	0603	TIME CARD SERVER	ADVANCED MANAGEMENT SOLUTIONS
1542	0606	IMAGE SERVER ADDRESS LOOK-UP	WATERMARK SOFTWARE
1543	0607	ARTS RLOGIND SERVER	AMERICAN REAL TIME, REUTERS CO
1544	0608	ARTS GENERIC SERVER	AMERICAN REAL TIME, REUTERS CO
1545	0609	ARTS RUUPD - ARE YOU UP DAEMON	AMERICAN REAL TIME, REUTERS CO
1546	060A	MONEY CENTER DATA SERVER	KNIGHT RIDDER FINANCIAL INC
1548	060C	AXIS PRINTER SERVER	AXIS COMMUNICATIONS AB
1549	060D	UNIX MAIL SERVER	FELPAUSCH
1552	0610	SCSI MANAGEMENT	ADAPTEC
1553	0611	STEALTH EVENT CAPTURE ENGINE	INTELLIGENCE QUOTIENT INTL LTD
1555	0613	UNIVERSAL DATA TRANSPORTER	CONSEIL FORMATION ET DEVELOPPE
1556	0614	COMMUNICATION SOFTWARE (NLM)	NEC - NIPPON ELECTRIC COMPANY

SAP-DEC	SAP-HEX	Service	Company
1557	0615	COMMUNICATION SOFTWARE (NLM)	NEC - NIPPON ELECTRIC COMPANY
1558	0616	FAX GATEWAY SERVER	RUSSELL CONSULTING
1564	061C	PRINT SERVER	EMULEX CORPORATION
1565	061D	PRINT SERVER	EMULEX CORPORATION
1566	061E	PRINT SERVER	EMULEX CORPORATION
1567	061F	PRINT SERVER	EMULEX CORPORATION
1568	0620	PRINT SERVER	EMULEX CORPORATION
1573	0625	SETUP MANAGER ACCESS CONTROL	MAXIMIZED SOFTWARE
1583	062F	METERING PROGRAM	SECURE DESIGN
1590	0636	COMET TERMINAL SERVER	GOODALL SOFTWARE
1591	0637	COMET FILE SERVER	GOODALL SOFTWARE
1600	0640	NT SERVER-RPC SERVICES, GATEWAY SERVICES FOR NW OR WIN95-USERLEVELSECURITY	MICROSOFT
1614	064E	NT SERVER-INTERNET INFORMATION SERVER	MICROSOFT
1853	073D	CONNECT:DIRECT NLM SERVER	STERLING SOFTWARE
1854	073E	CONNECT:DIRECT NLM SERVER	STERLING SOFTWARE
1855	073F	CONNECT:DIRECT NLM SERVER	STERLING SOFTWARE
1856	0740	CONNECT:DIRECT NLM SERVER	STERLING SOFTWARE
1857	0741	CONNECT:DIRECT NLM SERVER	STERLING SOFTWARE
1858	0742	CONNECT:DIRECT NLM SERVER	STERLING SOFTWARE
1859	0743	MAXSERV 3270	MAXSERV
1860	0744	MAXSERV PRICE	MAXSERV
1861	0745	MAXSERV MAIL	MAXSERV
1862	0746	MAXSERV MACS	MAXSERV
1863	0747	MAXSERV RESERVED	MAXSERV
1864	0748	MAXSERV RESERVED	MAXSERV
1865	0749	TASKING IPX/LWSI GATEWAY	TASKING SOFTWARE BV
1869	074D	NLM APPLICATIONS	KNX LTD
1870	074E	NLM APPLICATIONS	KNX LTD
1871	074F	NLM APPLICATIONS	KNX LTD
1872	0750	NLM APPLICATIONS	KNX LTD
1875	0753	STP PROTOCOL SERVICE AGENT (CSA_FTP)	DIGITAL TECHNOLOGIES
1876	0754	IND$FILE FILE TRANASFER AGT (CSA_INDFILE)	DIGITAL TECHNOLOGIES
1877	0755	ALL KALPANA SWITCHES	KALPANA
1878	0756	CHAT SERVER	MICROSOFT
1879	0757	TITANIUM DATABASE ENGINE	MICRO DATA BASE SYSTEMS INC
1880	0758	TCS COMMUNICATION SERVER	MICRO TEMPUS INC
1881	0759	PC COMMUNICATION PARTNER SAP	FUJITSU LTD
1882	075A	PC COMMUNICATION PARTNER SAP	FUJITSU LTD
1883	075B	PC COMMUNICATION PARTNER BINDERY	FUJITSU LTD
1884	075C	DPC SERVER SAP	FUJITSU LTD
1885	075D	DPC SERVER BINDERY	FUJITSU LTD
1886	075E	CAM HOST	TMD CONSULTING
1887	075F	SERVER MNITORING APPLICATION	SOFTWORK GMBH
1898	076A	ALARM MANAGER NLM	CONNER PERIPHERALS
1899	076B	EVENT MANAGER NLM	CONNER PERIPHERALS
1901	076D	DESKTOP MANAGEMENT NLM	ADVANCED MODULAR SOLUTIONS INC
1904	0770	REAL-TIME INTEGRATION SERVICES	INDUSTRIAL PEER TO PEER
1905	0771	LANDEEP SERVER MONITOR	DEERFIELD COMPUTER SOLUTIONS
1907	0773	HITECSOFT API MANAGER	HITECSOFT CORP
1908	0774	HITECSOFT PUBLIC LIBRARY	HITECSOFT CORP
1909	0775	HITECSOFT PHONE SERVER	HITECSOFT CORP
1915	077B	ADVANTAGE X-BASE SERVER	EXTENDED SYSTEMS
1919	077F	SECURITY AUDITOR	SECURE DESIGN
1920	0780	U OF WISCONSIN UTILITIES	U OF WISCONSIN - MADISON, CAE
1921	0781	U OF WISCONSIN UTILITIES	U OF WISCONSIN - MADISON, CAE
1922	0782	U OF WISCONSIN UTILITIES	U OF WISCONSIN - MADISON, CAE
1923	0783	U OF WISCONSIN UTILITIES	U OF WISCONSIN - MADISON, CAE
1931	078B	A NON-NAME-PIPE PIPE SERVER	TEAM DEVELOPMENT CORP
1944	0798	NETHOPPER ROUTER	ROCKWELL NETWORK SYSTEMS
1945	0799	NETHOPPER CLIENT ROUTER	ROCKWELL NETWORK SYSTEMS
1946	079A	NETHOPPER CLIENT	ROCKWELL NETWORK SYSTEMS
1951	079F	TIMBUKTU ADDRESS RESOLUTIONS	FARALLON COMPUTING
1952	07A0	REMOTE ACCESS SERVER	MERIDIAN TECHNOLOGY CORP
1959	07A7	BACKUP EXEC JOB QUEUE	ARCADA SOFTWARE
1960	07A8	BACKUP EXEC JOB MANAGER	ARCADA SOFTWARE

SAP-DEC	SAP-HEX	Service	Company
1961	07A9	BACKUP EXEC JOB SERVICE	ARCADA SOFTWARE
1964	07AC	SERVICE NW_CONNECT_PW_SERVER	FLAGSTAR
1967	07AF	SITE METER SERVER	MCAFEE ASSOCIATES
1968	07B0	PROXY AGENT SITE MTER	MCAFEE ASSOCIATES
1971	07B3	DISC DISASTER RECOVERY S/W SCHEDULER	COLUMBIA DATA PRODUCTS
1972	07B4	CUBIX COMM. SERVER	CUBIX CORP
1976	07B8	IDENTIFY SERVER VENDOR FOR NMS	SIEMENS-NIXDORF INFOSYSTEME
1978	07BA	FILE TRANSFER BETWEEN LAN/MAINFRAME	PROGINET CORP
1983	07BF	DATABASE SERVER	COMP SOFT PLC
1984	07C0	APPLICATION ADVERTISEMENT	COMP SOFT PLC
1985	07C1	INTERNET APPS ON NW PCS	MANGO SYSTEMS
1998	07CE	COMPUTER SUPPORTED TELEPHONY APPS	GLOBAL COMMUNICATIONS LTD
2001	07D1	A SYNCHRONOUS COMMUNICATION	LAN ACCESS
2002	07D2	RM AUDITOR-MONITORING NW USAGE	RESEARCH MACHINES
2003	07D3	GENERIC NW MANAGEMENT PRODUCT	3COM CORPORATION
2004	07D4	SCSI MANAGEMENT	ADAPTEC
2007	07D7	FAXSERVE	CHEYENNE
2009	07D9	DISTRIBUTION JOB SERVER	EMOTION INC
2010	07DA	DISTRIBUTION QUE	EMOTION INC
2011	07DB	DISTRIBUTION FEEDBACK SERVER	EMOTION INC
2012	07DC	VOICE PROCESSING SERVER	INTERNATIONAL VOICE EXCHANGE
2013	07DD	TELEPHONY SERVER MONITOR	INTERNATIONAL VOICE EXCHANGE
2016	07E0	PRINT SERVER	CREATIVE CONTROLLERS
2017	07E1	BOX ROUTER SUPPORTING IP, IPX & APPLE TALK	FURUKAWA ELECTRIC CO LTD
2018	07E2	TEL SERVER	FUJITSU NS CENTER
2019	07E3	TELEMARKETING LIBRARY	FUJITSU NS CENTER
2020	07E4	MULTI-SERVER METERING	MCAFEE ASSOCIATES
2022	07E6	STRESS MAGIC INSTALLATION UTILITY HANDLE	NETMAGIC SYSTEMS INC
2026	07EA	AVM-ISDN CALL CONTROL AGENT (ISDNCCA.NLM)	AVM
2028	07EC	FINANCIAL DATA SERVER	TELEMET AMERICA INC
2029	07ED	ISDN SOFTWARE SHARING OF ISDN BOARDS	HIGH SOFT TECH
2030	07EE	COMGATE-COMM GATEWAY FOR IBM AS 400	COMWARE INTERNATIONAL
2032	07F0	ZIP CODE SERVER	THIRD PLANET SOFTWARE
2034	07F2	NORTON ANTI-VIRUS CLIENT SERVER COMM	SYMANTEC PETER NORTON GROUP
2040	07F8	SOFTWARE ENTERTAINMENT	ORIGIN SYSTEMS
2041	07F9	REMOTE MANAGEMENT OF REMOTE ACCESS SERVER	MERIDIAN TECHNOLOGY CORP
2044	07FC	FTMANAGER-CONTROLS NETWARE SERVER	MITSUBISHI DENKI COMPUTER
2045	07FD	SAP FOR AUDITTRACK VS. 2.0	E.G. SOFTWARE INC
2047	07FF	SOFT-TUB AGENT PROXY FOR TRI-COM ADAPTOR	3COM CORPORATION
2049	0801	SAP FOR ONGUARD	E.G. SOFTWARE INC
2061	080D	IFONY FLOW FOR NETWARE	FUJITSU SOCIAL SCIENCE LAB
2062	080E	PRINT SERVER FOR NW 3.X	ISHIGAKI
2064	0810	ELAN LICENSE SERVER DEMO	ELAN COMPUTER GROUP
2065	0811	ElAN LICENSE SERVER	ELAN COMPUTER GROUP
2067	0813	BVNCS ENTERPRISE WAN	LAN SUPPORT GROUP
2068	0814	EVERGREEN WINDCAP MANAGEMENT AGENT	GOODALL SOFTWARE
2070	0816	POLICY ENGINE	EMERALD SYSTEMS
2071	0817	POLICY ENGINE	EMERALD SYSTEMS
2072	0818	POLICY ENGINE	EMERALD SYSTEMS
2073	0819	POLICY ENGINE	NCSU
2074	081A	POLICY ENGINE	NCSU
2075	081B	POLICY ENGINE	NCSU
2076	081C	POLICY ENGINE	NCSU
2077	081D	OFF-LINE COPYING	ON TECHNOLOGIES
2087	0827	PALINDROME SERVICE BROKER	PALINDROME
2089	0829	LAN LENS SERVER	DIGITAL EQUIPMENT CORP
2090	082A	LAN LENS PROBE	DIGITAL EQUIPMENT CORP
2091	082B	JOB QUEUE	ARCADA SOFTWARE
2095	082F	DE SERVER 3.X	SIMULTAN AG
2096	0830	PIN64	ASHI ELECTRONICS CO LTD
2098	0832	APPMETER 2-LICENSING PRODUCT	FUNK SOFTWARE
2099	0833	COMMUNICATION SERVER	DIGITAL PRODUCTS INC
2100	0834	FTANALYZER	MITSUBISHI ELECTRIC CORP

SAP-DEC	SAP-HEX	Service	Company
2101	0835	INSTANT INTERNET	PERFORMANCE TECHNOLOGY
2102	0836	HIGH SPEED COMMUNICATION SERVER	C SPEC CORP
2103	0837	ALANTA/3 PRINT SERVER	KELLY COMPUTER SYSTEMS
2109	083D	APPLICATION SERVER FOR NT	CITRIX SYSTEMS
2110	083E	CUSTOM AUTHENTICATION/LOAD BALANCING	UNIVERSITY OF PITTSBURGH
2112	0840	IP TO IPX GATEWAY	INTERNET JUNCTION
2113	0841	RDB/7000	FUJITSU LTD
2124	084C	BOX-TYPE HDWR ROUTER SUPPTS APPLETALK	BUG INC
2125	084D	SNAP-IN TOOL FOR NMS	NEC CORPORATION
2128	0850	APLINKTERM	MIKASA SYSTEM ENGINEERING
2131	0853	APLINKGATE	MIKASA SYSTEM ENGINEERING
2132	0854	AUTOMATION PROTOCOL SERVER	ARCADA SOFTWARE
2133	0855	TAPE ID SERVER	ARCADA SOFTWARE
2136	0858	NETWARE TO INTERNET GATEWAY	ON TECHNOLOGIES
2138	085A	RDB/7000 SERVER FOR NW	FUJITSU LTD
2140	085C	DB-EXPRESS	FUJITSU KOBE ENGINEERING LTD
2143	085F	AGENT ROUTER PROXY	ARCADA SOFTWARE
2144	0860	CATALOG SERVER	ARCADA SOFTWARE
2145	0861	INTETIX GATEWAY	MICRO COMPUTER SYSTEMS
2152	0868	SOUNDBYTE SERVER	ENVISION TELEPHONY INC
2153	0869	PEER TO PEER FILE REDIRECTION	ALEX LLOYD
2159	086F	VISION 20 MULTI-FUNCTIONAL PRODUCT	RICOH CORP
2160	0870	DPS-SV	PFU LIMITED
2166	0876	AUDIT SERVER	LAN SUPPORT GROUP
2167	0877	ALERT SERVER	LAN SUPPORT GROUP
2168	0878	NETSQUEEZE	LAN SUPPORT GROUP
2169	0879	EWAM	LAN SUPPORT GROUP
2170	087A	BINDING SERVER	LAN SUPPORT GROUP
2171	087B	AUTHENTICATION SERVER	LAN SUPPORT GROUP
2173	087D	MODEM POOL SERVER	ATTACHMATE
2174	087E	FLORA MANAGER FOR NETWARE	HITACHI LTD SOFTWARE DEVT CTR
2182	0886	MULTI-USER DOS GOLD	CONCURRENT CONTROLS INC
2187	088B	SPX TELENET AGENT	TELEBIT
2194	0892	UPS CONTROL SOFTWARE UPSCON	AKIRA SAITO
2199	0897	SHARED LAN CACHE	MEASUREMENT TECHNIQUES
2202	089A	"CLEAR IT"	DELL COMPUTER CORP
2205	089D	FAX SERVER - PERFECT FIT	NEWTECH SYSTEMS
2207	089F	NET TUNE PRO LICENSE	HAWKNET INC
2208	08A0	NET TUNE PRO ALERT SERVICE	HAWKNET INC
2209	08A1	POWERVISOR, POWER MONITOR	ALFATECH INC
2212	08A4	INETIX GATEWAY ADMIN SERVICE	MICRO COMPUTER SYSTEMS
2216	08A8	RAPPORT 112	SHIVA CORP
2218	08AA	CPA "CORP PRINT ACCOUNTING"	BLUE LANCE
2220	08AC	GROUPEASE FOR WIN-NEIGHBORHOOD WATCH	ETHOSOFT, INC
2221	08AD	CS-CARE NETWORK	COMPU-SHACK PRAGUE
2222	08AE	FIRECALL	TEAM DEVELOPMENT CORP
2224	08B0	NETWARE ENHANCED INTEGRATION FOR AS400	IBM - ROCHESTER
2226	08B2	OVERDRIVE	STAMPEDE TECHNOLOGY INC
2227	08B3	IIR	ANSWERSOFT INC.
2228	08B4	WANDERLINK DIALOUT	FUNK SOFTWARE
2230	08B6	F-PROT PROFESSIONAL FOR NETWARE COMMAND	SW
2231	08B7	WEB MANAGEMENT	VIRTUAL MOTION INC
2232	08B8	CAN DISTRIBUTED TRAFFIC MONITOR	CHEVIN SOFTWARE ENGINEERING LT
2233	08B9	DS GATE	DATASTREAM INTERNATIONAL
2234	08BA	VTD/X GATEWAY	MICRO MATIC RESEARCH
2235	08BB	SALIENT CORP	SALIENT TECHNOLOGIES INC
2236	08BC	IP ACCESS	AMERICAN INTERNET CORPORATION
2238	08BE	ELECTRONIC ARTS GAMES	ELECTRONIC ARTS INC
2239	08BF	IPX/IP GATEWAY	UKIAH SOFTWARE SOLUTIONS INC
2240	08C0	NORTON ENTERPRISE FRAMEWORK (nef)	SYMANTEC CORP
2241	08C1	XYRATEX VIRTUALSCSI NLM	XYRATEX
2242	08C2	8235 DIALS SWITCH FOR TOKEN RING	IBM - RESEARCH TRIANGLE PARK
2243	08C3	STAMPEDE OVERDRIVE	STAMPEDE TECHNOLOGY INC
2244	08C4	INTEGATE	NETSPHERE COMMUNICATIONS CORP
2246	08C6	PICCOLO	CORNERSTONE
2248	08C8	SERVERTRAK	INTRAK INC
2249	08C9	CDSERVE	M4COM LTD
2252	08CC	POWERBURST/INTEGRATED	SHIVA CORP

SAP-DEC	SAP-HEX	Service	Company
2253	08CD	SURFCONTROL	WONDERWARE CORPORATION
2254	08CE	TREND TRAK	INTRAK INC
2255	08CF	SURFCONTROL	JSB COMPUTER SYSTEMS LTD
2256	08D0	INTERNET LANBRIDGE	VIRTUAL MOTION INC
2257	08D1	HYPERDRIVE	PAPERWISE
2258	08D2	GOLF SCORING SYSTEM	INFORMATION AND DISPLAY SYST
2259	08D3	SURFCONTROL	JSB COMPUTER SYSTEMS LTD
2260	08D4	SURFCONTROL	JSB COMPUTER SYSTEMS LTD
2261	08D5	SURFCONTROL	JSB COMPUTER SYSTEMS LTD
2262	08D6	SURFCONTROL	JSB COMPUTER SYSTEMS LTD
2263	08D7	SURFCONTROL	JSB COMPUTER SYSTEMS LTD
2264	08D8	SURFCONTROL	JSB COMPUTER SYSTEMS LTD
2265	08D9	SURFCONTROL	JSB COMPUTER SYSTEMS LTD
2266	08DA	SURFCONTROL	JSB COMPUTER SYSTEMS LTD
2267	08DB	SURFCONTROL	JSB COMPUTER SYSTEMS LTD
2268	08DC	SURFCONTROL	JSB COMPUTER SYSTEMS LTD
2269	08DD	SURFCONTROL	JSB COMPUTER SYSTEMS LTD
2270	08DE	SURFCONTROL	JSB COMPUTER SYSTEMS LTD
2271	08DF	SURFCONTROL	JSB COMPUTER SYSTEMS LTD
2272	08E0	HTTP-WEBMAIL, EXPRESSIT!_2000	INFINITE SOFTWARE
2273	08E1	SMPT-EXPRESSIT!2000, WEBMAIL, & INFINITE INTERCHANGE	INFINITE SOFTWARE
2274	08E2	POP3-EXPRESSIT!2000 & INFINITE INTERCHANGE	INFINITE SOFTWARE
2275	08E3	IMAP4-EXPRESSIT!2000 & INFINITE INTERCHANGE	INFINITE SOFTWARE
2276	08E4	NNTP-EXPRESSIT!2000 & INFINITE INTERCHANGE	INFINITE SOFTWARE
2277	08E5	MPR FOR ISDN	AVM GMBH
3113	0C29	LICENSE PROFILE	INTEGRITY SOFTWARE
3114	0C2A	VIRUS FILE LIST	INTEGRITY SOFTWARE
3116	0C2C	GLOBAL LICENSE SAP	INTEGRITY SOFTWARE
3121	0C31	LICENSE REPORT DEFINITION	INTEGRITY SOFTWARE
9100	238C	MEETING MAKER XP	ON TECHNOLOGY
9101	238D	NLM	ON TECHNOLOGY
9102	238E	NLM	ON TECHNOLOGY
18440	4808	SITELOCK	
32770	8002	INTEL NETPORT (PRINT SERVER)	INTEL
34952	8888	WORDPERFECT NETWORK VERSION	WORDPERFECT
36864	9000	REMOTE NLM SPAWN /RSPAWN UTILITY	NOVELL - SAN JOSE
38432	9620	LICENSE PROFILE ADMINISTRATOR	INTEGRITY SOFTWARE
62723	F503	MICROSFT SQL SERVER	MICROSOFT
65535	FFFF	ALL TYPES	NOVELL - PROVO CORP HQ

Appendix D – Novell® Private Socket Numbers

Socket Decimal	Socket Hex	Service	Company
32768	8000		NATIONAL ADVANCED SYSTEMS
32769	8001		NATIONAL ADVANCED SYSTEMS
32770	8002		NATIONAL ADVANCED SYSTEMS
32771	8003	COMM DRIVER	SPERRY CORP. COMPUTER SYSTEMS
32772	8004		KTA
32773	8005		KTA
32774	8006		KTA
32775	8007		KTA
32776	8008		NOVELL - PROVO
32777	8009	SPERRY TERM EMULATOR	TURNBULL AUTOMATIONS
32778	800A	PRINT SERVER	COMMUNICATION HORIZONS
32779	800B		DATA LANGUAGE CORP
32783	800F	BATRAM	SANTA CLARA SYSTEMS
32784	8010	OFFICE WARE	CENTURY ANALYSIS
32785	8011	UPS	ELGAR CORP
32786	8012	UPS	ELGAR CORP
32787	8013		CHI CORP
32788	8014		INTEL - AMERICAN FORK
32789	8015		COMPASS COMPUTING
32790	8016		COMPASS COMPUTING
32791	8017		COMPASS COMPUTING
32792	8018		COMPASS COMPUTING
32793	8019		COMPASS COMPUTING
32794	801A		COMPASS COMPUTING
32795	801B		COMPASS COMPUTING
32796	801C		COMPASS COMPUTING
32797	801D		COMPASS COMPUTING
32798	801E		COMPASS COMPUTING
32799	801F		NOVELL - SUNNYVALE LATIN AMERI
32800	8020		NOVELL - SUNNYVALE LATIN AMERI
32801	8021		NOVELL - SUNNYVALE LATIN AMERI
32802	8022		NOVELL - SUNNYVALE LATIN AMERI
32803	8023		MCAFEE ASSOCIATES
32804	8024		BLUE LANCE NETWORK INFO SYS
32807	8027		GATEWAY COMMUNICATIONS INC
32808	8028		GATEWAY COMMUNICATIONS INC
32809	8029		GATEWAY COMMUNICATIONS INC
32810	802A	FILE SHARING	NETLINE INC
32811	802B	FILE SHARING	NETLINE INC
32812	802C		INTEL - AMERICAN FORK
32813	802D		INTEL - AMERICAN FORK
32814	802E		ICM
32815	802F	C-TREE VAP	FAIR COM
32816	8030		MICROMIND
32817	8031		MICROMIND
32818	8032		NORTH STAR COMPUTERS
32819	8033		NORTH STAR COMPUTERS
32820	8034	X.25 GATEWAY	RSJ SOFTWARE
32821	8035		SANYO ICON INC
32822	8036		DATA ACCESS CORP
32825	8039	NET MANAGEMENT	NOVELL - AUSTIN
32826	803A		BETA SOFT
32827	803B		PHASER SYSTEMS
32828	803C		PHASER SYSTEMS
32829	803D		PHASER SYSTEMS
32830	803E		PERFORMANCE GROUP
32831	803F		PERFORMANCE GROUP
32832	8040		HORIZON TECHNOLOGY INC
32833	8041	CD-ROM SERVER	MERIDIAN DATA CORP
32834	8042		NATIONWIDE COMPUTER SERVICES
32835	8043	COMM SERVER	COMPUTER LANGUAGE RESEARCH INC
32839	8047	3274 CONTROLLER EMULATORS	SOFTWARE DYNAMICS
32840	8048	3274 CONTROLLER EMULATORS	SOFTWARE DYNAMICS
32841	8049	3274 CONTROLLER EMULATORS	SOFTWARE DYNAMICS
32842	804A	3274 CONTROLLER EMULATORS	SOFTWARE DYNAMICS
32843	804B	MIC SNA DFV SERVER	COMPUTERLAND - BONN
32844	804C	MIC SNA DFV SERVER	COMPUTERLAND - BONN
32845	804D	DATABASE SERVER	MIGENT SOFTWARE INC
32846	804E		SWITCH LINK SYSTEMS
32847	804F		INTERACTIVE FINANCIAL SOL INC
32848	8050		INTERACTIVE FINANCIAL SOL INC
32849	8051		INTERACTIVE FINANCIAL SOL INC
32850	8052		INTERACTIVE FINANCIAL SOL INC
32851	8053		UNKNOWN CUSTOMER
32852	8054		UNKNOWN CUSTOMER
32853	8055	E-MAIL CHAT	NICHE CO
32854	8056	MONEY TRANSFER	IPI INC
32855	8057	E-MAIL CHAT	NICHE CO
32862	805E		TELEBASE SYSTEMS
32863	805F		TELEBASE SYSTEMS
32864	8060	PRINT SERVER	NOVELL - PROVO
32865	8061	T-NET, LAN BRIDGES	BRITISH TELECOM
32866	8062		WOLLONGONG GROUP
32868	8064		MICRO DATA BASE SYSTEMS INC
32869	8065		MICRO DATA BASE SYSTEMS INC
32870	8066		NORTON LAMBERT CORP
32871	8067		NORTON LAMBERT CORP
32872	8068		NORTON LAMBERT CORP
32873	8069	NETWARE TO HP LAN GATEWAY	HEWLETT PACKARD - SUNNYVALE
32874	806A	REMOTE PC SOFTWARE	ALM
32876	806C	CHAT PROGRAM	DIGITAL AV INC
32877	806D	PC DEX	MERGENT INTERNATIONAL
32878	806E	DART	COLLEGE HILL SYSTEMS
32880	8070	NETWORK COURIER	MICROSOFT WORKGROUP CANADA
32881	8071	PIPES	PEER LOGIC
32882	8072		WORDPERFECT CORP
32883	8073	DATABASE & 4GL	PROGRESS SOFTWARE CORP
32884	8074	RIGHT HAND MAN	FUTURESOFT
32885	8075	RIGHT HAND MAN	FUTURESOFT
32886	8076	LASER DISK PROGRAM	U OF WISCONSIN - SUPERIOR
32887	8077	FAX LINK & VAX MANAGER	OPTUS INFORMATION SYSTEMS
32888	8078	FAX LINK & VAX MANAGER	OPTUS INFORMATION SYSTEMS
32889	8079	TELECOMMUNICATIONS	VAN AUTO PARTS
32890	807A	TELECOMMUNICATIONS	VAN AUTO PARTS
32891	807B	TELECOMMUNICATIONS	VAN AUTO PARTS
32892	807C	TELECOMMUNICATIONS	VAN AUTO PARTS
32893	807D	TELECOMMUNICATIONS	VAN AUTO PARTS
32894	807E	R21PX	CROSSTALK
32895	807F		ORACLE CORP
32896	8080		ORACLE CORP
32897	8081		ORACLE CORP
32898	8082		ORACLE CORP
32899	8083		ORACLE CORP
32900	8084		ORACLE CORP
32901	8085		ORACLE CORP
32902	8086		ORACLE CORP
32903	8087		ORACLE CORP
32904	8088		ORACLE CORP
32905	8089	REMOTE PC SOFTWARE	ALM
32906	808A		UNKNOWN CUSTOMER
32907	808B		UNKNOWN CUSTOMER
32908	808C		UNKNOWN CUSTOMER
32909	808D		UNKNOWN CUSTOMER
32910	808E		UNKNOWN CUSTOMER
32911	808F		UNKNOWN CUSTOMER
32912	8090	VAP	PILLSBURY CO
32913	8091	VAP	PILLSBURY CO
32914	8092	VAP	PILLSBURY CO
32915	8093	VAP	PILLSBURY CO
32916	8094	VAP	PILLSBURY CO
32917	8095	VAP	PILLSBURY CO
32918	8096	VAP	PILLSBURY CO
32919	8097	VAP	PILLSBURY CO
32920	8098	VAP	PILLSBURY CO
32921	8099	VAP	PILLSBURY CO
32922	809A		MARSHFIELD CLINIC
32923	809B		MARSHFIELD CLINIC
32924	809C		MARSHFIELD CLINIC
32925	809D		MARSHFIELD CLINIC
32926	809E		MARSHFIELD CLINIC
32927	809F		MARSHFIELD CLINIC
32943	80AF		DSI DYNAPRO SYSTEMS INC
32944	80B0		DSI DYNAPRO SYSTEMS INC
32945	80B1		DSI DYNAPRO SYSTEMS INC
32946	80B2		DSI DYNAPRO SYSTEMS INC
32947	80B3		DSI DYNAPRO SYSTEMS INC
32948	80B4		CHICAGO RESEARCH & TRADING
32949	80B5		STREETWISE SYSTEMS INC
32950	80B6		STREETWISE SYSTEMS INC
32951	80B7	FAX SERVER	DIGITAL VISIONS CORP

Socket Decimal	Socket Hex	Service	Company
32952	80B8	VOICE SERVER	DIGITAL VISIONS CORP
32953	80B9		DIGITAL VISIONS CORP
32954	80BA	NETWARE MANAGEMENT APPLICATION	FRYE COMPUTER SYSTEMS
32955	80BB	MAJOR BBS SOFTWARE	GALACTICOMM INC
32956	80BC	MAJOR BBS SOFTWARE	GALACTICOMM INC
32957	80BD	MAJOR BBS SOFTWARE	GALACTICOMM INC
32958	80BE	MAJOR BBS SOFTWARE	GALACTICOMM INC
32959	80BF	MAJOR BBS SOFTWARE	GALACTICOMM INC
32960	80C0	MAJOR BBS SOFTWARE	GALACTICOMM INC
32961	80C1	MAJOR BBS SOFTWARE	GALACTICOMM INC
32962	80C2	MAJOR BBS SOFTWARE	GALACTICOMM INC
32963	80C3	CHAT PROGRAM/IPX TALK	FELSINA SOFTWARE
32964	80C4	IPX/SPX COMM PROTOCOL APPLICATIONS	MAGEE ENTERPRISES INC
32965	80C5		WATCOM
32966	80C6		NEWPORT SYSTEMS SOLUTIONS INC
32967	80C7		NEWPORT SYSTEMS SOLUTIONS INC
32968	80C8		NEWPORT SYSTEMS SOLUTIONS INC
32969	80C9		NEWPORT SYSTEMS SOLUTIONS INC
32970	80CA		NEWPORT SYSTEMS SOLUTIONS INC
32971	80CB		NEWPORT SYSTEMS SOLUTIONS INC
32972	80CC		NEWPORT SYSTEMS SOLUTIONS INC
32973	80CD		NEWPORT SYSTEMS SOLUTIONS INC
32974	80CE	SPX DRIVER/NOVA FOCUS REMOTE PC ACCESS	DRIGGS CORPORATION
32975	80CF	APPL IPX/SPX COMMUNICATIONS	MAGEE ENTERPRISES INC
32976	80D0	APPL IPX/SPX COMMUNICATIONS	MAGEE ENTERPRISES INC
32977	80D1	APPL IPX/SPX COMMUNICATIONS	MAGEE ENTERPRISES INC
32978	80D2	APPL IPX/SPX COMMUNICATIONS	MAGEE ENTERPRISES INC
32979	80D3	APPL 1PX/SPX COMMUNICATIONS	MAGEE ENTERPRISES INC
32981	80D5	DATABASE SERVER	GUPTA TECHNOLOGIES
32982	80D6	DATABASE SERVER	GUPTA TECHNOLOGIES
32983	80D7	LAN/FILE TRANSFER, COMM SERVERS, REMOTES	US ROBOTICS SOFTWARE
32984	80D8	POWERCHUTE— VAP/NLM SERVER POWER SUPPLY	AMERICAN POWER CONVERSION
32985	80D9	COREL DRIVER PRODUCT UNDER NOVELL 386	COREL SYSTEMS, OPTICAL DIV
32986	80DA	ARCHIVE SERVER	GIGATREND INC
32987	80DB	GATEWAY PRODUCT	ATLANTA TECHNOLOGIES
32988	80DC	GATEWAY PRODUCT	ATLANTA TECHNOLOGIES
32989	80DD	OFFICE ORGANIZER NLM	UNISYS - CAMARILLO
32990	80DE		UNIVERSAL NETWORK SYSTEMS
32991	80DF	APPLICATION SERVER	NATIONSBANK APPL SYSTEMS SUPP
32993	80E1		MODULAR SOFTWARE CORPORATION
32994	80E2	CLIENT SERVER APPLICATION	SOFTWARE AG
32996	80E4	LANPORT VIRTUAL EXTENSION OF PORTS	MICROTEST INC
32997	80E5	WORK STATION PEER-TO-PEER	CONVEYANT SYSTEMS INC
32998	80E6	DESKVIEW X - IPX SOCKET INTERFACE	QUARTERDECK OFFICE SYSTEMS
32999	80E7	DESKVIEW X - IPX SOCKET INTERFACE	QUARTERDECK OFFICE SYSTEMS
33000	80E8	DESKVIEW X - IPX SOCKET INTERFACE	QUARTERDECK OFFICE SYSTEMS
33001	80E9	DESKVIEW X - IPX SOCKET INTERFACE	QUARTERDECK OFFICE SYSTEMS
33002	80EA	DESKVIEW X - IPX SOCKET INTERFACE	QUARTERDECK OFFICE SYSTEMS
33003	80EB	DESKVIEW X - IPX SOCKET INTERFACE	QUARTERDECK OFFICE SYSTEMS
33004	80EC		MERIDIAN DATA CORP
33005	80ED		BIZTECH
33006	80EE		RATIONAL DATA SYSTEMS
33007	80EF		RATIONAL DATA SYSTEMS
33008	80F0		RATIONAL DATA SYSTEMS
33009	80F1		RATIONAL DATA SYSTEMS
33012	80F4	GARP SERVER COMMUNICATION	NET RESEARCH PTY LTD
33013	80F5	NETWORK MANAGEMENT PRODUCT	NCR - COLUMBIA
33014	80F6	NETWORK MANAGEMENT PRODUCT	NCR - COLUMBIA
33015	80F7	NETWORK MANAGEMENT PRODUCT	NCR - COLUMBIA
33016	80F8	NETWORK MANAGEMENT PRODUCT	NCR - COLUMBIA
33017	80F9	NETWORK MANAGEMENT PRODUCT	NCR - COLUMBIA
33018	80FA	NETWORK MANAGEMENT PRODUCT	NCR - COLUMBIA
33019	80FB		PROFESSIONAL PROGRAMMING SVCS
33020	80FC		PROFESSIONAL PROGRAMMING SVCS
33021	80FD		PEER LOGIC
33022	80FE		WALL DATA
33023	80FF	DISTRIBUTED APPLICATION	FOLIO CORPORATION

Socket Decimal	Socket Hex	Service	Company
33024	8100		MARSHFIELD CLINIC
33025	8101		MARSHFIELD CLINIC
33026	8102		MARSHFIELD CLINIC
33027	8103		MARSHFIELD CLINIC
33029	8105		VIA
33039	810F	NET 3270	MCGILL UNIVERSITY COMPUTING CT
33040	8110		PROFESSIONAL PRODUCTIVITY CORP
33041	8111	TODS	TELETRAK
33042	8112	LUFTHANSA BS	LUFTHANSA GERMAN AIRLINES
33043	8113	LANPORT	MICROTEST INC
33044	8114	LANPORT	MICROTEST INC
33045	8115	LANPORT	MICROTEST INC
33046	8116	LANPORT	MICROTEST INC
33047	8117	LANPORT	MICROTEST INC
33048	8118	LANPORT	MICROTEST INC
33049	8119	LANPORT	MICROTEST INC
33050	811A	LANPORT	MICROTEST INC
33051	811B	LANPORT	MICROTEST INC
33052	811C	LANPORT	MICROTEST INC
33053	811D	IMAGE SERVER	FILE NET CORPORATION
33055	811F	RTK OPERATING SYSTEM	OWL MICRO SYSTEMS
33056	8120	TXD	THOMAS CONRAD CORP
33057	8121	SPECIAL REQUEST	SPECTRAFAX
33058	8122	NET MONITOR	ARTEFACT NETWORK SUPPORT
33059	8123	NET MONITOR	ARTEFACT NETWORK SUPPORT
33060	8124	NET MONITOR	ARTEFACT NETWORK SUPPORT
33061	8125	NET MONITOR	ARTEFACT NETWORK SUPPORT
33062	8126	NET MONITOR	ARTEFACT NETWORK SUPPORT
33063	8127	NET MONITOR	ARTEFACT NETWORK SUPPORT
33064	8128	NET MONITOR	ARTEFACT NETWORK SUPPORT
33065	8129	NET MONITOR	ARTEFACT NETWORK SUPPORT
33066	812A	NET MONITOR	ARTEFACT NETWORK SUPPORT
33067	812B	NET MONITOR	ARTEFACT NETWORK SUPPORT
33068	812C	NET MONITOR	ARTEFACT NETWORK SUPPORT
33069	812D	NET MONITOR	ARTEFACT NETWORK SUPPORT
33080	8138	LANSIGHT	LAN SYSTEMS
33081	8139	PC CHALKBOARD	INTEL - AMERICAN FORK
33083	813B	REAL TIME BACK-UP	EMERALD SYSTEMS
33084	813C	NETWORK MANAGEMENT	PURE DATA INC
33086	813E	FILE TRANSFER	AAC SYSTEMS
33087	813F	IMAGE SERVER	WANG LABORATORIES
33088	8140	IMAGE SERVER	WANG LABORATORIES
33089	8141	IMAGE SERVER	WANG LABORATORIES
33090	8142	IMAGE SERVER	WANG LABORATORIES
33091	8143		NETWORK DESIGNERS LTD
33092	8144	MAYNESTREAM	ARCADA SOFTWARE
33093	8145	MAYNESTREAM	ARCADA SOFTWARE
33094	8146	NETWORK MANAGEMENT SYSTEM	ACCUNETICS
33095	8147	NETWORK MANAGEMENT SYSTEM	ACCUNETICS
33096	8148	NETWORK MANAGEMENT SYSTEM	ACCUNETICS
33097	8149	NETWORK MANAGEMENT SYSTEM	ACCUNETICS
33098	814A		BR COMPUTING
33099	814B	NLM	AUTOMATED DESIGN SYSTEMS
33100	814C		BR COMPUTING
33101	814D		BR COMPUTING
33102	814E		BR COMPUTING
33103	814F		BR COMPUTING
33104	8150		BR COMPUTING
33105	8151		BR COMPUTING
33106	8152		BR COMPUTING
33107	8153		BR COMPUTING
33108	8154		BR COMPUTING
33109	8155		BR COMPUTING
33110	8156		BR COMPUTING
33111	8157		BR COMPUTING
33112	8158		BR COMPUTING
33113	8159		BR COMPUTING
33114	815A		BR COMPUTING
33115	815B		BR COMPUTING
33116	815C		BR COMPUTING
33117	815D		BR COMPUTING
33118	815E		BR COMPUTING
33119	815F		BR COMPUTING
33120	8160	PRINTQLAN	SOFTWARE DIRECTIONS INC
33121	8161	PRINTQLAN	SOFTWARE DIRECTIONS INC
33122	8162	PRINTQLAN	SOFTWARE DIRECTIONS INC
33123	8163	PRINTQLAN	SOFTWARE DIRECTIONS INC
33124	8164	PRINTQLAN	SOFTWARE DIRECTIONS INC
33125	8165	PRINTQLAN	SOFTWARE DIRECTIONS INC
33126	8166	PRINTQLAN	SOFTWARE DIRECTIONS INC

Socket Decimal	Socket Hex	Service	Company
33127	8167	PRINTQLAN	SOFTWARE DIRECTIONS INC
33128	8168	PRINTQLAN	SOFTWARE DIRECTIONS INC
33129	8169	PRINTQLAN	SOFTWARE DIRECTIONS INC
33130	816A	PRINTQLAN	SOFTWARE DIRECTIONS INC
33131	816B	PRINTQLAN	SOFTWARE DIRECTIONS INC
33132	816C	PRINTQLAN	SOFTWARE DIRECTIONS INC
33133	816D	PRINTQLAN	SOFTWARE DIRECTIONS INC
33134	816E	PRINTQLAN	SOFTWARE DIRECTIONS INC
33135	816F	PRINTQLAN	SOFTWARE DIRECTIONS INC
33136	8170	CDROM	ONLINE COMPUTER SYSTEMS
33137	8171	CDROM	ONLINE COMPUTER SYSTEMS
33138	8172		BENTLEY SYSTEMS
33139	8173		AVALAN
33140	8174		AVALAN
33141	8175		AVALAN
33142	8176		AVALAN
33143	8177		AVALAN
33144	8178		AVALAN
33145	8179		AVALAN
33146	817A		AVALAN
33147	817B	ACCOUNTING APD	SUPERNET
33148	817C		ACS
33149	817D	SERVER DB	XDB SYSTEMS
33150	817E	PIPES	PEER LOGIC
33151	817F	PIPES	PEER LOGIC
33152	8180		NETWARE SOFTWARE ASSOCIATES
33153	8181	PEER-TO-PEER	LODGISTIX INC
33154	8182	PEER-TO-PEER	LODGISTIX INC
33155	8183	PEER-TO-PEER	LODGISTIX INC
33156	8184	PEER-TO-PEER	LODGISTIX INC
33157	8185	PEER-TO-PEER	LODGISTIX INC
33158	8186		DANWARE DATA AS
33159	8187		DANWARE DATA AS
33160	8188		DANWARE DATA AS
33161	8189		DANWARE DATA AS
33162	818A		DANWARE DATA AS
33163	818B	NETFRAME NW 386	NETFRAME
33164	818C	NETFRAME NW 386	NETFRAME
33165	818D	NETFRAME NW 386	NETFRAME
33166	818E	NETFRAME NW 386	NETFRAME
33167	818F	MAXIBACK	SYSGEN INC
33168	8190	MAXIBACK	SYSGEN INC
33169	8191	MAXIBACK	SYSGEN INC
33170	8192	MAXIBACK	SYSGEN INC
33171	8193	DCA IPX COMM PRODUCT	DIGITAL COMMUNICATIONS ASSOC (DCA)
33172	8194	DCA IPX COMM PRODUCT	DIGITAL COMMUNICATIONS ASSOC (DCA)
33173	8195	DCA IPX COMM PRODUCT	DIGITAL COMMUNICATIONS ASSOC (DCA)
33174	8196	QUICKCHART	HEALTHWARE
33175	8197	QUICKCHART	HEALTHWARE
33176	8198	QUICKCHART	HEALTHWARE
33177	8199	QUICKCHART	HEALTHWARE
33178	819A		U OF OTAGO
33179	819B		U OF OTAGO
33180	819C	MINI SQL	ISICAD
33184	81A0	DEALING ROOM SYSTEMS	ART & SCIENCE LTD
33185	81A1	DEALING ROOM SYSTEMS	ART & SCIENCE LTD
33186	81A2	DEALING ROOM SYSTEMS	ART & SCIENCE LTD
33187	81A3	DEALING ROOM SYSTEMS	ART & SCIENCE LTD
33188	81A4	DEALING ROOM SYSTEMS	ART & SCIENCE LTD
33189	81A5	DEALING ROOM SYSTEMS	ART & SCIENCE LTD
33190	81A6	DEALING ROOM SYSTEMS	ART & SCIENCE LTD
33191	81A7	DEALING ROOM SYSTEMS	ART & SCIENCE LTD
33192	81A8	DEALING ROOM SYSTEMS	ART & SCIENCE LTD
33193	81A9	DEALING ROOM SYSTEMS	ART & SCIENCE LTD
33194	81AA	DEALING ROOM SYSTEMS	ART & SCIENCE LTD
33195	81AB	DEALING ROOM SYSTEMS	ART & SCIENCE LTD
33196	81AC	DEALING ROOM SYSTEMS	ART & SCIENCE LTD
33197	81AD	DEALING ROOM SYSTEMS	ART & SCIENCE LTD
33198	81AE	DEALING ROOM SYSTEMS	ART & SCIENCE LTD
33199	81AF	DEALING ROOM SYSTEMS	ART & SCIENCE LTD
33200	81B0	DEALING ROOM SYSTEMS	ART & SCIENCE LTD
33201	81B1	DEALING ROOM SYSTEMS	ART & SCIENCE LTD
33202	81B2	DEALING ROOM SYSTEMS	ART & SCIENCE LTD
33203	81B3	DEALING ROOM SYSTEMS	ART & SCIENCE LTD
33204	81B4	DEALING ROOM SYSTEMS	ART & SCIENCE LTD
33205	81B5	DEALING ROOM SYSTEMS	ART & SCIENCE LTD
33206	81B6	DEALING ROOM SYSTEMS	ART & SCIENCE LTD
33207	81B7	DEALING ROOM SYSTEMS	ART & SCIENCE LTD
33208	81B8	DEALING ROOM SYSTEMS	ART & SCIENCE LTD
33209	81B9	DEALING ROOM SYSTEMS	ART & SCIENCE LTD
33210	81BA	DEALING ROOM SYSTEMS	ART & SCIENCE LTD
33211	81BB	DEALING ROOM SYSTEMS	ART & SCIENCE LTD
33212	81BC	DEALING ROOM SYSTEMS	ART & SCIENCE LTD
33213	81BD	DEALING ROOM SYSTEMS	ART & SCIENCE LTD
33214	81BE	DEALING ROOM SYSTEMS	ART & SCIENCE LTD
33215	81BF	DEALING ROOM SYSTEMS	ART & SCIENCE LTD
33216	81C0	DEALING ROOM SYSTEMS	ART & SCIENCE LTD
33217	81C1	DEALING ROOM SYSTEMS	ART & SCIENCE LTD
33218	81C2	DEALING ROOM SYSTEMS	ART & SCIENCE LTD
33219	81C3	DEALING ROOM SYSTEMS	ART & SCIENCE LTD
33220	81C4	DEALING ROOM SYSTEMS	ART & SCIENCE LTD
33221	81C5	DEALING ROOM SYSTEMS	ART & SCIENCE LTD
33222	81C6	DEALING ROOM SYSTEMS	ART & SCIENCE LTD
33223	81C7	DEALING ROOM SYSTEMS	ART & SCIENCE LTD
33224	81C8	DEALING ROOM SYSTEMS	ART & SCIENCE LTD
33225	81C9	DEALING ROOM SYSTEMS	ART & SCIENCE LTD
33226	81CA	DEALING ROOM SYSTEMS	ART & SCIENCE LTD
33227	81CB	DEALING ROOM SYSTEMS	ART & SCIENCE LTD
33228	81CC	DEALING ROOM SYSTEMS	ART & SCIENCE LTD
33229	81CD	DEALING ROOM SYSTEMS	ART & SCIENCE LTD
33230	81CE	DEALING ROOM SYSTEMS	ART & SCIENCE LTD
33231	81CF	DEALING ROOM SYSTEMS	ART & SCIENCE LTD
33232	81D0	DEALING ROOM SYSTEMS	ART & SCIENCE LTD
33233	81D1	DEALING ROOM SYSTEMS	ART & SCIENCE LTD
33234	81D2	DEALING ROOM SYSTEMS	ART & SCIENCE LTD
33235	81D3	DEALING ROOM SYSTEMS	ART & SCIENCE LTD
33236	81D4	DEALING ROOM SYSTEMS	ART & SCIENCE LTD
33237	81D5	DEALING ROOM SYSTEMS	ART & SCIENCE LTD
33238	81D6	DEALING ROOM SYSTEMS	ART & SCIENCE LTD
33239	81D7	DEALING ROOM SYSTEMS	ART & SCIENCE LTD
33240	81D8	DEALING ROOM SYSTEMS	ART & SCIENCE LTD
33241	81D9	DEALING ROOM SYSTEMS	ART & SCIENCE LTD
33242	81DA	DEALING ROOM SYSTEMS	ART & SCIENCE LTD
33243	81DB	DEALING ROOM SYSTEMS	ART & SCIENCE LTD
33244	81DC	DEALING ROOM SYSTEMS	ART & SCIENCE LTD
33245	81DD	DEALING ROOM SYSTEMS	ART & SCIENCE LTD
33246	81DE	DEALING ROOM SYSTEMS	ART & SCIENCE LTD
33247	81DF	DEALING ROOM SYSTEMS	ART & SCIENCE LTD
33248	81E0	DEALING ROOM SYSTEMS	ART & SCIENCE LTD
33249	81E1	DEALING ROOM SYSTEMS	ART & SCIENCE LTD
33250	81E2	DEALING ROOM SYSTEMS	ART & SCIENCE LTD
33251	81E3	DEALING ROOM SYSTEMS	ART & SCIENCE LTD
33252	81E4	DEALING ROOM SYSTEMS	ART & SCIENCE LTD
33253	81E5	DEALING ROOM SYSTEMS	ART & SCIENCE LTD
33254	81E6	DEALING ROOM SYSTEMS	ART & SCIENCE LTD
33255	81E7	DEALING ROOM SYSTEMS	ART & SCIENCE LTD
33256	81E8	DEALING ROOM SYSTEMS	ART & SCIENCE LTD
33257	81E9	DEALING ROOM SYSTEMS	ART & SCIENCE LTD
33258	81EA	DEALING ROOM SYSTEMS	ART & SCIENCE LTD
33259	81EB	DEALING ROOM SYSTEMS	ART & SCIENCE LTD
33260	81EC	DEALING ROOM SYSTEMS	ART & SCIENCE LTD
33261	81ED	DEALING ROOM SYSTEMS	ART & SCIENCE LTD
33262	81EE	DEALING ROOM SYSTEMS	ART & SCIENCE LTD
33263	81EF	DEALING ROOM SYSTEMS	ART & SCIENCE LTD
33264	81F0	DEALING ROOM SYSTEMS	ART & SCIENCE LTD
33265	81F1	DEALING ROOM SYSTEMS	ART & SCIENCE LTD
33266	81F2	DEALING ROOM SYSTEMS	ART & SCIENCE LTD
33267	81F3	DEALING ROOM SYSTEMS	ART & SCIENCE LTD
33268	81F4	DEALING ROOM SYSTEMS	ART & SCIENCE LTD
33269	81F5	DEALING ROOM SYSTEMS	ART & SCIENCE LTD
33270	81F6	DEALING ROOM SYSTEMS	ART & SCIENCE LTD
33271	81F7	DEALING ROOM SYSTEMS	ART & SCIENCE LTD
33272	81F8	DEALING ROOM SYSTEMS	ART & SCIENCE LTD
33273	81F9	DEALING ROOM SYSTEMS	ART & SCIENCE LTD
33274	81FA	DEALING ROOM SYSTEMS	ART & SCIENCE LTD
33275	81FB	DEALING ROOM SYSTEMS	ART & SCIENCE LTD
33276	81FC	NETWORK SUPPORT MGR/PC REMOTE CNTRL PRGM	PCI LTD
33277	81FD	NETWORK SUPPORT MGR/PC REMOTE CNTRL PRGM	PCI LTD
33278	81FE		IWI
33279	81FF		MARTELLO & ASSOCIATES
33283	8203		NETWORK COMPUTING INC - NCI
33284	8204		NETWORK COMPUTING INC - NCI
33285	8205		NETWORK COMPUTING INC - NCI
33286	8206		NETWORK COMPUTING INC - NCI
33287	8207		NETWORK COMPUTING INC - NCI

Socket Decimal	Socket Hex	Service	Company
33288	8208		NETWORK COMPUTING INC - NCI
33289	8209		NETWORK COMPUTING INC - NCI
33290	820A		NETWORK COMPUTING INC - NCI
33291	820B		NETWORK COMPUTING INC - NCI
33292	820C		NETWORK COMPUTING INC - NCI
33293	820D		DATA VOICE SOLUTIONS CORP
33294	820E	ID 5001 WEATHER STATION	ZENITH DATA SYSTEMS
33295	820F	WRITE SERVER	ARC CALCULON
33296	8210	WRITE SERVER	QUANTUM CONSULTING
33297	8211	SYSTEM 9	HBF GROUP
33298	8212	SYSTEM 9	HBF GROUP
33299	8213	SYSTEM 9	HBF GROUP
33300	8214	SYSTEM 9	HBF GROUP
33301	8215	SYSTEM 9	HBF GROUP
33302	8216	SYSTEM 9	HBF GROUP
33303	8217	SYSTEM 9	HBF GROUP
33304	8218	SYSTEM 9	HBF GROUP
33305	8219	ARGUS	TRITICOM
33306	821A	ARGUS	TRITICOM
33307	821B	TCP/IP GATEWAY	COMPUTERVISION SERVICES
33308	821C	PICKIT (COMM SERVER)	INTEL CORP
33309	821D		PEER LOGIC
33310	821E		PEER LOGIC
33311	821F	DATA FACE NET BATCH	DATA FACE
33312	8220	DATA FACE NET BATCH	DATA FACE
33313	8221	LUMINAR OPTICAL SERVER	COREL SYSTEMS CORP
33314	8222		UNKNOWN CUSTOMER
33315	8223		UNKNOWN CUSTOMER
33316	8224		UNKNOWN CUSTOMER
33317	8225		UNKNOWN CUSTOMER
33318	8226		UNKNOWN CUSTOMER
33319	8227	X-BRIDGE	ADVANCED POLICY COMMUNICATIONS
33320	8228	X-BRIDGE	ADVANCED POLICY COMMUNICATIONS
33321	8229	FLEXCOM	EVERGREEN SYSTEMS
33322	822A	FLEXCOM	EVERGREEN SYSTEMS
33323	822B	FLEXCOM	EVERGREEN SYSTEMS
33324	822C	FLEXCOM	EVERGREEN SYSTEMS
33325	822D	FLEXCOM	EVERGREEN SYSTEMS
33326	822E	GATEWAYS & WKST PROCESSOR	TEKNOS SYSTEMS LTD
33327	822F	GATEWAYS & WKST PROCESSOR	TEKNOS SYSTEMS LTD
33328	8230	GATEWAYS & WKST PROCESSOR	TEKNOS SYSTEMS LTD
33329	8231	GATEWAYS & WKST PROCESSOR	TEKNOS SYSTEMS LTD
33330	8232	GATEWAYS & WKST PROCESSOR	TEKNOS SYSTEMS LTD
33331	8233	GATEWAYS & WKST PROCESSOR	TEKNOS SYSTEMS LTD
33332	8234	GATEWAYS & WKST PROCESSOR	TEKNOS SYSTEMS LTD
33333	8235	GATEWAYS & WKST PROCESSOR	TEKNOS SYSTEMS LTD
33334	8236	GATEWAYS & WKST PROCESSOR	TEKNOS SYSTEMS LTD
33335	8237	GATEWAYS & WKST PROCESSOR	TEKNOS SYSTEMS LTD
33336	8238	LANWARE	HORIZON TECHNOLOGY INC
33337	8239	LANWARE	HORIZON TECHNOLOGY INC
33338	823A	LANWARE	HORIZON TECHNOLOGY INC
33339	823B	LANWARE	HORIZON TECHNOLOGY INC
33340	823C	LANWARE	HORIZON TECHNOLOGY INC
33341	823D		UNKNOWN CUSTOMER
33342	823E	TEAM 286	IWI
33343	823F	DBMS LOCK MANAGER	RAIMA CORP
33344	8240	DBMS LOCK MANAGER	RAIMA CORP
33345	8241		CENTRAL POINT SOFTWARE
33346	8242	REMOTE COMPUTING	CENTRAL POINT SOFTWARE
33347	8243	REMOTE COMPUTING	CENTRAL POINT SOFTWARE
33348	8244	REMOTE COMPUTING	CENTRAL POINT SOFTWARE
33349	8245	REMOTE COMPUTING	CENTRAL POINT SOFTWARE
33350	8246	REMOTE COMPUTING	CENTRAL POINT SOFTWARE
33351	8247	REMOTE COMPUTING	CENTRAL POINT SOFTWARE
33352	8248	REMOTE COMPUTING	CENTRAL POINT SOFTWARE
33353	8249	TOKEN RING RPL	NCR - DAYTON
33354	824A	TOKEN RING RPL	NCR - DAYTON
33355	824B	DEALING ROOM SYSTEMS	ART & SCIENCE LTD
33356	824C	DEALING ROOM SYSTEMS	ART & SCIENCE LTD
33357	824D	DEALING ROOM SYSTEMS	ART & SCIENCE LTD
33358	824E	DEALING ROOM SYSTEMS	ART & SCIENCE LTD
33359	824F	DEALING ROOM SYSTEMS	ART & SCIENCE LTD
33360	8250	DEALING ROOM SYSTEMS	ART & SCIENCE LTD
33361	8251	DEALING ROOM SYSTEMS	ART & SCIENCE LTD
33362	8252	DEALING ROOM SYSTEMS	ART & SCIENCE LTD
33363	8253	TOTAL AUTOMATION SYSTEMS	DYNATECH UTAH SCIENTIFIC
33364	8254	TOTAL AUTOMATION SYSTEMS	DYNATECH UTAH SCIENTIFIC
33365	8255	VANTAGE POINT	CONNECT COMPUTER

Socket Decimal	Socket Hex	Service	Company
33366	8256	NETARC SCHEDULER	EMERALD SYSTEMS
33367	8257	DISTRIBUTED PROCESSING	BRIGHAM YOUNG UNIVERSITY
33368	8258	XTREE NET	CENTRAL POINT SOFTWARE
33369	8259	XTREE NET	CENTRAL POINT SOFTWARE
33370	825A	SYSM/LAN2	H&W COMPUTER SYSTEMS
33371	825B	VANTAGE POINT	CONNECT COMPUTER
33372	825C	VANTAGE POINT	CONNECT COMPUTER
33373	825D	VANTAGE POINT	CONNECT COMPUTER
33374	825E	TIME OUT	NORDRA INC
33375	825F	MULIT-PROCESSOR CONTROLLER	MBAC
33376	8260	MULIT-PROCESSOR CONTROLLER	MBAC
33377	8261	MULIT-PROCESSOR CONTROLLER	MBAC
33378	8262	MULIT-PROCESSOR CONTROLLER	MBAC
33379	8263	MULIT-PROCESSOR CONTROLLER	MBAC
33380	8264	TIME OUT	NORDRA INC
33381	8265	INGRES DATABASE	INGRES CORPORATION
33382	8266	EASY STREET PREM FINANCE	STREETWISE SYSTEMS INC
33383	8267	EASY STREET PREM FINANCE	STREETWISE SYSTEMS INC
33385	8269	APT NET	AUTOMATED PROGRAMMING TECH
33386	826A	APT NET	AUTOMATED PROGRAMMING TECH
33387	826B	APT NET	AUTOMATED PROGRAMMING TECH
33388	826C	APT NET	AUTOMATED PROGRAMMING TECH
33389	826D	TOTAL AUTOMATION SYS ED	LUTHERAN SOCIAL SERVICES
33390	826E	TOTAL AUTOMATION SYS MC	LUTHERAN SOCIAL SERVICES
33391	826F	TOTAL AUTOMATION SYS ML	DYNATECH UTAH SCIENTIFIC
33392	8270	TOTAL AUTOMATION SYS - ADI	DYNATECH UTAH SCIENTIFIC
33393	8271	TOTAL AUTOMATION SYS - FAX	DYNATECH UTAH SCIENTIFIC
33394	8272	SERVICE POINT	INTERPOINT SOFTWARE
33395	8273	SERVICE POINT	INTERPOINT SOFTWARE
33396	8274	PRODIGY GATEWAY	COMPUTEREASE SOFTWARE
33397	8275	PRODIGY GATEWAY	COMPUTEREASE SOFTWARE
33398	8276	NEWSMANAGER	VSS INC
33399	8277	NEWSMANAGER	VSS INC
33400	8278	NEWSMANAGER	VSS INC
33407	827F		MARTELLO & ASSOCIATES
33408	8280		MARK HURST
33409	8281		MARK HURST
33410	8282		MARK HURST
33411	8283		MARK HURST
33412	8284	BARR GATE/PC-TO-MAINFRAME COMMUNICATION	BARR SYSTEMS INC
33413	8285	BARR GATE/PC-TO-MAINFRAME COMMUNICATION	BARR SYSTEMS INC
33414	8286		PURE DATA RESEARCH LTD
33415	8287	VOICE MAIL	PLAN COMMUNICATIONS
33416	8288		CENTERS FOR DISEASE CONTROL
33417	8289	RPC CALLS	INTEGRATED DATA SYSTEMS
33418	828A		INTEGRATED DATA SYSTEMS
33419	828B		FOLIO CORPORATION
33420	828C		MULTITECH
33421	828D		MULTITECH
33422	828E		MULTITECH
33423	828F		MULTITECH
33424	8290		MULTITECH
33425	8291		BUS TECH INC
33426	8292		AMERICAN AIRLINES DECISION TEC
33427	8293	REQ BY DORELOWITZ - CONSULTING	MICROCOM INC
33428	8294		GATEWAY USA
33429	8295		GATEWAY USA
33430	8296		GATEWAY USA
33431	8297		GATEWAY USA
33432	8298		GATEWAY USA
33433	8299		GATEWAY USA
33434	829A		COMPUTERVISION SERVICES
33435	829B		COMPUTERVISION SERVICES
33436	829C		COMPUTERVISION SERVICES
33437	829D		SHIVA CORP
33438	829E		SHIVA CORP
33439	829F		TODD WEISS
33440	82A0		TODD WEISS
33441	82A1		TODD WEISS
33442	82A2	TAPE NLM	ARCADA SOFTWARE
33443	82A3	LANLORD PRODUCT	MICROCOM CLIENT SERVER TECHNOL
33444	82A4	LANLORD PRODUCT	MICROCOM CLIENT SERVER TECHNOL
33445	82A5	X25 AUTOMATED BRIDGE MONITOR & CONTROL	AUTOMATED INTERACTIONS DIV OF
33446	82A6	REQ BY DORELOWITZ - CONSULTING	MICROCOM INC
33447	82A7	PEER-TO-PEER COMMUNICATIONS	WITNESS SYSTEMS
33448	82A8		MICRO INTEGRATION
33449	82A9		MICRO INTEGRATION

Socket Decimal	Socket Hex	Service	Company
33450	82AA		IBM - POUGHKEEPSIE
33451	82AB		IBM - POUGHKEEPSIE
33452	82AC		IBM - POUGHKEEPSIE
33453	82AD		IBM - POUGHKEEPSIE
33454	82AE		IBM - POUGHKEEPSIE
33455	82AF		IBM - POUGHKEEPSIE
33456	82B0		J&L INFORMATION SYSTEMS
33457	82B1		J&L INFORMATION SYSTEMS
33458	82B2		J&L INFORMATION SYSTEMS
33459	82B3		J&L INFORMATION SYSTEMS
33460	82B4		J&L INFORMATION SYSTEMS
33461	82B5		J&L INFORMATION SYSTEMS
33462	82B6		J&L INFORMATION SYSTEMS
33463	82B7		J&L INFORMATION SYSTEMS
33464	82B8		J&L INFORMATION SYSTEMS
33465	82B9		J&L INFORMATION SYSTEMS
33466	82BA		J&L INFORMATION SYSTEMS
33467	82BB		J&L INFORMATION SYSTEMS
33468	82BC		J&L INFORMATION SYSTEMS
33469	82BD		J&L INFORMATION SYSTEMS
33470	82BE		J&L INFORMATION SYSTEMS
33471	82BF		J&L INFORMATION SYSTEMS
33472	82C0		J&L INFORMATION SYSTEMS
33473	82C1		J&L INFORMATION SYSTEMS
33474	82C2		J&L INFORMATION SYSTEMS
33475	82C3		J&L INFORMATION SYSTEMS
33476	82C4		J&L INFORMATION SYSTEMS
33477	82C5		J&L INFORMATION SYSTEMS
33478	82C6		J&L INFORMATION SYSTEMS
33479	82C7		J&L INFORMATION SYSTEMS
33480	82C8		J&L INFORMATION SYSTEMS
33481	82C9		J&L INFORMATION SYSTEMS
33482	82CA		J&L INFORMATION SYSTEMS
33483	82CB		J&L INFORMATION SYSTEMS
33484	82CC		J&L INFORMATION SYSTEMS
33485	82CD		J&L INFORMATION SYSTEMS
33486	82CE		J&L INFORMATION SYSTEMS
33487	82CF		J&L INFORMATION SYSTEMS
33488	82D0		J&L INFORMATION SYSTEMS
33489	82D1		J&L INFORMATION SYSTEMS
33490	82D2		J&L INFORMATION SYSTEMS
33491	82D3		J&L INFORMATION SYSTEMS
33492	82D4		J&L INFORMATION SYSTEMS
33493	82D5		J&L INFORMATION SYSTEMS
33494	82D6		J&L INFORMATION SYSTEMS
33495	82D7		J&L INFORMATION SYSTEMS
33496	82D8		LEGATO SYSTEMS
33497	82D9		LEGATO SYSTEMS
33498	82DA		LEGATO SYSTEMS
33499	82DB		LEGATO SYSTEMS
33500	82DC		LEGATO SYSTEMS
33501	82DD		LEGATO SYSTEMS
33502	82DE	VALUE ADDED SERVER	SKYLINE TECHNOLOGY
33503	82DF	VALUE ADDED SERVER	SKYLINE TECHNOLOGY
33504	82E0	FCP FOR OS/2 SUPPORT (FOUNDTN COOP PROC)	ANDERSEN CONSULTING - CHICAGO
33505	82E1	FCP FOR OS/2 SUPPORT (FOUNDTN COOP PROC)	ANDERSEN CONSULTING - CHICAGO
33507	82E3		SYTRON CORP
33508	82E4		AMERICAN AIRLINES DECISION TEC
33509	82E5		IMAGE RETRIEVAL INC
33510	82E6		CONNECT COMPUTER
33511	82E7		CONNECT COMPUTER
33512	82E8		CONNECT COMPUTER
33513	82E9		CONNECT COMPUTER
33514	82EA		CONNECT COMPUTER
33515	82EB		CONNECT COMPUTER
33516	82EC		CONNECT COMPUTER
33517	82ED		CONNECT COMPUTER
33518	82EE		CONNECT COMPUTER
33519	82EF		CONNECT COMPUTER
33520	82F0		CONNECT COMPUTER
33521	82F1		CONNECT COMPUTER
33522	82F2		CONNECT COMPUTER
33523	82F3		CONNECT COMPUTER
33524	82F4		CONNECT COMPUTER
33525	82F5		CONNECT COMPUTER
33526	82F6	HELLO-1	JOSHIN DENKI CO LTD J&P DIV
33527	82F7	HELLO-1	JOSHIN DENKI CO LTD J&P DIV
33528	82F8	HELLO-1	JOSHIN DENKI CO LTD J&P DIV

Socket Decimal	Socket Hex	Service	Company
33529	82F9	HELLO-1	JOSHIN DENKI CO LTD J&P DIV
33530	82FA	HELLO-1	JOSHIN DENKI CO LTD J&P DIV
33533	82FD	NETWARE JUKEBOX	COREL SYSTEMS, OPTICAL DIV
33534	82FE	FCP FOR WINDOWS SUPPORT (FNDTN COOP PROC)	ANDERSEN CONSULTING - CHICAGO
33535	82FF	FCP FOR WINDOWS SUPPORT (FNDTN COOP PROC)	ANDERSEN CONSULTING - CHICAGO
33536	8300		SMITH MICRO SOFTWARE INC
33537	8301		SMITH MICRO SOFTWARE INC
33538	8302		SMITH MICRO SOFTWARE INC
33539	8303		SMITH MICRO SOFTWARE INC
33540	8304		SMITH MICRO SOFTWARE INC
33541	8305		SMITH MICRO SOFTWARE INC
33542	8306		SMITH MICRO SOFTWARE INC
33543	8307		SMITH MICRO SOFTWARE INC
33544	8308		SMITH MICRO SOFTWARE INC
33545	8309		SMITH MICRO SOFTWARE INC
33546	830A		SMITH MICRO SOFTWARE INC
33547	830B		SMITH MICRO SOFTWARE INC
33548	830C		SMITH MICRO SOFTWARE INC
33549	830D		SMITH MICRO SOFTWARE INC
33550	830E		SMITH MICRO SOFTWARE INC
33551	830F		SMITH MICRO SOFTWARE INC
33552	8310		SMITH MICRO SOFTWARE INC
33553	8311		SMITH MICRO SOFTWARE INC
33554	8312		SMITH MICRO SOFTWARE INC
33555	8313		SMITH MICRO SOFTWARE INC
33556	8314		SMITH MICRO SOFTWARE INC
33557	8315		SMITH MICRO SOFTWARE INC
33558	8316		SMITH MICRO SOFTWARE INC
33559	8317		SMITH MICRO SOFTWARE INC
33560	8318		SMITH MICRO SOFTWARE INC
33561	8319		SMITH MICRO SOFTWARE INC
33562	831A		SMITH MICRO SOFTWARE INC
33563	831B		SMITH MICRO SOFTWARE INC
33564	831C		SMITH MICRO SOFTWARE INC
33565	831D		SMITH MICRO SOFTWARE INC
33566	831E		SMITH MICRO SOFTWARE INC
33567	831F		SMITH MICRO SOFTWARE INC
33568	8320		SMITH MICRO SOFTWARE INC
33569	8321		SMITH MICRO SOFTWARE INC
33570	8322		SMITH MICRO SOFTWARE INC
33571	8323		SMITH MICRO SOFTWARE INC
33572	8324		SMITH MICRO SOFTWARE INC
33573	8325		SMITH MICRO SOFTWARE INC
33574	8326		SMITH MICRO SOFTWARE INC
33575	8327		SMITH MICRO SOFTWARE INC
33576	8328		SMITH MICRO SOFTWARE INC
33577	8329		SMITH MICRO SOFTWARE INC
33578	832A		SMITH MICRO SOFTWARE INC
33579	832B		SMITH MICRO SOFTWARE INC
33580	832C		SMITH MICRO SOFTWARE INC
33581	832D		SMITH MICRO SOFTWARE INC
33582	832E		SMITH MICRO SOFTWARE INC
33583	832F		SMITH MICRO SOFTWARE INC
33584	8330		SMITH MICRO SOFTWARE INC
33585	8331		SMITH MICRO SOFTWARE INC
33586	8332		SMITH MICRO SOFTWARE INC
33587	8333		SMITH MICRO SOFTWARE INC
33588	8334		SMITH MICRO SOFTWARE INC
33589	8335		SMITH MICRO SOFTWARE INC
33590	8336		SMITH MICRO SOFTWARE INC
33591	8337		SMITH MICRO SOFTWARE INC
33592	8338		SMITH MICRO SOFTWARE INC
33593	8339		SMITH MICRO SOFTWARE INC
33594	833A		SMITH MICRO SOFTWARE INC
33595	833B		SMITH MICRO SOFTWARE INC
33596	833C		SMITH MICRO SOFTWARE INC
33597	833D		SMITH MICRO SOFTWARE INC
33598	833E		SMITH MICRO SOFTWARE INC
33599	833F		SMITH MICRO SOFTWARE INC
33600	8340		SMITH MICRO SOFTWARE INC
33601	8341		SMITH MICRO SOFTWARE INC
33602	8342		SMITH MICRO SOFTWARE INC
33603	8343		SMITH MICRO SOFTWARE INC
33604	8344		SMITH MICRO SOFTWARE INC
33605	8345		SMITH MICRO SOFTWARE INC
33606	8346		SMITH MICRO SOFTWARE INC
33607	8347		SMITH MICRO SOFTWARE INC
33608	8348		SMITH MICRO SOFTWARE INC

Socket Decimal	Socket Hex	Service	Company
33609	8349		SMITH MICRO SOFTWARE INC
33610	834A		SMITH MICRO SOFTWARE INC
33611	834B		SMITH MICRO SOFTWARE INC
33612	834C		SMITH MICRO SOFTWARE INC
33613	834D		SMITH MICRO SOFTWARE INC
33614	834E		SMITH MICRO SOFTWARE INC
33615	834F		SMITH MICRO SOFTWARE INC
33616	8350	POWER GRID SERVER	COGNOS INC
33617	8351	DATA SERVICE TO WORKSTATION/ SCHOOL ADMIN	CHANCERY SOFTWARE LTD
33618	8352	TRANSMITTING	UNISYNC
33619	8353	RECEIVING	UNISYNC
33620	8354	MULTICOM NET - MODEM SHARING SOFTWARE	MULTICOM
33621	8355	MULTICOM NET - MODEM SHARING SOFTWARE	MULTICOM
33622	8356	MULTICOM NET - MODEM SHARING SOFTWARE	MULTICOM
33623	8357	CD CONNECTIONS	CBIS INC
33625	8359		RIVERVIEW SYSTEMS
33626	835A	TOTAL AUTOMATION SYSTEMS	DYNATECH UTAH SCIENTIFIC
33627	835B	TOTAL AUTOMATION SYSTEMS	DYNATECH UTAH SCIENTIFIC
33628	835C		ITAC INC
33629	835D		ASP COMPUTER PRODUCTS INC
33630	835E		ASP COMPUTER PRODUCTS INC
33631	835F		ASP COMPUTER PRODUCTS INC
33633	8361	FAX SERVER	TRANSFAX CORPORATION
33634	8362	FAX PRINT SERVER	TRANSFAX CORPORATION
33635	8363	FAX MERGE SERVER	TRANSFAX CORPORATION
33636	8364	NETWORK MANAGEMENT SERVER	TRANSFAX CORPORATION
33637	8365		FUNK SOFTWARE
33638	8366		MICRO INTEGRATION
33639	8367		MICRO INTEGRATION
33640	8368		MICRO INTEGRATION
33641	8369		MICRO INTEGRATION
33642	836A		MICRO INTEGRATION
33643	836B		TRIPLE A MOTOR CLUB
33644	836C	SOCKET # FOR LAN TIMES JAPAN ARTICLE	LAN TIMES JAPAN, SOFTBANK CORP
33645	836D		WATCHTOWER
33646	836E	CLIENT SERVER COMM SYSTEMS, SPX PROTOCOL	MICRORIM
33647	836F		NORTON LAMBERT CORP
33648	8370		NORTON LAMBERT CORP
33649	8371		NORTON LAMBERT CORP
33650	8372		NORTON LAMBERT CORP
33651	8373		NORTON LAMBERT CORP
33652	8374		CENTRAL POINT SOFTWARE
33653	8375	DRIVERS	PRESOFT ARCHITECTS
33654	8376	NLM FOR REMOTE VOLUME MOUNT	INTECK CORPORATION
33655	8377	NLM FOR REMOTE VOLUME MOUNT	INTECK CORPORATION
33656	8378		SYMANTEC PETER NORTON GROUP
33657	8379		SYMANTEC PETER NORTON GROUP
33658	837A		DIGITAL EQUIPMENT - MERRIMACK
33659	837B	CBS FACILITIES ASSIGNMENT - REQUEST	DYNATECH UTAH SCIENTIFIC
33660	837C	CBS FACILITIES ASSIGNMENT - REPLY	DYNATECH UTAH SCIENTIFIC
33661	837D	MULTI-PROTOCOL ROUTER INSIDE IPX	RESEARCH MACHINES PLC
33662	837E	SHAREWARE	CHERRY TREE SOFTWARE
33663	837F	ENTERPRISE ECS	INTEL CORP
33664	8380	ENTERPRISE MMT	INTEL CORP
33665	8381	STOCK TICKER BROADCAST SERVER	NCOMPASS DEVELOPMENT INTL
33666	8382	QUERY UNIQUE USERS	US ROBOTICS SOFTWARE
33667	8383	CBS ADA SERVER	DYNATECH UTAH SCIENTIFIC
33668	8384		PACE SOFTWARE SYSTEMS INC
33669	8385		ANDERSEN CONSULTING - CHICAGO
33670	8386	GATEWAY MANAGEMENT	WALL DATA
33671	8387	GATEWAY MANAGEMENT	WALL DATA
33675	838B	POWERCHUTE ALERT - UPS MONITORING	AMERICAN POWER CONVERSION
33677	838D		AVAIL SYSTEMS CORP
33678	838E	QA+ FOR WINDOWS/REMOTE DIAGNOSTICS	DIAGSOFT INC
33679	838F	POWERCHUTE ADMINSTRATIVE SOCKET	AMERICAN POWER CONVERSION
33680	8390		DATAMEDIC
33681	8391	COREL DRIVER	COREL SYSTEMS, OPTICAL DIV
33682	8392	ID FOR LASERMASTER PRINTER PRODUCTS	LASER MASTER CORP
33683	8393	TFTP TRIVIAL FILE TRANSFER PROTOCOL	HEWLETT PACKARD - ROSEVILLE
33684	8394	FTP FILE TRANSFER PROTOCOL	HEWLETT PACKARD - ROSEVILLE
33685	8395		HEWLETT PACKARD - ROSEVILLE
33686	8396		HEWLETT PACKARD - ROSEVILLE
33687	8397		SITA
33688	8398		SITA
33689	8399		TECHGNOSIS INC
33690	839A	QA+ FOR WINDOWS/REMOTE DIAGNOSTICS	DIAGSOFT INC
33691	839B	MAIL SYSTEMS	SYNECTIC SYSTEMS LTD
33692	839C	MAIL SYSTEMS	SYNECTIC SYSTEMS LTD
33693	839D	NLM/VIDEO FILES ON NOVELL SERVICE	PROTOCOMM CORPORATION
33694	839E	TURNAX EMULATION GATEWAY	IDE CORPORATION
33695	839F	CNF 16000 CONNECTION STATION	COROLLARY INC
33696	83A0	CNF 16000 CONNECTION STATION	COROLLARY INC
33697	83A1	CNF 16000 CONNECTION STATION	COROLLARY INC
33698	83A2	CNF 16000 CONNECTION STATION	COROLLARY INC
33699	83A3	CNF 16000 CONNECTION STATION	COROLLARY INC
33700	83A4	CNF 16000 CONNECTION STATION	COROLLARY INC
33701	83A5	WORKSTATION PEER-TO-PEER COMMUNICATIONS	IBM - RESEARCH TRIANGLE PARK
33702	83A6	WORKSTATION PEER-TO-PEER COMMUNICATIONS	IBM - RESEARCH TRIANGLE PARK
33703	83A7	WORKSTATION PEER-TO-PEER COMMUNICATIONS	IBM - RESEARCH TRIANGLE PARK
33704	83A8	WORKSTATION PEER-TO-PEER COMMUNICATIONS	IBM - RESEARCH TRIANGLE PARK
33705	83A9	WORKSTATION PEER-TO-PEER COMMUNICATIONS	IBM - RESEARCH TRIANGLE PARK
33706	83AA	WORKSTATION PEER-TO-PEER COMMUNICATIONS	IBM - RESEARCH TRIANGLE PARK
33707	83AB		DATANEX CORPORATION
33708	83AC	HP OPEN MAIL & PORTABLE NETWARE	HEWLETT PACKARD - BERKSHIRE
33709	83AD	COMMUNICATION/MAIL SERVER SOFTWARE	DATOR 3 SPOL SRO
33710	83AE	COMMUNICATION/MAIL SERVER SOFTWARE	DATOR 3 SPOL SRO
33711	83AF	COMMUNICATION/MAIL SERVER SOFTWARE	DATOR 3 SPOL SRO
33712	83B0	COMMUNICATION/MAIL SERVER SOFTWARE	DATOR 3 SPOL SRO
33713	83B1	POWER MANAGEMENT SERVER	ELGAR CORP
33714	83B2	POWER MANAGEMENT CLIENT	ELGAR CORP
33715	83B3	NETWORK PERIPHERALS - PRINT SERVER	CANON INFORMATION SYSTEMS
33716	83B4	NETWORK PERIPHERALS	CANON INFORMATION SYSTEMS
33717	83B5	NETWORK PERIPHERALS	CANON INFORMATION SYSTEMS
33718	83B6	NETWORK PERIPHERALS	CANON INFORMATION SYSTEMS
33719	83B7	NETWORK PERIPHERALS	CANON INFORMATION SYSTEMS
33720	83B8	NETWORK PERIPHERALS	CANON INFORMATION SYSTEMS
33721	83B9	FAX SERVER	FERRARI ELECTRONIC GMBH
33722	83BA	FAX SERVER	FERRARI ELECTRONIC GMBH
33723	83BB	FAX SERVER	FERRARI ELECTRONIC GMBH
33724	83BC	FAX SERVER	FERRARI ELECTRONIC GMBH
33725	83BD	CLIENT SERVER DATABASE ENGINE/SQL FOR NW	SYBASE INC
33726	83BE	PRINTER CONTROLLER BOARD/VAP OR NLM	DP TEK
33727	83BF	PRINTER CONTROLLER BOARD/VAP OR NLM	DP TEK
33728	83C0		IWI
33729	83C1		LEXMARK INTERNATIONAL INC
33730	83C2		LEXMARK INTERNATIONAL INC
33731	83C3		LEXMARK INTERNATIONAL INC
33732	83C4		LEXMARK INTERNATIONAL INC
33733	83C5		LEXMARK INTERNATIONAL INC
33734	83C6		LEXMARK INTERNATIONAL INC
33735	83C7		OKNA CORP
33736	83C8		OKNA CORP
33737	83C9		OKNA CORP
33738	83CA		OKNA CORP
33739	83CB		OKNA CORP
33740	83CC		OKNA CORP
33741	83CD	DEVELOPMENT/COMMUNICATIONS TOOLKIT	MICHAEL RICH
33742	83CE	RESET PRINT SERVERS	MOTOROLA
33743	83CF		NETWORK DESIGNERS
33744	83D0	REMOTE PRINTER SOCKET	INDUSTRIAL EXOTICA
33745	83D1	FILE MANAGEMENT SERVICES	SYSTEMS AXIS PLC
33746	83D2	QUEUE MANAGEMENT SERVICES	SYSTEMS AXIS PLC

Socket Decimal	Socket Hex	Service	Company
33748	83D4		LANTECH SERVICES
33749	83D5	CC MAIL GATEWAY 3.30/SPX TRANSPORT	CC MAIL
33750	83D6	REMOTE CONTROL - NODE-TO-NODE	DST - DISTRIBUTED SYSTEMS TECH
33751	83D7	REMOTE CONTROL - BANK OF MODEMS	DST - DISTRIBUTED SYSTEMS TECH
33753	83D9	GENERAL COMMUNICATION FORUM ON NOVELL NW	CREDIT LYONNAIS
33754	83DA	DATABASE ENGINES	SYBASE INC
33755	83DB	DATABASE ENGINES	SYBASE INC
33756	83DC	DATABASE ENGINES	SYBASE INC
33757	83DD	DATABASE ENGINES	SYBASE INC
33758	83DE	DATABASE ENGINES	SYBASE INC
33759	83DF	DATABASE ENGINES	SYBASE INC
33760	83E0	DATABASE ENGINES	SYBASE INC
33761	83E1	DATABASE ENGINES	SYBASE INC
33762	83E2	GATEWAY SERVER PRODUCT TO SPERRY HOST	ICC
33763	83E3	WAN CONNECTION SERVER	IDEASSOCIATION
33764	83E4	LAN SPOOL 3.5	INTEL - AMERICAN FORK
33766	83E6	REMOTE INTERNAL HUB DRIVER	INTEL PCED
33767	83E7	SOFTWARE ACCESS CONTROL SERVER	U OF PLYMOUTH
33769	83E9	COMMUNICATIONS SYSTEM	UNICABLES SA
33770	83EA	COMMUNICATIONS SYSTEM	UNICABLES SA
33771	83EB	COMMUNICATIONS SYSTEM	UNICABLES SA
33772	83EC	COMMUNICATIONS SYSTEM	UNICABLES SA
33773	83ED	COMMUNICATIONS SYSTEM	UNICABLES SA
33774	83EE	COMMUNICATIONS SYSTEM	UNICABLES SA
33775	83EF	COMMUNICATIONS SYSTEM	UNICABLES SA
33776	83F0	COMMUNICATIONS SYSTEM	UNICABLES SA
33777	83F1	COMMUNICATIONS SYSTEM	UNICABLES SA
33778	83F2	COMMUNICATIONS SYSTEM	UNICABLES SA
33779	83F3	COMMUNICATIONS SYSTEM	UNICABLES SA
33780	83F4	COMMUNICATIONS SYSTEM	UNICABLES SA
33781	83F5	COMMUNICATIONS SYSTEM	UNICABLES SA
33782	83F6	COMMUNICATIONS SYSTEM	UNICABLES SA
33783	83F7	COMMUNICATIONS SYSTEM	UNICABLES SA
33784	83F8	COMMUNICATIONS SYSTEM	UNICABLES SA
33785	83F9	COMMUNICATIONS SYSTEM	UNICABLES SA
33786	83FA	COMMUNICATIONS SYSTEM	UNICABLES SA
33787	83FB	COMMUNICATIONS SYSTEM	UNICABLES SA
33788	83FC	COMMUNICATIONS SYSTEM	UNICABLES SA
33789	83FD	COMMUNICATIONS SYSTEM	UNICABLES SA
33790	83FE	COMMUNICATIONS SYSTEM	UNICABLES SA
33791	83FF	COMMUNICATIONS SYSTEM	UNICABLES SA
33792	8400	COMMUNICATIONS SYSTEM	UNICABLES SA
33793	8401	COMMUNICATIONS SYSTEM	UNICABLES SA
33794	8402	GENERIC SERVER	GREENBAUM ASSOCIATES
33795	8403	OBJECT-STORE, DATABASE ACCESS PROTOCOL	OBJECT DESIGN
33796	8404	OBJECT-STORE, DATABASE ACCESS PROTOCOL	OBJECT DESIGN
33797	8405	OBJECT-STORE, DIRECTORY PROTOCOL	OBJECT DESIGN
33798	8406	OBJECT-STORE, DIRECTORY PROTOCOL	OBJECT DESIGN
33799	8407	OBJECT-STORE, CACHE COHERENCE PROTOCOL	OBJECT DESIGN
33800	8408	OBJECT-STORE, CACHE COHERENCE PROTOCOL	OBJECT DESIGN
33801	8409	VISINET NLM ID#	TECHNOLOGY DYNAMICS INC
33802	840A	WDAGR - VAP	JOSTENS LEARNING CORP
33803	840B	WKILL	JOSTENS LEARNING CORP
33804	840C	INTERNET LAN CONTROLLER FOR NETWARE SAA	BUS TECH INC
33805	840D	PEER-TO-PEER COMMUNICATIONS	M&M MARS INC
33806	840E	PEER-TO-PEER COMMUNICATIONS	M&M MARS INC
33807	840F	CONNECTION MANAGER (SOFTWARE HOUSE PROD)	TBR INTERNATIONAL CORP
33808	8410	CONNECTION MANAGER (SOFTWARE HOUSE PROD)	TBR INTERNATIONAL CORP
33809	8411	CONNECTION MANAGER (SOFTWARE HOUSE PROD)	TBR INTERNATIONAL CORP
33810	8412	CONNECTION MANAGER (SOFTWARE HOUSE PROD)	TBR INTERNATIONAL CORP
33811	8413	CONNECTION MANAGER (SOFTWARE HOUSE PROD)	TBR INTERNATIONAL CORP
33812	8414	CONNECTION MANAGER (SOFTWARE HOUSE PROD)	TBR INTERNATIONAL CORP
33813	8415	FOR PROXY HOST	FUNK SOFTWARE
33814	8416	WORKSTATION 3-LAN	ADVANCED TECHNICAL SOLUTIONS
33815	8417	DEVELOPING SERVER PERFORMANCE ANALIZER	BANYAN SYSTEMS INC
33816	8418	DEVELOPING SERVER PERFORMANCE ANALIZER	BANYAN SYSTEMS INC
33817	8419	DEVELOPING SERVER PERFORMANCE ANALIZER	BANYAN SYSTEMS INC
33818	841A	DEVELOPING SERVER PERFORMANCE ANALIZER	BANYAN SYSTEMS INC
33819	841B	DEVELOPING SERVER PERFORMANCE ANALIZER	BANYAN SYSTEMS INC
33820	841C	DEVELOPING SERVER PERFORMANCE ANALIZER	BANYAN SYSTEMS INC
33821	841D	REMOTE DATABASE SERVICES	INTERACTIVE DATA
33822	841E	REMOTE DATABASE SERVICES	INTERACTIVE DATA
33823	841F	ENVELOPE PRINTER FOR NETWORK	THURIDION SOFTWARE ENGINEERING
33824	8420	DACS OFFICE II	DOCUNET GMBH
33825	8421	TERMINAL EMULATOR FOR MDOS - TRANSMIT	INTELLIGENT MICRO SOFTWARE LTD
33826	8422	TERMINAL EMULATOR FOR MDOS - RECEIVE	INTELLIGENT MICRO SOFTWARE LTD
33827	8423		PRAXIS
33828	8424	REFLEX COMPLIANCE	PRODIGY SERVICES
33829	8425	REFLEX COMPLIANCE	PRODIGY SERVICES
33830	8426	REFLEX COMPLIANCE	PRODIGY SERVICES
33831	8427	REFLEX COMPLIANCE	PRODIGY SERVICES
33832	8428	REFLEX COMPLIANCE	PRODIGY SERVICES
33833	8429	REFLEX COMPLIANCE	PRODIGY SERVICES
33834	842A	IPX REMOTE CONTROL FUNCTION	NETWORTH INC
33835	842B	IPX REMOTE CONTROL FUNCTION	NETWORTH INC
33836	842C	SERVER-CLIENT COMMUNICATION VIA SPX	AT&T JENS CORP
33837	842D	TSR BROADCASTING VIA IPX	AT&T JENS CORP
33838	842E		SITA
33839	842F		SITA
33840	8430	INSTANT RECALL	DAYTIMER TECHNOLOGIES
33841	8431	APPLICATION SERVER FC1	THOMSON FINANCIAL
33842	8432	APPLICATION SERVER FC2	THOMSON FINANCIAL
33843	8433	APPLICATION SERVER FC3	THOMSON FINANCIAL
33844	8434	APPLICATION SERVER FC3	THOMSON FINANCIAL
33845	8435	ENVELOPE MANAGER SOFTWARE	PSI ASSOCIATES
33846	8436	ENVELOPE MANAGER SOFTWARE	PSI ASSOCIATES
33847	8437	ENVELOPE MANAGER SOFTWARE	PSI ASSOCIATES
33848	8438	ENVELOPE MANAGER SOFTWARE	PSI ASSOCIATES
33849	8439	LAA SERVER BINDARY SOCKET	SABER SOFTWARE
33850	843A	440 IPX COMMUNICATIONS	INFORMATION BUILDERS
33851	843B	VITAL SIGNS/LAN SERVER	BLUELINE SOFTWARE INC
33852	843C	VITAL SIGNS/LAN SERVER	BLUELINE SOFTWARE INC
33853	843D	ENVELOPE PRINTER FOR NETWORK	THURIDION SOFTWARE ENGINEERING
33854	843E	OS2 SEQUEL SERVER IPX/SPX SUPPORT	MICROSOFT
33855	843F	ASYNCHRONOUS SERIAL COMMUNICATIONS	BLACK CREEK INTEGRATED SYSTEMS
33856	8440	ASYNCHRONOUS SERIAL COMMUNICATIONS	BLACK CREEK INTEGRATED SYSTEMS
33857	8441	ASYNCHRONOUS SERIAL COMMUNICATIONS	BLACK CREEK INTEGRATED SYSTEMS
33858	8442	ASYNCHRONOUS SERIAL COMMUNICATIONS	BLACK CREEK INTEGRATED SYSTEMS
33859	8443	COMMUNICATION BETWEEN SAGES	AMERICAN AUTO MATRIX INC
33860	8444	TV BROADCAST AUTOMATION STATUS	UTAH SCIENTIFIC
33861	8445	TV BROADCAST AUTOMATION STATUS	UTAH SCIENTIFIC
33862	8446	TV BROADCAST AUTOMATION STATUS	UTAH SCIENTIFIC
33863	8447	CLIENT-SERVER VERSION OF CC MAIL	CC MAIL
33864	8448	TV BROADCAST AUTOMATION STATUS	UTAH SCIENTIFIC
33865	8449	NETSPRINT	DIGITAL PRODUCTS INC
33866	844A	WORKSTATION REMOTE CONTROL	CALIFORNIA FEDERAL
33867	844B	FULL TEXT RETRIEVAL CLIENT/SERVER DB ENG	IMPACT ITALIANA SRL
33868	844C	GATEWAY IPX	ICOT CORP
33869	844D	GATEWAY SPX	ICOT CORP
33870	844E	WORKSTATION IPX	ICOT CORP
33871	844F	WORKSTATION SPX	ICOT CORP
33872	8450	NETWORK SERVICES IPX	ICOT CORP
33873	8451	NETWORK SERVICES SPX	ICOT CORP
33874	8452	NETWORK LOGGER IPX	ICOT CORP
33875	8453	NETWORK LOGGER SPX	ICOT CORP
33876	8454	GATEWAY SOFTWARE	DATEV EG
33877	8455	GATEWAY SOFTWARE	DATEV EG
33878	8456		NOVELL - PROVO
33879	8457	RE:ACTION	CONCENTRIC TECHNOLOGIES
33880	8458	RE:ACTION	CONCENTRIC TECHNOLOGIES
33881	8459	CAD SERVER	ISICAD
33882	845A	CAD SERVER	ISICAD

Socket Decimal	Socket Hex	Service	Company
33883	845B	ICL PORTABLE NETWARE	ICL PERSONAL SYSTEMS OY
33884	845C	LOCATE	ZENITH DATA SYSTEMS
33885	845D	BOML	ZENITH DATA SYSTEMS
33886	845E	RHOTHEOS	ZENITH DATA SYSTEMS
33887	845F	CHAT PROGRAM	INTEL - AMERICAN FORK
33888	8460	MAILSLOTS	IBM - FRANKLIN LAKES
33889	8461	MAILSLOTS	IBM - FRANKLIN LAKES
33890	8462	FILETALK	MOUNTAIN NETWORK SOLUTIONS INC
33891	8463	FILETALK	MOUNTAIN NETWORK SOLUTIONS INC
33892	8464	FILETALK	MOUNTAIN NETWORK SOLUTIONS INC
33893	8465	FILETALK	MOUNTAIN NETWORK SOLUTIONS INC
33894	8466	FILETALK	MOUNTAIN NETWORK SOLUTIONS INC
33895	8467	FILETALK	MOUNTAIN NETWORK SOLUTIONS INC
33896	8468		MICROCOM INC
33897	8469		MICROCOM INC
33898	846A		MICROCOM INC
33899	846B		MICROCOM INC
33900	846C		MICROCOM INC
33901	846D		MICROCOM INC
33902	846E		MICROCOM INC
33903	846F		MICROCOM INC
33904	8470		MICROCOM INC
33905	8471		MICROCOM INC
33906	8472		MICROCOM INC
33907	8473		MICROCOM INC
33908	8474		MICROCOM INC
33909	8475		MICROCOM INC
33910	8476		MICROCOM INC
33912	8478	DOCUMENT MANGEMENT PACKAGE	PERFECT SOLUTIONS CORPORATION
33913	8479	LITIGATION SUPPORT	GIBSON OCHSNER & ADKINS
33914	847A	MONOTREX	PRIME COMPUTER
33915	847B	MONOTREX	PRIME COMPUTER
33916	847C	MONOTREX	PRIME COMPUTER
33917	847D	MONOTREX	PRIME COMPUTER
33918	847E	MONOTREX	PRIME COMPUTER
33919	847F	MONOTREX	PRIME COMPUTER
33920	8480	MONOTREX	PRIME COMPUTER
33921	8481	MONOTREX	PRIME COMPUTER
33922	8482	MONOTREX	PRIME COMPUTER
33923	8483	MONOTREX	PRIME COMPUTER
33924	8484	LITIGATION SUPPORT	GIBSON OCHSNER & ADKINS
33925	8485	LITIGATION SUPPORT	GIBSON OCHSNER & ADKINS
33926	8486	LITIGATION SUUPORT	GIBSON OCHSNER & ADKINS
33927	8487	ARGUS/N	TRITICOM
33928	8488	ARGUS/N	TRITICOM
33929	8489	ARGUS/N	TRITICOM
33930	848A	CHANNEL SWITCHER APPLICATION	DYNATECH UTAH SCIENTIFIC
33931	848B	CHANNEL SWITCHER APPLICATION	DYNATECH UTAH SCIENTIFIC
33932	848C	CHANNEL SWITCHER APPLICATION	DYNATECH UTAH SCIENTIFIC
33933	848D	CHANNEL SWITCHER APPLICATION	DYNATECH UTAH SCIENTIFIC
33934	848E	CHANNEL SWITCHER APPLICATION	DYNATECH UTAH SCIENTIFIC
33935	848F	CHANNEL SWITCHER APPLICATION	DYNATECH UTAH SCIENTIFIC
33936	8490	CHANNEL SWITCHER APPLICATION	DYNATECH UTAH SCIENTIFIC
33937	8491	CHANNEL SWITCHER APPLICATION	DYNATECH UTAH SCIENTIFIC
33938	8492		OXFORD INFORMATION TECHNOLOGY
33939	8493	GATEWAY INTEGRATION ARCHITECT	MORRISEY ASSOCIATES
33940	8494	BOOTWARE/MSD	LANWORKS
33941	8495	WORKGROUP COMPUTING TOOL	MEMOREX TELEX
33942	8496	WORKGROUP COMPUTING TOOL	MEMOREX TELEX
33943	8497	WORKGROUP COMPUTING TOOL	MEMOREX TELEX
33944	8498	WORKGROUP COMPUTING TOOL	MEMOREX TELEX
33945	8499	WORKGROUP COMPUTING TOOL	MEMOREX TELEX
33946	849A	WORKGROUP COMPUTING TOOL	MEMOREX TELEX
33947	849B	WORKGROUP COMPUTING TOOL	MEMOREX TELEX
33952	84A0	IPX/SPX SOCKETS	ARTEFACT NETWORK SUPPORT
33953	84A1	IPX/SPX SOCKETS	ARTEFACT NETWORK SUPPORT
33954	84A2	IPX/SPX SOCKETS	ARTEFACT NETWORK SUPPORT
33955	84A3	IPX/SPX SOCKETS	ARTEFACT NETWORK SUPPORT
33956	84A4	IPX/SPX SOCKETS	ARTEFACT NETWORK SUPPORT
33957	84A5	IPX/SPX SOCKETS	ARTEFACT NETWORK SUPPORT
33958	84A6	IPX/SPX SOCKETS	ARTEFACT NETWORK SUPPORT
33959	84A7	IPX/SPX SOCKETS	ARTEFACT NETWORK SUPPORT
33960	84A8	IPX/SPX SOCKETS	ARTEFACT NETWORK SUPPORT
33961	84A9	IPX/SPX SOCKETS	ARTEFACT NETWORK SUPPORT
33962	84AA	CLIENT-SERVER DRIVER FOR IPX/SPX	REFERENCE POINT SOFTWARE
33963	84AB		INTRAK INC
33964	84AC		INTRAK INC
33965	84AD		INTRAK INC
33966	84AE		INTRAK INC
33967	84AF		INTRAK INC

Socket Decimal	Socket Hex	Service	Company
33968	84B0		INTRAK INC
33969	84B1	DATABASE APPLICATIONS	DIGITAL EQUIPMENT - NASHUA
33970	84B2	DATABASE APPLICATIONS	DIGITAL EQUIPMENT - NASHUA
33971	84B3	LOADER SOCKET	NETWORK TECHNICAL SOLUTIONS
33972	84B4	FINDER SOCKET	NETWORK TECHNICAL SOLUTIONS
33973	84B5	AUTOMATED CONTROL SYSTEM	AIR PRODUCTS & CHEMICALS
33974	84B6	AUTOMATED CONTROL SYSTEM	AIR PRODUCTS & CHEMICALS
33975	84B7	AUDIT TRAIL PACKAGE	BLUE LANCE INC
33976	84B8	SBACKUP ENHANCEMENT PRODUCT	SYTRON CORP
33977	84B9	TAPE BACKUP SYSTEMS	COLORADO MEMORY SYSTEMS
33978	84BA	QA+ ATTENTION SOCKET	DIAGSOFT INC
33979	84BB	ADMINISTRATION SERVER	MCGILL UNIVERSITY FAC OF ENGIN
33980	84BC	ADMINISTRATION SERVER	MCGILL UNIVERSITY FAC OF ENGIN
33981	84BD	WORKSTATION COMMUNICATIONS	SYMANTEC PETER NORTON GROUP
33982	84BE	WORKSTATION COMMUNICATIONS	SYMANTEC PETER NORTON GROUP
33983	84BF	NETWORK DYNAMIC DATA EXCHANGE	NETLOGIC INC
33984	84C0	ASYNCHRONOUS COMMUNICATIONS SERVER	US ROBOTICS SOFTWARE
33985	84C1	MISCELLANEOUS SOFTWARE COMMUNICATIONS	TENTERA COMPUTER SERVICES
33986	84C2	FORUM SEND	TEXAS A&M UNIVERSITY
33987	84C3	FORUM RECEIVE	TEXAS A&M UNIVERSITY
33988	84C4	FORUM CONTROL	TEXAS A&M UNIVERSITY
33989	84C5	REMOTE PRINTER CONFIGURATION	NEWGEN SYSTEMS CORP
33990	84C6	AUDIT TRAIL PACKAGE	BLUE LANCE INC
33991	84C7	PEER-TO-PEER COMMUNICATIONS	FUJITSU NETWORKS INDUSTRY
33992	84C8	SNA GATEWAY	MICROSOFT
33993	84C9	SNA GATEWAY	MICROSOFT
33994	84CA	WORKSTATION TERMINAL ACCESS	HSD HARDWARE SOFTWARE DEVELOPM
33995	84CB		SERCOMM
33996	84CC		DE INTERNATIONAL LTD
33997	84CD	APPLICATION TRACKING SYSTEM	AUTOMATED INTERACTIONS DIV OF
33998	84CE	IBM HOST GATEWAY	IDEA COURIER
33999	84CF	CREDIT AUTHORIZATION GATEWAY	MERCHANTEC INTERNATIONAL
34000	84D0	GRAPHICAL HOTEL MANAGEMENT APPLICATION	INSURE INC
34001	84D1	GRAPHICAL HOTEL MANAGEMENT APPLICATION	INSURE INC
34002	84D2	NETWORK BACK-UP	DIGITAL EQUIPMENT - SHREWSBURY
34003	84D3	CLIENT SERVER APPLICATION	ALCON SYSTEMS
34004	84D4	CLIENT SERVER APPLICATION	ALCON SYSTEMS
34005	84D5	COMMUNICATIONS SERVER SDD SYNTHESIZER	SCANDINAVIAN AIRLINES DATA
34006	84D6	INFORMATION SYSTEMS PRODUCT	PROSOFTIA AB
34007	84D7	INFORMATION SYSTEMS PRODUCT	PROSOFTIA AB
34008	84D8	INFORMATION SYSTEMS PRODUCT	PROSOFTIA AB
34009	84D9	INFORMATION SYSTEMS PRODUCT	PROSOFTIA AB
34010	84DA	INFORMATION SYSTEMS PRODUCT	PROSOFTIA AB
34011	84DB	INFORMATION SYSTEMS PRODUCT	PROSOFTIA AB
34012	84DC	INFORMATION SYSTEMS PRODUCT	PROSOFTIA AB
34013	84DD	INFORMATION SYSTEMS PRODUCT	PROSOFTIA AB
34014	84DE	INFORMATION SYSTEMS PRODUCT	PROSOFTIA AB
34015	84DF	INFORMATION SYSTEMS PRODUCT	PROSOFTIA AB
34016	84E0	OBJECT ORIENTED DATABASE SYSTEM	ONTOS INC
34017	84E1	OBJECT ORIENTED DATABASE SYSTEM	ONTOS INC
34018	84E2	TAPE BACK-UP FOR NLM APPLICATION	MOUNTAIN NETWORK SOLUTIONS INC
34019	84E3	TAPE BACK-UP FOR NLM APPLICATION	MOUNTAIN NETWORK SOLUTIONS INC
34020	84E4	TAPE BACK-UP FOR NLM APPLICATION	MOUNTAIN NETWORK SOLUTIONS INC
34021	84E5	TAPE BACK-UP FOR NLM APPLICATION	MOUNTAIN NETWORK SOLUTIONS INC
34022	84E6		ATTACHMATE CORPORATION
34023	84E7	TAPE BACK-UP FOR NLM APPLICATION	MOUNTAIN NETWORK SOLUTIONS INC
34024	84E8	TAPE BACK-UP FOR NLM APPLICATION	MOUNTAIN NETWORK SOLUTIONS INC
34025	84E9	BRMSG NETWORK MAIL SERVER	SOFTBRIDGE INC
34026	84EA	CLIENT SERVER MONITORING UTILITY	DELL COMPUTER
34027	84EB	SYNECTICS FOR OS/2 VERSION 2.0	PARALLEL PCS INC
34028	84EC	SYNECTICS FOR OS/2 VERSION 2.0	PARALLEL PCS INC
34029	84ED	INFORMATION SYSTEMS PRODUCT	PROSOFTIA AB
34030	84EE	INFORMATION SYSTEMS PRODUCT	PROSOFTIA AB
34031	84EF	INFORMATION SYSTEMS PRODUCT	PROSOFTIA AB
34032	84F0	INFORMATION SYSTEMS PRODUCT	PROSOFTIA AB
34033	84F1	INFORMATION SYSTEMS PRODUCT	PROSOFTIA AB
34034	84F2	INFORMATION SYSTEMS PRODUCT	PROSOFTIA AB
34035	84F3	INFORMATION SYSTEMS PRODUCT	PROSOFTIA AB
34036	84F4	INFORMATION SYSTEMS PRODUCT	PROSOFTIA AB
34037	84F5	INFORMATION SYSTEMS PRODUCT	PROSOFTIA AB
34038	84F6	INFORMATION SYSTEMS PRODUCT	PROSOFTIA AB
34039	84F7	INFORMATION SYSTEMS PRODUCT	PROSOFTIA AB
34040	84F8	INFORMATION SYSTEMS PRODUCT	PROSOFTIA AB
34041	84F9	INFORMATION SYSTEMS PRODUCT	PROSOFTIA AB
34042	84FA	INFORMATION SYSTEMS PRODUCT	PROSOFTIA AB

Socket Decimal	Socket Hex	Service	Company
34043	84FB	NETSCRIBE	MERIDIAN DATA CORP
34044	84FC	NETSCRIBE	MERIDIAN DATA CORP
34045	84FD	NETSCRIBE	MERIDIAN DATA CORP
34046	84FE	NETSCRIBE	MERIDIAN DATA CORP
34047	84FF	FAX SERVER	FERRARI ELECTRONIC GMBH
34050	8502	FOURTH SHIFT MFG ADD-ON PACKAGE	NOVELL
34058	850A	MODEM-SHARING SOFTWARE - DOS	LANSOURCE TECHNOLOGIES
34059	850B	MODEM-SHARING SOFTWARE - WINDOWS	LANSOURCE TECHNOLOGIES
34060	850C	TELEPHONE ANSWERING SYSTEM	A&M COMMUNICATIONS
34061	850D	FILE/IPX-BASED RPC SYSTEM	AUTOMATED INTERACTIONS DIV OF
34062	850E	NETWORK DISC BACK-UP SOFTWARE	FORTUNET INC
34063	850F	ARCHIVES & MUSEUM MANAGEMENT	CACTUS SOFTWARE
34064	8510	FAX SERVER	EXTENDED SYSTEMS
34065	8511	TELETEXT SERVICE	U OF PLYMOUTH
34066	8512	NETWORK ERROR LOG	U OF PLYMOUTH
34067	8513	MEASURESERVERS AND MEASURECLIENTS	ADVANTECH BENELUX BV
34068	8514	3270 NETWARE FOR SAA EMULATOR	FORVUS RESEARCH INC
34069	8515	3270 NETWARE FOR SAA EMULATOR	FORVUS RESEARCH INC
34070	8516	WORKSTATION UTILIZATION DATA COLLECTION	NETWORK SECURITY SYSTEMS
34071	8517	WORKSTATION UTILIZATION DATABASE MGMT	NETWORK SECURITY SYSTEMS
34072	8518	DATABASE GATEWAY	INFORMATION BUILDERS
34073	8519	FAX SERVER	EXTENDED SYSTEMS
34074	851A	REMOTE PRINTER CONSOLE	PEERLESS GROUP
34075	851B	BATCHFILER IPX SOCKET	JOVANDI INTERNATIONAL INC
34076	851C	TIME SYNCHRONIZATION IPX SOCKET	JOVANDI INTERNATIONAL INC
34077	851D	FAX SERVER LOGIN SOCKET	ASCOM TELECOMMUNICATION LTD
34078	851E	FAX SERVER - CAS REQUEST SOCKET	ASCOM TELECOMMUNICATION LTD
34079	851F	FAX SERVER - WORKSTATION UTILITY REQUEST	ASCOM TELECOMMUNICATION LTD
34080	8520	FAX SERVER - FAXBIOS REQUEST SOCKET	ASCOM TELECOMMUNICATION LTD
34081	8521	FAX SERVER - NETWORKED ATTACHED FAX UNIT	ASCOM TELECOMMUNICATION LTD
34082	8522	DFDSM DATA FACILITIES DATA STORAGE MGMT	IBM - SAN JOSE
34083	8523	RELATIONAL DATABASE	GUPTA TECHNOLOGIES
34084	8524	REPORT SERVER STATS	CENTRAL POINT SOFTWARE
34085	8525	REPORT SERVER STATS	CENTRAL POINT SOFTWARE
34086	8526	COMMUNICATION BETWEEN CLIENT AND SERVER	CDC - OFFICE OF PROGRAM SUPPRT
34089	8529	DOCUMENT PROCESSING SERVER NLM	BOSS LOGIC INC
34090	852A	DOCUMENT PROCESSING SERVER NLM	BOSS LOGIC INC
34091	852B	DOCUMENT PROCESSING SERVER NLM	BOSS LOGIC INC
34092	852C	DOCUMENT PROCESSING SERVER NLM	BOSS LOGIC INC
34093	852D	DOCUMENT PROCESSING SERVER NLM	BOSS LOGIC INC
34094	852E	DOCUMENT PROCESSING SERVER NLM	BOSS LOGIC INC
34095	852F	FINANCIAL MARKETS INFORMATION SERVER	AT FINANCIAL
34096	8530	EMAIL NOTIFICATION	ON TECHNOLOGY
34097	8531	NETVIEW SUPPORT	MEMOREX TELEX
34098	8532	LANLORD PRODUCT	MICROCOM CLIENT SERVER TECHNOL
34099	8533	RTS TERMINAL EMULATION	DATA RESEARCH & APPLICATIONS
34100	8534	RSCF CLIENT-SERVER API	DATA RESEARCH & APPLICATIONS
34101	8535	CD NETWORKER IPX VERSION	UNKNOWN CUSTOMER
34102	8536	SQL SERVER IPX/SPX HIDDEN SERVER	MICROSOFT
34109	853D	DATABASE LOCK SERVER	HIGH ASPECT DEVELOPMENT
34110	853E	MESSAGE MANAGER	LANCORP PTY LTD
34111	853F	OBJECT MANAGER	LANCORP PTY LTD
34112	8540	OBJECT AGENT	LANCORP PTY LTD
34113	8541	REQUEST MANAGER	LANCORP PTY LTD
34116	8544	WORKSTATION 4-LAN	ADVANCED TECHNICAL SOLUTIONS
34117	8545	GENESYS NETWORK PRODUCT DOS	GENESYS DATA TECHNOLOGY
34118	8546	GENESYS NETWORK PRODUCT OS/2	GENESYS DATA TECHNOLOGY
34120	8548	INTERNET GATEWAY	METASCYBE SYSTEMS LTD
34121	8549	INTELLIGENT HOST GATEWAY	AMERICAN AIRLINES DECISION TEC
34124	854C	INTERNAL PROTOCOL/WIRELESS LAN PRODUCT	IBM - LA GAUDE
34125	854D	MULTI-SYSTEM MGR - ACCESS TO SERVER INFO	IBM - RESEARCH TRIANGLE PARK
34126	854E	NETPRINT CALLING CHANNEL	INTERLINK COMMUNICATIONS LTD
34127	854F	NETPRINT WORKING CHANNEL	INTERLINK COMMUNICATIONS LTD
34128	8550	REMOTE SESSION	DST - DISTRIBUTED SYSTEMS TECH
34129	8551	REMOTE SESSION	DST - DISTRIBUTED SYSTEMS TECH
34130	8552	REMOTE SESSION	DST - DISTRIBUTED SYSTEMS TECH
34131	8553	REMOTE SESSION	DST - DISTRIBUTED SYSTEMS TECH
34132	8554	PEER-TO-PEER COMMUNICATIONS - DESKTOP	FUJITSU NETWORKS INDUSTRY
34133	8555	DIAGNOSTIC UTILITY	FUJITSU NETWORKS INDUSTRY
34134	8556	NLM HEALTH MONITOR	PRESOFT ARCHITECTS
34135	8557	CRISLER MCGEE	CRISLER MCKEE SOFTWARE DEVELOP
34136	8558	CRISLER MCGEE	CRISLER MCKEE SOFTWARE DEVELOP
34137	8559	INFORMATION SHARING BETWEEN WORKSTATIONS	INTEL CORP
34138	855A	POLICY ENGINE	EMERALD SYSTEMS
34139	855B	POLICY ENGINE	EMERALD SYSTEMS
34140	855C	POLICY ENGINE	EMERALD SYSTEMS
34141	855D	POLICY ENGINE	EMERALD SYSTEMS
34142	855E	POLICY ENGINE	EMERALD SYSTEMS
34143	855F	POLICY ENGINE	EMERALD SYSTEMS
34144	8560	STAND-ALONE PRINT SERVER	BAY TECHNICAL ASSOCIATES
34145	8561	SERVICE DISTRIBUTION	US WEST ADVANCED TECHNOLOGIES
34146	8562	SERVICE DISTRIBUTION	US WEST ADVANCED TECHNOLOGIES
34147	8563	SERVICE DISTRIBUTION	US WEST ADVANCED TECHNOLOGIES
34148	8564	SERVICE DISTRIBUTION	US WEST ADVANCED TECHNOLOGIES
34149	8565	SERVICE DISTRIBUTION	US WEST ADVANCED TECHNOLOGIES
34150	8566	SERVICE DISTRIBUTION	US WEST ADVANCED TECHNOLOGIES
34152	8568	MPRST PEER-TO-PEER	US SPRINT
34153	8569	MPRST BROADCAST	US SPRINT
34154	856A	LAN ASSIST PLUS REMOTE CONTROL	MICROTEST INC
34155	856B	LAN ASSIST PLUS REMOTE CONTROL	MICROTEST INC
34156	856C	LAN ASSIST PLUS REMOTE CONTROL	MICROTEST INC
34157	856D	LAN ASSIST PLUS REMOTE CONTROL	MICROTEST INC
34158	856E	MAP ASSIST PEER-TO-PEER	MICROTEST INC
34159	856F	MAP ASSIST PEER-TO-PEER	MICROTEST INC
34160	8570	MAP ASSIST PEER-TO-PEER	MICROTEST INC
34161	8571	MAP ASSIST PEER-TO-PEER	MICROTEST INC
34162	8572	ASYNCHRONOUS COMMUNICATIONS SERVERS	US ROBOTICS SOFTWARE
34163	8573	DATABASE SERVER	FAIR COM
34164	8574	NLM-BASED DATABASE ENGINE	AUTO GRAPHICS INC
34165	8575	TIGER QUOTE INCOMING SERVER REQUESTS	JOSHUA GROUP LTD
34166	8576	TIGER QUOTE BROADCAST	JOSHUA GROUP LTD
34167	8577	USER SOCKET	NETWORK TECHNICAL SOLUTIONS
34168	8578	GHOST SOCKET	NETWORK TECHNICAL SOLUTIONS
34169	8579	REMOTE PROCEDURE PROTOCOL	FORTUNET INC
34170	857A	EICON INTERCONNECT SERVER	EICON TECHNOLOGY
34171	857B	EICON SECURITY AGENT	EICON TECHNOLOGY
34172	857C	COST RECOVERY SERVER	VINCENT LARSEN
34173	857D	COST RECOVERY SERVER	VINCENT LARSEN
34174	857E	PC-BASED SNA GATEWAY	UNGERMANN BASS
34175	857F	COMMUNICATION BETWEEN PRINT SERVER & W/S	NISSIN ELECTRIC CO LTD
34176	8580	PEER-TO-PEER MESSAGING	HANS SPATZIER
34177	8581	BANKING DEALING ROOMS	ART & SCIENCE LTD
34178	8582	BANKING DEALING ROOMS	ART & SCIENCE LTD
34179	8583	BANKING DEALING ROOMS	ART & SCIENCE LTD
34180	8584	BANKING DEALING ROOMS	ART & SCIENCE LTD
34190	858E	CLIENT UTILITY - WS TO PRINTER COMMUNICA	TOKYO DENSHI SEKKEI KK
34192	8590	NETWORK WORKSTATION CONTROL	WESTERN PACIFIC TECHNOLOGIES
34193	8591	TELETEXT SERVER	TEVESCOM
34194	8592	PRINT SERVER	FORESYTE TECHNOLOGIES
34201	8599	WAN NETWORKS	PROSOFTIA AB
34202	859A	WAN NETWORKS	PROSOFTIA AB
34203	859B	WAN NETWORKS	PROSOFTIA AB
34204	859C	WAN NETWORKS	PROSOFTIA AB
34205	859D	WAN NETWORKS	PROSOFTIA AB
34206	859E	WAN NETWORKS	PROSOFTIA AB
34207	859F	WAN NETWORKS	PROSOFTIA AB
34208	85A0	WAN NETWORKS	PROSOFTIA AB
34209	85A1	WAN NETWORKS	PROSOFTIA AB
34210	85A2	WAN NETWORKS	PROSOFTIA AB
34215	85A7	DATABASE SERVER	SOFTWRIGHT SYSTEMS
34216	85A8	CHANGE CONTROL PRODUCT	OCCIDENTAL PETROLEUM SVCS INC
34217	85A9	CHANGE CONTROL PRODUCT	OCCIDENTAL PETROLEUM SVCS INC
34218	85AA	STATISTIC MANAGEMENT	MULTITECH
34219	85AB	STATISTIC MANAGEMENT	MULTITECH
34220	85AC	REMOTE CONTROL SOFTWARE	MULTITECH
34221	85AD	REMOTE CONTROL SOFTWARE	MULTITECH
34222	85AE		MULTITECH
34223	85AF	REMOTE ACCESS SERVER	DIGITAL COMMUNICATIONS ASSOC (DCA)
34224	85B0	WINDOWS-BASED FAX SYSTEM	ICONOGRAPHIC SYSTEMS
34225	85B1	PRINT SERVER ADD-ON	INTEL - AMERICAN FORK
34226	85B2	AGV CONTROLLER COMMUNICATIONS	CONTROL ENGINEERING
34227	85B3	INDEX SEQUENTIAL ACCESS NLM	INFOPOINT SYSTEMS
34228	85B4	ASSOCIATIVE INDEX SERVER	INFOPOINT SYSTEMS
34229	85B5	STRESSMAGIC SERVER UTILITY	NET MAGIC SYSTEMS INC

Socket Decimal	Socket Hex	Service	Company
34230	85B6	NETWORK POWER TOOLS	NET MAGIC SYSTEMS INC
34231	85B7	DOCUMENT MANAGEMENT SVC	IMAGERY SOFTWARE INC
34232	85B8	IMAGE MANAGEMENT SVC	IMAGERY SOFTWARE INC
34233	85B9	MASS STORAGE SVC	IMAGERY SOFTWARE INC
34234	85BA	CITRIX APPLICATION SERVER	CITRIX SYSTEMS
34235	85BB	CITRIX APPLICATION SERVER	CITRIX SYSTEMS
34236	85BC		KLOS TECHNOLOGIES INC
34237	85BD	HOSPITAL MANAGEMENT PACKAGE	SOFTWORK GMBH
34238	85BE	ROUTER MANAGEMENT APPLICATION	CISCO SYSTEMS
34239	85BF	NETWORK MODEM	NANAGRAM
34240	85C0	NETWORK MODEM	NANAGRAM
34244	85C4	5250 GATEWAY COMMUNICATIONS	MICRO INTEGRATION
34245	85C5	SOFTWARE DISTRIBUTION SUITE	CENTERA PTY LTD
34246	85C6	SOFTWARE DISTRIBUTION SUITE	CENTERA PTY LTD
34247	85C7	SOFTWARE DISTRIBUTION SUITE	CENTERA PTY LTD
34248	85C8	SOFTWARE DISTRIBUTION SUITE	CENTERA PTY LTD
34249	85C9	TOBIT !TEAM - REMOTE CONTROLLING	TOBIT SOFTWARE GMBH
34250	85CA	FAXWARE 3.0 - API COMMUNICATION	TOBIT SOFTWARE GMBH
34251	85CB	TOBIT PLZ5 - DATABASE SERVER	TOBIT SOFTWARE GMBH
34252	85CC	AVL NLM DATABASE	AETNA LIFE & CASUALTY
34253	85CD	LU6.2 GATEWAY	AETNA LIFE & CASUALTY
34255	85CF	TELI-LINK VOICE SERVER	COMPUTER & COMMUNICATIONS CO
34256	85D0	REMOTE DOWNLOAD SOFTWARE	ASANTE TECHNOLOGIES
34261	85D5	LAN EXPANDERS MANAGEMENT & DATA TRANSFER	GATEWAY COMMUNICATIONS INC
34262	85D6	CALENDAR SERVER	CAMPBELL SERVICES INC
34263	85D7	CA UNICENTER	COMPUTER ASSOCIATES
34264	85D8	CA UNICENTER	COMPUTER ASSOCIATES
34265	85D9	CA UNICENTER	COMPUTER ASSOCIATES
34266	85DA	CA UNICENTER	COMPUTER ASSOCIATES
34267	85DB	CA UNICENTER	COMPUTER ASSOCIATES
34268	85DC	CA UNICENTER	COMPUTER ASSOCIATES
34269	85DD	NET MODEM SPX SOCKET	PRACTICAL PERIPHERALS INC
34270	85DE	NET MODEM IPX SOCKET	PRACTICAL PERIPHERALS INC
34271	85DF	ALERT SERVER NLM	CENTRAL POINT SOFTWARE
34272	85E0	MESSAGE ROUTER	CENTRAL POINT SOFTWARE
34273	85E1	OPTICAL FILE SERVER COMMUNICATIONS	PEGASUS DISK TECHNOLOGIES INC
34274	85E2	OPTICAL FILE SERVER LOGIN/LOGOUT	PEGASUS DISK TECHNOLOGIES INC
34277	85E5	PRINT SERVER	RASTEROPS PRINTER TECH DIV
34278	85E6	QUARK EXPRESS	QUARK INC
34279	85E7	SECURITY NLM	NOVELL - OREM
34280	85E8	ENDPOINT MAPPER FOR RPC	MICROSOFT
34281	85E9	PC ANYWHERE/NETWARE LITE	SYMANTEC CORP
34282	85EA	INTERSERVER FILE COPYING	BANKERS TRUST CO
34283	85EB	CONNECTION SERVICES	RABBIT SOFTWARE CORP
34284	85EC	DISCOVERY SERVICES	RABBIT SOFTWARE CORP
34285	85ED	NETWORK MONITOR SERVICES	RABBIT SOFTWARE CORP
34286	85EE	CA-DATACOM/PC	COMPUTER ASSOCIATES
34287	85EF	CA-IDMS/PC	COMPUTER ASSOCIATES
34289	85F1	INDUSTRIAL CONTROL AUTOMATION NETWORK	TELE DENKEN
34290	85F2	PRINT SERVER	RINGDALE UK LTD
34291	85F3	COMMUNICATIONS SERVER	CSB SYSTEMS GMBH
34294	85F6	DATABASE DATAGRAM SOCKET	LYNC INC
34295	85F7	NETPORT EXPRESS STATUS RESPONDER	INTEL - AMERICAN FORK
34296	85F8	CADENCE TIME SYNCHRONIZATION PRODUCT	POLYGON INC
34297	85F9	CADENCE TIME SYNCHRONIZATION PRODUCT	POLYGON INC
34298	85FA	REMOTE VIRUS SCANNING	MCAFEE ASSOCIATES
34299	85FB	REMOTE MEMORY CONTROL	MCAFEE ASSOCIATES
34300	85FC	NORTON BACK-UP DEVICE SHARING PROTOCOL	ASTORA SOFTWARE INC
34301	85FD	SAMS:EXPERT	STERLING TEFEN LAB
34302	85FE	SAMS:CONTROL	STERLING TEFEN LAB
34303	85FF	SAMS:VANTAGE	STERLING TEFEN LAB
34304	8600	SAMS:SAVE	STERLING TEFEN LAB
34305	8601	SAMS:DISPATCHER	STERLING TEFEN LAB
34306	8602	RENDEZVOUS IPX	GREYHOUSE TECHNOLOGIES
34307	8603	RENDEZVOUS IPX	GREYHOUSE TECHNOLOGIES
34308	8604	RENDEZVOUS SPX	GREYHOUSE TECHNOLOGIES
34309	8605	RENDEZVOUS SPX	GREYHOUSE TECHNOLOGIES
34310	8606	VOICE/FAX RESPONDING MACHINE	SYSTEM SOPHIA
34312	8608	SHARE MODE BROADCAST	NOVELL - OREM
34313	8609	IPX ENCAPSULATED RM3 PACKETS	CAYMAN SYSTEMS INC
34314	860A	DATABASE SERVICE	TRIFOX INC
34315	860B	IMAGESOLVE_OFS	IMAGESOLVE INTERNATIONAL
34316	860C	TECHRA: CLIENTSERVER RDBMS	KVATRO AS
34317	860D	DOCRA: CLIENTSERVER DOC MGMT SYSTEM	KVATRO AS

Socket Decimal	Socket Hex	Service	Company
34318	860E	NETWORK MANAGEMENT APPLICATION	WANDEL & GOLTERMANN
34319	860F	NETWORK MANAGEMENT APPLICATION	WANDEL & GOLTERMANN
34320	8610	NETWORK MANAGEMENT APPLICATION	WANDEL & GOLTERMANN
34321	8611	NETWORK MANAGEMENT APPLICATION	WANDEL & GOLTERMANN
34322	8612	NETWORK MANAGEMENT APPLICATION	WANDEL & GOLTERMANN
34323	8613	NETWORK MANAGEMENT APPLICATION	WANDEL & GOLTERMANN
34324	8614	CONNECTION ACCEPTANCE SOCKET	CHANCERY SOFTWARE LTD
34325	8615	OPTIDRIVER-NET	OPTISYS
34326	8616	EDM CLIENT/PC	COMPUTER VISION
34327	8617	VIDEO CONFERENCING	LLOYD ALLAN CORPORATION
34328	8618	PRINTER GATEWAY/PEER-TO-PEER COMMUNICATIONS	ADACOM GROUP
34332	861C	COMMUNICATIONS UTILITY	LEXMARK INTERNATIONAL INC
34333	861D	COMMUNICATIONS UTILITY	LEXMARK INTERNATIONAL INC
34334	861E	PRINT SERVER	LEXMARK INTERNATIONAL INC
34335	861F	PRINT SERVER	LEXMARK INTERNATIONAL INC
34336	8620	PHONE SYSTEM CONTROL	DASH OPEN PHONE SYSTEMS
34337	8621	PHONE SYSTEM CONTROL	DASH OPEN PHONE SYSTEMS
34338	8622	PHONE SYSTEM CONTROL	DASH OPEN PHONE SYSTEMS
34339	8623	PHONE SYSTEM CONTROL	DASH OPEN PHONE SYSTEMS
34340	8624	OVERSIGHT AGENT	NETWORK UTILITIES SOFTWARE LTD
34341	8625	OVERSIGHT MASTER	NETWORK UTILITIES SOFTWARE LTD
34342	8626	ERL DATABASE SERVER	SILVER PLATTER INFORMATION LTD
34343	8627	ERL DIRECTORY SERVER	SILVER PLATTER INFORMATION LTD
34344	8628	IPX BROADCAST	INTERTECH IMAGING CORPORATION
34345	8629	SPX CONNECT	INTERTECH IMAGING CORPORATION
34346	862A	ZIFF PROPRIETARY SERVICES	ZIFF INFORMATION SERVICES
34347	862B	GAMES	LOOKING GLASS
34348	862C	LPT PORTS	LEXMARK INTERNATIONAL INC
34349	862D	LPT PORTS	LEXMARK INTERNATIONAL INC
34350	862E	LPT PORTS	LEXMARK INTERNATIONAL INC
34351	862F	LPT PORTS	LEXMARK INTERNATIONAL INC
34352	8630	LPT PORTS	LEXMARK INTERNATIONAL INC
34353	8631	LPT PORTS	LEXMARK INTERNATIONAL INC
34354	8632	LPT PORTS	LEXMARK INTERNATIONAL INC
34355	8633	LPT PORTS	LEXMARK INTERNATIONAL INC
34356	8634	TIME SERVER BROADCAST SOCKET	MEINBERG FUNKUHREN
34357	8635	ACCELERATION DATA	SABLE TECHNOLOGY CORP
34358	8636	CALLPATH	IBM - RESEARCH TRIANGLE PARK
34359	8637	CALLPATH	IBM - RESEARCH TRIANGLE PARK
34360	8638	COMMUNICATION INTEGRATOR - SPX	COVIA CORP
34361	8639	COMMUNICATION INTEGRATOR - IPX	COVIA CORP
34363	863B	NETWARE AWARE INFORMATION MGMT SYSTEM	INTUITIVE SOLUTIONS
34364	863C	NETWARE AWARE INFORMATION MGMT SYSTEM	INTUITIVE SOLUTIONS
34365	863D	NETWARE AWARE INFORMATION MGMT SYSTEM	INTUITIVE SOLUTIONS
34366	863E	NETWARE AWARE INFORMATION MGMT SYSTEM	INTUITIVE SOLUTIONS
34367	863F	ARTS RLOGIN APPLICATION	AMERICAN REAL TIME, REUTERS CO
34368	8640	ARTS GENERIC SERVER	AMERICAN REAL TIME, REUTERS CO
34369	8641	NETOP PROGRAM	DANWARE DATA AS
34370	8642	NETOP PROGRAM	DANWARE DATA AS
34371	8643	NETOP PROGRAM	DANWARE DATA AS
34372	8644	NETOP PROGRAM	DANWARE DATA AS
34373	8645	NETOP PROGRAM	DANWARE DATA AS
34374	8646	DOCUMENT MANAGEMENT SYSTEM	SR ASSOCIATES/CYBERMEDIA
34375	8647	SECURITY CHECK	MCAFEE ASSOCIATES
34376	8648	NEWSWIRE NOTIFICATION	GENERATION TECHNOLOGIES CORP
34377	8649	P-NET GATEWAY	PROCES DATA SILKEBORG APS
34378	864A	VIRTUAL MANUFACTURING DEVICE	PROCES DATA SILKEBORG APS
34379	864B	SALES APPLICATION	PROXIM INC
34380	864C	BROADCASTS	REMUERA CORP
34381	864D	COMMUNICATIONS	REMUERA CORP
34382	864E	LAN/CD ROM SERVER	LOGICRAFT
34384	8650	SERVER SOCKET	KNIGHT RIDDER FINANCIAL INC
34385	8651	PING SOCKET	KNIGHT RIDDER FINANCIAL INC
34386	8652	BROADCAST DATAGRAM SOCKET	KNIGHT RIDDER FINANCIAL INC
34387	8653	EMPOWER LINK APPLICATION LOADER	NETWORK SECURITY SYSTEMS
34388	8654	NAME RESOLUTION	INTELEC SYSTEMS CORPORATION
34389	8655	MESSAGE LINE	NORMAN DATA DEFANSE SYSTEMS
34390	8656	CLIENT MESSAGE LINE	NORMAN DATA DEFANSE SYSTEMS
34391	8657	CD SHARING ON NOVELL WORKSTATION	CROSS INTERNATIONAL CORP
34392	8658	NETTALK LAN COMMUNICATIONS SOFTWARE	CROSS INTERNATIONAL CORP
34393	8659	TELEPHONE COMMUNICATIONS SOFTWARE	CROSS INTERNATIONAL CORP
34394	865A	LAN CHATTING	CROSS INTERNATIONAL CORP
34395	865B	NETWORK MANAGEMENT SERVER	POLE POSITION SOFTWARE GMBH

Socket Decimal	Socket Hex	Service	Company
34396	865C	REMOTE ACCESS SOCKET #1	TRAVELING SOFTWARE
34397	865D	REMOTE ACCESS SOCKET #2	TRAVELING SOFTWARE
34398	865E	NET TRAX ADMINISTRATION	NET X CORP
34399	865F	NET TRAX AGENT	NET X CORP
34400	8660	NET TRAX ALARM MONITOR	NET X CORP
34401	8661	AO CLIENT	ICL PERSONAL SYSTEMS OY
34402	8662	AO SERVER FOR CLIENT	ICL PERSONAL SYSTEMS OY
34403	8663	AO DIRECTORY SERVER	ICL PERSONAL SYSTEMS OY
34404	8664	AO SERVER ALARMER	ICL PERSONAL SYSTEMS OY
34405	8665	AO CLIENT ALARMER	ICL PERSONAL SYSTEMS OY
34406	8666	AO LAN RTS	ICL PERSONAL SYSTEMS OY
34407	8667	AO REMOTE CMD SERVER	ICL PERSONAL SYSTEMS OY
34408	8668	AO REMOTE CMD CLIENT	ICL PERSONAL SYSTEMS OY
34409	8669	AO DIR JOIN SERVER	ICL PERSONAL SYSTEMS OY
34410	866A	AO STORAGE SERVER FOR CLIENT	ICL PERSONAL SYSTEMS OY
34411	866B	SAVE UTILTIY/2	IBM - CARY
34412	866C	SAVE UTILTIY/LIBRARIAN	IBM - CARY
34413	866D	SAVE UTILTIY/CURATOR	IBM - CARY
34414	866E	SAVE UTILTIY/JANITOR	IBM - CARY
34415	866F	SAVE UTILTIY/ARCHIVES I	IBM - CARY
34416	8670	SAVE UTILTIY/ARCHIVES II	IBM - CARY
34417	8671	SAVE UTILTIY/ARCHIVES III	IBM - CARY
34418	8672	SAVE UTILTIY/ARCHIVES IV	IBM - CARY
34419	8673	SAVE UTILTIY/ARCHIVES V	IBM - CARY
34420	8674	SAVE UTILTIY/ARCHIVES VI	IBM - CARY
34421	8675	SAVE UTILTIY/ARCHIVES VII	IBM - CARY
34422	8676	SAVE UTILTIY/ARCHIVES VIII	IBM - CARY
34423	8677	SAFESERVER REQUEST SOCKET	OMNITECH CORPORATE SOLUTIONS
34424	8678	CLIENT NLM COMMUNICATIONS	SOFTWARE SECURITY INC
34425	8679	NLM SERVER TO SERVER	SOFTWARE SECURITY INC
34426	867A	FILE TRANSFER APPLICATION	URS ZURBUCHEN
34427	867B	SD ROM JUKEBOX COMMAND SERVER	TODD ENTERPRISES INC
34428	867C	COURSEWARE SERVER	FIRST CLASS SYSTEMS
34429	867D	UDP OVER IPX TRANSMIT	SYNOPTICS
34430	867E	UDP OVER IPX RECEIVE	SYNOPTICS
34431	867F	REALTIME VOICE COMMUNICATION SOFTWARE	VOCALTEC INC
34433	8681	PERSON TO PERSON PRODUCT	IBM - HAMPSHIRE UK
34434	8682	NET TUNE	HAWKNET INC
34435	8683	NET TUNE	HAWKNET INC
34436	8684	NET TUNE	HAWKNET INC
34437	8685	NET TUNE	HAWKNET INC
34442	868A	ACCESS CONTROL & LICENSE MANAGEMENT	DALLAS SEMICONDUCTOR
34443	868B	STAND-ALONE PRINT SERVER	SERCOMM
34444	868C	CD-VINE PEER-TO-PEER COMMUNICATIONS	INFO LINE
34445	868D	FTP	CENTRAL POINT SOFTWARE
34446	868E	TFTP	CENTRAL POINT SOFTWARE
34447	868F	BOOT PS	CENTRAL POINT SOFTWARE
34448	8690	BOOT PC	CENTRAL POINT SOFTWARE
34450	8692	CLIENT TO SERVER COMMUNICATION	NBS SYSTEMS INC
34451	8693	SERVER TO SERVER COMMUNICATION	NBS SYSTEMS INC
34452	8694	NOTIFICATION PURPOSES	NBS SYSTEMS INC
34453	8695	FUTURE EXPANSION	NBS SYSTEMS INC
34454	8696	FUTURE EXPANSION	NBS SYSTEMS INC
34455	8697	SNA SERVICES	NETWORK CONTROLS INTERNATIONAL
34456	8698	SNA SERVICES	NETWORK CONTROLS INTERNATIONAL
34458	869A	PROTOCOL IPX	OST-OUEST STANDARD TELEMATIQUE
34459	869B	PROTOCOL IPX	OST-OUEST STANDARD TELEMATIQUE
34460	869C	MULTI-PLAYER GAME - DOOM	ID SOFTWARE
34461	869D	NETWORK MANAGEMENT	HEWLETT PACKARD - GRENOBLE
34462	869E	NETWORK MANAGEMENT	HEWLETT PACKARD - GRENOBLE
34463	869F	DISTRIBUTION SERVICES DISCOVERY	IBM - ROME
34464	86A0	DOCUMENT MANAGEMENT PACKAGE	NOVELL - OREM
34465	86A1	DOCUMENT MANAGEMENT PACKAGE	NOVELL - OREM
34466	86A2	DOCUMENT MANAGEMENT PACKAGE	NOVELL - OREM
34467	86A3	DOCUMENT MANAGEMENT PACKAGE	NOVELL - OREM
34472	86A8	CENTRAL MONITORING SYSTEM	TALX CORP
34473	86A9	CENTRAL MONITORING SYSTEM	TALX CORP
34474	86AA	CENTRAL MONITORING SYSTEM	TALX CORP
34475	86AB	CENTRAL MONITORING SYSTEM	TALX CORP
34476	86AC	CENTRAL MONITORING SYSTEM	TALX CORP
34477	86AD	END POINT MAPPER ID'S	MICROSOFT
34478	86AE	END POINT MAPPER ID'S	MICROSOFT
34479	86AF	END POINT MAPPER ID'S	MICROSOFT
34480	86B0	CALENDAR SERVER	CAMPBELL SERVICES INC
34490	86BA	NETWORK MANAGEMENT APPLICATION	XIRCOM

Socket Decimal	Socket Hex	Service	Company
34491	86BB	NETWORK MANAGEMENT APPLICATION	XIRCOM
34492	86BC	NETWORK MANAGEMENT APPLICATION	XIRCOM
34493	86BD	NETWORK MANAGEMENT APPLICATION	XIRCOM
34494	86BE	NETWORK MANAGEMENT APPLICATION	XIRCOM
34495	86BF	NETWORK MANAGEMENT APPLICATION	XIRCOM
34510	86CE	SERVICE LOCATION PROTOCOL	EICON TECHNOLOGY
34511	86CF	TWINSCOPE FOR IPX	NIPPON SYSTEM KAIHUTSU
34512	86D0	MAJOR BBS SOFTWARE	GALACTICOMM INC
34513	86D1	MAJOR BBS SOFTWARE	GALACTICOMM INC
34514	86D2	MAJOR BBS SOFTWARE	GALACTICOMM INC
34515	86D3	MAJOR BBS SOFTWARE	GALACTICOMM INC
34516	86D4	MAJOR BBS SOFTWARE	GALACTICOMM INC
34517	86D5	MAJOR BBS SOFTWARE	GALACTICOMM INC
34518	86D6	MAJOR BBS SOFTWARE	GALACTICOMM INC
34519	86D7	MAJOR BBS SOFTWARE	GALACTICOMM INC
34520	86D8	MAJOR BBS SOFTWARE	GALACTICOMM INC
34521	86D9	MAJOR BBS SOFTWARE	GALACTICOMM INC
34524	86DC	FAXWARE 3.0 (C-REQ)	TOBIT SOFTWARE GMBH
34525	86DD	FAXWARE 3.0 (HS-COMM)	TOBIT SOFTWARE GMBH
34526	86DE	FAXWARE 3.0 (TLD)	TOBIT SOFTWARE GMBH
34527	86DF	HIGH PERFORMANCE COMMUNICATION SERVER	TOBIT SOFTWARE GMBH
34528	86E0	HIGH PERFORMANCE COMMUNICATION SERVER	TOBIT SOFTWARE GMBH
34529	86E1	HIGH PERFORMANCE COMMUNICATION SERVER	TOBIT SOFTWARE GMBH
34530	86E2	HIGH PERFORMANCE COMMUNICATION SERVER	TOBIT SOFTWARE GMBH
34531	86E3	HIGH PERFORMANCE COMMUNICATION SERVER	TOBIT SOFTWARE GMBH
34532	86E4	HIGH PERFORMANCE COMMUNICATION SERVER	TOBIT SOFTWARE GMBH
34533	86E5	ELECTRONIC SPELLING BOOK (SPOOK!)	TOBIT SOFTWARE GMBH
34534	86E6	ELECTRONIC SPELLING BOOK (SPOOK!)	TOBIT SOFTWARE GMBH
34535	86E7	ELECTRONIC SPELLING BOOK (SPOOK!)	TOBIT SOFTWARE GMBH
34536	86E8	ELECTRONIC SPELLING BOOK (SPOOK!)	TOBIT SOFTWARE GMBH
34545	86F1	OFFICE EXTEND SERVER	FRANSEN KING
34546	86F2	WINDOWS NT FACSYS SERVER	OPTUS INFORMATION SYSTEMS
34547	86F3	WINDOWS NT FACSYS SERVER	OPTUS INFORMATION SYSTEMS
34548	86F4	WINDOWS NT FACSYS SERVER	OPTUS INFORMATION SYSTEMS
34551	86F7	KEYFILE NAME SERVICE	KEYFILE CORP
34552	86F8	EVERGREEN MANAGEMENT AGENT	GOODALL SOFTWARE
34553	86F9	EVERGREEN MANAGEMENT AGENT	GOODALL SOFTWARE
34554	86FA	EVERGREEN MANAGEMENT AGENT	GOODALL SOFTWARE
34555	86FB	EVERGREEN MANAGEMENT AGENT	GOODALL SOFTWARE
34556	86FC	EVERGREEN MANAGEMENT AGENT	GOODALL SOFTWARE
34557	86FD	EVERGREEN MANAGEMENT AGENT	GOODALL SOFTWARE
34558	86FE	FILE SYNCHRONIZITION	NOMADIC SYSTEMS
34559	86FF	WINDOWS BULLETIN BOARD SYSTEM	PACER SOFTWARE
34562	8702	LAN NETVIEW MANAGEMENT UTILITIES	IBM - RESEARCH TRIANGLE PARK
34563	8703	LAN NETVIEW MANAGEMENT UTILITIES	IBM - RESEARCH TRIANGLE PARK
34564	8704	LAN NETVIEW MANAGEMENT UTILITIES	IBM - RESEARCH TRIANGLE PARK
34565	8705	LAN NETVIEW MANAGEMENT UTILITIES	IBM - RESEARCH TRIANGLE PARK
34566	8706	LAN NETVIEW MANAGEMENT UTILITIES	IBM - RESEARCH TRIANGLE PARK
34567	8707	LAN NETVIEW MANAGEMENT UTILITIES	IBM - RESEARCH TRIANGLE PARK
34568	8708	DOCUMENT MANAGEMENT PACKAGE	NOVELL - OREM
34569	8709	DOCUMENT MANAGEMENT PACKAGE	NOVELL - OREM
34570	870A	DOCUMENT MANAGEMENT PACKAGE	NOVELL - OREM
34571	870B	DOCUMENT MANAGEMENT PACKAGE	NOVELL - OREM
34572	870C	DOCUMENT MANAGEMENT PACKAGE	NOVELL - OREM
34573	870D	GOODALL VIRTUAL PROTOCOL ADAPTOR	GOODALL SOFTWARE
34574	870E	GOODALL VIRTUAL PROTOCOL ADAPTOR	GOODALL SOFTWARE
34575	870F	GOODALL VIRTUAL PROTOCOL ADAPTOR	GOODALL SOFTWARE
34576	8710	GOODALL VIRTUAL PROTOCOL ADAPTOR	GOODALL SOFTWARE
34589	871D	PINNACLE RELATIONAL ENGINE	VERMONT DATABASE CORP
34590	871E	FAULT TOLERANCE	CLONE STAR SOFTWARE
34596	8724	UNIX MAIL SERVER	FELPAUSCH
34599	8727	INDUSTIRAL TEST & HANDLING EQUIPMENT	Q CORP
34600	8728	INDUSTIRAL TEST & HANDLING EQUIPMENT	Q CORP
34601	8729	INDUSTIRAL TEST & HANDLING EQUIPMENT	Q CORP
34602	872A	TELI-LINK VOICE SERVER	COMPUTER & COMMUNICATIONS
34603	872B	SECURE FAX CLIENT SOCKET	RUSSELL CONSULTING
34611	8733	BRIDGE ROUTER MENU CONNECTION	NETWORKS NORTHWEST INC
34612	8734	BRIDGE ROUTER ERROR LOG	NETWORKS NORTHWEST INC
34613	8735	IMAGE SERVER CONNECTION	WATERMARK SOFTWARE
34615	8737	EXSEKEY INTERFACE FOR PROTECTED PROGRAMS	CLOVER INFORMATICA SNC

Socket Decimal	Socket Hex	Service	Company
34616	8738	METERING PROGRAM	SECURE DESIGN
34620	873C	COMET TERMINAL SERVER	GOODALL SOFTWARE
34621	873D	COMET TERMINAL SERVER	GOODALL SOFTWARE
34622	873E	COMET TERMINAL SERVER	GOODALL SOFTWARE
34623	873F	COMET TERMINAL SERVER	GOODALL SOFTWARE
34624	8740	COMET TERMINAL SERVER	GOODALL SOFTWARE
34625	8741	COMET TERMINAL SERVER	GOODALL SOFTWARE
34626	8742	COMET FILE SERVER	GOODALL SOFTWARE
34627	8743	COMET FILE SERVER	GOODALL SOFTWARE
34628	8744	COMET FILE SERVER	GOODALL SOFTWARE
34629	8745	COMET FILE SERVER	GOODALL SOFTWARE
34630	8746	COMET FILE SERVER	GOODALL SOFTWARE
34631	8747	COMET FILE SERVER	GOODALL SOFTWARE
34634	874A	MAXSERV COMMUNICATIONS	MAXSERV
34635	874B	MAXSERV COMMUNICATIONS	MAXSERV
34636	874C	NCP COMMUNICATION - JOB SCHEDULER & NLM	SIMWARE
34638	874E	TRACE ROUTE	3COM CORPORATION
34639	874F	CLIENT COLLABORATIVE DATA SHARE PRODUCT	AT&T
34651	875B	TVI DESKTOP SERVER	TARGET VISION
34652	875C	MASTERSHOW	TARGET VISION
34653	875D	USER-ACCESS SERVER	TMD CONSULTING
34654	875E	TCP GATEWAY	TMD CONSULTING
34655	875F	CAM SERVER	TMD CONSULTING
34656	8760	CAM SECURE DATABASE	TMD CONSULTING
34657	8761	CAM RESOURCE MANAGER	TMD CONSULTING
34658	8762	CAM BACK-UP	TMD CONSULTING
34659	8763	FIREFOX NLMS AND CLIENT SOFTWARE	FIREFOX COMMUNICATIONS LTD
34660	8764	FIREFOX NLMS AND CLIENT SOFTWARE	FIREFOX COMMUNICATIONS LTD
34661	8765	FIREFOX NLMS AND CLIENT SOFTWARE	FIREFOX COMMUNICATIONS LTD
34662	8766	FIREFOX NLMS AND CLIENT SOFTWARE	FIREFOX COMMUNICATIONS LTD
34663	8767	FIREFOX NLMS AND CLIENT SOFTWARE	FIREFOX COMMUNICATIONS LTD
34664	8768	FIREFOX NLMS AND CLIENT SOFTWARE	FIREFOX COMMUNICATIONS LTD
34665	8769	FIREFOX NLMS AND CLIENT SOFTWARE	FIREFOX COMMUNICATIONS LTD
34666	876A	FIREFOX NLMS AND CLIENT SOFTWARE	FIREFOX COMMUNICATIONS LTD
34667	876B	FIREFOX NLMS AND CLIENT SOFTWARE	FIREFOX COMMUNICATIONS LTD
34668	876C	FIREFOX NLMS AND CLIENT SOFTWARE	FIREFOX COMMUNICATIONS LTD
34669	876D	DATA TRANSACTIONS	TENFORE RESEARCH & DEVELOPMENT
34670	876E	FLOW CONTROL	TENFORE RESEARCH & DEVELOPMENT
34672	8770	REMOTE CONTROL PRODUCT	IBM - RALEIGH
34673	8771	REMOTE CONTROL PRODUCT	IBM - RALEIGH
34674	8772	REMOTE CONTROL PRODUCT	IBM - RALEIGH
34675	8773	REMOTE CONTROL PRODUCT	IBM - RALEIGH
34684	877C	HITECSOFT SEND SOCKET	HITECSOFT CORP
34685	877D	HITECSOFT RECEIVE SOCKET	HITECSOFT CORP
34686	877E	LAN SCHOOL FOR WINDOWS	LAN FAN TECHNOLOGIES
34687	877F	REMOTE ACCESS PROTOCOL FOR DESKTOP MGMT	INTEL - AMERICAN FORK
34695	8787	VIDEO SERVER	COREL SYSTEMS CORP
34696	8788	LAN PERFORMANCE MNGMT TOOL IPX/SPX	DIGITAL EQUIPMENT CORP
34697	8789	LAN PERFORMANCE MNGMT TOOL TCP	DIGITAL EQUIPMENT CORP
34699	878B	STATUS SOCKET	NATIONAL SOFTWARE DEVELOPMENT
34700	878C	PRODUCTION SOCKET	NATIONAL SOFTWARE DEVELOPMENT
34702	878E	U OF WISCONSIN UTILITIES	U OF WISCONSIN - MADISON, CAE
34703	878F	U OF WISCONSIN UTILITIES	U OF WISCONSIN - MADISON, CAE
34704	8790	MARC SOFTWARE SYSTEMS	QDATA
34705	8791	TIMBUKTU APPLICATION	FARALLON COMPUTING
34706	8792	LED DISPLAY SERVER	INOVA CORPORATION
34708	8794	BORLAND INTERFACE SERVER FOR NW	BORLAND INTERNATIONAL
34709	8795	NPTN SPX	IBM - RTP
34710	8796	NPTN IPX	IBM - RTP
34711	8797	REMOTE ACCESS SOCKET 3	TRAVELING SOFTWARE
34712	8798	SERVERBENCH	NOVELL
34713	8799	INITIAL CONNECTION TO NLM	FLAGSTAR
34715	879B	SITE METER SECURITY	MCAFEE ASSOCIATES
34716	879C	SITE METER COMMAND	MCAFEE ASSOCIATES
34717	879D	PROXY AGENT SITE METER	MCAFEE ASSOCIATES
34718	879E	DATABASE SERVER/DB2 INSTANCES	IBM - CANADA LTD.
34719	879F	DATABASE SERVER/DB2 INSTANCES	IBM - CANADA LTD.
34720	87A0	DATABASE SERVER/DB2 INSTANCES	IBM - CANADA LTD.
34721	87A1	DATABASE SERVER/DB2 INSTANCES	IBM - CANADA LTD.
34729	87A9	FILE TRANSFER BETWEEN LAN/MAINFRAME	PROGINET CORP
34730	87AA	CONNECTION SERVICE AND SECURITY	PROGINET CORP
34733	87AD	ENTERTAINMENT PRODUCT-BILLIARD POOL	CELERIS INC.
34734	87AE	GAME DEVELOPMENT	GOLDSTEIN GOLLUB KESSLER & CO
34738	87B2	NOVELL SERVER MNGMT BY MAESTROVISION	CALYPSO SOFTWARE
34746	87BA	HEALTHCARE VALIDATIONS APP FOR DOS	OPENVISION
34747	87BB	HEALTHCARE VALIDATIONS APP FOR UNIX	OPENVISION
34749	87BD	A SYNCHRONOUS COMMUNICATION	LAN ACCESS
36864	9000	NP/SQL SERVER	NOVELL - PROVO
36865	9001	WIDE AREA ROUTER	NOVELL - SUNNYVALE (CPD)
36866	9002	WIDE AREA ROUTER	NOVELL - SUNNYVALE (CPD)
36867	9003	WIDE AREA ROUTER	NOVELL - SUNNYVALE (CPD)
36868	9004	WIDE AREA ROUTER	NOVELL - SUNNYVALE (CPD)
36869	9005	WIDE AREA ROUTER	NOVELL - SUNNYVALE (CPD)
36870	9006	WIDE AREA ROUTER	NOVELL - SUNNYVALE (CPD)
36871	9007	WIDE AREA ROUTER	NOVELL - SUNNYVALE (CPD)
36872	9008	WIDE AREA ROUTER	NOVELL - SUNNYVALE (CPD)
36873	9009	WIDE AREA ROUTER	NOVELL - SUNNYVALE (CPD)
36874	900A	WIDE AREA ROUTER	NOVELL - SUNNYVALE (CPD)
36875	900B	WIDE AREA ROUTER	NOVELL - SUNNYVALE (CPD)
36876	900C	WIDE AREA ROUTER	NOVELL - SUNNYVALE (CPD)
36879	900F	SMNP OVER IPX	NOVELL - SAN JOSE
36880	9010	SMNP OVER IPX	NOVELL - SAN JOSE
36882	9012	SOFTWARE DISTRIBUTION	PHASER SYSTEMS
36883	9013	SOFTWARE DISTRIBUTION	PHASER SYSTEMS
36884	9014	SOFTWARE DISTRIBUTION	PHASER SYSTEMS
36885	9015	SOFTWARE DISTRIBUTION	PHASER SYSTEMS
36886	9016	SOFTWARE DISTRIBUTION	PHASER SYSTEMS
36887	9017	CHAT - WINDOWS	NOVELL - PROVO CORP HQ
36889	9019	RPC BIND	NOVELL - AUSTIN
36890	901A	RPC BIND	NOVELL - AUSTIN
36894	901E		NOVELL - PROVO
36895	901F	NETWARE 'SLURPY'	NOVELL - PROVO
36897	9021	SPX CONNECTION	NOVELL - PROVO
36898	9022	JOB SERVER	NOVELL - PROVO
36899	9023	NETWARE 'SLURPY'	NOVELL - PROVO
36900	9024	NETWARE 'SLURPY'	NOVELL - PROVO
36901	9025	NETWARE 'SLURPY'	NOVELL - PROVO
36902	9026	NETWARE 'SLURPY'	NOVELL - PROVO
36903	9027		NOVELL - PROVO
36904	9028	NETWORK MANAGEMENT	NOVELL - PROVO
36905	9029		NOVELL - SALT LAKE CITY
36906	902A		NOVELL - SALT LAKE CITY
36907	902B		NOVELL - SALT LAKE CITY
36908	902C	NETWORK MANAGEMENT CLIENT INFO	NOVELL - PROVO
36909	902D		NOVELL - PROVO
36910	902E	DOS TARGET SERVICE AGENT	NOVELL - PROVO
36911	902F		NOVELL - PROVO
36912	9030	SUPERLAB AUTOMATION SERVER	NOVELL - PROVO
36913	9031	RPC BIND	NOVELL - AUSTIN
36914	9032	IPX BIFF	UNIVEL
36915	9033	IPX BOOTPC	UNIVEL
36916	9034	IPX BOOTPS	UNIVEL
36917	9035	IPX CHARGEN	UNIVEL
36918	9036	IPX DAYTIME	UNIVEL
36919	9037	IPX DISCARD	UNIVEL
36920	9038	IPX ECHO	UNIVEL
36921	9039	IPX EPRC	UNIVEL
36922	903A	IPX MONITOR	UNIVEL
36923	903B	IPX NAME	UNIVEL
36924	903C	IPX NAMESERVER	UNIVEL
36925	903D	IPX NETSTAT	UNIVEL
36926	903E	IPX NEW-RWHO	UNIVEL
36927	903F	IPX NFSD	UNIVEL
36928	9040	IPX NTP	UNIVEL
36929	9041	IPX QOTD	UNIVEL
36930	9042	IPX RMONITOR	UNIVEL
36931	9043	IPX ROUTE	UNIVEL
36932	9044	IPX SYSLOG	UNIVEL
36933	9045	IPX SYSTAT	UNIVEL
36934	9046	IPX TALK	UNIVEL
36935	9047	IPX TIME	UNIVEL
36936	9048	IPX WHO	UNIVEL
36937	9049	IPX WHOIS	UNIVEL
36938	904A	SPX APFS	UNIVEL
36939	904B	SPX APTS	UNIVEL
36940	904C	SPX AUTH	UNIVEL
36941	904D	SPX BFTP	UNIVEL
36942	904E	SPX CHARGEN	UNIVEL
36943	904F	SPX CMIP-AGENT	UNIVEL

Socket Decimal	Socket Hex	Service	Company
36944	9050	SPX CMIP-MANAGE	UNIVEL
36945	9051	SPX COURIER	UNIVEL
36946	9052	SPX CSNET-NS	UNIVEL
36947	9053	SPX DAYTIME	UNIVEL
36948	9054	SPX DISCARD	UNIVEL
36949	9055	SPX ECHO	UNIVEL
36950	9056	SPX EXEC	UNIVEL
36951	9057	SPX FINGER	UNIVEL
36952	9058	SPX FTP	UNIVEL
36953	9059	SPX FTP-DATA	UNIVEL
36954	905A	SPX HOSTNAMES	UNIVEL
36955	905B	SPX INGRESLOCK	UNIVEL
36956	905C	SPX ISO-IP	UNIVEL
36957	905D	SPX ISO-TP0	UNIVEL
36958	905E	SPX ISO-TSAP	UNIVEL
36959	905F	SPX LINK	UNIVEL
36960	9060	SPX LISTEN	UNIVEL
36961	9061	SPX LOGIN	UNIVEL
36962	9062	SPX NAME	UNIVEL
36963	9063	SPX NAMESERVER	UNIVEL
36964	9064	SPX NETSTAT	UNIVEL
36965	9065	SPX NNTP	UNIVEL
36966	9066	SPX NTP	UNIVEL
36967	9067	SPX PCSERVER	UNIVEL
36968	9068	SPX POP-2	UNIVEL
36969	9069	SPX PRINT-SRV	UNIVEL
36970	906A	SPX PRINTER	UNIVEL
36971	906B	SPX QOTD	UNIVEL
36972	906C	SPX RJE	UNIVEL
36973	906D	SPX SFTP	UNIVEL
36974	906E	SPX SHELL	UNIVEL
36975	906F	SPX SMTP	UNIVEL
36976	9070	SPX SUPDUP	UNIVEL
36977	9071	SPX SYSTAT	UNIVEL
36978	9072	SPX TELNET	UNIVEL
36979	9073	SPX TIME	UNIVEL
36980	9074	SPX TTYMON	UNIVEL
36981	9075	SPX UUCP	UNIVEL
36982	9076	SPX UUCP-PATH	UNIVEL
36983	9077	SPX WHOIS	UNIVEL
36984	9078	SPX X400	UNIVEL
36985	9079	SPX X400-SND	UNIVEL
36986	907A	SPX XSERVER0	UNIVEL
36987	907B	SMS	NOVELL - PROVO
36988	907C	SMS	NOVELL - PROVO
36989	907D	QUEUE SERVER FOR IBM PSF/2	NOVELL - PROVO CORP HQ
36990	907E	IPX SOCKET FOR BTRIEVE REQUESTER	NOVELL - AUSTIN
36991	907F	NETWARE FOR SAA	NOVELL - SUNNYVALE
36992	9080	ADDRESS SERVER	NOVELL - PROVO
36993	9081	NOVELL MHS DS GATEWAY FOR OCE	NOVELL - WALNUT CREEK
36994	9082	NDS GATEWAY FOR OCE	NOVELL - WALNUT CREEK
36995	9083	X.400 PROTOCOL ACCESS MODULE	NOVELL - RICHMOND HILL
36996	9084	SNADS PROTOCOL ACCESS MODULE	NOVELL - RICHMOND HILL
36997	9085	REMOTE PROGRAM SPAWNING	NOVELL - SAN JOSE
36998	9086	IPX PING	NOVELL - SAN JOSE
36999	9087	ENHANCED NCP COMMUNICATIONS	NOVELL - PROVO
37000	9088	ROAMING CLIENT SUPPORT	NOVELL - FORTUNE
37001	9089	NETWARE FOR SAA - LOAD BALANCING	NOVELL - SUNNYVALE
37011	9093	MONITOR SOCKET	NOVELL - MONTERREY
37012	9094	DATALINK SWITCHING (DLSW)	NOVELL - SAN JOSE
37013	9095	REMOTE CONTROL SOFTWARE PROGRAM	NOVELL - PROVO CORP HQ
37015	9097	NETWORK INSTALL OF UNIXWARE #2	NOVELL
37016	9098	NEST DEVICE	NOVELL - MONTERREY
37017	9099	DATABASE SERVER DOC. MNGMT. PKG.	NOVELL - OREM
37018	909A	DATABASE SERVER DOC. MNGMT. PKG.	NOVELL - OREM
37019	909B	DATABASE SERVER DOC. MNGMT. PKG.	NOVELL - OREM
37020	909C	DATABASE SERVER DOC. MNGMT. PKG.	NOVELL - OREM
37021	909D	DATABASE SERVER DOC. MNGMT. PKG.	NOVELL - OREM
37022	909E	DATABASE SERVER DOC. MNGMT. PKG.	NOVELL - OREM
37023	909F	DATABASE SERVER DOC. MNGMT. PKG.	NOVELL - OREM
37024	90A0	DATABASE SERVER DOC. MNGMT. PKG.	NOVELL - OREM
37025	90A1	DATABASE SERVER DOC. MNGMT. PKG.	NOVELL - OREM
37026	90A2	DATABASE SERVER DOC. MNGMT. PKG.	NOVELL - OREM

Appendix E – Router Port Queue Buffer Sizes

Link Speed	Bits per Second	Bytes per Second	Seconds in Buffer				
			0.1	0.30	0.5	1	2
56kbps	56,000	7,000	700	2,100.00	3,500	7,000	14,000
112kbps	112,000	14,000	1,400	4,200.00	7,000	14,000	28,000
128kbps	128,000	16,000	1,600	4,800.00	8,000	16,000	32,000
256kbps	256,000	32,000	3,200	9,600.00	16,000	32,000	64,000
512kbps	512,000	64,000	6,400	19,200.00	32,000	64,000	128,000
1.544Mb T1	1,536,000	192,000	19,200	57,600.00	96,000	192,000	384,000
4Mb TR	4,000,000	500,000	50,000	150,000.00	250,000	500,000	1,000,000
10Mb EN	10,000,000	1,250,000	125,000	375,000.00	625,000	1,250,000	2,500,000
16Mb TR	16,000,000	2,000,000	200,000	600,000.00	1,000,000	2,000,000	4,000,000
45Mb T3	45,000,000	5,625,000	562,500	1,687,500.00	2,812,500	5,625,000	11,250,000
100Mb FDDI or 100Mb Ethernet	100,000,000	12,500,000	1,250,000	3,750,000.00	6,250,000	12,500,000	25,000,000
Sonet OC3	155,520,000	19,440,000	1,944,000	5,832,000.00	9,720,000	19,440,000	38,880,000
Sonet OC12	622,080,000	77,760,000	7,776,000	23,328,000.00	38,880,000	77,760,000	155,520,000
Gb Ethernet	1,000,000,000	125,000,000	12,500,000	37,500,000.00	62,500,000	125,000,000	250,000,000

Each router interface can be configured with an amount of memory (Queue Buffer) which is used to store the packets which are to be transmitted onto the line. The memory is used when the router is receiving data destined for a specific interface at a faster rate than the data can be transmitted onto the interface. The router will sequentially store the packets in the Queue Buffer until they can be sent. Pine Mountain Group recommends that the Queue Buffer Size should be no more than two times (in seconds) the interface speed. On a 256kbps interface, this would equate to 512kbit (64kbyte) of Queue Buffer. This table shows the Queue Buffer Size based on time intervals of .1, .3, .5, 1, or 2 second increments.

Appendix F – Packet Transmit Times

Packet Size (Bytes)		64	128	256	512	1024	1514	2048	4096
Line Type	**Link Speed**	**Transmission Times**							
POTS/SER	9600	0.053333	0.106667	0.213333	0.426667	0.853333	1.261667	1.706667	3.413333
POTS/SER	19200	0.026667	0.053333	0.106667	0.213333	0.426667	0.630833	0.853333	1.706667
4 Wire/Frac T1	56000	0.009143	0.018286	0.036571	0.073143	0.146286	0.216286	0.292571	0.585143
4 Wire/Frac T1	128000	0.004000	0.008000	0.016000	0.032000	0.064000	0.094625	0.128000	0.256000
4 Wire/Frac T1	256000	0.002000	0.004000	0.008000	0.016000	0.032000	0.047313	0.064000	0.128000
4 Wire/Frac T1	512000	0.001000	0.002000	0.004000	0.008000	0.016000	0.023656	0.032000	0.064000
4 Wire/Frac T1	768000	0.000667	0.001333	0.002667	0.005333	0.010667	0.015771	0.021333	0.042667
4 Wire/Frac T1	1024000	0.000500	0.001000	0.002000	0.004000	0.008000	0.011828	0.016000	0.032000
T1	1536000	0.000333	0.000667	0.001333	0.002667	0.005333	0.007885	0.010667	0.021333
4Mb TR	4000000	0.000128	0.000256	0.000512	0.001024	0.002048	0.003028	0.004096	0.008192
10Mb EN	10000000	0.000051	0.000102	0.000205	0.000410	0.000819	0.001211	0.001638	0.003277
16Mb TR	16000000	0.000032	0.000064	0.000128	0.000256	0.000512	0.000757	0.001024	0.002048
25Mb ATM	25000000	0.000020	0.000041	0.000082	0.000164	0.000328	0.000484	0.000655	0.001311
45Mb T3	45000000	0.000011	0.000023	0.000046	0.000091	0.000182	0.000269	0.000364	0.000728
100Mb FDDI 100Mb EN	100000000	0.000005	0.000010	0.000020	0.000041	0.000082	0.000121	0.000164	0.000328
Sonet OC3	155520000	0.0000033	0.0000066	0.0000132	0.0000263	0.0000527	0.0000779	0.0001053	0.0002107
Sonet OC12	622080000	0.0000008	0.0000016	0.0000033	0.0000066	0.0000132	0.0000195	0.0000263	0.0000527
1Gb EN	1000000000	0.0000005	0.0000010	0.0000020	0.0000041	0.0000082	0.0000121	0.0000164	0.0000328

This table shows the amount of time that is required to transmit a stream of bytes (packet size) onto different transmission interfaces. For example, it would take .001211 seconds (1.2 milliseconds) for a station to transmit a 1,514 byte packet onto an Ethernet LAN interface. It would take .216286 seconds (216 milliseconds) for a router to transmit the same packet onto a 56kbps line. The formulas used to calculate these values are as follows:

Transmission Time = (Packet Size * 8) / Transmission Speed (bps)

Ethernet 10BaseT LAN (1514 bytes * 8) / 10Mbps = .00121 seconds

56kbps Sync Line (1514 bytes * 8) / 56kbps = .216 seconds

Appendix G – 4 & 8 bit Hex-Bin-Dec Conversion Table

Hex	Bin	Dec
0	0000	0
1	0001	1
2	0010	2
3	0011	3
4	0100	4
5	0101	5
6	0110	6
7	0111	7
8	1000	8
9	1001	9
A	1010	10
B	1011	11
C	1100	12
D	1101	13
E	1110	14
F	1111	15

Hex	Bin	Dec
01	00000001	1
02	00000010	2
03	00000011	3
04	00000100	4
05	00000101	5
06	00000110	6
07	00000111	7
08	00001000	8
09	00001001	9
0A	00001010	10
0B	00001011	11
0C	00001100	12
0D	00001101	13
0E	00001110	14
0F	00001111	15
10	00010000	16
11	00010001	17
12	00010010	18
13	00010011	19
14	00010100	20
15	00010101	21
16	00010110	22
17	00010111	23
18	00011000	24
19	00011001	25
1A	00011010	26
1B	00011011	27
1C	00011100	28
1D	00011101	29
1E	00011110	30
1F	00011111	31
20	00100000	32
21	00100001	33
22	00100010	34
23	00100011	35
24	00100100	36
25	00100101	37
26	00100110	38
27	00100111	39
28	00101000	40
29	00101001	41
2A	00101010	42
2B	00101011	43
2C	00101100	44
2D	00101101	45
2E	00101110	46
2F	00101111	47
30	00110000	48
31	00110001	49
32	00110010	50
33	00110011	51
34	00110100	52
35	00110101	53
36	00110110	54
37	00110111	55
38	00111000	56
39	00111001	57
3A	00111010	58
3B	00111011	59
3C	00111100	60
3D	00111101	61
3E	00111110	62
3F	00111111	63
40	01000000	64
41	01000001	65
42	01000010	66
43	01000011	67
44	01000100	68
45	01000101	69
46	01000110	70
47	01000111	71
48	01001000	72
49	01001001	73
4A	01001010	74
4B	01001011	75
4C	01001100	76
4D	01001101	77
4E	01001110	78
4F	01001111	79
50	01010000	80
51	01010001	81
52	01010010	82
53	01010011	83
54	01010100	84
55	01010101	85
56	01010110	86
57	01010111	87
58	01011000	88
59	01011001	89
5A	01011010	90
5B	01011011	91
5C	01011100	92
5D	01011101	93
5E	01011110	94
5F	01011111	95
60	01100000	96
61	01100001	97
62	01100010	98
63	01100011	99
64	01100100	100
65	01100101	101
66	01100110	102
67	01100111	103
68	01101000	104
69	01101001	105
6A	01101010	106
6B	01101011	107
6C	01101100	108
6D	01101101	109
6E	01101110	110
6F	01101111	111
70	01110000	112
71	01110001	113
72	01110010	114
73	01110011	115
74	01110100	116
75	01110101	117
76	01110110	118
77	01110111	119
78	01111000	120

79 01111001 121
7A 01111010 122
7B 01111011 123
7C 01111100 124
7D 01111101 125
7E 01111110 126
7F 01111111 127
80 10000000 128
81 10000001 129
82 10000010 130
83 10000011 131
84 10000100 132
85 10000101 133
86 10000110 134
87 10000111 135
88 10001000 136
89 10001001 137
8A 10001010 138
8C 10001100 140
8D 10001101 141
8E 10001110 142
8F 10001111 143
90 10010000 144
91 10010001 145
92 10010010 146
93 10010011 147
94 10010100 148
95 10010101 149
96 10010110 150
97 10010111 151
98 10011000 152
99 10011001 153
9A 10011010 154
9B 10011011 155
9C 10011100 156
9D 10011101 157
9E 10011110 158
9F 10011111 159
A0 10100000 160
A1 10100001 161
A2 10100010 162
A3 10100011 163
A4 10100100 164
A5 10100101 165
A6 10100110 166
A7 10100111 167
A8 10101000 168
A9 10101001 169
AA 10101010 170
AB 10101011 171
AC 10101100 172
AD 10101101 173
AE 10101110 174
AF 10101111 175
B0 10110000 176
B1 10110001 177
B2 10110010 178
B3 10110011 179
B4 10110100 180
B5 10110101 181
B6 10110110 182
B7 10110111 183
B8 10111000 184
B9 10111001 185
BA 10111010 186
BB 10111011 187
BC 10111100 188
BD 10111101 189
BE 10111110 190
BF 10111111 191
C0 11000000 192
C1 11000001 193
C2 11000010 194
C3 11000011 195
C4 11000100 196
C5 11000101 197
C6 11000110 198
C7 11000111 199
C8 11001000 200
C9 11001001 201
CA 11001010 202
CB 11001011 203
CC 11001100 204
CD 11001101 205
CE 11001110 206
CF 11001111 207
D0 11010000 208
D1 11010001 209
D2 11010010 210
D3 11010011 211
D4 11010100 212
D5 11010101 213
D6 11010110 214
D7 11010111 215
D8 11011000 216
D9 11011001 217
DA 11011010 218
DB 11011011 219
DC 11011100 220
DD 11011101 221
DE 11011110 222
DF 11011111 223
E0 11100000 224
E1 11100001 225
E2 11100010 226
E3 11100011 227
E4 11100100 228
E5 11100101 229
E6 11100110 230
E7 11100111 231
E8 11101000 232
E9 11101001 233
EA 11101010 234
EB 11101011 235
EC 11101100 236
ED 11101101 237
EE 11101110 238
EF 11101111 239
F0 11110000 240
F1 11110001 241
F2 11110010 242
F3 11110011 243
F4 11110100 244
F5 11110101 245
F6 11110110 246
F7 11110111 247
F8 11111000 248
F9 11111001 249
FA 11111010 250
FB 11111011 251
FC 11111100 252
FD 11111101 253
FE 11111110 254
FF 11111111 255

Appendix H – Hex & Decimal Integer Conversion Table

0.......	0	.0......	0	..0.....	0	...0....	0
1.......	268,435,456	.1......	16,777,216	..1.....	1,048,576	...1....	65,536
2.......	536,870,912	.2......	33,554,432	..2.....	2,097,152	...2....	131,072
3.......	805,306,368	.3......	50,331,648	..3.....	3,145,728	...3....	196,608
4.......	1,073,741,824	.4......	67,108,864	..4.....	4,194,304	...4....	262,144
5.......	1,342,177,280	.5......	83,886,080	..5.....	5,242,880	...5....	327,680
6.......	1,610,612,736	.6......	100,663,296	..6.....	6,291,456	...6....	393,216
7.......	1,879,048,192	.7......	117,440,512	..7.....	7,340,032	...7....	458,752
8.......	2,147,483,648	.8......	134,217,728	..8.....	8,388,608	...8....	524,288
9.......	2,415,919,104	.9......	150,994,944	..9.....	9,437,184	...9....	589,824
A.......	2,684,354,560	.A......	167,772,160	..A.....	10,485,760	...A....	655,360
B.......	2,952,790,016	.B......	184,549,376	..B.....	11,534,336	...B....	720,896
C.......	3,221,225,472	.C......	201,326,592	..C.....	12,582,912	...C....	786,432
D.......	3,489,660,928	.D......	218,103,808	..D.....	13,631,488	...D....	851,968
E.......	3,758,096,384	.E......	234,881,024	..E.....	14,680,064	...E....	917,504
F.......	4,026,531,840	.F......	251,658,240	..F.....	15,728,640	...F....	983,040

....0...	0	.....0..	0	......0.	0	0	0
....1...	4096	.....1..	256	......1.	16	1	1
....2...	8192	.....2..	512	......2.	32	2	2
....3...	12288	.....3..	768	......3.	48	3	3
....4...	16384	.....4..	1024	......4.	64	4	4
....5...	20480	.....5..	1280	......5.	80	5	5
....6...	24576	.....6..	1536	......6.	96	6	6
....7...	28672	.....7..	1792	......7.	112	7	7
....8...	32768	.....8..	2048	......8.	128	8	8
....9...	36864	.....9..	2304	......9.	144	9	9
....A...	40960	.....A..	2560	......A.	160	.......A	10
....B...	45056	.....B..	2816	......B.	176	.......B	11
....C...	49152	.....C..	3072	......C.	192	C	12
....D...	53248	.....D..	3328	......D.	208	.......D	13
....E...	57344	.....E..	3584	......E.	224	.......E	14
....F...	61440	.....F..	3840	......F.	240	.......F	15

Examples:

0000000D = 13

000000F0 = 240

00000C00 = 3072

00001000 = 4096

000B0000 = 720896

00000510 = 1280 + 16 = 1296

0000A0E1 = 40960 + 224 + 1 = 41185

Appendix I – Popular RFCs

	Major IP RFCs
760,791	IP
777,792	ICMP
781	IP Timestamp Option
815	IP Datagram Reassembly Algorithms
1063,1191	Path MTU Discovery
1071	Computing the Internet Checksum
	ICMP RFCs
777,792	ICMP
1016,1018	Source Quench Introduced Delay (SQuID)
1256	ICMP Router Discovery Messages
1788	ICMP Domain Name Messages
2521	ICMP Security Failures Messages
	ARP/RARP RFCs
826	ARP
903	RARP
925	Multi-LAN ARP
1027	Using ARP to Implement Transparent Subnet Gateways
1293	Inverse ARP
1329	Address Resolution for Dual MAC FDDI Networks
1433	Directed ARP
1735	NBMA ARP (NARP)
	Major Routing and Gateway RFCs
823	Gateway-Gateway Protocol
985	Requirements for Internet Gateways
1058	Routing Information Protocol (RIP)
1075	Distance Vector Multicast Routing Protocol
1105	Border Gateway Protocol (BGP)
1131,1247,1583	OSPF Version 2
1142	OSI IS-IS Intra-Domain Routing Protocol
1163 & 1267	BGP 3
1164,1268,1655	Application of BGP in the Internet
1245	OSPF Protocol Analysis
1265	BGP Protocol Analysis
1338,1519	Classless Inter-Domain Routing (CIDR)
1370	Applicability Statement for OSPF
1387,1721	RIP II Protocol Analysis
1517	Applicability Statement for the Implementation of CIDR
1519	Classless Inter-Domain Routing (CIDR)
1654	BGP 4

Continued on next page

TCP RFCs

793	Transmission Control Protocol
1006	ISO Transport Service
1122	Requirements for Internet Hosts – Communications Layers
1144	Compressing TCP Headers
1323	TCP Extensions for High Performance
2001	TCP Slow Start, Congestion Avoidance, Fast Retransmit, Fast Recovery
2018	TCP Selective Acknowledgements

Application Layer RFCs

821,788	SMTP (Simple Mail Transport Protocol)
854	Telnet
959	FTP (File Transfer Protocol)
1034,1035,1591	DNS
1089	SNMP (Simple Network Management Protocol)
1350	TFTP (Trivial File Transport Protocol)
1352	SNMP Security Protocols
2131,2132	DHCP
3011	DHCP Subnetting

Get all RFCs on CD ROM, e-mail: freerfc@pmg.com

Appendix J – DOS Commands

arp

This diagnostic command displays and modifies the IP-to-Ethernet or IP-to-token ring address translation tables used by the address resolution protocol (ARP).

Syntax

arp -a [*inet_addr*] [**-N** [*if_addr*]]
arp -d *inet_addr* [*if_addr*]
arp -s *inet_addr ether_addr* [*if_addr*]

Parameters

-a - Displays current ARP entries by querying TCP/IP. If *inet_addr* is specified, only the IP and physical addresses for the specified computer are displayed.

-d - Deletes the entry from the ARP cache table that is specified by *inet_addr*.

-s - Adds an entry in the ARP cache to associate the IP address *inet_addr* with the physical address *ether_addr*. The physical address is given as 6 hexa decimal bytes separated by hyphens. The IP address is specified using dotted decimal notation. The entry is permanent, that is, it will not be automatically removed from the cache after the timeout expires.

-N [*if_addr*] - Displays the ARP entries for the network interface specified by *if_addr*.

ether_addr - Specifies a physical address.

if_addr - Specifies, if present, the IP address of the interface for which the address translation table should be modified. If not present, the first applicable interface will be used.

inet_addr - Specifies an IP address in dotted decimal notation.

nbtstat

This diagnostic command displays protocol statistics and current TCP/IP connections using NetBIOS over TCP/IP.

Syntax

nbtstat [**-a** *remotename*] [**-A** *IPaddress*] [**-c**] [**-n**] [**-R**] [**-r**] [**-S**] [**-s**] [*interval*]

Parameters

-a *remotename* - Lists the remote computer's name table using the computer's name.

-A *Ipaddress* - Lists the remote computer's name table using the computer's IP address.

-c - Lists the contents of the NetBIOS name cache, giving the IP address of each name.

-n - Lists local NetBIOS names.

-R - Reloads the LMHOSTS file after purging all names from the NetBIOS name cache.

-r - Lists name resolution statistics for Windows networking. On a Windows NT computer configured to use WINS, this option returns the number of names resolved and registered via broadcast or via WINS.

-S - Displays both workstation and server sessions, listing the remote hosts by IP address only.

-s - Displays both workstation and server sessions. It attempts to convert the remote host IP address to a name using the HOSTS file.

Interval - Redisplays selected statistics, pausing *interval* seconds between each display. Press Ctrl+C to stop redisplaying statistics. If this parameter is omitted, **nbtstat** prints the current configuration information once.

Notes

The column headings generated by the **nbtstat** utility have the following meanings.

In- Number of bytes received.

Out - Number of bytes sent.

In/Out - Whether the connection is from the computer (outbound) or from another system to the local computer (inbound).

Life - The remaining time that a name table cache entry will live before it is purged.

Local Name - The local NetBIOS name associated with the connection.

Remote Host - The name or IP address associated with the remote host.

Type - This refers to the type of name. A name can either be a unique name or a group name.

<03> - Each NetBIOS name is 16 characters long. The last byte often has special significance, because the same name can be present several times on a computer. This notation is simply the last byte converted to hexadecimal. For example, <20> is a space in ASCII.

State

The state of NetBIOS connections. The possible states are shown in the following list:

State	Meaning
Connected	The session has been established
Associated	A connection endpoint has been created and associated with an IP address
Listening	This endpoint is available for an inbound connection
Idle	This endpoint has been opened but cannot receive connections
Connecting	The session is in the connecting phase where the name-to-IP address mapping of the destination is being resolved
Accepting	An inbound session is currently being accepted and will be connected shortly
Reconnecting	A session is trying to reconnect if it failed to connect on the first attempt
Outbound	A session is in the connecting phase where the TCP connection is currently being created
Inbound	An inbound session is in the connecting phase
Disconnecting	A session is in the process of disconnecting
Disconnected	The local computer has issued a disconnect, and it is waiting for confirmation from the remote system

netstat

This diagnostic command displays protocol statistics and current TCP/IP network connections.

Syntax

netstat [**-a**] [**-e**][**-n**][**-s**] [**-p** *protocol*] [**-r**] [*interval*]

Parameters

-a - Displays all connections and listening ports; server connections are usually not shown.

-e - Displays Ethernet statistics. This can be combined with the **-s** option.

-n - Displays addresses and port numbers in numerical form (rather than attempting name lookups).

-s - Displays per-protocol statistics. By default, statistics are shown for TCP, UDP, ICMP, and IP; the **-p** option can be used to specify a subset of the default.

-p *protocol* - Shows connections for the protocol specified by *proto*; *proto* (See notes) can be **tcp** or **udp**. If used with the **-s** option to display per-protocol statistics, *protocol* can be **tcp**, **udp**, **icmp**, or **ip**.

-r - Displays the contents of the routing table.

Interval - Redisplays selected statistics, pausing *interval* seconds between each display. Press Ctrl+C to stop redisplaying statistics. If this parameter is omitted, **netstat** prints the current configuration information once.

Notes

The **netstat** utility provides statistics on the following network components.

Statistic	Purpose
Foreign Address	The IP address and port number of the remote computer to which the socket is connected. The name corresponding to the IP address is shown, instead of the number, if the HOSTS file contains an entry for the IP address. In cases where the port is not yet established, the port number is shown as an asterisk (*).
Local Address	The IP address of the local computer, as well as the port number the connection is using. The name corresponding to the IP address is shown, instead of the number, if the HOSTS file contains an entry for the IP address. In cases where the port is not yet established, the port number is shown as an asterisk (*).
Proto	The name of the protocol used by the connection.
(state)	Indicates the state of TCP connections only. The possible states are: CLOSED FIN_WAIT_1 SYN_RECEIVED CLOSE_WAIT FIN_WAIT_2 SYN_SEND ESTABLISHED LISTEN TIMED_WAIT LAST_ACK

route

This diagnostic command manipulates network routing tables.

Syntax
route [**-f**] [*command* [*destination*] [**MASK** *netmask*] [*gateway*] [METRIC *metric*]]

Parameters

-f - Clears the **routing** tables of all gateway entries. If this parameter is used in conjunction with one of the commands, the tables are cleared prior to running the command.

-p - When used with the **route add** command, makes a **route** persistent across restarts of the computer. By default, **routes** are not preserved when the computer is restarted. When used with the **route print** command, displays the list of registered persistent **routes**. Ignored for all other commands.

command
Specifies one of four commands.

Command	Purpose
print	Prints a **route**
add	Adds a **route**
delete	Deletes a **route**
change	Modifies an existing **route**

destination - Specifies the host to send *command.*
MASK - Specifies, if present, that the next parameter be interpreted as the *netmask* parameter.
netmask - Specifies, if present, the subnet mask value to be associated with this **route** entry. If not present, this parameter defaults to 255.255.255.255.
gateway - Specifies the gateway.
METRIC - Specifies the **route** metric (cost) for the destination.

Notes
The **route** utility does not accept a subnet mask value of 255.255.255.255 on the command line. To specify a subnet mask with this value, you must accept the default.

On a multihomed computer on which a network is available from more than one adapter card, all network traffic is passed over the first gateway defined. If you add a second gateway to the same network, the entry is added to the **route** table, but it is never used.

The **route** utility uses the Networks file to convert *destination* names to addresses. For the **route** utility to work correctly, the network numbers in the Networks file must specify all four octets in dotted decimal notation. For example, a network number of 284.122.107 must be specified in the Networks file as 284.122.107.0, with trailing zeroes appended.

All symbolic names used for *destination* or *gateway* are looked up in the network and host name database files NETWORKS and HOSTS, respectively. If the command is **route print** or **route delete**, wildcards may be used for the *destination* and *gateway*, or the *gateway* argument may be omitted.

ping

This diagnostic command verifies connections to one or more remote hosts.

Syntax

ping [**-t**] [**-a**] [**-n** *count*] [**-l** *length*] [**-f**] [**-i** *ttl*] [**-v** *tos*] [**-r** *count*] [**-s** *count*] [[**-j** *host-list*] | [**-k** *host-list*]] [**-w** *timeout*] *destination-list*

Parameters

-t - **Pings** the specified host until interrupted.

-a - Specifies not to resolve addresses to host names.

-n *count* - Sends the number of echo packets specified by *count*. The default is 4.

-l *length* - Sends echo packets containing the amount of data specified by *length*. The default is 64 bytes; the maximum is 8192.

-f - Sends a Do Not Fragment flag in the packet. The packet will not be fragmented by gateways on the route.

-i *ttl* - Sets the Time To Live field to the value specified by *ttl*.

-v *tos* - Sets the Type Of Service field to the value specified by *tos*.

-r *count* - Records the route of the outgoing packet and the returning packet in the Record Route field. A minimum of 1 to a maximum of 9 hosts must be specified by *count*.

-s *count* - Specifies the time stamp for the number of hops specified by *count*.

-j *host-list* - Routes packets by means of the list of hosts specified by *host-list*. Consecutive hosts may be separated by intermediate gateways (loose source routed). The maximum number allowed by IP is 9.

-k *host-list* - Routes packets by means of the list of hosts specified by *host-list*. Consecutive hosts may not be separated by intermediate gateways (strict source routed). The maximum number allowed by IP is 9.

-w *timeout* - Specifies a time-out interval in milliseconds.

destination-list - Specifies the remote hosts to **ping**.

Note on Ping

The **ping** command verifies connections to remote host or hosts, by sending Internet Control Message Protocol (ICMP) echo packets to the host and listening for echo reply packets. The **ping** command waits for up to 1 second for each packet sent and prints the number of packets transmitted and received. Each received packet is validated against the transmitted message. By default, four e cho packets containing 64 bytes of data (a periodic uppercase sequence of alphabetic characters) are transmitted.

You can use the **ping** utility to test both the host name and the IP address of the host. If the IP address is verified but the host name is not, you may have a name resolution problem. In this case, be sure that the host name you are querying is in either the local HOSTS file or in the DNS database.

tracert

This diagnostic utility determines the route taken to a destination by sending Internet Control Message Protocol (ICMP) echo packets with varying Time-To-Live (TTL) values to the destination. Each router along the path is required to decrement the TTL on a packet by at least 1 before forwarding it, so the TTL is effectively a hop count. When the TTL on a packet reaches 0, the router is supposed to send back an ICMP Time Exceeded message to the source system. The **tracert** command determines the route by sending the first echo packet with a TTL of 1 and incrementing the TTL by 1 on each subsequent transmission until the target responds or the maximum TTL is reached. The route is determined by examining the ICMP Time Exceeded messages sent back by intermediate routers. Notice that some routers silently drop packets with expired TTLs and are invisible to **tracert**.

Syntax

tracert [**-d**] [**-h** *maximum_hops*] [**-j** *host-list*] [**-w** *timeout*] *target_name*

Parameters

-d - Specifies not to resolve addresses to host names.

-h *maximum_hops* - Specifies maximum number of hops to search for target.

-j *host-list* - Specifies loose source route along *host-list*.

-w *timeout* - Waits the number of milliseconds specified by *timeout* for each reply.

target_name - Specifies the host name of the destination computer.

Notes on Tracert

The following shows sample output for **tracert**. The first column is the hop number, which is the Time To Live (TTL) value set in the packet. Each of the next three columns contains the round-trip times in milliseconds for an attempt to reach the destination with that TTL value. An asterisk (*) means that the attempt timed out. The fourth column is the host name (if it was resolved) and IP address of the responding system.

```
F:\>tracert www.pmg.com

Tracing route to www.pmg.com [207.141.76.86]
over a maximum of 30 hops:

  1   <10 ms   <10 ms    10 ms  192.168.0.1
  2    10 ms    10 ms    40 ms  10.85.180.1
  3    10 ms    30 ms    20 ms  r1-fe1-7.sttls1.wa.home.net [24.9.59.209]
  4    10 ms    10 ms    20 ms  10.0.220.149
  5    10 ms    20 ms    20 ms  c1-pos6-2.sttlwa1.home.net [24.7.76.185]
  6    10 ms    20 ms    10 ms  home-gw.st6wa.ip.att.net [192.205.32.249]
  7    10 ms    10 ms    20 ms  gbr3-p40.st6wa.ip.att.net [12.123.44.130]
  8    30 ms    20 ms    30 ms  gbr4-p10.sffca.ip.att.net [12.122.2.61]
  9    31 ms    30 ms    30 ms  gbr1-p60.sffca.ip.att.net [12.122.5.249]
 10    20 ms    30 ms    30 ms  ar2-a300s4.sffca.ip.att.net [12.127.1.141]
 11    40 ms    50 ms    50 ms  12.127.195.182
 12    40 ms    50 ms    50 ms  www.pinemountaingroup.com [207.141.76.86]

Trace complete.
```

Appendix K – DHCP

Default DHCP Options

Code	Option name	Meaning
1	Subnet mask	Specifies the subnet mask of the client subnet. This option is defined in the DHCP Manager **Create Scope** or **Scope Properties** dialog box. It cannot be set directly in the DHCP **Options** dialog box.
3	Router	Specifies a list of IP addresses for routers on the client's subnet. Multihomed computers can have only one list per computer, not one per network adapter.
6	DNS servers	Specifies a list of IP addresses for DNS name servers available to the client.
15	Domain name	Specifies the DNS domain name that the client should use for DNS computer name resolution.
44	WINS/NBNS servers	Specifies a list of IP addresses for NetBIOS name servers (NBNS).
46	WINS/NBT node type	Allows configurable NetBIOS over TCP/IP (NetBT) clients to be configured as described in RFC 1001/1002, where 1 = b-node, 2 = p-node, 4 = m-node, and 8 = h-node. On multihomed computers, the node type is assigned to the entire computer, not to individual network adapters.
47	NetBIOS scope ID[1]	Specifies a text string that is the NetBIOS over TCP/IP scope ID for the client, as specified in RFC 1001/1002.
51	Lease time	Specifies the time, in seconds, from address assignment until the client's lease on the address expires. Lease time is specified in the DHCP Manager **Create Scope** or **Scope Properties** dialog box, and can be set directly in the DHCP **Options** dialog box.
58	Renewal (T1) time value	Specifies the time in seconds from address assignment until the client enters the Renewing state. Renewal time is a function of the lease time option, which is specified in the DHCP Manager **Create Scope** or **Scope Properties** dialog box and can be set directly in the DHCP Options dialog box.
59	Rebinding (T2) time value	Specifies the time, in seconds, from address assignment until the client enters the Rebinding state. Rebinding time is a function of the lease time option, which is specified in the DHCP Manager **Create Scope** or **Scope Properties** dialog box and can be set directly in the DHCP Options dialog box.

DHCP Message Fields

Field Name	Octets	Description
op	1	Message op code/message type. 1 = BOOTREQUEST, 2 = BOOTREPLY
htype	1	Hardware address type.
hlen	1	Hardware address length.
hops	1	Client sets to zero; optionally used by relay agents when booting via a relay agent.
xid	4	Transaction ID, a random number chosen by the client, used by the client and server to associate messages and responses between a client and a server.
secs process.	2	Filled in by client; seconds elapsed since client began address acquisition or renewal
flags	2	Flags. To work around some clients that cannot accept IP unicast datagrams before the TCP/IP software is configured, DHCP uses the flags field. The leftmost bit is defined as the BROADCAST (B) flag. The remaining bits of the flags field are reserved for future use. They must be set to zero by clients and ignored by servers and relay agents.
ciaddr	4	Client IP address; only filled in if client is in BOUND, RENEW, or REBINDING state and can respond to ARP requests.
yiaddr	4	Client IP address.
siaddr	4	IP address of next server to use in bootstrap; returned in DHCPOffer, DHCPAck by server.
giaddr	4	Relay agent IP address; used in booting via a relay agent.
chaddr	16	Clients hardware address.
sname	64	Optional server host name, null terminated string.
file	128	File name, null terminated string; "generic" name or null in DHCPDiscover, fully qualified path name in DHCPOffer.
options	variable	Optional parameters field. For more information about DHCP options, see "DHCP Options" on previous page.

Appendix L – NetBIOS

NetBIOS 16th Character Suffix

<computername>	<00>	Unique	Workstation Service
<computername>	<01>	Unique	Messenger Service
<\\--__MSBROWSE__>	<01>	Group	Master Browser
<computername>	<03>	Unique	Messenger Service
<computername>	<06>	Unique	RAS Server Service
<computername>	<1F>	Unique	NetDDE Service
<computername>	<20>	Unique	File Service Service
<computername>	<21>	Unique	RAS Client Service
<computername>	<22>	Unique	MS Exchange Interchange
<computername>	<23>	Unique	MS Exchange Store
<computername>	<24>	Unique	MS Exchange Directory
<computername>	<30>	Unique	Modem Sharing Server Service
<computername>	<31>	Unique	Modem Sharing Client Service
<computername>	<43>	Unique	SMS Clients Remote Control
<computername>	<44>	Unique	SMS Admin Remote Control Tool
<computername>	<45>	Unique	SMS Clients Remote Chat
<computername>	<46>	Unique	SMS Clients Remote Transfer
<computername>	<4C>	Unique	DEC Pathworks TCPIP Service on WINNT
<computername>	<52>	Unique	DEC Pathworks TCPIP Service on WINNT
<computername>	<87>	Unique	MS Exchange MTA
<computername>	<6A>	Unique	MS Exchange IMC
<computername>	<BE>	Unique	Network Monitor Agent
<computername>	<BF>	Unique	Network Monitor Application
<username>	<03>	Unique	Messenger Service
<domain>	<00>	Group	Domain Name
<domain>	<1B>	Unique	Domain Master Browser
<domain>	<1C>	Group	Domain Controllers
<domain>	<1D>	Unique	Master Browser
<domain>	<1E>	Group	Browser Service Elections
<INet-Services>	<1C>	Group	IIS
<IS-computername>	<00>	Unique	IIS
<computername>	[2B]	Unique	Lotus Notes Server Service
IRISMULTICAST	[2F]	Group	Lotus Notes
IRISNAMESERVER	[33]	Group	Lotus Notes
Forte_$ND800ZA	[20]	Unique	DCA IrmaLan Gateway Server Service

About Pine Mountain Group

Network Analysis Training
Network Analysis Services
Certified NetAnalyst — the standards-based certification
Network Analysis Video Library

PMG NetAnalyst
Pine Mountain Group, Inc.

Portfolio of PMG NetAnalyst Services

Synergy Management Methods™ are designed for the complex enterprise network that requires innovative methods and services to attain reliability and performance levels for mission critical applications. Every dollar spent on IT systems is dependent on the enterprise network to deliver its return on investment.

Critical Problem Resolution – Fast, definitive, on-site diagnosis of application, network operating system and network infrastructure problems.

Network Health Check – Network architecture is reviewed, performance is assessed and inconsistencies located.

DocuNet – Effective layered approach to network documentation. Provides the architectural visualization needed to design, operate, troubleshoot, support and maintain control of enterprise networks.

SafetyNet – Cost-effective remote analysis services when immediate response is required. Or when intermittent problems require ongoing analysis. Provided from our Network Analysis Center.

Application Impact Analysis – Characterize applications to determine infrastructure and WAN link speeds required for acceptable performance.

Infrastructure Validation Testing – Evaluate network infrastructure using a variety of stress testing tools. Make recommendations for improved response times.

Network Guardian – Synergy Management clients enjoy the services of a Network Guardian, providing consistent on-site visits with Synergy teams to fulfill charters and master plan development.

Training and Certification – of network support personnel through our public courses or on-site training and the Certified NetAnalyst™ Professional Certification Program.

Network Analysis Video Library

A cost-effective way to receive the very best of our training programs–*when you need them!* Recorded during live presentations at national conferences and our Network Training classes.

Order No. Title

S100 Ethernet Troubleshooting Set

V101 — Ethernet Troubleshooting Methods & Practices
Covers the basic elements of network troubleshooting to include audits, documentation, monitoring, value added reporting and proactively addressing priorities.

V102 — Ethernet Operation & Analysis
Focuses on the essential details of using protocol analysis to solve Ethernet data link problems.

V103 — Switched Network Analysis & VLANs
Various switching technologies explained and how to analyze networks employing them.

S200 WAN Analysis Set

V201— WAN, HDLC & Frame Relay
Learn the essentials of passing data across WANs with Frame Relay services.

V202 — High Speed Media–ATM & FDDI.
How can voice, video and data traffic be transferred over a network? The answer is ATM.

V203 — Link State Protocol Analysis
RIP2, OSPF & NLSP — Alternatives for conserving IP addresses and how overhead traffic can be reduced between routers.

S300 Network Performance Set

V301 — Throughput & Latency Analysis
Tricks for using your analyzer to measure the throughput of popular protocols including NCP, Packet Burst, TCP and SMB.

V302 — Packet Burst & TCP Sliding Window Operation
Discusses advantages of these techniques in LAN & WAN environments.

V303 — TCP/IP Troubleshooting
Covers the details of IP and TCP headers through trace file analysis.

S400 TCP/IP Tools Set

V401 — DNS & DHCP
Crash course in DNS concepts and analysis methods plus DHCP operation as seen in packets on the wire.

V402 — TCP/IP Tools, RMON, SNMP & DNS
Covers the details of IP and TCP headers through trace file analysis.

V403 — SNMP & RMON Operation
Covers the essentials of popular network management protocols and how they can be used to help manage and baseline your network.

S500 Application Analysis Set

V501 — Client-Server Analysis
Overview of fundamental concepts of client-server architecture with case study examination.

V502 — Application Analysis & Preparation of an "Application Network Impact Statement"
Forum Tutorial. Includes session notes, 4 tapes.

Order No. **Title**

S600 Troubleshooting NetWare Set

V601 — IPX, RIP, SAP & NetWare Protocols
Details on commonly encountered IPX/NetWare problems.

V602 — IPX Troubleshooting Methods & Case Studies
Looks at the key analysis issues associated with applications, workstations, media, protocol, Internet and server.

S700 Microsoft Protocol Analysis Set

V701 — Microsoft's use of IP, IPX, NetBIOS
Implementation of protocols, retransmit time-outs, segment sizes, MTU discovery, multiple default router entry pitfalls and what you'll see on the wire.

V702 — Microsoft TCP/IP Implementation & Analysis
Overview and operation of the MS Network Monitor, NT Protocol Stack, DHCP, NetBIOS Naming, WINS and case studies.

V703 — LLC2, NetBIOS & SMB
Deployed in many mission-critical networks, NetBIOS and SMB are an integral part of IBM, Windows for Workgroups, and NT Server LANs.

S800 Documentation & Baselining Set

V801 — Network Architecture & Documentation
Learn productive documentation methods.

V802 — Establishing a Minimum Essential Baseline
Why gather more baseline data than you'll ever examine?

V803 — IP Operation & Analysis
Case studies show how to perform analysis on your Internet sessions.

LB100 Video Library Set

S100 through S800. ***Our best value!***
22 titles, over 50 hours of training, plus free Certified NetAnalyst testing.

S900 Additional Selected Titles Set

V901 — Token Ring 802.5 Theory
Overview of the fundamentals plus productive troubleshooting methods.

V902 — IP Troubleshooting & TCP
Don't have an analyzer? We will show you how to debug TCP/IP using UNIX and Windows NT/95.

V903 — Network Analysis Essentials
OSI layers, IEEE addressing, encapsulation and complexity analysis

S1000 Network Troubleshooting at its Best!

5-title set , Includes: V102, V103, V301, V803, V903

Order form on next page

Pricing plans:

Single title (order no. starts with V) $199
2-title Set (order no. starts with S) $349
3-title Set (order no. starts with S) $495
5-title Set (order no. S1000 only) $795
Video Library Set LB100 (Includes free Certified NetAnalyst testing) $3495

Or order on-line at www.NetworkTraining.com or call 800-645-8486

Video Library Order Form (Descriptions/prices on previous page)

Order No	Title	Qty.	Price each	Total
Subtotal				
Sales Tax Amount (Calif. Residents)				
Shipping, US only (5 videos or more) $25				
Total				

SHIPPING INFORMATION

Name ______________________ Job Title ______________

Company ______________________________

Street Address ______________________________

City ______________________ State ________ Zip ________

Phone ______________ Fax ______________ Email ______________

BILLING INFORMATION (If Different From Shipping Information)

Name ______________________ Job Title ______________

Company ______________________________

Address ______________________________

City ______________________ State ________ Zip ________

Phone ______________ Fax ______________ Email ______________

METHOD OF PAYMENT:
You may fax us this form with your credit card information, call us or mail us with check payment.

____
Company Check Number

____
Credit Card Number

____
Expiration Date

How to Reach Us

Phone
800-645-8486
or 209-962-6247

Fax
209-962-5914

Web
www.NetworkTraining.com

Mail
Pine Mountain Group, Inc.
Box 849, 20950 Ferretti Rd.
Groveland, CA 95321

Reference Guide Order Form

NetAnalyst™ Reference Guide	Qty.	Total
Price each: Qty: 1-4 $65.00 (includes shipping in US only) Qty: 5-9 $55.25 Qty: 10 or more $45.50		
Subtotal		
Sales Tax Amount (Calif. Residents)		
Total		

SHIPPING INFORMATION

Name ______________________ Job Title ____________

Company ______________________

Street Address ______________________

City ______________________ State ________ Zip ________

Phone ______________ Fax ______________ Email ______________

BILLING INFORMATION (If Different From Shipping Information)

Name ______________________ Job Title ____________

Company ______________________

Address ______________________

City ______________________ State ________ Zip ________

Phone ______________ Fax ______________ Email ______________

METHOD OF PAYMENT:
You may fax us this form with your credit card information, call us or mail us with check payment.

____
Company Check Number

____
Credit Card Number

____
Expiration Date

How to Reach Us

Phone
800-645-8486
or 209-962-6247

Fax
209-962-5914

Web
www.NetworkTraining.com

Mail
Pine Mountain Group, Inc.
Box 849, 20950 Ferretti Rd.
Groveland, CA 95321